The Saga of Rocky Flats

THE SAGA OF ROCKY FLATS:

LESSONS LEARNED FOR A SAFER NUCLEAR WORLD IN THE 21ST CENTURY

John M. Whiteley

ISBN 978-0-578-67134-5

Dedication

For Future Stewards of a Safer Nuclear World In the 21st Century and Especially:

Abigail Dahlgren
Amelia Williams
Cash Whiteley
Elian Whiteley
Emmalyn Dahlgren
Isabela Whiteley
Jack McLean
Mia McLean
Noah Whiteley
Penelope Dahlgren

About the Covers

The upper frame of the Front Cover is from Closure Legacy which is presented in Chapter Twelve. There was an imperative for action to reduce toxicity at Rocky Flats for the U.S. government after the end of the Cold War and the closure for production. There were a number of sources for the action imperative. In 1994 DOE informed the public that the two most dangerous buildings in the entire U.S. nuclear weapons complex were at Rocky Flats. The aftermath of the FBI raid, the guilty plea by Rockwell International to serious criminal violations of the Resource Conservation and Recovery Act and the Clean Water Act, the findings of the Congressional Wolpe subcommittee, and the DOE (1995) estimates that it would cost $37 billion and take 65 years to cleanup the site combined to force the imperative for remedial action. And visually it was very clear that Denver suburbs were both downwind and downhill. The choices made by the U.S. government had a number of elements. The Clinton-Gore administration had instituted reforms in contracting to include contractor incentives. The contractor chosen to conduct the cleanup introduced a concept of engaging stakeholders across the Legislative and Executive branches including regulators. The Government Accountability Office estimated in 2006 that the total cost to taxpayers through 2080 for that extended period would be $10 billion. The work which was accomplished took 7 years, not 65 years as forecast by DOE in 1995, and ended in 2005 in that time frame at a cost of $7.3 billion.

The lower frame of the Front Cover is a field of agriculture irrigators. But these fields at Rocky Flats were being used to commit environmental crime. Starting in 1979 Rockwell International had been confronted with an excess of contaminated water from the operational requirements of the industrial enterprise. One of the original site selection requirements in 1951 had been for a reliable water supply of 1 million gallons a day, or 365 million gallons a year. By the end of the Rockwell International era they had to dispose of sometimes 250,000 gallons a day, and up to 80 million gallons a year. The way they chose to do this, in order to avoid direct discharge into the community (from a comparatively small site which dropped 528 feet in the 2 miles from West to East), was to utilize Spray Irrigation year around in all conditions of Colorado weather. There were (at least) five separate Spray Fields which had been established: North, South, East, West, and Pond. There were severe problems with all of them. At one point the Rockwell International Spray Field site manager asked permission of Plant management to stop the Spray Irrigation practice in the winter months but was turned down. Once Rockwell International was confronted with a human error caused toxic crisis. A worker had accidently left a hose running. This caused discharge of a distinctive green chromic acid (containing sulfuric acid among other toxic substances). Rockwell management choice was to Spray Irrigate the chromic acid spill onto the South Spray Field and not tell the community including drinking water users. The Special Federal Grand Jury detailed the broad elements of the environmental crime committed by Spray Irrigation in Chapter Eleven with the notation [Indict].

The Back Cover is an exhibit from the plaintiff's case in the Cook class action lawsuit and in its original conception was by ChemRisk (1992). It was augmented for The Saga of Rocky Flats by the addition of arrows to the major waste disposal locations identified in Chapters Five and Six almost all within the 334-acre high security production area.

TABLE OF CONTENTS

EPOCH THREE:
ENVIRONMENTAL JUSTICE DELAYED AND THE CONTINUING STRUGGLE OVER THE FUTURE OF THE ROCKY FLATS SITE: 1990-CONTINUING

Foreword

On January 20, 1961, President John F. Kennedy announced that:

> "Let every nation know, whether it wishes us well or ill, that we shall pay any price, bear any burden, meet any hardship, support any friend, oppose any foe to assure the survival and success of liberty."

Rocky Flats is a testament to the commitment of the government of the United States "to assure the survival and the success of liberty." The irony of Rocky Flats is that it fabricated successfully the first stage plutonium fission igniters for thermonuclear bombs for the War Reserve in the staggering number of 70,000, and each at 20 kilotons the size of the bomb which destroyed Nagasaki. Rocky Flats additionally renewed and recycled plutonium igniters which had been deployed previously on bombers, missiles and submarines in defense of the country.

The Saga of Rocky Flats also chronicles and analyzes tragedies which occurred as part of the willingness to "pay any price" and "bear any burden" in defense of liberty. The lessons learned as a consequence are the source of the aspiration for this book to contribute to a safer 21st Century nuclear world.

Earlier in this Inaugural Address President Kennedy stated that the world was very different now than when the first President of the United States took the original Oath of Office nearly a century and three-quarters ago:

> "For man holds in his mortal hands the power to abolish all forms of human poverty and all forms of human life. "

Three days after the Inauguration on January 23, 1961, a B-52 bomber of the United States Air Force loaded with two 3.8-megaton thermonuclear bombs (the explosive power of 7.6 million tons of TNT) disintegrated in flight over Goldsboro, North Carolina. The disintegration was caused by mechanical failure after 10 hours of flight.

The explosive power of all the bombs used in World War II, including the atomic bombs used on Hiroshima and Nagasaki, is variously estimated at between 5 megatons and 9 megatons. That means that the United States dropped on itself 7.6 megatons of explosive power, an amount within the estimated range of all of the explosive power used in World War II. In commenting on the Goldsboro tragedy described in Chapter Two, humorist John Oliver said: "That is why we have South Carolina, so we will always have a Carolina."

Closer to home for me, my companion in life's journey was, in January of 1961, a high school student living in Wayne County, North Carolina. This location is only 20 miles from the farmer's field where the two thermonuclear bombs landed. The "near miss' of an unthinkable catastrophe in North Carolina could have profoundly impacted negatively my life and work to come, but of course I did not know it at the time.

Policymakers and decision-makers alike in the Cold War were under duress of preserving the very fabric and survival of their societies. A consequence at Rocky Flats was the use of evolving technology to gain an advantage in weaponizing the atom at whatever the cost to environment, health and safety.

America's principal adversaries, the Soviet Union and China, possessed nuclear weapons with a stated intent to impose their model of the relationship of the individual to an authoritarian state on the people of the United States. Nuclear weapons in the Cold War were the weapon of choice to protect the three societies from preemptive attack.

The Saga of Rocky Flats is a lesson learned book based on insights toward creating a safer nuclear world for the 21st Century. The source of the insights is not only from a detailed examination of the sixty-nine years of Rocky Flats as an integral part of the nuclear weapons development complex of the United States, but also the broader Cold War global context.

My greatest hope is that in this Foreword I can "hook" readers to motivate them to really want to get into this work for what can be a transformative intellectual and personal experience. This book is very relevant to environmental science, social ecology, political science, and peace studies. The future of our country and world as we know it can be positively influenced by actions some of our readers may take.

This Foreword is written by a professor about the work of a fellow professor, albeit from different universities and from different disciplinary perspectives. This will definitely be academic writing! I especially want to communicate with college and university students, and perhaps even, hopefully, some secondary school students. When any professor prepares a presentation the basic question of "who is the audience?" must be addressed. I remember this from taking Speech 101 in my first year of college, in which my grade was a D!

As I thought about who would be the readers and consumers of Professor Whiteley's saga, I predicted many different types of readers and audiences. Surely this work will be read, studied, debated, argued about, hopefully emulated and replicated by:

*undergraduate students
*graduate students
*law students
*public policy students and practitioners
*prosecutors
*defense attorneys
*Congressional and state legislative staff
*corporation planners and counsel
*investigators
*journalists
*law enforcement personnel

*military planners
*governmental regulatory agency staff
*diplomats
*and more!

I found that even with the above speculation on potential types of readers I could not possibly catalogue all the possible occupations and roles of potential readers. There is just so much to be learned and applied from Professor Whiteley's research.

Because my own work has focused especially on undergraduate students, I decided to write a Foreword especially for them. I want to focus on them because if we are to prevent future Rocky Flats disasters which could destroy our county and planet, then by virtue of the age of traditional undergraduates, they have the most time left to prevent thermonuclear disaster resulting from the continuing arms race. They are not in their professions yet. They are developing both knowledge and a set of values and ethics that could combine to prevent the kinds of disasters chronicled in this study.

This work will motivate undergraduate students to want to learn more about threats to the future of our species because of unintended consequences of the thermonuclear arms race and how citizens in multiple roles can be more engaged to prevent future catastrophe. This work can motivate undergraduate students to see how enormously multifaceted and complex the arms race and threat of nuclear disaster is, and that it cannot possibly be understood and addressed from the perspective of any one discipline. Hence this book makes a wonderful case for the intellectual need and illuminating power of going beyond one discipline (e.g. a "major" for undergraduates), which often works in a kind of self-imposed academic isolation, and instead pursue multi-disciplinary and interdisciplinary reading, research, thinking and action.

This work will help undergraduate students see and understand what is all around them but which they do not see let alone understand; namely, the integration of the university/military/industrial/political/regulatory/judicial complex which has produced this hugely complex ecosystem designed to protect in our case, the United States, from its real and imagined enemies. Because of the ways this complex ecosystem actually functions, it has created enormous risks to our survival as a human society. The technologies we have devised and the cultural forces that govern and monitor them have outstripped the moral capacities of our citizenry to protect us from the unintended consequences of this ecosystem we have built since the 1940s. Technological thermonuclear defenses we have created to protect and preserve us now threaten to destroy us. As the American essayist, Ralph Waldo Emerson wrote in the 1840s "things are in the saddle and ride mankind!"

Undergraduate students will see for themselves how selfless citizens can make a difference to pursue persistently and ferret out, and then publish the information we must have to not only preserve our democracy but our very human existence itself. This work then may be viewed as a primer for how to conduct research on the ecosystem created by humans for very good, noble even, legal, practical purposes, but which contains inherent flaws, self-contained contradictions, and threats to our health and societal survival.

Undergraduate students can learn intellectual habits of disciplined informational literacy and research skills—how to find the information they seek, evaluate its credibility, and then do/create something with that information. They can learn how to create information that society needs to preserve itself, and how to disseminate and act upon that information.

My writing in this vein is not meant to be alarmist or defeatist. I have to be more hopeful. I am more hopeful, especially when I see researchers like Whiteley taking what risks may be involved to pursue the kind of information we need to have better chances of preserving our planetary life. What I am trying to say is that for undergraduate students there are risks all around them that they do not see, know, understand, think about, but which really are very close to home.

I have another reason for thinking of undergraduates as readers for this work. I believe some of them will find this book reads more like a novel. It truly is replete with suspense and drama. They will find plenty of crime: individual, corporate, governmental, and environmental. And deception, falsehoods and other misdeeds. I believe some undergraduates could eventually convert this to a novel and/or screenplay—I am sure the furthest possible outcome that its author might have intended. But I am having a fantasy about what an impact that could have on public awareness and demands for proactive interventions.

So what was the author trying to do? What was his intent? Why would he devote the majority of his professional time for over a decade to studying the environmental disaster of Rocky Flats? Of course, Professor Whiteley had many reasons but fundamentally it all comes down to an overriding sense of purpose. And this gets back to why I especially want to see this book read by undergraduates. It is my contention that the most important outcome of undergraduate education is the development, pursuit, and attainment of purpose—in order to be an adult citizen of our country leading a purpose driven life. Whiteley certainly has had a purpose driven life. That is what I would want for all undergraduates. And from my experience I have seen many of them discover this all-important sense of purpose during and because of undergraduate education. I believe then that this work can be inspirational for the development of purpose on behalf of some readers.

Whiteley's purpose is inextricable from something that undergraduates need to come to understand; namely, what are the purposes of universities and their faculties? Stated quite simply then, students need to understand and appreciate what is it that university professors really do, especially at our country's greatest research universities like the University of California Irvine, where Whiteley has been on the faculty for nearly 50 years.

University professors are expected to discover and create new knowledge. They are expected to become leading experts in the world on some subject or subjects. They are expected to spend their entire productive lives working on the intersectional developments that arise from the synergies of their continuing discoveries and creations. That is definitely the story here. I am saying then, that part of the story behind Rocky Flats is also the story of what university professors really do. And I recommend to my readers here a relevant brief, inspirational, and intellectual exercise to more fully appreciate the importance of this question: very simply, Google Nobel Prize Laureates for any given year and read their biographies. These are men, and now much more recently women, the majority of them university researchers, who developed an early and initial focus and spent a life pursuing it—only to be finally recognized much later in life for their life purpose as translated into this unique, pathfinding and commendable research.

In one word, this work has been all about *justice*, albeit in multiple forms and contexts, but especially justice for protection and preservation of the environment, both that created by human actions and the natural environment is which humans must live. Whiteley has practiced this pursuit of justice in the intellectual context of a 20th century new discipline known as social ecology. Social ecology is an academic discipline that is relevant to solving complex problems in a globalized world. Social ecology is a form of "convergent science" where the contributions for solving complex human problems converge from a form of "team science" whereby researchers, scientists, thinkers, problem solvers come from multi- and transdisciplinary disciplines and work together to integrate their findings and ideas to be more effective than they could be working alone.

Another form of fundamental justice for Whiteley became that of the pursuit of world peace. What could one university professor and researcher do in his respective sphere of influence to support the pursuit of many for world peace? One answer was the creation of a series of 180 interviews, known as Quest for Peace, nationally broadcast video interviews of world leaders and thinkers. These conversations were viewed by many thousands of Whiteley's fellow human beings, surely many of them influenced to increase their own involvement and commitments towards world peace and justice.

This brings me full circle to my own connection(s) to Professor Whiteley's work. I will start by returning to the question I raised above; namely, what is it that college and university professors really do? Like Professor Whiteley, my professional and personal life has also been related to the same life purpose: the pursuit of justice. I am focused on this in multiple contexts, including civil rights and educational equity, but one in particular that is a central focus of my own academic work is the need for greater justice for an often exploited and neglected population of undergraduate students: first-year students. It turned out that Whiteley also shared this interest as had been evident by actions he took to support first-year students during his tenure as Vice Chancellor for Student Affairs at UCI. He learned about my interest in this population after I had founded what became a very influential series of so-called Conferences on the First-Year Experience in 1982, now 39 years in the offering and participated in by many thousands of educators around the globe. John (Whiteley) reached out to me in the early 1980s and proposed that we meet sometime and see how our work might intersect. I will make a long story somewhat shorter here and say that we did indeed meet at John's initiative. He provided me with a vision for expanding my still nascent work and crusade on behalf of justice for first year students to establish a national center for this purpose at the University of South Carolina where I held my professorship. I did so and became an accidental self-appointed spokesperson for an international movement to improve the attention paid to struggling first-year students.

One of the first things we did together was to get his University in partnership with mine to host a series of conferences on the first-year experience in California with UCI as the co-host. And now some 35 years later we continue to pursue a number of partnerships for shared work in pursuit of justice for people more at risk—in my case for multiple sub-populations of undergraduate college students, and in John Whiteley's case, literally on the planet whose prospects for a full and natural life are threatened daily by the negative consequences of the thermonuclear arms race. Because I know Whiteley well I have been following his work on this book for more than a decade and now am truly pleased to see it finally "out" with its prospect of great impact on undergraduates, but also many other categories of learned persons of influence and power.

I will close with this final personal observation on how this book is really all about purpose. On one level it is about the purpose of a scholar working with other professionals and gathering information to report to the human community what really happened at Rocky Flats and why this should be of concern and action for many more of us. But this work is also about the well-intentioned purpose(s) of the US government and its policy to protect and defend our citizenry during the Cold War. It is also about the purposes, some of them well intentioned, of for-profit corporations and employees, whose talents and resources were necessary to implement the policy decisions of the US government to produce weapons of mass destruction---and who almost destroyed millions of us.

I have previously stated my belief that colleges and universities are all about helping people discover both new knowledge and legitimate purposes of socially redeeming value. In my own case, as a college student, the single most important outcome for me was the development of an abstract vision of purpose—a purpose identical to Whiteley's: the pursuit of justice. I did not have any idea upon graduation from college how I was going to pursue this. All I knew was that I was at risk of being drafted to serve in the Vietnam War. But it was thanks to the US military and my first commanding officer in the US Air Force who gave me a direct order to: "Gardner, you will perform community service!" I had no idea what he meant. But he immediately informed me what HE meant and that was "college teaching." He explained that South Carolina, where I had been involuntarily stationed, was desperate for teachers and that I was to do so immediately in my off-duty time.

And as events unfolded I discovered how to implement a life based on purpose in the pursuit of justice, a career which brought me into contact with the author of this book, who has helped me greatly give more substance and meaning to my version of a career and life based on a pursuit of justice.

In the second part of President Kennedy's Inaugural Address, with which I opened this Foreword, humankind now has the capacity to abolish "all forms of human life." Rocky Flats and the nuclear age in the conception of sociologist Kai Erikson (1991, 1994) is a frame used by Professor Whiteley to convey that "the all clear" never sounds with nuclear technological disasters. Erikson observes that as human beings test their competence with technology, being that they are of human manufacture, disasters are in principle preventable. The result is there is a story with a moral to tell about them, and a share of blame to be assigned. The "new species of trouble" is a recurrent theme in The Saga of Rocky Flats.

I hope by now you can't wait to read this book. I urge you to be open to letting this book move you, disturb you, challenge your sense of purpose and your role in our human eco-system, such as that studied, researched and taught about by inspiring professors of social ecology such as John M. Whiteley. Reading and digesting this book will be the closest thing to actually being in class with him—and that I will leave to your imagination and inspiration.

John N. Gardner
Pisgah Forest, North Carolina
December 2019.

Distinguished Professor Emeritus and Senior Fellow
Founding Executive Director, University 101 Programs and National Resource Center for The First-Year Experience and Students in Transition
University of South Carolina

Co-Founder, Chief Executive Officer, and Chair
John N. Gardner Institute for Excellence in Undergraduate Education
Brevard, N.C.

Preface:
The Saga of Rocky Flats: A Study of Exceptional Contributions To National Security and the Ironic and Tragic Clash Over Secrecy, the Release of Plutonium Into the Environment, the Rule of Law and the Public Trust

This book is about the intractable conflicts which arose during the Cold War and its aftermath over the imperative of the government of the United States to protect American society against communist adversaries possessing nuclear weapons and emerging societal concerns over threats to environmental protection, safety and the health of surrounding communities and workers. Intractable conflicts which developed were both within and between "separate but equal" institutions of U.S. constitutional government and with significant constituencies of organized civil society.

The Saga of Rocky Flats is a sixty-nine year (and counting) study of strivings for national security and weaponization of the atom, the global context driving the nuclear arms race, the release of plutonium into the environment, the commission of environmental crime and its aftermath, evolving regimes of law where the relationship between the regimes had not been settled, the current intractable struggle over the future of the site itself, and the public trust.

Pristine Cattle Lands Chosen In Secret As a Nuclear Weapons Production Site

The once pristine cattle lands of rolling prairie, woodlands, and wetlands nestled close to the Front Range of the Rocky Mountains and only sixteen miles from an expanding Denver, Colorado were chosen in secret as a nuclear weapons production site known as Rocky Flats. The year was 1951.

The Saga of Rocky Flats has unfolded across its history governed and regulated by very different regimes of law each of which had been established by bipartisan consensus. Environmental consciousness in the United States evolved over the long history of Rocky Flats. This evolution has been reflected in the dominant purposes of the major legal regimes, and since the late 1970s, in the pursuit of environmental justice.

When conflicts have arisen over citizen attempts at influence over Rocky Flats as a major national security imperative of the U.S. government, the conflicts seldom were resolved fundamentally, but instead were carried over into the years which followed. When conflicts have arisen within or between the very different regimes of law and branches of government, it has taken an extended number of years to resolve them.

Unique Contribution To National Security For Weaponizing the Atom
At its inception contributing to nuclear weapons production behind a cloak of Cold War secrecy, Rocky Flats operated under a regime of law which had as a core purpose weaponizing the atom. Nuclear weapons were the weapon of choice in the Cold War. The United States had tested a nuclear bomb at Alamogordo, New Mexico in July of 1945. On direct orders of President Harry S. Truman, nuclear weapons were exploded over the cities of Hiroshima and Nagasaki on August 6 and August 9, 1945, killing close to an estimated 200,000 people, and followed shortly by Japan's surrender to mark the end of World War II.

The Saga of Rocky Flats begins just after World War II when the U.S. government in a time of great duress moved quickly to strengthen the nation's defenses by achieving nuclear superiority through the production of hydrogen bombs. As part of that process, and under the aegis of the Atomic Energy Commission (AEC), a site was hastily chosen to produce plutonium igniters (or "pits") for detonating new super bombs.

Rushing to stay ahead of the Soviet Union and protect America during an escalating arms race, the selected site was inexplicably located and constructed only 16 miles from downtown Denver on a windy plateau. Prudent and long-term planning was obviously compromised by expediency and, as the pages of this book detail, workers, managers, and first- responders were placed in harm's way during four decades of operation while the health and well being of residents in Denver and surrounding communities were placed at risk.

"How could that happen?" In retrospect, there is no good answer in the public interest. But tremendous pressures in the moment to make America more secure may partially explain seriously bad decisions to locate the plant so close to a major city, and to proceed without having adequate policies and protocol in place to prevent contamination of air and soil by toxic wastes discharged through burning, waste-water runoff, or seepage of cooling oil contaminated with radionuclides often with cancer causing chemicals from barrels which corroded over time.

By way of explanation (but not justification), the pressure was great for stewards of the Rocky Flats project to rush ahead. Understanding the geopolitical context that gave rise to such great urgency sheds some light on how a faulty decision-making process of this magnitude could evolve.

Competition with the Soviet Union greatly intensified when the Soviets tested their first atomic (fission) weapon in 1949. President Truman authorized the creation and stockpiling of thermonuclear (fusion) weapons in the form of hydrogen bombs. This move was intended to deter possible Soviet aggression by creating a force so enormous that it could instantly wipe out any adversary. Hydrogen bombs had that power. Some hydrogen bombs had explosive capability equivalent to 10,000,000 tons of TNT (compared to the atomic bombs used on Hiroshima (16 tones of TNT) and Nagasaki (which had equivalency of 20,000 tons.)

As the United States sped forward in weaponizing the atom, the Soviet Union was also pulling out all stops to achieve matching nuclear capability. In response to the Soviet threat, Truman (on January 31, 1950) announced massive funding to rapidly produce hydrogen bombs. Within this context of importance and urgency, Dow Chemical was chosen to create hastily the Rocky Flats manufacturing facility for production of essential plutonium igniters and other components as quickly as could be achieved. Assigned engineering and production teams worked swiftly and efficiently. Igniters were successfully produced at Hanford Nuclear Reservation and Rocky Flats to enable detonation of the first hydrogen bomb on November 1, 1952. (Detonation occurred on the Eniwetok Atoll in the Pacific Marshall Islands, completely vaporizing the island and leaving a crater one mile wide.) Nikita Khrushchev of the Soviet Union instructed Soviet bomb designers to create a nuclear weapon so powerful that it would demonstrate the superiority of Soviet technology: a 50 million ton of TNT equivalent bomb was exploded in December of 1961.

Rocky Flats' fulfilled its crucial mission. Eventually, 70,000 igniters were produced from 1951 to 1989, each of the power of the bomb which destroyed Nagasaki. As time passed, Dow Chemical was replaced by Rockwell International in 1975. EG&G replaced Rockwell in 1990. In 1977, oversight of the facility was transferred from the AEC to the Department of Energy (DOE). Many mistakes in planning and oversight were made along the way by each party. Each successor inherited those mistakes with narrowed options for correction. Eventually, some of the practices were so egregious and dangerous that, based on inside information and surveillance, the FBI conducted its first criminal raid on another U.S. government facility on June 6, 1989. Termed "Operation Desert Glow," the raid revealed many violations of environmental laws and weapons production ceased by the end of 1989 and President George H.W. Bush declared Rocky Flats to be officially closed in 1992.

At its core, <u>The Saga of Rocky Flats</u> is an account of a project that successfully fulfilled a mission to make all inhabitants of the country safer while (ironically and in some cases tragically) creating great risk to workers and local communities. But this history is much more than that. It is also a detailing of conflicts and cross-purposes between and among government agencies, Congress, the Executive branch, the Judicial branch, and citizen and environmental groups.

As the reader will discover, this Saga is also a disturbing portrayal of governmental malfeasance and in some cases criminal activity, some of it willful and some of it as a result of Catch-22s caused by inherited mistakes. The result, however, is an on-going legacy of residual contamination and health impairment for those affected by contamination. Nearly seventy years from the inception of Rocky Flats, many environmental health and justice issues remain unresolved.

My intent has been to produce a lessons-learned study to contribute to a safer nuclear world in future years. Much can be learned from studying the dysfunctions that contributed to such a troubling debacle and how this important national defense project turned into a dangerous and still unsettled fiasco. The nuclear age is unforgiving of mistakes.

John M. Whiteley
Irvine, California

EPOCH ONE

STRIVINGS FOR NATIONAL SECURITY AND WEAPONIZATION OF THE ATOM, EXTENSIVE GENERATION OF ENVIRONMENTAL POLLUTION, AND THE ORIGINS OF THE CONTINUING CONTROVERSY OVER OFF-SITE CONTAMINATION: 1951-1989

Introduction To Part I: Framing the Irony and the Tragedy of Rocky Flats

Key Elements of the Chapter One Narrative

- Rocky Flats was founded under a legal regime to weaponize the atom and transformed by a bipartisan change in environmental consciousness and a legal regime grounded in protecting the environment.
- Strivings for national security and weaponization of the atom by the Atomic Energy Commission and Dow Chemical essentially governing and regulating themselves was the system from 1951 through 1974.
- The flawed siting decision is introduced which directly led to subsequent unbelievable tragedies.
- Three prisms for examining Rocky Flats as a scientific inquiry are introduced:
  - questions which arise when direct concerns of some citizens and some elements of organized civil society come into conflict with national security priorities and obligations of the U.S. government,
  - evaluating decision-making by stewards at Rocky Flats in the frame of the categories for understanding why some societies in history have made disastrous decisions and subsequently collapsed, and
  - assessing the functioning of U.S. constitutional democracy: where the "system" worked, where it did not work, where it may yet work.
- Distinctive dimensions to the legacy of Rocky Flats:
  - siting errors created immense difficulties for solving evolving public health, waste management and safety challenges,
  - vulnerability to risk of fire,
  - wellsprings and transformation of environmental crime,
  - reconfiguration of the Rocky Flats site into a Wildlife Refuge and a Superfund Site,
  - prism for viewing clashes and controversies over national security and environmental crime and over responsiveness in a democratic society to citizen concerns, and
  - overarching conflict over off-site contamination.

Key Elements of the Chapter Two Narrative

- Nuclear weapons were the weapon of choice in the Cold War.
- Unstable plutonium is "a new species of trouble" for humankind.
- Dwight D. Eisenhower said in 1951, "Because, therefore, we are defending a way of life, we must be respectful of that way of life...We must not violate its principles and precepts, and we must not destroy from within what we are trying to defend from without."
- U.S. disasters with nuclear weapons delivery systems posed the gravest danger that a thermonuclear explosion could have occurred over North Carolina, in Arkansas, and over Greenland and Spain.
- Half a world away geographically, and on a different planet ideologically, the Soviet Union and China both learned how dangerous nuclear weapons are to their creators and possessors.
- A major lesson from the dangerous global context of weaponizing the atom is that on numerous occasions the world came very close to another nuclear explosion after Hiroshima and Nagasaki, this time by accident.

Key Elements of the Chapter Three Narrative

- Rocky Flats had two historic missions: to create the first-stage plutonium fission bombs which ignited the second-stage fusion explosions of thermonuclear bombs, and to recycle nuclear warheads which had been deployed previously in support of America's national security.
- A Federal District Court Judge declared in 1982 the Rocky Flats industrial enterprise in support of the missions to be "ultrahazardous."
- There were at least twenty-five "materials of concern" at Rocky Flats for both off-site and on-site contamination, most were cancer causing, nineteen were not monitored at all, and the remaining six not monitored effectively.

Key Elements of the Chapter Four Narrative

- There was a major failure to anticipate industrial enterprise requirements in site selection criteria, building replacement specifications, and waste management.
- The Hanford Nuclear Reservation was sited on 670,000 square miles with a hazardous production area of 200 square miles and far from metropolitan areas.
- In contrast, Rocky Flats was sited on 4 square miles by choice, with a hazardous production area of 334 acres and only 16 miles in 1951 from an expanding Denver, Colorado.

- As late as 1955, a Dow Chemical report stated: "waste was being generated when the ultimate disposition had not yet been decided upon."
- A Dow Chemical Report in 1969 said that in the early years waste management was a low priority because "it is desirable to do as little as possible to get rid of any waste streams."
- With an unassailable belief that science, engineering and technology can solve technical problems, the Atomic Energy Commission, the Department of Energy, and its military-industrial contractors acted outside of the normal safeguards and oversight typical of private-sector nuclear power facilities in the United States.
- Events at Rocky Flats which were unbelievable and unthinkable reflected a compelling necessity to use science, engineering, and new and untried technology to increase the scope of the possible, and in doing so, had the unintended tragic consequence of exceeding the limits of the prudent. In global Cold War context, the Soviets and the Chinese did the same.

The Saga of Rocky Flats is a study of distinctive accomplishments in weaponizing the atom in the context of a Cold War. Communist adversaries possessed nuclear weapons. The 70,000 plutonium first-stage igniters fabricated at Rocky Flats for thermonuclear weapons were noteworthy for their quality and reliability. After 1965, Rocky Flats was the mass-production provider of the igniters in the quantities essential for the War Reserve, and for subsequent deployment in nuclear weapons on bombers, submarines, and ground launched ballistic missiles.

Yet the production era at Rocky Flats ended with the facility containing the two most dangerous buildings in the entire U.S. nuclear weapons production complex. Production of igniters ceased at Rocky Flats just over six months after the first criminal raid ever by the Federal Bureau of Investigation on another facility belonging to the government of the United States.

The four chapters which comprise Part I provide the context for understanding the background of why such an extraordinary contributor to national security became mired in a burden of long-lived radioactive and chemical pollution which resulted in 1,300 acres of the now closed Rocky Flats designated as a Superfund site under provisions of the Comprehensive Environmental Response, Compensation, and Liability Act of 1980.

Chapter One presents the evaluative framework for analyzing this Rocky Flats study. A first frame of evaluation is determining viable approaches within American democracy by which citizens can maintain at least some level of democratic influence over such a sensitive area of national security as nuclear weapons production.

A second frame of evaluation is assessing decision-making by stewards at Rocky Flats using categories created by Jared Diamond (2005) for understanding why some societies in history made disastrous decisions and subsequently collapsed at the height of their power and glory.

A third frame of evaluation is inspired by Gelb and Betts (1979) in The Irony of Vietnam: The System Worked. They distinguished between the foreign policy of the United States in Vietnam which failed, and the ultimate decision-making "system" which worked. The irony of Rocky Flats is that it was highly successful in contributions to weaponizing the atom within the legal framework of the Atomic Energy Acts of 1946 and 1954 and hence to national security. But Rocky Flats did not (or could not) adapt to performing the fabrication of new plutonium igniters in its industrial enterprise within the new requirements of a legal regime based on environmental protection.

Chapter Two places weaponization of the atom initiatives at Rocky Flats in their global context. Plutonium itself was (and is) unstable and prone to technological disasters. Disasters grew in numbers at Rocky Flats and around the world in what sociologist Kai Erikson called "a new species of trouble" for humankind.

The dynamics of the nuclear arms race are explained by physicist John Holdren along with how countries use the rhetoric of fear to intimidate enemies internationally and to acquire support domestically for resource acquisition.

The Kennedy and succeeding administrations confronted challenges in melding the science, engineering and technology of nuclear weapons creation with the reliabilities and safety of delivery systems. Chapter Two chronicles four disasters between 1961 and 1980 involving nuclear weapons of the United States and their delivery systems which came close to being truly catastrophic.

A problem which Rocky Flats never successfully solved over its thirty-eight-year production era was disposal of the types, the longevity and the consequences of long-lived toxic waste which was generated in the production of plutonium igniters. The challenge proved to be insurmountable on a site which did not have either the size, or the topography, to solve the problem of toxic waste disposal.

The global nuclear arms race context is presented by juxtaposing the U.S. experience with that of the Soviet and Chinese approaches to nuclear weapons. The snapshots of the Soviet and Chinese experiences lack the depth of information available about Rocky Flats, but the known dramatic differences are informative about how dangerous nuclear weapons are to their possessors.

Chapter Three presents the missions of Rocky Flats and its ultrahazardous industrial enterprise in the historical context of accelerating the weaponization of the atom. A theme running through this chapter, and indeed the entire study, is that nuclear weapons are dangerous to their creators.

In accomplishing its industrial missions, Rocky Flats utilized 8,000 substances in three broad categories: solvents, metals and "others" (mostly toxic chemicals). A contractor after the close of production, ChemRisk (1992), identified twenty-five "materials of concern" for off-site contamination. With extensive access to classified records at Rocky Flats, ChemRisk determined that only 6 of 25 substances had received at least some level of monitoring for off-site pollution.

Chapter Four presents the flawed siting decisions. Rocky Flats had been established in 1951 on only 2,500 acres within 4 square miles at a distance then of only 16 miles from an expanding Denver, Colorado. The high security production area was a highly compacted 334 acres. The site chosen was thought in error to be downwind and downhill from Denver when in fact it was both uphill and upwind with winds from the Front Range of the Rocky Mountains occasionally exceeding 100 miles an hour.

In stark contrast with Rocky Flats, the Hanford Nuclear Reservation had been originally sited on 670,000 square miles. The nine reactors to produce plutonium were located miles away from the five chemical separation facilities to remove plutonium from spent nuclear fuel, and from the massive tanks constructed to store the highly toxic chemical and radioactive "witches brew" which resulted from the process of separating the plutonium in spent nuclear fuel from the other toxic and highly radioactive elements. The siting consequences of these very different choices are visually compelling. The siting decisions which were made in secret at Rocky Flats were ultimately to foreclose the possibility of operating safely the Rocky Flats industrial enterprise without creating an on-site burden of toxic contamination and extensive off-site pollution.

Rocky Flats encountered challenges in common with other U.S. nuclear production sites. But it also encountered significant challenges which were not reflected in site selection criteria, or in original building design and construction. Large scale production processes in one facility did not translate directly into the design of a major replacement building. The new building never opened for production.

The original site selection criteria, and small site size chosen, combined to create very negative consequences caused by the subsequent failure to anticipate the sheer volume, dangerous volatility, and complexity of toxic waste. Early waste management had been established as a "low priority." Management of waste was largely by crisis. The ad-hoc nature of early waste disposal decisions created dangerous conditions for workers, and led to significant off-site contamination from not considering the topography of the site in the location chosen for what became the industrial enterprise known as Rocky Flats.

Chapter One
Evolving Legal Regimes and Environmental Consciousness: Clashes Within and Between Institutions of Shared Constitutional Governance, and the Quest For Environmental Justice Across the Epochs of Rocky Flats

The framework for organizing this study of Rocky Flats is three overlapping and interrelated Epochs which span the time period from 1951 to the present. This chronicle of the interrelated Epochs is followed by three prisms for evaluating the case of Rocky Flats: decision-making by stewards, responsiveness to citizen concerns by the U.S. government and the military-industrial contractors, and the functioning of the "system" of U.S. constitutional democracy.

Epoch One: Strivings For National Security and Weaponization of the Atom, Extensive Generation of Environmental Pollution, and the Origins of the Continuing Controversy Over Off-Site Contamination: 1951-1989

The production era in Epoch One describes occurrences at Rocky Flats which were not disclosed fully at the time when they happened, the 1957 and 1969 fires for example, or the waste disposal practices within a site which was not able to accommodate the sheer volume of nuclear and chemical waste and contaminated water generated by plutonium igniter production and its accompanying industrial processes. Also not disclosed by the U.S. government was the circumstance behind the acquisition of the approximately 4,000 acre "buffer" zone in 1974. The continuing controversies over the extent of off-site contamination which began to be addressed by independent scientists Martell and Poet in 1970 (CCEI, 1970a; 1970b) are just that: continuing controversies for forty-six years from 1970 until resolved by legal settlement in 2016. The clash over the Wildlife Refuge as documented in Chapter Thirteen reveals the continuing significance of on-site and off-site plutonium contamination, which escalated in late 2019.

The central elements of the history of the production era from 1951 until 1974, with Dow Chemical as the contractor, through the departure of Rockwell International at the end of 1989, have been covered extensively by the Lamm-Wirth Task Force (1975), by Abas (1989), by regular coverage in Westword and the Bulletin of the Atomic Scientists, and by Ackland (1999). Since then, more primary source documents have become available about the production era from many initiatives. The archives of the Church cluster of lawsuits (Holme, 1978), The Ambushed Grand Jury (McKinley and Balkany, 2004), and the Citizens Investigation of 2005 (Lipsky, McKinley, Brever, and Balkany, 2005) have made an exemplary contribution to primary source document identification, as did the Cook class action litigation during its twenty-six year history, and as did the U.S. House of Representatives (Wolpe, 1993) when it relied on its power of subpoena when the Executive branch was not cooperating sufficiently with its investigation.

Flawed Siting Decision In 1951 and Its On-Going Negative Consequences

The series of siting decisions leading to the choice of where to locate Rocky Flats were made in secret without independent expert review. An independent scientific review would surely have discovered that the location chosen for Rocky Flats was not downwind from Denver, but in fact both upwind and uphill. This fundamental error in 1951 was to haunt the future of Rocky Flats until its closure for production in 1989 and beyond.

A major siting flaw was in not considering the drainage flow problem with two "navigable waters of the United States" (under the definition of the Clean Water Act of 1972), Walnut Creek and Woman Creek, flowing through the site and into Great Western Reservoir and Standley Lake compounded by the drop in elevation of 528 feet from West to East. Of course the Clean Water Act of 1972 had not been passed when the siting decision was finalized in 1951. Storage, treatment and disposal of the toxic byproducts of the industrial enterprise had not been considered in site selection. There is no evidence that provisions of the 1948 Federal Water Pollution Control Act were considered relevant to the siting decision.

Epoch Two: Discovery of Environmental Crime, and the Transformation of Rocky Flats By the Criminal Justice System and By the "Separate But Equal" Constitutional "Checks and Balances" Between the Executive, the Legislative, and the Judicial Branches of the U.S. Government: 1986-1994

By the start of Epoch Two in 1986, the legal underpinnings of governance and regulation at Rocky Flats were two primary evolving legal regimes. The authorities between them had not been litigated to the point where there was settled law by the time that the Rockwell International era ended, and the production of plutonium igniters ceased.

The Genesis of Environmental Crime

The genesis of environment crime as defined by the legal regimes of the 1970s was unexpectedly the disposal of contaminated water not the storage, treatment or disposal of radionuclides. The discovery of environmental crime began with a 1986 secret briefing memo prepared by an attorney within the Department of Energy for the purpose of a high level meeting to inform Assistant Secretary Mary L. Walker and others of conditions at Rocky Flats.

The phrase "patently 'illegal'" is the most famous part of this briefing memo (DOE, 1986a). A year later the memo was released routinely to the U.S. Congress and to the Criminal Enforcement Division of the

Environmental Protection Agency and subsequently to a Special Agent of the Federal Bureau of Investigation (FBI) based in Denver, Colorado. After a two-year joint criminal investigation, an FBI Search Warrant was approved by a U.S. Magistrate. This authorized the first FBI raid ever on another U.S. government entity which began on June 6, 1989. The Department of Justice in late 1989 asked Federal District Judge Sherman Finesilver in Denver, Colorado to empanel a Special Federal Grand Jury which heard testimony for over two years.

Decisive Choices By the Department of Justice

The Justice Department made several choices which influenced decisively the course of subsequent events and determined the direction taken by the criminal justice system in the Second Epoch. These choices had the unintended consequence of precipitating a clash between two "separate but equal" branches of the U.S. Government, in this case the Legislative branch and the Executive branch.

After the Grand Jury had heard over 200 witnesses and received over 2,000 exhibits, and the U.S. Attorney had issued "target letters" to officials of the Department of Energy at Rocky Flats and to officials of the contractor from the military-industrial complex, Rockwell International, the Grand Jury was disbanded in 1992 without the indictment of the individuals whom the Grand Jury held accountable for committing environmental crimes.

Federal Judge Sherman Finesilver accepted a plea bargain which required Rockwell International to pay a fine of $18.5 million and to plead guilty as a corporation to ten felony and misdemeanor criminal violations of the Resource Conservation and Recovery Act and the Clean Water Act. Rockwell International retained its eligibility as a contractor for future federal contracts.

The Leaked and Subpoenaed Grand Jury Reports

Judge Finesilver determined that the first Grand Jury Report did not meet "necessary standards." He released only a heavily redacted version of it which did not include such judgments of the Grand Jurors as Rocky Flats was "an on-going criminal enterprise" which "should be closed." A copy of the first Grand Jury Report was "leaked" by person or persons unknown and published by a Denver newsmagazine, Westword. The Constitution Society made available on the World Wide Web the "leaked" first Grand Jury Report which included all of the redactions highlighted in blue. Most of the redactions minimized the responsibility of the Department of Energy for complicity in environmental crime, as well as that of the new contractor, EG&G, after Rockwell International ceased as contractor on December 31, 1989.

A second even harsher and more detailed version of the Grand Jury Report was uncovered in 2013 in the archives of the Congressional Wolpe subcommittee of 1992. It lacked a cover page so had gone unnoticed. But the "Second Citizen Special Federal Grand Jury Report," as it is known in this volume, cites specific exhibits as the basis for findings of "probable cause" that named individuals had engaged in environmental crime in the judgment of the Grand Jurors. The Grand Jurors in this second report (see Chapter Eleven) were highly critical of the U.S. Attorney in Denver, Colorado for granting immunity in return for their testimony to individuals who in their judgment should have been indicted.

The Wolpe subcommittee of the U.S. Congress held extensive hearings in 1992. The subcommittee issued a three volume report (Wolpe, 1992, 1993) which concluded that the Justice Department had bargained away the right of the American people to "know the truth" about Rocky Flats. The Justice Department spent two years compiling a rebuttal to the Wolpe subcommittee which was endorsed by Attorney General Janet Reno in 1994. A key conclusion of the Justice Department was that in such a high profile case, the prosecutors had used "appropriate discretion" in their pursuit of justice.

Epoch Three: Environmental Justice Delayed and the Continuing Struggle Over the Future of the Rocky Flats Site: 1990-Continuing

The "justice delayed" portion of the Third Epoch refers in part to the Cook class action lawsuit filed in 1990 which was not resolved until 2016. The lawsuit did not come to trial before a jury until 2005. The result of the trial, when it finally occurred, was a jury verdict in 2006 in favor of the plaintiffs of $926 million with accumulated interest, penalties, and fines. The 10th Circuit Court of Appeals vacated the jury verdict and the award based on provisions of the Price-Anderson Act. In 2012, the U.S. Supreme Court declined to accept the case, and returned it to the original trial court for further disposition. The plaintiffs re-filed an amended complaint in the Federal District Court of original jurisdiction. The Federal District Judge dismissed the re-filed complaint. The 10th Circuit Court of Appeals subsequently determined in 2015 that Colorado state tort law could be the basis for a finding that the plaintiffs had a valid judgment based on the 2006 jury verdict. Lawyers for the defendants appealed this determination to the U.S. Supreme Court. In May of 2016 the case was settled for $375 million just over one week before the Justices on the U.S. Supreme Court were to meet in private to decide whether or not they would accept the appeal.

Major Change In Congressional Funding Philosophy

The "clash over the public interest" portion of Epoch Three centers around the evolution of a major change in funding philosophy by the U.S. Congress on the subject of environmental stabilization and restoration of former nuclear weapons facilities, and two major interrelated developments at Rocky Flats. The evolution of funding philosophy is symbolized by the title of a report commissioned by a committee of the U.S. Senate: Train Wreck Along the River of Money (Blush and Heitman, 1995). The clear bipartisan consensus within the U.S. Congress was that as a country the United States has more significant and urgent national funding priorities than environmental stabilization and restoration at nuclear weapons plants dating from the Cold War.

The Clash Over the Public Interest

Two interrelated developments at Rocky Flats were the "Accelerated Cleanup" and the controversial creation of the Rocky Flats National Wildlife Refuge. The Department of Energy with their "Accelerated Cleanup" either imploded or dismantled and hauled away production facilities and ultrahazardous radioactive and chemical waste from mostly the high security production area. The second development is that portions of the site have been designated as the Rocky Flats National Wildlife Refuge administered by the United States Fish and Wildlife Service. An on-going clash over these developments has not resolved the concerns of many citizens over the adequacy of the cleanup and the safety of the Wildlife Refuge, and is the subject of continuing civil litigation.

Unique Hazards Encountered By Nuclear Workers

There are unique hazards encountered by workers within a nuclear weapons production facility. This obvious fact was known at Los Alamos during World War II, and at the Hanford Nuclear Reservation starting in 1943 where plutonium was produced in large quantities over the years. And the unique hazards to workers were well known at Rocky Flats. But there was not a federal law passed until 2000 which would provide compensation to former workers whose health had been negatively affected by work place unique hazards. The characterization of Epoch Three as containing "justice delayed" refers importantly to the former nuclear workers at Rocky Flats and elsewhere.

Prisms For Examining Rocky Flats

These intertwined and overlapping Epochs have in common aspects of what sociologist Kai Erikson (1991, 1994) called a "new species of trouble" for humankind. Referring to the nuclear era, the "all clear" never sounds. Nor have the conflicts and unresolved struggles of one Epoch at Rocky Flats failed to generate on-going and determined inter-related struggles over many of the same issues in subsequent Epochs.

There are three prisms for examining Rocky Flats across all three Epochs: the functioning of American democracy when citizen concerns clash with national security imperatives; examination of basic U.S. government and contractor decision-making derived from the categories which help understand failures of successful past societies which ultimately collapsed; and identification of where the U.S. "system" of constitutional government worked, where it failed to work, and where it may yet work.

Government Response To Citizen Concerns

Adapted from Lovrich and Whiteley (1999, pp. 287-332) and Dalton, Garb, Lovrich, Pierce and Whiteley (1999, pp. 9-12) inquiry centers on the following questions which arise when direct concerns of some citizens and some elements of organized civil society come into conflict with national security priorities and obligations of the U.S. government:

1. What are the viable approaches within American democracy by which citizens can maintain democratic influence over such a sensitive area of national security as nuclear weapons production when there is a curtain of secrecy covering the government's actions? This question is addressed in Chapters Fourteen through Eighteen.
2. Was it necessary in order to preserve national security to engage the Judicial branch to seal the settlement of a successful citizen lawsuit (the Church cluster of lawsuits of 1975), the work of twenty-three citizens on a Special Federal Grand Jury and the first FBI raid on another government facility? Was it necessary for national security to withhold the Appendix containing evidence underpinning conclusions in the Justice Department rebuttal to the U.S. Congress? Was it necessary for national security to destroy in secret Rocky Flats records? Was it necessary for national security to restrict access to the known evidence from the 1970s of the extent of soil contamination with plutonium? These questions are addressed in the Epilogue.

Three contexts at Rocky Flats influenced this inquiry. The first context is that in order to defend American society from adversaries which possessed nuclear weapons, there was an imperative to produce large quantities of plutonium igniters. The second context is related to an under-assessment of risk by stewards of Rocky Flats based on an unassailable belief that science, technology and engineering can solve challenging problems which were encountered over the production era from 1951 through 1989. The third context was the choice by the Atomic Energy Commission (and as much as possible in a different legal context by the Department of Energy and it military-industrial contractors) proceeding under the authorities of the Atomic Energy Acts of 1946 and 1954 to act largely outside of the normal safeguards and oversight typical of the private-sector nuclear power facilities in the United States.

The case of Rocky Flats informs the study of these questions related to the workings of American democracy in different arenas. One arena is the government response to citizen concerns by providing different sources of evidence from the nuclear weapons production era, the environmental crime era and the clashes within and between branches of the U.S. government era. A second arena is the clashes within complex civil lawsuits, over environmental justice delayed, and over the public interest in determining the future of the Rocky Flats site.

Functioning of Constitutional Democracy: Evaluating Decision-Making By Stewards At Rocky Flats In the Frame of Categories For Understanding Why Some Societies In History Have Made Disastrous Decisions and Subsequently Collapsed

This approach adapts elements of Jared Diamond's (2005) typology of categories of why some societies have made disastrous decisions (and subsequently failed) the different Epochs of Rocky Flats. Each of the Epochs has been a time when disastrous decisions were made on a number of the following dimensions:

Failure to anticipate a problem before it arrived.

In the Diamond framework, this may be because a society had no prior experience with such problems and so may not have been sensitive to the possibility of potential danger. Another reason a society may fail to anticipate a problem is that when faced with an unfamiliar situation it falls back on analogies to old familiar situations.

Perceiving or failing to perceive a problem that has actually arrived.

Diamond offers a number of reasons for such situations. For example, the origins of some problems are imperceptible. The problem may be in the form of a slow trend hidden by wide up-and-down fluctuations. A second reason is "landscape amnesia" where people forget how things used to be because change is so gradual. A classic example in contemporary society is environmental "surprise" by which two known problems combine to become more than twice as significant: super problems Hurricane Katrina or Super Storm Sandy.

Societies often fail even to attempt to solve a problem once it has been perceived.

Diamond calls this the most frequent and the most surprising. Economists and other social scientists term this "rational behavior" (or in this situation "rational bad behavior") arising between clashes of interests between people. This is called "rational" because it employs correct reasoning even though it may be morally reprehensible. Diamond says a frequent type of bad behavior is "good for me, bad for you and everybody else." Another form of clashes of interests is called the "tragedy of the commons." The tragedy is that many common resources become over harvested or destroyed.

Disastrous values

People often cling to values when they no longer make sense for the society as a whole. For example, a commitment to the values of individual freedom and independent self-sufficiency make it difficult to accept that new conditions may require some form of government planning and curbing of individual rights.

An example Diamond cited was China rejecting environmental protection values because they were perceived as alien western values and errors of capitalism—until they had serious environmental problems such as the massive air pollution engulfing many urban areas.

The problem may be beyond the capacity of a society to solve it.

The reasons may be lack of the necessary technology, knowledge, or financial resources. Forest fires are good examples. It is prohibitively expensive to reduce new growth occurring under large trees and to remove dead timber. Doing this on a large scale in the millions of acres of forest land vulnerable to forest fires is beyond the existing capacity of a society.

Efforts may be too little, too late.

Greenhouse gas global warming is a clear example. There is already so much momentum in the generation of greenhouse gases that global warming will continue for years ahead even if there is bold action now.

Some attempted solutions backfire and make the problem worse.

An example in the Diamond framework is adding chemicals to high-level waste storage tanks at the Hanford Nuclear Reservation. These chemicals led unexpectedly to the buildup of explosive gases. It was a classic example of how trying to fix one problem made another problem worse.

The Diamond framework provides a prism for highlighting and understanding disastrous decisions which were made within each Epoch, and sometimes continued across Epochs. But across the Epochs the inter-relationship and the carryover of unsolved problems, and the effects of disastrous decisions from one era to the next, underscore some of the reasons why there are such distinctive dimensions to the legacy of Rocky Flats.

"Getting Inside the Minds" of Stewards of Rocky Flats

In a most valuable section of his Introductory chapter in Collapse, Diamond (Diamond, 2005) identified an approach to "getting inside the minds" of stewards of past societies on whose watch fateful decisions were made (or not made) which became disastrous ones leading to the collapse of their societies. Diamond conveyed his idea of "getting inside the minds" of stewards whose decisions brought down their societies with a memorable turn of phrase and question about Easter Island: "...what were Easter Islanders thinking as they cut down the last tree on their island?"

In the "getting inside the minds" section, Diamond posed five questions:

> How could a society fail to have seen the dangers that seem so clear to us in retrospect?
>
> Can we say that their end was the inhabitants' own fault?
>
> Were they instead tragic victims of unsolvable problems?

How much past environmental damage was unintentional and imperceptible?

How much past environmental damage was perversely wrought by people acting in full awareness of the consequences?

These questions are addressed in the Epilogue to this volume.

Functioning of Constitutional Democracy: Where the "System" Worked, Where It Did Not Work, and Where It May Yet Work

The idea of using the evaluative frame of whether or not the "system" of U.S. constitutional democracy worked at Rocky Flats was inspired by the approach taken by Gelb and Betts (1979) in The Irony of Vietnam: The System Worked. They argued that in Vietnam the foreign policy of the United States had failed but the decision-making "system" had worked. An irony of Rocky Flats is that it was highly successful in contributions to weaponizing the atom within the legal framework of the Atomic Energy Acts of 1946 and 1954 but it did not (or could not) adapt to the new requirements of a legal regime based on environmental protection.

This prism for evaluation examines both the decision-making and policy choices which were made in the broad context of the thirty-eight years of production. During this extended time frame both the legal framework and the national consciousness evolved in ways that ultimately led to the ignominy of facility closure mired in long-lived pollution and environmental crime (see Chapters Eight through Eleven, Sixteen, Seventeen and Eighteen).

Evolution of Legal Regimes and the Protection of Human Health and Environment

Legal authority for the founding of the Atomic Energy Commission and under its direction the framework for the Nation's nuclear weapons development complexes, flowed from the Atomic Energy Acts of 1946 and 1954. President Eisenhower's Atoms for Peace program was authorized in 1954 along with the establishment of the Nuclear Regulatory Commission for the development and regulation of nuclear power.

The Price-Anderson Nuclear Industries Indemnity Act of 1957 was enacted as Section 170 of the Atomic Energy Act. It provides compensation to members of the public from a nuclear or radiological "incident." The legal definition of "incident" under this law and the role of state tort laws, especially nuisance and trespass, were to become important in a civil case filed against military-industrial contractors at Rocky Flats in 1990 and not settled until 2016.

At the founding of Rocky Flats in 1951, utilizing essentially new and unproven industrial practices, there was not yet an understanding that short-term facility infrastructure design choices and initial operational practices would have long-term negative consequences for off-site contamination, risk to human health, workplace safety, and negative impacts on the surrounding ecosystem with no known technology for remediation.

Reflecting bipartisan consensus, there had evolved by the 1970s a recognition that the extraordinary contributions to human welfare of the post World War II synthetic chemical revolution carried with it an imperative to create another legal regime emphasizing protection of human and environmental health across domains of clean air, clean water, and prevention of the release of toxics into the environment. The new suite of environmental laws from the 1970s had not been created in response to nuclear weapons production. But the new laws included provisions for felony and misdemeanor criminal violations reflecting the seriousness in creating them of the bipartisan commitment of the U.S. Congress.

The National Environmental Protection Act of 1969 (NEPA) required environmental impact statements before a federal project with potential human health impacts could be undertaken. The establishment of the Environmental Protection Agency in 1970 created a powerful Executive branch organization with a mandate totally different from that of the Atomic Energy Commission and its successor Department of Energy. Under the Clean Water Act of 1972 the two creeks passing through Rocky Flats became legally "navigable waters of the United States" with established protections. The Resource Conservation and Recovery Act of 1976 sets standards for toxic waste management and disposal.

Decisions about long-term infrastructure and decision-making about waste disposal made in the early years of the production era at Rocky Flats resulted in levels of toxic contamination which by the 1970s and 1980s did not conform to the new suite of environmental laws. This basic circumstance is why there has been such a continuing clash at Rocky Flats over criminal and civil law. The Judicial branch often has exercised a decisive role in settling conflicts between branches of the U.S. government and an empowered civil society.

The 1980s Were a Period of Regulatory and Governance Transition and the Escalation of Spirited Clashes Within and Between Institutions of the U.S. Government

The 1980s were a period of struggle between the application of legal regimes and over the authority for governance by the Department of Energy on the one hand, and on the other hand by the institutions of law associated with the creation of the Environmental Protection Agency in 1970, the legal provisions of the Clean Water Act of 1972, and the Resource Conservation and Recovery Act of 1976.

The regulatory transition struggle over which legal regimes applied to Rocky Flats escalated into spirited clashes within and between the "separate but equal" branches of the U.S. government between 1986 and 1994. During that critical period of time, Rocky Flats closed forever to the production of essential nuclear weapons components, the Soviet Union dissolved, and the Cold War ended.

Environmental Justice and Unresolved Clashes Over the Future of the Ultrahazardous Site

The challenge arose concerning what to do with the Rocky Flats site over the long term after the end of the Cold War and the close of production. A Federal District Court Judge had declared Rocky Flats in 1982 to be "ultrahazardous." The Denver metropolitan area had by 1989 expanded very close to the Rocky Flats site.

There is a continuing clash over the future of the site which will not be resolved anytime soon. An Act of Congress in 2001 created the Rocky Flats National Wildlife Refuge surrounding the 1,300 acres declared to be a Superfund site under the Comprehensive Environmental Response, Compensation, and Liability Act of 1980. The site is to be assessed every five years by the Department of Energy with the primary involvement of the Environmental Protection Agency.

The concept of justice became an established part of the dialogue about Rocky Flats, and in particular strivings to realize environmental justice for workers at Rocky Flats and for residents of surrounding communities. The American Friends Service Committee hired organizers in 1973. The Lamm-Wirth Task Force began work in 1974 (see Chapter Fourteen). The Rocky Flats Truth Force occupied the railroad in 1978-1979. Protesters from 50 countries joined together to encircle Rocky Flats. Other organizations with justice goals were soon to follow. The Rocky Mountain Peace and Justice Center was founded in 1983. Among principal goals of the Rocky Mountain Peace and Justice Center are "to restore and protect Earth and human rights," "to create a culture of justice and peace," and "to seek a healthy, sustainable relationship between people and the planet."

Facilitating Classroom Instruction In Interdisciplinary Inquiry and an Introduction To the Science of Team Science

Each chapter in the volume contains an introduction to essential elements contained within the chapter followed by a narrative summary of how the essential elements interconnect. The essential elements portion also aids a reader in determining if this is a chapter which contains content they wish to focus upon.

For law students, by way of example, Chapters Ten and Eleven present the work of a Special Federal Grand Jury. Chapter Fourteen demonstrates the power of discovery in a civil lawsuit which led to a 520-page presentation. The U.S. government made a determination to settle the lawsuit for $9 million provided the settlement was sealed.

Chapter Fifteen presents a class action lawsuit which took an unbelievable 26 years before the U.S. government agreed to a settlement of $375 million. There were ten Memorandum Opinions and Orders by Federal District Court Judges and two Writs of Certiorari to the U.S. Supreme Court.

The Saga of Rocky Flats lends itself to classroom instruction in interdisciplinary inquiry and contains significant content across the sixty-nine years where analysis and understanding is facilitated by the science of team science. So many instances of flawed decision-making occurred at Rocky Flats when stewards made choices based on one discipline or perspective, and the choices foreclosed reaching the intended long term goal.

The Compelling Future For the Rocky Flats Site Was Foretold In a Secret Meeting of Dow Chemical Senior Management and Mostly Leadership of the Steelworkers Union On April 3, 1970

There is a transcript of the meeting discovered by Holme (1978) and retained in Dow Chemical Archives. The secret meeting was of Dow Chemical senior managers and representatives of Labor Unions. Among the memorable characterizations of Rocky Flats in the production era was by Dow Chemical Senior Manager Bill Lee when he said:
"I think we have the world's barrel capitol in Rocky Flats."

And another presentation on the future of the Rocky Flats site was (at least partially predicted) for 1,300 acres and now an open question for 5,237 acres of Wildlife Refuge as of 2020 when Dow Chemical Senior Manager Jim Willging said: "You see if you start using a place for a burial ground, you can't ever use that place again for anything else."

Continuing clashes over the future of the Rocky Flats site are inevitable!

Distinctive Dimensions To the Legacy of Rocky Flats

There has been social and health harm and long-lived nuclear pollution at other nuclear weapons production facilities. But the Rocky Flats legacy is distinctive on at least six dimensions:

- **Original Siting Errors Curtailed the Capacity To Solve Evolving Public Health, Waste Management and Safety Challenges**
  There are a number of intertwined reasons why Rocky Flats was the most dangerous Cold War nuclear weapons production facility in the United States. It was located uphill and upwind from one of America's major metropolitan areas. It had two creeks which ran through it into downstream drinking water reservoirs. Concerns for preventing risks to public health and the environment were not taken into account in the original siting decisions. There was either a gross miscalculation of risk, or a major flaw in planning, concerning what to do with the sheer volume of nuclear and toxic chemical waste which would be (and was) generated at Rocky Flats by the nature of the industrial enterprise, and how to dispose of that waste in an environmentally safe and responsible manner.

- **Vulnerability To Risk of Fire**
  The special vulnerability to risk of fire in the plutonium waste generated by plutonium igniter fabrication was not sufficiently understood by the planners when the physical infrastructure of Rocky Flats was designed in the early 1950s. The lessons available

from Los Alamos and Hanford either were not known, or a choice was made not to follow them.

- **Wellsprings and Transformation of Environmental Crime**
  Understanding the wellsprings of environmental crime at Rocky Flats, and what was done about the crimes themselves, has been transformed by the clash between the Legislative, the Executive, and the Judicial branches of the U.S. government and by the workings of the criminal and civil justice systems of the United States.

- **Reconfiguration of Significant Portions of a Nuclear Weapons Production Site Into a Wildlife Refuge**
  Large portions of Rocky Flats have been reconfigured into the Rocky Flats National Wildlife Refuge administered by the U.S. Fish and Wildlife Service. This reconfiguration has galvanized civil society over issues of potential long-term threats to public health and safety.

- **Prism For Viewing Clashes and Controversies Over National Security and Environmental Crime and Over Responsiveness In a Democratic Society To Citizen Concerns**
  Rocky Flats, in part because of previously unreleased archival material, is a prism for viewing the clashes and controversies within and between segments of the U.S. government, Dow Chemical and Rockwell International as contractors from the military-industrial complex, FBI and EPA investigators of environmental crime, twenty-three citizens empanelled as a Special Federal Grand Jury, the Chief Federal Judge and the U.S. Attorney's office in Denver, the roles and actions of the Department of Energy and the Department of Justice in Washington, D.C., and the investigative role of the U.S. Congress. Civil lawsuits using the power of subpoena of the Judicial branch often have been decisive in settling divisive conflicts.

- **The Overarching Conflict Over Off-Site Contamination**
  It is clear that in their initial operational planning for Rocky Flats, the Atomic Energy Commission and Dow Chemical did not anticipate the magnitude of the nuclear and chemical waste that would be produced in normal operations, nor were there established effective monitoring systems for the flow of contamination off-site of Rocky Flats.

This chapter introduced the clash between strivings for national security, unresolved citizen concerns, and the commission of environmental crime. The notion of "clash" is an accurate characterization for Rocky Flats probably indefinitely.

Chapter Two
"A New Species of Trouble" For Humankind: The Dangerous Cold War Global Context Which Drove Weaponization of the Atom

Weaponization of the atom at Rocky Flats occurred in a dangerous Cold War global context where policy makers and decision-makers alike were under duress of preserving the very fabric and survival of their societies. Choices made at Rocky Flats were not made in isolation from developments in the rest of the world. Soviet and Chinese leaders were responding to their concerns for societal survival. The following are illustrative quotes from themes which were recurrent across the Cold War nuclear arms race in which Rocky Flats had such an indispensable role in preserving and enhancing national security for the United States.

Imperative of National Security

"For history does not long trust the care of freedom to the weak or timid."

Dwight D. Eisenhower, First Inaugural Address January 20, 1953

"The building of such a peace is a bold and solemn purpose. To proclaim it is easy. To serve it will be hard. And to attain it, we must be aware of its full meaning—and ready to pay its full price."

Dwight D. Eisenhower, Second Inaugural Address January 21, 1957

A "New Species of Trouble" For Humanity: The Nuclear Age Is Unforgiving of Mistakes

"The irony, though, is that the technological advances which have afforded us this degree of protection from natural disasters have created a whole new category of what specialists have come to call technological disasters—things that can go wrong when systems fail, humans err, designs prove faulty, engines misfire, and so on. Technological disasters have clearly grown in numbers as human beings test the limits of their competence...Natural disasters are almost always experienced as acts of God or caprices of nature. They happen *to us*. They *visit* us, as if from afar. Technological disasters, however, being of human manufacture, are at least in principle preventable, so there is a story to be told about them, a moral to be drawn from them, a share of blame to be assigned. They provoke outrage rather than acceptance or resignation."

Kai Erikson "Radiation's Lingering Dread."
Bulletin of the Atomic Scientists, March, 1991

We Could Or Should Have Acted Differently Had We Known What We Know Now

"It was pointed out that our plant [Rocky Flats] was about 20 years old working in a technical field that, for all practical purpose, was only about 25 years old. Our learning rate was approximately equal to the rate of the growth of knowledge in this "new" field...As we learned, we obviously could look on problems we had faced in the past in light of current knowledge and clearly decide we could or should have acted differently had we known what we know now."

James R. Seed, Dow Chemical Scientist,
1971 (in a report only released in 2018)

Decision-Making Half a World Away: The Soviet Union and China

"Ironically, the Russian nuclear weapons complex now has less waste in storage than the United States because large quantities of the high-level waste (as much as 1.7 billion curies) were poured into rivers and lakes or injected deep underground rather than stored in tanks."

Office of Strategic Planning and Analysis,
Closing the Circle on the Splitting of the Atom, (1995, p. 75)

"...create a new nuclear device which will make the world tremble at the might of Soviet technology."

First Secretary of the Community Party of the Soviet Union
Nikita Khrushchev's direction to Soviet nuclear weapons designers in 1961

"...the weapons [Chinese nuclear weapons] were made to fend off American and Soviet aggressors encircling China."
Inscription on a museum still closed to foreigners in remote Qinghai Province

"At the time, China's social conditions and international position were a bit like North Korea now...For many people it was purely in the spirit of 'I'll do whatever Chairman Mao tells me to'."

Liao Tianli, Chinese writer who interviewed scientists who created the bomb
"Where China Built the Bomb, Dark Memories Haunt the Ruins."
The New York Times, January 20, 2018

Wise Decisions Are Vital In the Nuclear Age: "The Man Who Saved the World"

"The slightest false move can lead to colossal consequence. That hasn't changed...That's when things can go terribly wrong...One way or another, you still need a person to order a launch of one of those weapons, and a person can always make a mistake."

Stanislav Petrov, Soviet Scientist
"Stanislav Petrov, the Russian Officer Who Averted a Nuclear War, Feared History Repeating Itself." Time Magazine September 19, 2017

Imperative of Preventing Nuclear War

> "Not even the most accomplished ideologue can tell the difference between the ashes of capitalism and the ashes of communism."
>
> John Kenneth Galbraith, Harvard Economist,
> Quest for Peace Interview with John M. Whiteley, 1984

Nuclear Weapons Were the Cold War Weapon of Choice

In the immediate post-World War II context, nuclear weapons became the weapon of choice for both the United States and the Soviet Union to deter the other from a preemptive attack. China became the third country to contribute to this dangerous Cold War global context with the explosion of its first nuclear weapon in 1964 followed by the detonation of a hydrogen bomb in 1967. The imperative in all three societies was to utilize evolving technology to gain an advantage in weaponizing the atom at whatever the cost to environment, health and safety within their respective societies.

The utilization of evolving and unproven technology to gain advantage in the Cold War contributed to the depth of psychological tension and distress within the leadership of the three competing societies. Other critical contributors to tension and distress were the addition to "acts of God" as threats to a society of emerging technological disasters as a "new species of trouble" for humankind, the dynamics of the nuclear arms race and the rhetoric of fear which compelled stewards of the national security of the United States, for example, to require Rocky Flats to fabricate the astounding number of 70,000 first stage igniters for thermonuclear bombs.

The production era at Rocky Flats unfolded in a global context which created the imperative to accelerate the weaponization of the atom. This circumstance raised the risk of encountering disastrous and truly catastrophic outcomes. The industrial processes of weaponizing the atom are not forgiving of errors.

The stakes in the global context in the early years of the Cold War became preserving freedom and the American way of life against implacable Communist adversaries who were intent on imposing their ideologies and social systems on others. Communist foes were weaponizing the atom at a breakneck pace. This disquieting global situation developed in tandem with the creation of delivery systems for nuclear weapons that were not only to provide deterrence, but also to possess the capability of launching a preemptive nuclear strike.

Increasing the Scope of the Possible and Exceeding the Prudent

Events at Rocky Flats which proved to be unbelievably fraught with danger reflected a compelling necessity to use science, engineering and technology to increase the scope of the possible and in so doing had the tragic consequence of exceeding the limits of the prudent. This chapter introduces the broad outlines of the dangerous global system and its dynamics which shaped importantly the production imperative at Rocky Flats. In so doing, it illuminates the pressures behind valuing the production of plutonium igniters at Rocky Flats above all other considerations. The U.S. government was required to proceed in a manner that assured the preservation of national security as a paramount objective in a very dangerous period of human history. This chapter is organized into nine sections:

- The unstable nature of plutonium and the implications of this "new species of trouble" for humankind, and the encounter with "radiations lingering dread" as conceptualized by sociologist Kai Erikson;
- The dynamics of the nuclear arms race and the explanation of the use of the rhetoric of fear by physicist John Holdren, the Science Advisor for President Barak Obama and a recipient of the MacArthur "Genius" Award;
- Understanding the sources of tension and psychological distress in American society from 1945-1961: Dwight D. Eisenhower on the folly of war and the imperative to pay the price to prevent war;
- The Kennedy and succeeding administrations 1961-1980s who were willing to "bear any burden" and "oppose any foe" to preserve freedom (and four examples of disasters involving delivery systems for nuclear weapons which occurred as a result of bearing the burden which could easily have been catastrophic);
- The types, the longevity and the consequences of the long-lived toxic waste which was generated in the production of nuclear weapons;
- Snapshots of occurrences within the Soviet and Chinese nuclear weapons programs which were the direct result of their approaches to addressing the challenges of nuclear weapons production;
- Lessons derived from the dangerous Cold War global context;
- A dramatic lesson for humanity from disobeying the rules: "The man who saved the world;" and
- Weaponizing the atom in this historical era.

Plutonium As "A New Species of Trouble For Humankind" and "Radiations Lingering Dread"

Plutonium was first created in December of 1940 at the University of California, Berkeley by Glenn Seaborg, Arthur Wahl, Joseph Kennedy and Edwin McMillan. They created it by bombarding uranium-238 with deuterium nuclei. This in turn created neptunium-238 with a half-life of two days, and this in turn decayed into plutonium. By 1945 there was enough plutonium to create the bomb which destroyed Nagasaki.

Plutonium was, and is, a "new species of trouble" for humankind. In Erikson's (1991, 1994) conceptualization, technological disasters have clearly grown in number as humans have tested the limits of their competence. Technological disasters at Rocky Flats and in the broader initiative to weaponize the atom have occurred, for example, when systems failed, human beings made errors, or key apparatus did not function in actual operations as it was designed to do.

A key insight of Erikson (1991, 1994) is that technological disasters are perceived as preventable in contrast to natural disasters which are perceived as "acts of God." And because technological disasters are in theory preventable, there is a story to tell about them, a moral lesson to be drawn, and a share of blame to be assigned.

Radiation scares people in "new and special ways." Emergencies involving radiation exposure "violate the rules of plot." For victims, the "all clear never sounds." Erikson observed that radiation contaminates rather than merely damages, and pollutes, befouls, and taints.

The Dynamics Driving the Nuclear Arms Competition and the Use of the Rhetoric of Fear

In a videotaped interview with John M. Whiteley in the Quest for Peace Series, Professor John Holdren explained the peculiar dynamics of the Cold War nuclear arms race. In his conception, four phrases capture the elements of why the world found itself with 50,000 nuclear weapons, each with a delivery system and a target. Destruction of the environment was driven by these dynamics which powered the massive effort to weaponize the atom in the Soviet Union, China, and the United States:

1. The "fallacy of the last move." The "fallacy of the last move" is the supposition by either side that one more weapons system or other strengthening of military forces will give it the margin of safety - or other advantage - that it seeks, as if this move will be the last. It is a fallacy because, although each side would like to have the advantage of the last move, neither will concede that advantage to the other; thus there is never a "last move."

2. The "worst-case analysis." "Worst-case analysis" means that each side tends to assume the worst about both the military capabilities and the intentions of the other side, and to try to develop its own capability to cope with this "worse case," preferably with a further margin of safety for the sake of prudence in assuring national security. The United States government argues that, given its responsibilities not only for the defense of the nation but also for the defense of democracy, to do any less would be irresponsible. The government of the Soviet Union, of course, makes the same argument, but with the substitution of "communism" for "democracy" to explain what is being defended. Both sides cannot simultaneously enjoy the "margin of safety" in military capability.

3. The "action-reaction syndrome." The "action-reaction syndrome" means simply that each time one side develops a new weapon or refinement, or takes a major geopolitical initiative, the other side feels obliged to emulate or counter it. The reaction may be in kind

(for example, one side deploys a particular type of missile because its adversary is doing so); it may be a specific countermeasure (for example, deployment of more air defenses by one side to counter more bombers on the other); or it may be a general increase in weapons development and deployment in response to a broadly threatening move by the adversary (for example, the U.S. military buildup starting in 1950, in reaction to the Communist invasion of South Korea). This syndrome may be characterized as "Monkey-See-Monkey-Do."

4. The "military-industrial complex." The concept of the "military-industrial complex" embraces not only the designers and manufacturers of weapons, whose influence on the arms race has been documented by Solly Zuckerman and many others, but also the armed forces. The branches of the armed forces compete with one another - as well as with their counterparts on the opposing side - in upgrading their arsenals. The governmental defense bureaucracies maintain and expand their influence by guarding jealously the weapons programs under their jurisdiction, and by campaigning tirelessly for new ones. On the Western side, defense-oriented legislators stay in office with the help of the steady flow of money they steer into military facilities and defense industries in their districts.

The Use of the Rhetoric of Fear

Throughout the history of the nuclear arms race, leaders on both sides of the Cold War conflict have used rhetoric about nuclear forces - and about intentions relating to their use - for two main purposes. The first purpose has been to intimidate one's adversaries, with the aim of reducing their inclination to counter or initiate attempts to alter the geopolitical status quo. The second purpose has been to frighten one's own society into willingness to bear the economic burdens of high military spending.

An illustration of the utilization of the rhetoric of fear in the nuclear arms race was an initiative in 1961 by Nikita Khrushchev, the First Secretary of the Communist Party of the Soviet Union. Khrushchev directed Soviet nuclear weapons designers to create a new nuclear device which will "make the world tremble at the might of Soviet technology" (as quoted by the British Broadcasting Corporation in 2017) or "to show the imperialists what we can do" (as quoted by Adamsky and Smirnov in 1994 in the Cold War International History Project). The new device was a "fission-fusion-fission" creation with the capability of 100 megatons (100 million tons of TNT) which is 3,000 times more powerful than the two bombs which destroyed Hiroshima and Nagasaki. When the actual device was detonated on October 30, 1961 it was configured so the explosive power was "only" 50 million tons of TNT or 1,500 times more powerful than the two bombs used on Japan.

The timing was approximately one year before the Cuban Missile Crisis erupted as a threat to world peace between the United States and the Soviet Union. The total explosive power of all the bombs in World War II including the atomic bombs is variously estimated at between 5 megatons and 9 megatons in comparison with the 50 megatons of the Khrushchev initiated device. The reasons for introducing this largest device ever exploded in the atmosphere (or anywhere else on the planet for that matter) in this narrative on the dynamics of the Cold War context was that nuclear weapons were the weapon of choice for national survival.

Exceeding the limits of the prudent in the search for national security had the opposite consequence of increasing the risk of miscalculation from seeking strategic advantage relying on unproven technology.

Understanding the Sources of Tension and Psychological Distress In American Society 1945-1961: Dwight D. Eisenhower On the Folly of War and the Imperative To Pay the Price To Prevent War

Dwight D. Eisenhower left a very precise record of the progression of his thinking from 1945 through 1961. Representative excerpts are presented in chronological order. Collectively they reveal the complexity of his thinking on the folly and destructiveness of war, and on the imperative to pay the price to prevent war:

> "War is a grim, cruel business, a business justified only as a means of sustaining the forces of good against those of evil." Transcription Made for the National War Fund at the Request of Luther Hill, September 11, 1945
>
> "For Americans, only the threat to our way of life justified resort to conflict." Graduation Exercises, U.S. Military Academy, June 3, 1947
>
> "Because, therefore, we are defending a way of life, we must be respectful of that way of life as we proceed to the solution of our problem. We must not violate its principles and precepts, and we must not destroy from within what we are trying to defend from without." Speech Before NATO Council, November 26, 1951, Pre-Presidential Papers, Box 197
>
> "For history does not long trust the care of freedom to the weak or timid." First Inaugural Address, January 20, 1953
>
> "Since the advent of nuclear weapons, it seems clear that there is no longer any alternative to peace, if there is to be a happy and well world." Remarks to the Department of State Honors Awards Ceremony, October 19, 1954
>
> "...our readiness to meet and defeat this kind of possible attack is forced upon us, both as a potent preventive of actual war and to ensure survival in event of attack. This alertness to danger had to be translated into specific policies and activities in the several parts of our world where our rights--our way of life--can be seriously

damaged. Work of this kind occupies my days and nights." Letter from DDE to Hallock Brown Hoffman, February 7, 1955

"Today there is a great ideological struggle going on in the world. One side upholds what it calls the materialistic dialectic. Denying the existence of spiritual values, it maintains that man responds only to materialistic influences and consequently he is nothing. He is an educated animal and is useful only as he serves the ambitions–desires–of a ruling clique; though they try to make this finer–sounding than that, because they say their dictatorship is that of the proletariat, meaning they rule in the people's name–for the people... Now, on our side, we recognize right away that man is not merely an animal, that his life and his ambitions have at the bottom a foundation of spiritual values." Remarks at 11th Annual Washington Conference of the Advertising Council, March 22, 1955

"The building of such a peace is a bold and solemn purpose. To proclaim it is easy. To serve it will be hard. And to attain it, we must be aware of its full meaning--and ready to pay its full price." Second Inaugural Address, January 21, 1957

"Having established as our goals a lasting world peace with justice and the security of freedom on this earth, we must be prepared to make whatever sacrifices are demanded as we pursue this path to its end."

Remarks at the Fort Pitt Chapter, Association of the U.S. Army, May 31, 1961

The idea in his Second Inaugural Address on January 21, 1957 of being ready "to pay the full price" is a metaphor for understanding the challenges and perils of weaponization of the atom.

The Kennedy and Succeeding Administrations 1961-1980: Bearing Any Burden and Opposing Any Foe To Preserve Freedom

This section presents only four of the disasters with nuclear weapons delivery systems which occurred in the years between the start of the Kennedy administration in January of 1961 and toward the end of the Carter administration in 1980. These four disasters have in common that they did not result in catastrophic thermonuclear explosions. They did, however, create the gravest of danger that a thermonuclear explosion could have occurred:

- when a B-52 disintegrated after 10 hours of flight dropping two 3.8 megaton nuclear bombs onto a North Carolina farm;
- when a B-52 collided with a tanker when refueling after 17 hours of flight dropping three nuclear bombs in or near Palomares in southern Spain, and one into the Mediterranean Sea;
- when a B-52 crashed at 600 miles an hour into the Greenland ice 700 miles north of the Arctic Circle in the dead of Winter; and

- when a 9-megaton nuclear warhead was hurled 1,000 feet into the air by an explosion in a Titan ll missile silo and landed 200 yards from its destroyed silo and only 50 miles from Little Rock, Arkansas.

The Goldsboro, North Carolina Crash of January 1961
The crash at Goldsboro occurred on January 23, 1961, only three days after the inauguration of President Kennedy.

<u>The Cause of the Goldsboro B-52 Crash Was Equipment Malfunction</u>

The basic cause of the Goldsboro crash was equipment malfunction. During the second mid-air refueling, someone on the tanker noticed that jet fuel was cascading out of one of the wings of the B-52.

<u>During the Second Refueling After 10 Hours of Flight Equipment Failed</u>

The B-52 had taken off from Seymour Johnson Air Force Base in Goldsboro, North Carolina for what Schlosser (2013, p. 245) called "a long circular route along the East Coast." It was during refueling after ten hours of flight that the boom operator of the tanker observed a fuel leak from the right wing. An imbalance in weight occurred when jet fuel would not drain from the left wing after the tanks on the right wing had lost 40,000 gallons in about 2 minutes. The disintegration of the bomber caused the release of two hydrogen bombs over the farming village of Faro, North Carolina and the death of three crewmen on the B-52.

<u>The Fate of the B-52 and Its Crew</u>

Schlosser (2013, p. 245) reported that the plane went into an uncontrolled spin followed by an explosion. The plane started to break apart at an altitude of 10,000 feet. Major Eugene Shelton, Major Eugene H. Richards and Technical Sergeant Frank R. Barnish were put into harms way as the plane disintegrated, and lost their lives in defense of the United States.

<u>The Fate of the Two 3.8-Megaton Hydrogen Bombs</u>

Nuclear weapons in the arsenals of the United States are equipped with a variety of different safety devices to prevent either their unauthorized use or their accidental detonation. In the case of the two 3.8 megaton Mark 39 hydrogen bombs on this B-52, the following is the explanation for the bomb whose parachute opened:

> Every safety mechanism had failed, except one: the ready/safe switch in the cockpit. The switch was in the SAFE position when the bomb dropped. Had the switch been set to GROUND or AIR, the x-unit would've charged, the detonators would have triggered, and a thermonuclear weapon would have exploded…"
>
> Schlosser, 2013, p. 246

It was only 1 of 7 safety switches which had not fired: the one which had to be set manually by the pilot.

Figure 2.1 Fate of One of the Two 3.8-Megaton Hydrogen Bombs Dropped Onto an Agricultural Field In Rural North Carolina

The parachute of this hydrogen bomb deployed as it should have which prevented major damage to the bomb itself and the detonation of the high explosives which were part of the first stage plutonium igniter system.

The status of the authorization to explode mechanism for the second Mark 39 3.8 megaton bomb has now been disclosed with the declassification of a 1961 study by deMontmollin and Hoagland released by the National Security Archive in 2014. The parachute on the second Mark 39 bomb did not open. When the bomb hit the ground the explosive package penetrated the field it landed in to a depth of 18 feet. The high explosive package did not detonate. The primary plutonium igniter was not damaged.

The impact separated the secondary fusion portion from the rest of the bomb. An enduring mystery is why the impact caused the "authorization to detonate" mechanism to switch mechanically to the "fire-authorized to explode upon landing," and the device did not explode. It may be that the nuclear experts themselves from Sandia National Laboratory do not know.

The deMontmollin and Hoagland (1961) report provides significant new information on Weapon 2 which did not have its parachute even partially open:

> It landed in free-fall. Without the parachute opening, the timer did not initiate the bomb's high voltage battery ("trajectory arming") a step in the timing sequence. While the Arm/Safe switch was in the "safe" position, it had become virtually armed because the impact of the crash had rotated the indicator drum to the "armed" position. But the shock also damaged the switch contacts, which had to be intact for the weapon to detonate. While Weapon 2 was not close to detonation, the fact that the physical impact of a crash could activate the same arming mechanism that had kept Weapon 1 safe showed the danger of such accidents.

One Low Voltage Switch Prevented a Major Catastrophe

Schlosser quotes Parker F. Jones, a safety engineer at Sandia, as stating in an engineering report that "one simple, dynamo-technology, low-voltage switch stood between the United States and a major catastrophe!" (Schlosser, 2013, p. 247).

Figure 2.2 Authorization To Detonate Mechanism For a Hydrogen Bomb Controlled By the Pilot of a B-52 Bomber

This is the authorization to detonate mechanism for a hydrogen bomb on a B-52 bomber. Under the control of the pilot, there are three choices: SAFE, not authorized to explode; AIR, authorizing detonation above the target; or GROUND, authorizing an explosion upon landing. In this disaster, six authorization switches had fired by the time the bomb hit the ground. As the B-52 started to disintegrate, the pilot had left the detonate mechanism in the SAFE position.

Palomares B-52 Collision With a KC-135 Tanker On January 17, 1966

According to Seed (1971) the collision between the KC-135 tanker based at Morón Air Base in southern Spain and the B-52 based at Seymour Johnson Air Force Base in Goldsboro, North Carolina occurred at 32,800 feet on January 17, 1966. The B-52 had been on an Operation Chrome Dome Mission with a flight plan requiring it to fly across the Atlantic Ocean and the Mediterranean Sea to the European border of the Soviet Union before returning back to Seymour Johnson Air Force Base. The flight plan called for two mid-air refuelings over Spain. The collision occurred during the second refueling.

Schlosser (2013, p. 315) reported that the reason for the collision was that after 17 hours of flying time, the B-52 approached the tanker at about 10:14 A.M. "too quickly, flew into the fuel boom and started to break apart." Flames traveled straight through the boom into the tanker.

There are different emphases in the Seed (1971) and Schlosser (2013) reports. The Seed report, compiled by a committee chaired by James R. Seed (Seed, 1971), had a main focus on the contamination of plutonium deposited in the soil at Palomares. After service in the Korean War, Dr. Seed earned a Ph.D. from the University of California, Davis in Physical Chemistry in 1964, and developed a specialty in the field of radiation chemistry. Eric Schlosser is an investigative scholar who made in-depth use of both primary source documents and detailed reports from informed individuals which he compiled and reported in 2013.

This report on Palomares is vitally informed by both experts. It will begin with Schlosser's recounting of the immediate aftermath of the collision, the fates of the crews of both planes, the frantic search for nuclear weapons in the Spanish countryside, and the ultimate recovery of three of the four Mark 28 1.4 megaton nuclear weapons (or their remains). Subsequently, based on the assumption that the fourth Mark 28 had been lost somewhere in the Mediterranean Sea, a search offshore of Palomares began while a Soviet Union ship monitored the search area.

Fates of the Crews On Both Planes

The four crewmen on the KC-135 tanker based at Morón Air Base perished when fire from the B-52 accelerated up the boom into the tanker. The immediate aftermath for the seven crewmen on board the B-52 was both harrowing for the four who survived, and tragic for the three who lost their lives in the defense of the United States: the gunner, Sergeant Ronald Snyder; the electronic warfare officer, Lieutenant George Glesner; and the navigator and youngest crewman at 23 years, Lieutenant Stephan Montanus.

Three of the four B-52 crewmen who survived the collision were carried out to sea by the prevailing winds and rescued by Spanish fishing boats at distances ranging from 3 to 8 miles off of the coast: the copilot who was flying the plane at the time of the collision, Major Larry G. Messinger; copilot Michael J. Rooney; and the pilot Captain Charles J. Wendoff. The parachute of the radar navigator, Captain Ivan Buchanan, initially failed to open. Also, he was stuck in his ejection seat. He survived a hard crash landing because he had the presence of mind to unpack his parachute by hand. The landing knocked him unconscious, broke his shoulder, and hurt his back. Villagers found him (Schlosser, 2013, p. 315).

The Fate of Three of the Four Mark 28 Hydrogen Bombs

It took a period of several days to learn the fate of three of the four Mark 28 hydrogen bombs. A Disaster Control Team from the Strategic Air Command Base in Torrejón, Spain arrived to search for the bombs. Spanish Federal Police found one bomb southeast of Palomares. One of its parachutes had opened so it was mostly intact. The parachute of a second bomb had failed to open. Its package of high explosives partially detonated in the hills near a local cemetery. The high explosive package of a third bomb partially exploded near a farmhouse and tomato field. After a fruitless 6-week search for the fourth bomb, the search of the countryside was called off (Schlosser, 2013, p. 317).

Recovery of the Hydrogen Bomb Lost At Sea

The narrative of the recovery, created by Schlosser's (2013) access to primary sources, began with the assignment of Randall C. Mayhew to the recovery effort. He was head of the Aerodynamics Department at Sandia National Laboratory in New Mexico. The Aerodynamics Department had designed the parachutes and casings for the Mark 28.

Mayhew consulted with Bob Peurifoy, a leading advocate for the safety of nuclear weapons within Sandia and nationally. Whatever their independent contributions, the result of this consultation was the use of "reverse trajectory calculations" based on where the engines and the three recovered bombs hit the ground. The outcome of these calculations was that the collision was reconstructed to have happened (assumptions of altitude were not further explained) at an altitude of 15,000 feet two miles from the coast within a circular one-mile area. They factored in the prevailing winds, the fact that the tail plate of the lost bomb had been found on the beach, and the assumption that the parachute had opened. Several days later a report by a fisherman in the area collaborated this general location.

There were four submersibles searching for the lost hydrogen bomb. Schlosser (2013, p. 219) quoted Rear Admiral William S. Guest, the commander of the operation, that: "It isn't like looking for a needle in a haystack…It's like looking for the eye of a needle in a field full of haystacks in the dark." On March 15, 1966 the submersible *Alvin* found the bomb at a depth of approximately half-a-mile. Nine days later, while the bomb was being raised, the recovery line broke and the bomb was lost again for another week until *Alvin* located it a second time. This time cables were attached at a depth of 80 feet. The bomb supported by cables was safely hoisted onto the deck of a U.S. Navy ship.

Figure 2.3 The Mark 28 1.4-Megaton Hydrogen Bomb Which Was Recovered Largely Intact From the Mediterranean Sea

This is an originally classified U.S. Navy photograph of the hydrogen bomb (a tail plate from the bomb washed up on the shore near Palomares) which was recovered from the Mediterranean Sea. It was largely intact because one of the two parachutes had opened. The U.S. Navy Admiral supervising the recovery said "It's like looking for the eye of a needle in a field full of haystacks in the dark." He was referring to the exploration at a depth of approximately half-a-mile in a circular one-mile area by four submersibles.

Radioactive Contamination of Palomares

According to Seed (1971), 558 acres were contaminated by the non-nuclear explosions of two of the four hydrogen bombs. A total of 54 acres had an alpha contamination of 700,000 d/m/100 cm^2. The top 2" to 3" of soil totaling 1,100 cubic yards were removed along with 400 cubic yards of vegetation and shipped to the United States for burial at Savannah River in South Carolina. Lower amounts of alpha contamination on 42 acres were deep plowed to a depth of 13 inches, watered, and some vegetation removed. A remaining 511 acres were deep plowed and watered.

Sickness Among U.S. Airmen Involved In the Palomares Cleanup

On June 19, 2016 Dave Phillips of The New York Times in "Decades Later, Sickness Among Airmen After a Hydrogen Bomb Accident" quoted then 22-year old (in 1966) Frank B. Thompson that: "They told us it was safe, and we were dumb enough, I guess, to believe them...There was no talk about radiation or plutonium or anything else..."

Mr. Thompson spent days searching contaminated fields without protective equipment. Dave Phillips reported that Mr. Thompson, now age 72, has cancer of the liver, kidney, and lung. The thrust of the Phillips story is that the position of the U.S. government has been, and continues to be, that there was no harmful radiation at the crash site, the danger of contamination was minimal, and strict safety measures ensured that the 1,600 troops were protected.

There has been no mortality study done to determine whether or not there is an elevated incidence of disease in U.S. troops from the Palomares "cleanup." The New York Times was able to locate and subsequently contact 40 of the 1,600 troops: 21 had cancer and 9 had already died from cancer-related causes.

Abandoning a B-52 By Parachute Over Thule, Greenland On January 21, 1968

With access to primary source documents and knowledgeable human sources, Schlosser (2013, pp. 320-325) provides a definitive account of what caused the B-52 crew to have to abandon a B-52 near Thule, Greenland by parachute, and reported more complete information about the "lost" Mark 28 bomb than appears elsewhere in the unclassified literature about the incident.

The national security context was that by 1968 having twelve B-52s on airborne alert at all times had been augmented as a deterrent of the Soviet Union by 1,600 ballistic missiles. Schlosser (2013, p. 319) reported that both the Joint Chiefs of Staff and the new Commander of the Strategic Air Command believed in that era that continuing the B-52s on airborne alert was critical for national defense.

Cause of the Thule Disaster

The B-52 which crashed near Thule was flying a "bowtie" pattern at 35,000 feet above the ballistic missile early warning complex in Western Greenland. The cause of the on-board fire was the ignition of cloth-covered foam-rubber cushions on the chair of the instructor navigator. Because of coldness in the aircraft, the pilot brought into the cabin heated air from an engine manifold at 428 degrees Fahrenheit. The cushions were blocking a vent under the seat of the instructor navigator. The fire which ignited in the cushions could not be extinguished by the crew, and soon became out-of-control.

Captain John Haug wanted to increase the odds that the crew would survive in the -23 degree Fahrenheit weather (-44 Fahrenheit wind chill) when they bailed out. After the crew was all out, Captain Haug ejected from the plane 4 miles short of the Thule runway. The B-52 crashed and exploded on Bylot Sound. Co-Pilot Captain Leonard Svitenko died leaving the plane. Navigator Captain Cliss could not be found until midnight, and lost part of both feet to frost bite.

The Fate of the Hydrogen Bombs

The high explosives on the four Mark 28s and approximately 225,000 pounds of jet fuel burned for five or six hours. The patch of blackened ice totaled 160 yards wide and 720 yards long. An area of three square miles contained pieces of the plane and bomb remains. Schlosser (2013, p. 323) reported that the bombs did not detonate because during the Eisenhower administration they had been made "one point" safe (not further explained).

The Challenge To Decontaminate In the Cold and the Dark

The problem for the U.S. Air Force was to decontaminate three square miles of ice and snow in the dark during the middle of Winter 700 miles North of the Arctic Circle. The ice at the crash site had refrozen after the crash. The water under the crash site was 600 feet deep.

A new technology, the Field Instrument for the Detection of Low-Energy Radiation (Schlosser, 2013, p. 323) searched not for alpha particles, but for x-rays and gamma rays. It was not until Summer of 1968 that a Navy submersible could search 600 feet under the ice for the parts remaining of the missing bombs. The first stage igniters (undoubtedly fabricated at Rocky Flats) had been "blown to bits." Most of the uranium was recovered. The radioactive waste which filled 147 railway cars was sent to South Carolina for permanent disposal.

The B-52 was estimated to have hit the ice at 600 miles-per-hour. What Schlosser (2013, pp. 323-324) called "a crucial piece" of a bomb was still missing at the end of the search of the floor of Bylot Sound. That lost piece was an "enriched spark plug" (not more fully explained) and it was never found. There never had been an entire intact bomb missing. The U.S. airborne alert program was immediately cancelled after the crash in Thule in favor of B-52s on ready alert parked next to active runways. Deterrence continued to be bolstered by the 1,600 ballistic missiles which by then had been deployed.

Figure 2.4 The Challenging Cleanup of a B-52 Crash Site and Its Nuclear Debris In the Dark During the Middle of the Winter 700 Miles North of the Arctic Circle

The volume of the debris which was recovered spread over three square miles filled 147 railway cars. The first stage plutonium igniters had been destroyed. The severity of the incident danger led to a major U.S. change in deterrence policies. The continuous B-52 airborne alert was replaced by reliance on the deployed ballistic missiles in hardened silos.

The Damascus Titan ll Missile Accidental Puncture and Explosion On September 18-19, 1980

The Titan ll missile was 10 feet in diameter and 103 feet tall. This is the height of a 9-story building. The W-53 bomb mounted on top of the Titan ll had a yield of 9 megatons. The two liquid fuel propellants were extremely toxic. When mixed together they provided the explosive power to launch the missile. Aerozine-50 is a carcinogen which if inhaled can cause breathing difficulties, vomiting, convulsions, tremors, and death. The oxidizer, Nitrogen Tetroxide, is classified in the "Poison A" deadliest group. The Titan ll weighed 330,000 pounds. It had a range of over 6,000 miles and an accuracy within one mile. Compared to the explosive power of all the bombs used in World War II including both atomic bombs, the explosive power of the W-53 9 megaton warhead was greater than all of them combined. The Titan ll stayed in service for over 25 years.

<u>The Titan ll As "Flawed and Unreliable"</u>

Schlosser (2013) reviewed evidence that the Titan ll was widely understood to be a deeply flawed national security asset. Those flaws were fully understood by the top national security policy makers of the United States. During the Nixon administration, National Security Adviser Henry Kissinger explained that the Titan ll was "inaccurate and unreliable." The "Pentagon had been wanting to scrap [it] for years and I had kept it in service for trading purposes." In 1972 Kissinger offered a deal to the Soviets: the U.S. would decommission the Titan ll if the Soviets in return would decommission their SS-9 missiles. The Soviets declined the offer.

Figure 2.5 The Danger Posed By a Titan II and Its Warhead Was In Their Extraordinary Power and Potential for Disaster

The Titan II missile was massive: the height of a 9-story building with a diameter of 10 feet. The power generated by the mixing of the two liquid propellants could power the 330,000-pound missile to a reported range of over 6,000 miles. The explosive power of its W-53 warhead was greater than all of the bombs used in World War II.

The Ultimate Cause of the Explosion

The ultimate cause of the explosion was human error. A member of Propellant Transfer Team A used the wrong tool during a routine technical procedure. He used a ratchet wrench rather than a torque wrench to work on a fuel cap. The 9-pound socket of the ratchet wrench fell and hit a thrust mount, which deflected the socket into the skin of the rocket. The skin of the rocket only had the thickness of a nickel. A leak immediately began in the Aerozine-50 tank.

According to Schlosser (2013), it was not unusual for some implement to be dropped accidentally into the silo. It would land harmlessly at the bottom of the silo. The difference in this case was the deflection of such a heavy part by a thrust mount after it had fallen down the silo to the level of the Aerozine-50 tank which was immediately above the Nitrogen Tetroxide tank.

Chaos Followed the Puncture

Colonel John T. Moser, the commander of the 308th Strategic Missile Wing based in Little Rock, Arkansas, was leaving for a concert when notified that there might be a fire at Launch Complex 374-7. On his way to the Launch Complex, he ordered the convening of the Missile Potential Hazard Group. Schlosser (2013, p. 209) reported that flashing red lights in the Control Center were warning of a fuel leak, an oxidizer leak, or a fire in the Titan ll silo: "three things that could not be happening at once." (Colonel Moser had previously worked on solid fuel Minuteman Missiles and was new to the challenges and dangers posed by liquid fuel rockets such as the Titan ll). The Missile Potential Hazard Group joined the consultation as a vital expert resource. This was a "new species of trouble" for the command structure of the Titan ll in Damascus.

Two critical concerns quickly emerged after the electric power to the missile was shut off. The first critical concern was to prevent an electric spark from igniting an explosion. The second critical concern was whether they could have "faith in the steel blast doors" to contain an explosion if it occurred (Schlosser, 2013, p. 215). The pressure in the Stage 1 Aerozine-50 tank dropped to -2 psi raising the fear it could collapse. A collapse would cause a massive explosion from mixing the Aerozine-50 in the silo with the oxidizer Nitrogen Tetroxide.

Overall command authority was assumed by Lieutenant General Lloyd Levitt from the headquarters of the Strategic Air Command. He determined that it was not in the national security interest of the United States to open the steel silo door to release the toxic fumes. The fumes were to stay contained in the silo. He did not want to authorize the opening of the steel silo door which would increase the risk of losing control over a thermonuclear weapon (Schlosser, 2013, p. 215-216). General Levitt was especially credible as the overall commander: he had flown 100 combat missions in the Korean War, 152 combat missions in Vietnam, been a U-2 pilot from 1957 to 1960, and flown B-52s for 1 year. Schlosser (2013, p. 218) characterized the situation as "grim" when a representative of Martin Marietta, the manufacturer of the Titan ll, gave estimates of the range of tank pressure in Tank 1 which established the parameters for when Tank 1 was likely to collapse.

The Explosion Occurred When a Fan Was Ordered To Be Turned On

Later in the night following the evening that the socket had punctured a hole in the Aerozine-50 tank, Schlosser (2013, p. 232) quoted Senior Airman David Livingston as saying to Senior Airman Greg Devlin "I've got a bad feeling about this...somebody's going to die out there tonight."

An instruction from General Levitt at SAC Headquarters was: "no electrical switch was to be turned on or off without permission of SAC headquarters" (Schlosser, 2013, p. 239). Colonel Moser ordered that two men in Rocket Fuel Handler's Clothing Outfit (REFCO) protective suits should enter the Control Center. The men chosen for this hazardous assignment were Greg Devlin and Rex Hukle. They struggled with a crowbar for 15 minutes to open a heavy steel door. Once inside, the first blast door they encountered would not open with a hand pump. The air supply in their protective suits was low by then. They had to evacuate the complex.

Jeff Kennedy and David Livingston volunteered to complete the mission to gain access to the Control Center. When they entered the second blast door the level of the Aerozine-50 fuel vapor exceeded 250 parts per million, and inside the silo it was in excess of 21,000 parts per million (Schlosser, 2013, p. 391-392).

The explosion occurred after a crew chief had "an idea: maybe they should turn on the ventilation fan to clear out some of the fuel vapor" (Schlosser, 2013, p. 392). It is not clear if the "idea" had been approved by General Levitt at the Strategic Air Command. David Livingston volunteered to return back inside, turn on the switch for the fan, and had nearly exited the Titan ll when it exploded. Kennedy was catapulted against the security fence, knocked unconscious, and suffered a broken leg. When he awoke, he heard Senior Airman Livingston calling for help.

Figure 2.6 The Total Obliteration of the Silo Which Was Destroyed By Only the Non-Nuclear Elements of a System For Deterrence Provides a Visualization of the Danger To Their Possessors of Nuclear Weapons and Their Delivery Systems
This dramatic aerial view is reprinted with the permission of the Arkansas Democrat-Gazette. The 740-ton silo door had been blown two hundred feet into the air and landed 600 feet northeast of the missile complex.

<u>Extraordinary Bravery and Sacrifice Under Conditions of Extreme Danger</u>

Security Police Officers, Technical Sergeants Donald V. Green and Jimmy E. Roberts, approached the Titan ll complex to aid in the rescue. Jeff Kennedy was rescued by Richard L. English of the Disaster Preparedness Unit, Technical Sergeant David G. Rossborough who was a first responder, Captain George Short of PTS Maintenance, and Colonel James L. Morris. It was Technical Sergeant Jimmy E. Roberts who found the critically injured David Livingston, and first responder Technical Sergeant David G. Rossborough who helped move Livingston away from near the silo. Kennedy and Livingston were transported by helicopter to the Baptist Medical Center in Little Rock. David Livingston did not survive the severity of his wounds.

I'm the Vice President of the United States

Vice President of the United States Walter Mondale happened to be in Little Rock for the state convention of the Arkansas Democratic Party. Vice President Mondale called Secretary of Defense Harold Brown to ask if there had been a nuclear weapon on top of the Titan ll. According to Schlosser (2013, p. 424), at first Secretary Brown declined to tell him the answer. Mondale's response was "Goddamn it, Harold, I'm the Vice President of the United States." He got the answer. There was a 9-megaton warhead mounted on top of the Titan ll. The warhead had reached an altitude of 1,000 feet before falling back to Earth and landing in a ditch 200 yards from the silo.

"A New Species of Trouble": The Long Term Disposal of Transuranic Waste

The generation of nuclear waste and its disposal became a defining issue of Rocky Flats. Linking Legacies (Office of Environmental Management, 1997, p. 228) provides definitions of transuranic elements and transuranic waste. Since these man-made substances are such a "new species of trouble" for humanity in the formulation of sociologist Kai Erikson (1991, 1994), the definitions are as follows:

> Transuranic elements-All elements beyond uranium on the periodic table including *neptunium, plutonium, americium,* and *curium.* All transuranic elements are man-made.
>
> Transuranic waste-Waste contaminated with *uranium-233* or *transuranic elements* having half-lives of over 20 years in concentrations more than 1 ten-millionth of a *curie* per gram of waste.

Waste From Nuclear Weapons Production

Linking Legacies uses the term "waste" to refer to solids and liquids that are "radioactive, hazardous, or both" (Office of Environmental Management, 1997, p. 31). Waste in this use of the term had been disposed of in the past "by shallow burial, sea burial, or deep underground injection" (Office of Environmental Management, 1997, p. 31).

Without specifying a date, Linking Legacies indicated that "*hydrofracture* (an underground injection disposal technology) and sea disposal of radioactive waste have been discontinued" (Office of Environmental Management, 1997, p. 31). Waste from nuclear weapons production managed by DOE totals 24 million cubic meters by volume containing 900 million curies. Linking Legacies notes that by way of comparison, spent nuclear fuel from nuclear power generation contains 29 billion curies.

Definitions and Categories of Waste From Nuclear Weapons Production

Linking Legacies describes seven categories of waste from nuclear weapons production.

High-Level Waste

This category results from the chemical processing of spent nuclear fuel. At Hanford, 123 of the 149 single shelled tanks containing high-level waste had to be emptied and consolidated into double-shelled tanks in the first decade of the 21st century because of leaking into the environment. Linking Legacies observed that "under federal law, DOE high-level waste will eventually be disposed of in geologic repositories after it has been treated to produce solid waste forms acceptable for disposal, and repository facilities become available" (1997, p. 38). No such permanent repository facilities exist for high-level waste. Rocky Flats did not generate this category of waste.

Transuranic (TRU) Waste

This category applies directly to Rocky Flats. TRU waste contains alpha-emitting transuranic elements and can remain "radioactive for hundreds of thousands of years" (Office of Environmental Management, 1997, p. 40) and includes plutonium-239, -240, -241, -238 and -242; americium-241; and curium-244.

On a seriously disquieting note, Linking Legacies reported that:

> AEC first managed TRU waste as a separate category of radioactive waste until 1970. Prior to that time [1970], TRU waste and low-level waste were usually combined and managed as a single waste type and were disposed of in shallow burial trenches. Recognizing the need to isolate TRU waste more permanently from the environment, AEC discontinued shallow burial of TRU waste in 1970.

Rocky Flats prior to 1970 had 14 shallow burial trenches, some of which were utilized as repositories for TRU waste.

Low-Level Waste

This category of waste was present at Rocky Flats. Low-level waste "is composed of all radioactive waste not classified as high-level waste, TRU waste, spent nuclear fuel, or natural uranium or thorium byproduct material defined under section 11e(2) of the AEA...Low-level waste containing hazardous waste or PCBs is categorized as mixed low-level waste and is presented separately..." (Office of Environmental Management, 1997, p. 44). At Rocky Flats, this category of waste as of 1997 by volume was 5,300 square meters (p. 47).

11e(2) Byproduct Material

This category is waste from extraction or concentration of uranium or thorium was not reported to be present at Rocky Flats.

Mixed Low-Level Waste

As of 1997, there were 14,000 square meters by volume of mixed low-level waste at Rocky Flats. Its management is "considered separately from other low-level waste because of the presence of RCRA-regulated constituents [which] is a major factor in determining how it is managed" (Office of Environmental Management, 1997, p. 54).

Hazardous constituents in low-level mixed waste include "toxic heavy metals, organic and halogenated organic chemicals, cyanides, inorganic chemicals and elements, explosive compounds, and corrosive chemicals and solutions" (Office of Environmental Management, 1997, p. 56).

Hazardous Waste

Material is classified as hazardous under RCRA only if it is a solid waste and either listed in the regulations or it "exhibits a characteristic of corrosivity, ignitability, reactivity, or toxicity" (Office of Environmental Management, 1997, p. 56).

Other Waste

This is a DOE waste category not described above because it does not fit into a previous category or because of its "chemical and radiological composition." Examples are PCBs and PCBs mixed with radioactivity which are subject to the Toxic Substances Control Act, and asbestos and low-level asbestos waste.

The Soviet Experience: Half a World Away Geographically and On a Different Planet Ideologically

The Office of Strategic Planning and Analysis (1995) of the U.S. Department of Energy reports in Closing the Circle on the Splitting of the Atom that "Ironically, the Russian nuclear weapons complex now has less waste in storage than the United States because large quantities of the high-level waste (as much as 1.7 billion curies) were poured into rivers and lakes or injected deep underground rather than stored in tanks" (1995, p. 75).

The disquieting legacy of the Soviet era at the Mayak plutonium plant near Chelyabinsk is summarized by the Office of Strategic Planning and Analysis as follows:

> Techa River - from 1949 through 1951 liquid high-level radioactive waste was pumped directly into the river which was the water supply for 124,000 people without telling them.
>
> Lake Karachai - In November of 1951, Lake Karachai started receiving the high-level waste previously dumped in the Techa River. One estimate has the curies dumped into the lake at 120 million. Lake Staroe Boloto received the medium-level waste previously dumped into the Techa River.

Figure 2.7 The Fence Along the Techa River In the Village of Muslyumovo To Keep Residents Away

The plutonium plant known as Mayak (Lighthouse) deposited the amount of radiation in the Hiroshima Bomb in the Techa River which was the main water supply for 124,000 people and did not tell them for reasons of Cold War secrecy. The villages between Muslyumovo and Mayak were evacuated as were some further villages downstream. There is a main railroad which runs through Muslyumovo which is a reason offered by townspeople for why the village was not evacuated. The fence which is pictured was put up without explanation in order to keep villagers away from the river. The guard who was hired by authorities to keep villagers away from the river died from cancer and according to his family was not listed in official town records.

There are an estimated 193,000 curies trapped behind earthen reservoirs. There is no estimate available of what was deposited into deep injection wells. In 1953 a waste storage facility capable of holding 80,000 gallons began receiving high-level waste. An operational requirement was that cooling water had to be circulated constantly. When a cooling water pump failed on September 29, 1957, 20 million curies were released in an explosion caused when the evaporating mixture overheated.

In 1967, a hot dry summer caused parts of Lake Karachai to dry up. A severe dust storm distributed in one estimate 600 curies over the countryside.

Figure 2.8 Track of the Nuclear Waste Explosion Known As the Kyshtym Disaster

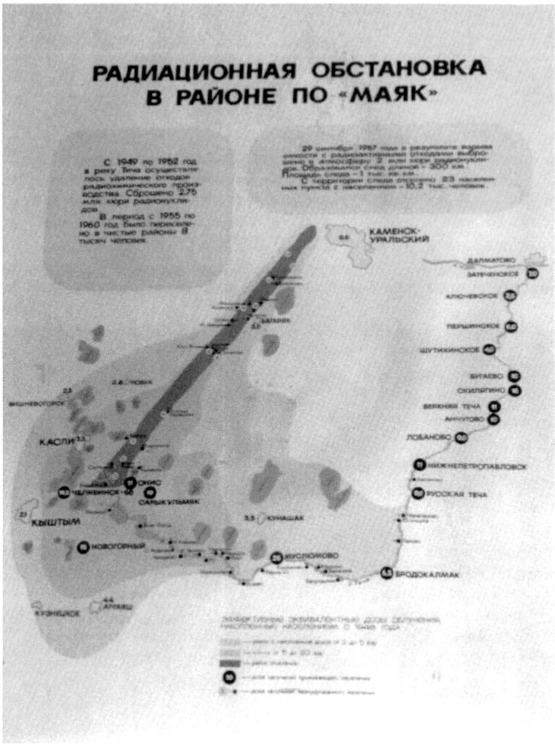

Starting from 1949 through 1952 radiochemical waste totaling 275 million curies of radionuclides was dumped into the Techa River. From 1955 to 1960 eight thousand people were resettled. On September 29, 1957 an explosion of a container with radioactive waste released 20 million curies of radionuclides into the atmosphere. The track of a length of about 200 kilometers was formed. The area of the track was 1000 square kilometers. 23 residential settlements with a population of 10,200 were relocated.

Figure 2.9 The Family of the Guard of the River Gather In Memorial of the Death of the Son Who Had Waded Into the Techa River To Obtain Water Samples For His Mother

Without explanation of the significance of her work, the wife of the guard for the river had been employed to obtain water samples from the river on a regular basis. These samples would be picked up by authorities without an explanation. While attending the memorial service, Professor Whiteley was told that the family had never been told of the need to take simple public health precautions. The family showed photojournalist Gail Fisher and Professor Whiteley the containers which the family had been given to use in taking water samples.

The China Experience Creating Nuclear Weapons
The broad outlines of the Chinese experience in creating nuclear weapons have been provided by the Nuclear Threat Initiative in their China Nuclear Chronology which was last updated in July of 2011. Two books report definitively on the early years of the Chinese nuclear program, China Builds the Bomb and Communist China's Strategy in the Nuclear Era. They reveal details such as that a prospecting team discovered uranium deposits in Guangxi Province in 1954. Further, the senior leadership of the Chinese Communist Party made a strategic decision in January of 1955 to pursue a nuclear weapons program (Lewis and Litai, 1988, pp. 35-72; Hsieh, 1962, pp. 20-21).

Cooperation Accelerated Between China and the Soviet Union

Cooperation between the Soviet Union and China accelerated in 1955 with a joint survey for uranium, an agreement that China would export surplus uranium to the Soviet Union, and that the Soviet Union would supply China with a nuclear reactor and a cyclotron (Lewis and Litai, 1988). China established in 1955 the deceptively named Bureau of Architectural Technology. The Chinese Politburo delegated authority to Chen Yun, Nie Rongzhen, and Bo Yibo to direct the Chinese nuclear weapons program (Lewis and Litai, 1988, pp. 46-47).

Meanwhile, the Nuclear Threat Initiative Chronology quotes Norris, Arkin, and Burr (1999, p. 30) that the United States began deployment in 1956 of nuclear weapons to bases in Guam, Hawaii and Okinawa. It cannot be known, of course, if the Soviets and the Chinese knew about these deployments in 1956, but they began intense collaboration in "building nuclear industry and research facilities" in China (Lewis and Litai, 1988, p. 41).

On October 15, 1957 China and the Soviet Union signed a "New Defense Technical Accord" in which the Soviet Union agreed to provide China with a prototype nuclear bomb, missiles and related data (Lewis and Litai, 1988, p. 41, pp. 62-64).

Chairman Mao: China Could Survive the Loss of 300 Million People

The Nuclear Threat Initiative Chronology quotes the Memoirs of Soviet Foreign Minister Andrei Gromyko that during an August 1958 visit to Beijing, Chairman Mao wanted the Soviet Union to use nuclear weapons on American soldiers if they invaded China over the Quemoy and Matsu Island dispute. The reason indicated by Foreign Minister Gromyko for the position of Chairman Mao (Taubman, 1988) was that China could survive a nuclear war even with the loss of 300 million people.

Soviet Politburo Subsequently Reneged On the Promise To Deliver a Prototype Bomb

The final relevant part of the most valuable Nuclear Threat Initiative Chronology is that on June 20, 1959 the Soviet Central Committee sent a letter to the Chinese Politburo saying that a prototype nuclear bomb would not be delivered citing as the reason the negotiations in Geneva over a test ban. The first Chinese nuclear bomb exploded in 1964. It was named "Device 596" after the month and date of this Soviet action perceived as a betrayal. The first Chinese hydrogen bomb was exploded in 1967.

Drawing on the reporting from the original Chinese bomb production site by Lam Yik Fei, Chris Buckley and Adam Wu (2018) in "Where China Built the Bomb, Dark Memories Haunt the Ruins" tell the chilling and tragic story of Plant 221, the secret city of Jinyintan in a remote plateau in Qinghai Province in Northwest China. Jinyintan disappeared from the map of China in 1958.

Figure 2.10 Isolation of Nuclear Weapons Production In a Desolate Area of China Parallels the Isolation of Plutonium Production at Hanford Nuclear Reservation In Washington State

The "hazardous production area" at Hanford closely paralleled in both size and isolation the Chinese site at Plant 221 near the secret city of Jinyintan. And as with Mayak in the former Soviet Union, Jinyintan located in a remote plateau in Qinghai Province in Northwest China is not on the map. The photo courtesy of The New York Times was taken by investigative reporter Lam Yik Fei. It pictures a museum still closed to foreigners which tells visitors that the weapons were made "to fend off American and Soviet aggressors encircling China." The rural character of the area is reflected in the background.

Liao Tianli, a writer who visits Jinyintan, interviewed scientists who worked on the project at its inception. He is quoted that: "At the time, China's social conditions and international position were a bit like North Korea now...For many people it was purely in the spirit of 'I'll do whatever Chairman Mao tells me to'."

The Tragic Chinese Backstory

The backstory is tragic, but has some parallels to the experiences of workers at the Soviet and American nuclear weapons development complexes: "Some veterans have said that nuclear workers were not adequately protected against radiation, or given effective care after they fell ill from cancer."

There are distinctive features in the narrative about the Chinese experience. In contrast to the quick but orderly relocation of residents of White Bluffs and Hanford and the Native American populations in the state of Washington, in China: "The herders and farmers who were moved for the project endured starvation, executions, and brutal expulsions," according to Buckley and Wu (2018).

Political Paranoia Engulfs Chinese Bomb Creation

Buckley and Wu (2018) quote Police Officer Yin Shusheng, then 80, who recalled the fate of the relocated herders:

> ...officials imprisoned about 700 herders...accusing them of joining counter revolutionary gangs. Seventeen died under brutal interrogation. Up to 9,000 herders were expelled in forced marches, given only a day or so to prepare and allowed to take just a few yaks per family. Hundreds died on the journey, beaten and abused by guards...

A surviving herder told a Mongolian researcher that "People were no better than beasts of burden...We weren't counted as humans."

The production area of Plant 221 was scattered across 220 square miles, roughly paralleling the size of the hazardous production area at Hanford Nuclear Reservation.

Much Sacrifice Was Unnecessary

Wei Shijie, then 76 and a retired physicist, said: "If nobody spoke out, then this episode in history would be buried...many people made agonizing sacrifices...Much of that sacrifice was unnecessary."

Political paranoia engulfed the project itself. Thousands of scientists and technicians were persecuted. Yin Shusheng, a police officer in Qinghai in 1963, said: "I wrote about this to sum up lessons from the past so we don't commit the same mistakes." Mr. Wei is the source quoted by Yak Fei Lam that he watched as one of Plant 221's top nuclear scientists, Qian Jin, was clubbed by interrogators and died several days later. Again, according to Mr. Wei, officials detained and interrogated about 4,000 workers and about 50 were executed. They were beaten to death or killed themselves under relentless interrogation.

Buckley and Wu (2018) sum up their narrative with the assessment that after 1966, when Chairman Mao launched his Cultural Revolution, Plant 221 was politically suspect because the by then hated Soviets had helped build the Plant.

Lessons of the Dangerous Cold War Global Context

A major lesson from the global context of weaponizing the atom is that on numerous occasions the world came very close to another nuclear explosion: this time an explosion of a device far more powerful than the bombs which destroyed Hiroshima and Nagasaki. A compelling observation about this historical era is the unknown danger to their own societies which confronted creators and deployers of nuclear weapons.

The dynamics of the nuclear arms race, and the accompanying rhetoric of fear, caused all of the possessors of nuclear weapons to extend too far the limits of human capability, to exceed the level of reliability of advanced technology, and to deploy new nuclear weapons delivery systems beyond their capacity to perform in a fail-safe manner. The lesson became known early: events with low probability of occurring actually did occur–often with disastrous results.

Technological disasters occurred too frequently. Technological disasters came close to resulting in unauthorized thermonuclear explosions which in every case would have been catastrophic. In the case of the technological disasters, in the framework of Erikson's "new species of trouble," there was plenty of blame to be assigned. The accidents or equipment failures which led to the disasters were clearly preventable in principle. Attention to risk had to be a constant top priority, and often was not.

The rapid weaponization of the atom, using new and unproven engineering practices and advanced technology not yet perfected in practice, did exceed the constraints of the precautionary principle: better safe than sorry. Since plutonium had only been created in December of 1940, and a plutonium bomb destroyed Nagasaki a little less than 5 years later, there may not have been a full realization of just how dangerous plutonium was to those who possess it making hydrogen bombs, or how difficult it was to deal with the disposal of toxic waste safely which is the byproduct of the industrial enterprise to create plutonium and fabricate it into reliable thermonuclear weapons. The perils of transuranic waste turned out to be a threat which was not learned satisfactorily and translated into safe disposal practices.

So much of the danger of disaster which is apparent in retrospect was exacerbated by the global context: nuclear weapons are overwhelmingly dangerous to the possessors of those weapons. This danger is ever present.

A Dramatic Lesson For Humanity From Disobeying the Rules: "The Man Who Saved the World"

A necessity of the nuclear age was that for both the leadership of the Soviet Union and the leadership of the United States it was imperative to be able to learn immediately if deterrence had failed with the consequence that their societies were under a preemptive attack from their sworn adversaries. The B-52s which crashed with nuclear weapons as part of Operation Chrome Dome chronicled earlier in this chapter were part of the

United States having B-52s airborne in order to retaliate if the Soviet leadership had determined that it was in their national interest to launch a preemptive strike on the United States. Likewise, the same circumstance applied if the Soviet top leadership had been informed that their country was under preemptive attack by incoming missiles. They would have had time, measured only in minutes, to authorize a retaliatory attack on the United States.

There are two relevant backstories. The first backstory is about the compelling logic of deterrence. This was expressed elegantly by famed Harvard University economist John Kenneth Galbraith in a Quest for Peace videotaped interview with John M. Whiteley: "Not even the most accomplished ideologue can tell the difference between the ashes of capitalism and the ashes of communism." There was a survival interest in recognizing the mutual national interest for both the Soviets and the Americans served by preserving the deterrence regime.

The second backstory was told to me by a former head of the Soviet Space Program: there was fallibility in early Soviet era computer hardware in integration with then available Soviet computer software. This second backstory, as well as the first, are very germane to the dilemma before Stanislav Petrov on the fateful day of September 26, 1983. Stanislav Petrov was an honors graduate of the Radio-Technical College in Kiev, Ukraine. After graduation he went to work at the secret Serpukhov-45 base in Moscow Region. His role in defense of the Soviet Union was to notify Soviet leadership if the country was under nuclear attack. The method for doing this was for Petrov to oversee the work of 200 computer operators who were monitoring surveillance by Oko satellites of six-missile launch areas of the United States where 1,000 missiles were known to be aimed at the Soviet Union.

At 12:30 A.M. on September 26, 1983, the warning screen at Serpukhov-45 reported that a missile was incoming from the United States. A warning siren in the facility started howling. The report he made to the leadership of the Soviet Politburo was that there was a fault in the system. The system which he had helped design subsequently reported that the probability of attack was 100 percent when it identified four more incoming missiles. This alert was not able to be confirmed by ground radar on the territory of the Soviet Union for another 10 minutes.

The following is how The Economist of September 30, 2017 reconstructed his dilemma:

> In 10 minutes ground radar could confirm it. But in 12 minutes the missiles, if they were incoming, would hit Russia. High command needed 12 minutes to organize their response.

Petrov called the leadership of the Soviet Politburo again to report a malfunction in the system only to determine that the officer at the other end of the line was drunk. The Economist reported that the next 15 minutes for Mr. Petrov were "unbearable." But there indeed had been a malfunction. The satellites (or their reporting software) had been fooled by clouds over North Dakota reflecting the rays of the sun. The Economist summarized his thinking:

> Some of the doubts were logical: the newness of the system, and the too-swift passage of the message through the 30 layers of verification he had set up himself. Other doubts were vaguer: a funny gut feeling, and a sense that he knew better than a machine. Even so, his decision to declare a false alarm was a 50-50 guess, no better.

The response of his superiors in retrospect was to sanction him for failing to fill in properly the operations log covering the incident. Mr. Petrov passed away in 2017 at the age of 77 years. The phrase used to describe him by The Economist in introducing his Obituary was: "The man who saved the world."

Figure 2.11. "The Man Who Saved the World"

This photograph was taken in Stanislav Petrov's simple apartment and is made available courtesy of the Associated Press and The Economist. Petrov's compelling life story is that over thirty years before this photograph was taken he twice exercised a personal judgment, not without extreme risk, to inform the leadership of the Politburo that there was most likely a fault in their warning system to alert that deterrence had failed and the Soviet Union was under nuclear attack by the United States. An observation he left for humanity was the warning that a nuclear standoff could kill millions "not by design but by accident." A further alert for humanity's future was that a person has to launch a nuclear weapon "and a person can always make a mistake."

In interviews with Time in 2015 (Shuster, 2017) he dismissed the hero designation with a curt: "Nonsense! I was just doing my job." Time documented the dangerous Cold War context. Three weeks before the events of September 26, 1983 a Soviet jet had shot down a Korean civilian airliner killing all 269 on board including a U.S. Congressman along with 60 other Americans. President Reagan had just announced plans for a missile defense system. The leader of the Soviet Union was former KGB head Yuri Andropov. Andropov was known to be worried about a preemptive attack by the United States. As Time noted, this series of events had both sides on "high alert." Time quoted Petrov in its 2015 interview that:

Responsibility For Starting World War III

> "But I didn't want to be the one responsible for starting a third world war."

Reason For the False Satellite Plus Software Report of a Preemptive Strike

> "Can you imagine? It was as though a child had been playing with a vanity mirror, throwing around the sun's reflection."
>
> "And by chance that blinding light landed right in the center of the system's eye."

A False Move By Accident Could Kill Millions

In his last years he worried that a nuclear standoff "not by design but by accident" could kill millions:

> "The slightest false move can lead to colossal consequence...That hasn't changed."

Inevitability of Human Error and Miscalculation

One thread in his 2015 interview with Time was the inevitability of human error and miscalculation:

> "That's when things can go terribly wrong...One way or another, you still need a person to order a launch of one of those weapons, and a person can always make a mistake."

The Economist raised a thoughtful issue. Petrov was a scientist first and foremost who understood at a technical level the design and functioning of the system to warn of a nuclear attack by the United States. His subsequent reflections and observations noted above are those of a most unusual person. While he did follow the rules governing his obligations that fateful early morning; namely, to notify the leadership of the Soviet Union that an alert of preemptive attack had been received, he took personal responsibility for an assessment that the alert should not be acted upon.

The magnitude of offering the judgment he did on two occasions in the early morning hours of September 26, 1983 is accurately reflected in how he is memorialized for posterity: "The man who saved the world."

Weaponizing the Atom In This Historical Era

Seed (1971) hosted a Problems Analysis Meeting at Rocky Flats on May 25, 1971. In preparation for that meeting he prepared a slide presentation that contained two observations very relevant to understanding this difficult period in the history of weaponizing the atom.

<u>The Technical Field Was Only 25 Years Old</u>

The first observation in the slide presentation by Seed was that:

> Our plant was approximately 20 years old–working in a technical field that, for all practical purposes, was only about 25 years old. Our learning rate is approximately equal to the rate of the growth of knowledge in this "new" field.

This observation is related to understanding and addressing technological challenges. In this circumstance, Seed highlighted that there was an absence of established roadmaps to successful practices.

<u>We Could Or Should Have Acted Differently</u>

The second observation by Seed was that:

> As we learned, we obviously could look on problems we had faced in the past in light of current knowledge and clearly decide we could or should have acted differently had we known what we know now.

The context is that the dynamics of the nuclear arms race as identified by Holdren did not provide any alternative to the production imperative for nuclear weapons in global context. At Rocky Flats there was no opportunity to redo previous decisions in light of new knowledge based on the "good lessons of bad experience" to paraphrase the title of an article by W. A. Rosenbaum (1999).

An example, which will be introduced in Chapter Four is the flawed siting decision. The siting criteria did not incorporate requirements mandated by the nature of the industrial enterprise. The original infrastructure investment in building the capacity to fabricate plutonium igniters in the required numbers made it impossible to start over, to act differently based on Seed's (1971) conception of "what we know now."

This reality gave rise to quite diminished choices available to subsequent stewards of Rocky Flats after the flawed siting decision. In 1971, the first recommendation by the Seed committee (Seed, 1971), based on their analysis of available data on plutonium levels in soil, was to expand the "buffer zone." Only 4 years later, ERDA (1975), in response to a major recommendation of the Lamm-Wirth Task Force (1975) that the "buffer zone" be expanded, firmly declared that the minimal risk posed by Rocky Flats mitigated any necessity for expansion of the site.

Moving forward 45 years, as documented in Chapter Fifteen, there was a settlement of $375 million in 2016 paid for by the taxpayers of the United States based on the presence of off-site plutonium in soil. The essence of the ERDA position was the absence of risk posed by Rocky

Flats. It cannot be known if the anonymous drafters of the ERDA position had access to the expert scientific analysis by the Seed committee.

The broader point is that the nuclear age is not forgiving. The Seed observations make clear that there is basically no chance to do it right the second time around. Mistakes made the first time a new challenge is encountered across domains of science, engineering, and emerging technology are most likely not to be correctable. Plutonium, after all, is both unstable and hazardous for 240,000 years.

Chapter Three
Missions of Rocky Flats and Its Ultrahazardous Industrial Enterprise

The historical missions of Rocky Flats and the five elements of the industrial enterprise to support the accomplishment of the missions will be described in a context that the processes of the creation of nuclear weapons are inherently both toxic and dangerous. There were two missions: to create the first-stage plutonium fission bombs which ignited the second-stage fusion explosions of thermonuclear bombs, and to recycle nuclear warheads which had been deployed previously in support of America's national security.

Nuclear Weapons Are Dangerous To Their Creators

The cumulative history at Rocky Flats has revealed that Rocky Flats was the most dangerous contributor to the national security of the United States based on its geographical location and the cumulative risks associated with its specialized missions. This chapter contains information that is highly disturbing about the overall lack of monitoring of dangerous radioactive and chemical substances both on-site, risking harm to workers, and off-site, bringing risk to the surrounding community.

The "gold standard" for the safe creation of nuclear weapons has been the Los Alamos National Laboratory in New Mexico during World War II where the first nuclear weapons were created. The contemporaneous history of its early years by Smythe (1945) made clear the efforts which management at Los Alamos exerted to create a culture of safety. With the closure of Rocky Flats for the production in 1989, the production function of fabricating plutonium igniters and recycling previously deployed warheads was transferred back to Los Alamos. Shockingly, Los Alamos was unable to produce plutonium igniters or recycle previously deployed warheads for several years. This was documented by the Center for Public Integrity (2017). The reason was that regulatory experts had determined that Los Alamos could not create igniters or recycle warheads without an unacceptable risk of a "criticality" incident. A "criticality" incident is when a chain reaction occurs in unstable plutonium causing it to fission uncontrollably thereby releasing energy and a burst of radiation which is fatal to persons in close proximity.

Los Alamos provides a chilling example of the dangers inherent in the industrial enterprise at Rocky Flats when it was operating. In 2017, the Center for Public Integrity released Nuclear Negligence (Center for Public Integrity, 2017), an in-depth report on safety weaknesses at U.S. nuclear weapons sites. The section on Los Alamos is disquieting. In 2013 the Acting Director of the National Nuclear Safety Administration in Washington, D.C. called the Director of Los Alamos National Laboratory with a recommendation that the laboratory be closed which fabricated new igniters and recycled previously deployed nuclear warheads. The reason was unresolved safety risks to workers. The major risk unresolved was not having enough personnel with expertise in preventing plutonium from going "critical." The Center for Public Integrity determined that between 2005 and 2016 there had been 40 reports of serious shortcomings in criticality safety by government oversight agencies, or teams of independent experts, or staff at Los Alamos.

There had been three preventable deaths from criticality accidents at Los Alamos between 1945 and 1958. The first death from a criticality accident occurred in 1945 when physicist Harry Daghlian lowered a large piece of plutonium into a cavity of tungsten bricks. The second death occurred in 1946 when scientist Louis Slotin positioned a shell of beryllium around a plutonium pit. Twelve years later chemist Cecil Kelley died from standing on a ladder to stir a vat which contained plutonium residue: the residue became too concentrated, and a criticality incident occurred leading to his death.

There are two points relevant to Rocky Flats. The first point is that the three deaths at Los Alamos were all preventable. The second point is that plutonium is unstable. Vigilance and proper practice in the extreme is essential. It is necessary to have highly trained experts to maintain a constant culture of safety. This was the dangerous context for workers at Rocky Flats as they undertook to operate the industrial enterprise with 8,000 substances and unpredictable and often unknown hazards. This is a prism for evaluating the monitoring of "materials of concern" both on-site during production and after they had migrated off-site.

Missions of Rocky Flats During the Production Era

Rocky Flats had two main core missions during its production era. The first mission was to produce the first-stage fission bombs which set off the second-stage fusion reaction in thermonuclear bombs. In this volume the first-stage fission bombs are referred to as "plutonium igniters." These first-stage fission bombs are also known elsewhere in the literature about Rocky Flats as "plutonium triggers" or "plutonium pits." The 70,000 igniters crafted as part of this first mission were formed and machined from some combination of plutonium, beryllium, stainless steel, and other materials only partially identified for continuing national security reasons.

The second mission was to process nuclear warheads which had been retired from deployment, recover the plutonium, and prepare another plutonium igniter. The task was usually to dismantle components which Rocky Flats itself had originally prepared for service in the defense of the country and "retrieve and recycle the materials" as ChemRisk phrased it (ChemRisk, 1992, p. 43).

There actually had been a third mission dating back to the time when the Atomic Energy Commission had a policy of redundancy (several sites performing the same function) across the nuclear weapons complex. Rocky Flats had worked earlier on enriched uranium until 1963 in tandem with Oak Ridge which specialized in processing enriched uranium for use in nuclear weapons. The overlap between the processing of the retired weapons mission and the enriched uranium mission is not known precisely. It is a reasonable assessment that at any one time Rocky Flats focused only on two core missions plus unspecified highly classified assignments such as for the weapons design laboratories.

Implementing the Missions of Rocky Flats: Fabrication of Nuclear and Nonnuclear Weapons Components

Linking Legacies (Office of Environmental Management, 1997) revealed the technical step-by-step processes followed by Rocky Flats in accomplishing its core missions in weaponizing the atom. The complexity of the technical processes is presented in terms of both fabrication of nuclear weapons components and plutonium recovery and purification. These separate technical processes and their corresponding visual representation flow-charts will be presented one at a time.

The reason for including the detailed technical representations is that despite the circumstance that the sophisticated processes originated in the very early years of weaponizing the atom, the quality and reliability of the plutonium igniters which were the end product were never in question.

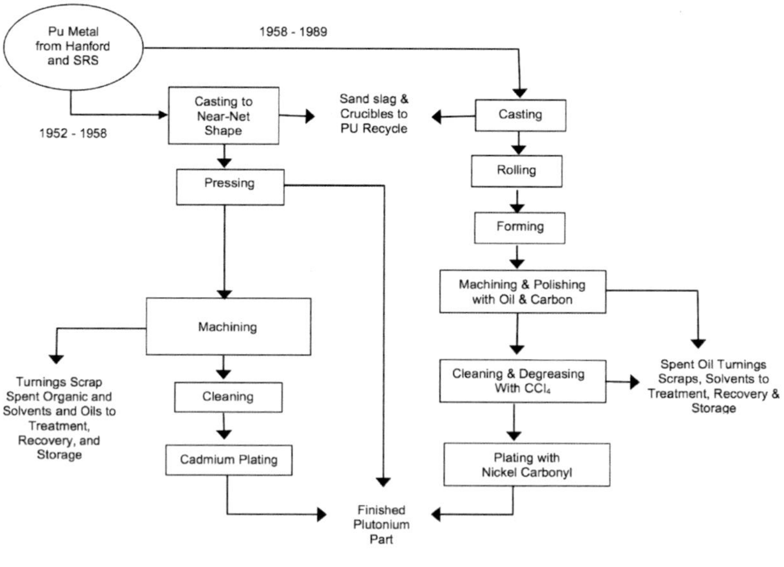

Figure 3.1 Rocky Flats Plutonium Component Manufacturing: From Receipt of Plutonium Metal From Hanford or Savannah River Through the Fabrication of a Finished Plutonium Part

The technical complexity of nuclear weapons "component fabrication" was presented by Linking Legacies in the form of a flow chart of manufacturing and assembly. The technical complexity necessitates presenting component manufacturing in two Figures 3.1 and 3.2.

Weapons Component Fabrication

Figure 3.1 and Figure 3.2 combine to present the visual representation for the process which was followed after the receipt of plutonium metal either from Hanford or Savannah River through the creation of a finished plutonium part.

This category of weapons component fabrication includes:

> ...the manufacturing, assembly, inspection, local testing, and verification of specialized parts and major weapons components.
> (Office of Environmental Management, 1997, p. 182)

Figure 3.2 Rocky Flats Plutonium Manufacturing: From Fabrication of a Finished Plutonium Part Through Assembly, Inspection and Testing To the Finished Pit

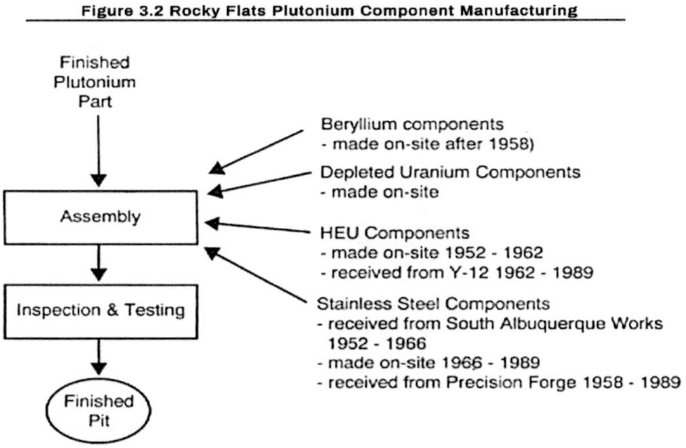

The assembly function of component manufacturing is diagrammed as involving beryllium components, depleted uranium (which declined overtime significantly in use at Rocky Flats), highly enriched uranium, and stainless steel. The inspection and testing functions were not elaborated. Across both Figures 3.1 and 3.2 the complexity and toxicity of manufacturing plutonium parts, and the technical skill involved in this phase of weaponizing the atom is very significant.

The irony of this major nuclear weapons accomplishment is that the component creation processes identified above were accompanied by companion industrial processes that the site was unable to accommodate because of its inability to safely dispose of the toxic wastes which were generated:

> Chemical processing to recover purify and recycle plutonium, tritium and lithium from retired warheads, and from component production scrap and residues, are included in this category...
> (Office of Environmental Management, p. 182)

The tragedy is that none of the site selection criteria anticipated the toxicity of the core industrial tasks which resulted from fabrication of nuclear weapons components.

Figure 3.3 Rocky Flats Plutonium Recovery and Purification

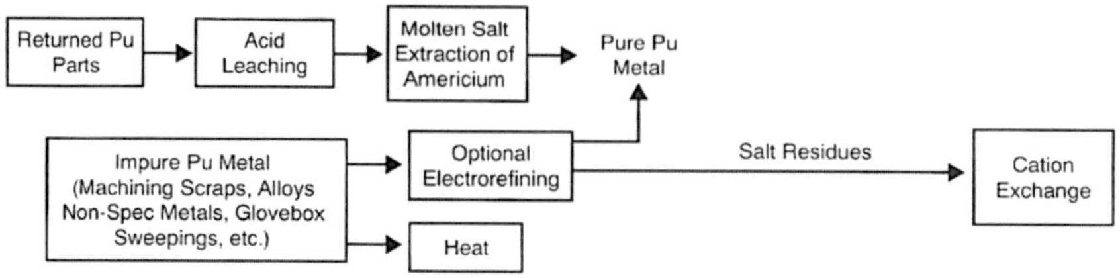

Figure 3.3 Plutonium Recovery and Purification: Identification of Processes Followed With Returned Plutonium Parts From Previously Deployed Nuclear Weapons and With Impure Plutonium Metal From Processing At Rocky Flats

The complexity of Plutonium Recovery and Purification is presented in both Figure 3.3 and Figure 3.4. Figure 3.3 diagrams the processes followed to create pure plutonium metal. The first process was to convert the impure plutonium from previously deployed nuclear weapons. The second process was to recover plutonium from machining scraps, alloys, non-specific metals, glovebox sweeping and related sources.

Weaponizing the Atom Successfully

Taken together Figures 3.1, 3.2, 3.3 and 3.4 reveal the magnitude of the national security accomplishment in weaponizing the atom successfully, and the toxicity of the underlying industrial enterprise.

Plutonium Recovery and Purification

Plutonium recovery and purification is a core element of fulfilling the missions of Rocky Flats. The complexity of visual representation by Linking Legacies is such that it is necessary to translate the original representations of the processing of returned plutonium parts (from previous deployed nuclear weapons), impure plutonium metal from several sources, and returned highly enriched uranium parts (Figure 3.3) separately from the technical processes involving aqueous dissolution with nitric acid (Figure 3.4).

The following Figure 3.4 illustrates the heavy toxic burden which was not anticipated in site selection criteria of construction of layout and the site itself. For example, nitric acid was used every year in the astounding amount of 400,000 pounds. Every plutonium igniter returned from prior deployment in defense of the United States required the now impure plutonium to under go aqueous dissolution with nitric acid.

Figure 3.4 Plutonium Recovery and Purification: Sources of Impure Plutonium Metal For the Significant Aqueous Dissolution With Nitric Acid Phases Resulting After Intermediate Stages In a Plutonium Metal Button

Figure 3.4 diagrams the processes followed in taking sources of impure metal through dissolution with nitric acid followed by the industrial processes which lead to a plutonium metal button. A key observation is that huge quantities of nitric acid were used every year at Rocky Flats. The aqueous dissolution stage was very important and highly toxic. This reality was not considered in site selection or design of the industrial enterprise.

Figure 3.4 Rocky Flats Plutonium Recovery and Purification

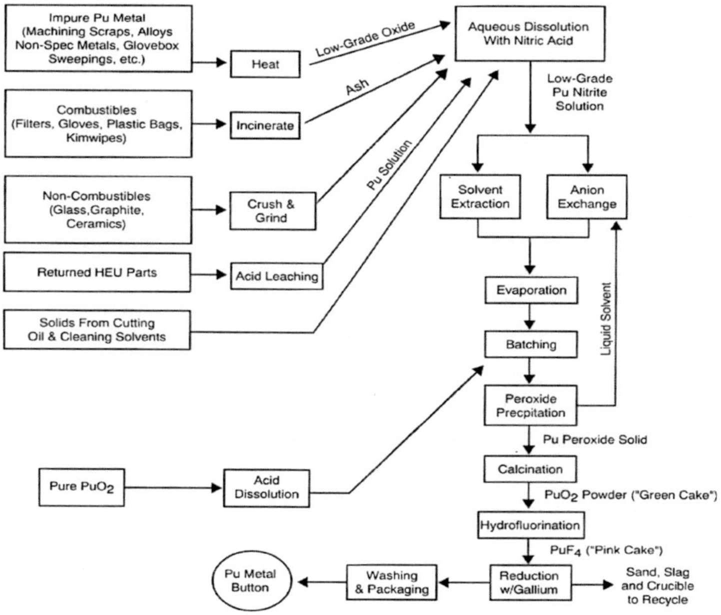

Main Industrial Support Functions At Rocky Flats

<u>Material Recovery and Purification</u>

The imperative to recover plutonium and uranium from lubricating and cleaning compounds reflected how exceptionally costly these radionuclides were to produce originally. Economically acceptable recovery methods for plutonium residues from used cooling oils and solvents were never devised. Unpermitted storage became an environmental crime under the suite of new environmental laws from the 1970s.

The purpose of "Recovery Operations" was to "recover and purify the fissionable material used in the weapons systems which are of strategic importance" (ChemRisk, 1992, p. 59). Building 771, the site of the catastrophic 1957 fire, was the location for plutonium recovery, plutonium purification, and plutonium component manufacturing. ChemRisk (1992) reports the comparable history for recovery and purification for uranium (pp. 67-69) and americium (pp. 69-71).

Component Manufacturing and Assembly

According to ChemRisk (1992, p. 72), in the early years of the production era "the majority of components were enriched uranium, depleted uranium, and plutonium." With the assignment of enriched uranium back to Oak Ridge, weapon designs at Rocky Flats called for use of more plutonium. The focus of this function was precision metal working of both fissionable materials and non-fissionable materials such as aluminum and stainless steel.

Building 776 became the plutonium machining facility. Building 777 became the center for assembly. In a fateful decision which increased risk that a serious fire could quickly spread, the wall between 776 and 777 was taken down, reportedly to facilitate operations. An unintended consequence of this decision was to escalate a serious fire in 1969 into a catastrophic fire.

As chronicled by ChemRisk (1992), in 1958 beryllium operations became an important part of the work at Rocky Flats. The center of that operation was Building 444. Beryllium became more prevalent in new weapons designs.

For readers interested in the manufacturing operations in general for stainless steel ChemRisk (1992, p. 77) provided a summary, as it does for components of depleted uranium (ChemRisk, 1992, p. 76), enriched uranium (ChemRisk, 1992, p. 76-77), and plutonium (ChemRisk, 1992, pp. 78-79).

Waste Processing At Rocky Flats

ChemRisk (1992) described waste processing at Rocky Flats separately for liquid process wastes, solid wastes, and sanitary wastes. Since waste processing became a focus of the investigation of environmental crime from 1986-1992, this section will provide capsule summaries of waste processing.

Liquid Process Wastes

ChemRisk (1992) indicated that liquid waste processing had undergone the fewest process changes over the years: "the general mission of the waste operations was to reduce the volume of wastes and put them in a form acceptable for transportation to off-site burial ground" (pp. 80-82). This section of ChemRisk (1992) describes one of the purposes of Building 771 as "treating radioactive wastes" (p. 80). The plan was to make adjustments for pH in liquids which were to be transferred to Building 774, "then go through a precipitation step to remove radionuclides" (p. 80). The resulting slurry was routed to vacuum filters, combined with cement (or another solidifying agent), and then sent to long-term storage.

This coverage of liquid process wastes introduces several of the limitations in practice, including discharging wastes into "B" pond which drained to the off-site Great Western Reservoir via Walnut Creek which traversed the Rocky Flats site. The lack of capacity of evaporators to treat liquids which accumulated in the Solar Evaporation Ponds became an area of subsequent commission of environmental crime.

A key source of off-site contamination was the plutonium in liquid organic wastes. Appearing in ChemRisk (1992, p. 81) was an initiative reported in Dow Chemical archives to use a centrifuge to process liquid organic waste to remove plutonium. This initiative did not meet with success.

Along with spent cooling oil with traces of radionuclides, carbon tetrachloride was in steel drums that leaked onto the ground of the 903 Pad. There were attempts made to extract carbon tetrachloride for reuse from the spent oil. (No measurement was made at the time of whether or not carbon tetrachloride became an off-site contaminant.)

Solid Wastes With Radioactivity

There were two categories of solid wastes containing radioactivity. The first category was retrievable wastes. Retrievable wastes, according to ChemRisk (1992), contained more than 10 nanocuries of radioactivity per gram of material:

> These wastes are packaged and stored to enable them to be repackaged if necessary, or if technology warrants, to enable their retrieval and the subsequent recovery or treatment of the contained radioactive and/or chemical toxins.
>
> (ChemRisk, 1992, p. 83)

There had been extensive efforts over the years to separate (retrieve) the contained radioactivity.

The second category of solid waste was non-retrievable wastes. Again, according to ChemRisk (1992), these wastes such as chairs, tables, and cabinets were reduced in size and placed in 55-gallon steel drums or wooden boxes. They contained less than 10 nanocuries of radioactivity per gram of material. This category of waste easily was shipped off-site to an approved repository.

Sanitary Wastes

Sanitary wastes were comprised of sewage resulting from the treatment of wastes from restrooms, food services, and cooling tower blowdown. Approximately 100 tons of sewage sludge was placed into trenches T-2 through T-8 from 1954 through 1968. After 1969, sludges were declared to be low-level radioactive waste and shipped off-site. The source of the radioactivity was drains in the production buildings which were designed not to flow into the system for sanitary wastes, but in error did so in some instances.

One focus of the Special Federal Grand Jury, according to its second report (see Chapter Eleven), was a chromic acid spill caused by a worker leaving a faucet running in a production area. At the direction of Rockwell International, the resulting spill was routed to the sanitary waste system. Chromic acid is Number 212 on the Priority List of Hazardous Substances maintained by the Agency for Toxic Substances and Disease Registry. The toxic spill was then Spray Irrigated off-site via the South Sprayfield to avoid detection.

Research and Development and Plant Support

The plant support missions were different from those at most industrial enterprises for three reasons. Rocky Flats was always run as an industrial enterprise by contractors from the military-industrial complex. The first difference was the extreme secrecy because of the need to protect the enterprise itself in Cold War context. The second difference was that the Atomic Energy Commission and then its successor, the Department of Energy, determined what the production goals were based upon changing national security priorities.

The third difference was that the acquisition of resources was a challenging multi-level process starting with the local Rocky Flats AEC/DOE office, then the regional AEC/DOE office in Albuquerque, then the Washington D.C. office of the AEC/DOE, and only then to the most meaningful, the Congressional appropriation process.

With the criminal prosecution of environmental crime, the interests diverged significantly and became adversarial between DOE on behalf of the U.S. government and that of Rockwell International as a business. And when the discovery process of civil lawsuits unfolded, it became clear that there had been major stresses in the relationship. Plant support of the industrial enterprise was only the "tip of the iceberg" of broader plant support issues and conflicts, including over the public trust.

There were two major challenges for the research and development mission. The first challenge was the pioneering of new industrial processes required by precision nuclear weapons component production. It was genuinely uncharted territory. The second challenge was the daunting task embedded in the fact that the size of the facility was both too small and too close to a major metropolitan area to accommodate the industrial enterprise itself.

The Extremely Toxic Nature of "Materials of Concern" At Rocky Flats and the Level of Monitoring For Off-Site Contamination

ChemRisk identified twenty-five substances used at Rocky Flats as problematic "materials of concern." Then they further identified twelve of the twenty-five substances which had the greatest potential for off-site contamination. Finally, and based on their access with "Q" security clearances, ChemRisk researchers disclosed the startling fact that six of the known twelve highly toxic substances with potential for off-site harm to health were not even monitored for off-site contamination.

Problematic "Materials of Concern" For Toxicological Review

One of the responsibilities of ChemRisk was to perform a toxicological review for the purpose of estimating the exposure dose to workers which accompanied Rocky Flats operations. The historical reconstruction of exposure dose was based in part on identifying twenty-five solvents, metals and "others." Most "others" were chemicals which were known carcinogens in 1992. And almost all of the twenty-five "materials of concern" are now on the Priority List of Hazardous Substances as maintained by the Agency for Toxic Substances and Disease Registry in 2017.

One thousand three hundred acres of Rocky Flats are now a Superfund site. Based on the level of hazard associated with the presence of the substance in a Superfund site, designation as a "Hazardous Substance" and its ranking on the List of Hazardous Substances is determined by general criteria established by the Comprehensive Environmental Response, Compensation, and Liability Act (CERCLA) and the Superfund Amendments and Reauthorization Act (SARA).

Figure 3.5 presents the 25 "Materials of Concern" selected by researchers for ChemRisk (1992) in three categories: solvents, metals, and "others."

Figure 3.5
"Materials of Concern" Selected By Researchers For ChemRisk (1992) From 8,000 Substances Used On the Rocky Flats Site

Seven Solvents	*Ten Metals*	*Eight Others*
Benzene	Americium-241	Benzidine
Carbon Tetrachloride	Beryllium	1,3-Butadiene
Chloroform	Cadmium	Ethylene Oxide
Methylene Chloride	Chromium	Formaldehyde
Tetrachloroethylene	Lead	Hydrazine
1,1,1-Trichloroethane	Mercury	Nitric Acid
Trichloroethylene	Nickel	Propylene Oxide
	Plutonium-238,239,240,241,242	Tritium
	Thorium-232, 230, 237, 238	
	Uranium-233, 234, 235 and 238	

Radionuclide "Materials of Concern"

There were more radionuclides used on the Rocky Flats site than appear as "Materials of Concern" with Neptunium as an example. The ChemRisk researchers used four categories of information to determine which radionuclides would be selected as a "Material of Concern":

> Radionuclides that have been included as materials of concern are all those which were handled in substantial quantity, were associated with production activities, were found in forms that are likely to be released, or were found to be present in plant effluents or in the environment.

ChemRisk reported that the presence of tritium was the only material which was an exception to the circumstance that "...monitoring data are consistent with the release of only the main production radionuclides from the facility" (ChemRisk, 1992, p. 3). Tritium was included because of the publicity given to the 1973 off-site release into Walnut Creek and Great Western Reservoir. Tritium turned out to be the only radioactive "material of concern" which was not on the ATSDR List of Priority Hazardous Substances as of 2017.

Figure 3.6
Seven Solvents Identified As "Materials of Concern" In the Rocky Flats Production Era and Their Listing On the Priority List of Hazardous Substances Compiled By the Agency For Toxic Substances and Disease Registry

Benzene #6 on the Priority List of Hazardous Substances
The EPA has classified benzene as a known carcinogen from all routes of exposure. Its use at Rocky Flats was (apparently) as a solvent. Negative reproductive effects are reported for women who are exposed at high levels. Chronic exposure causes blood disorders including the reduction of red blood cells and aplastic anemia. Inhalation exposure can cause neurological symptoms. Ingestion of large amounts of Benzene can cause dizziness, headaches, and convulsions.

*Carbon Tetrachloride #50 on the Priority List of Hazardous Substances

*Chloroform #11 on the Priority List of Hazardous Substances

*Methylene Chloride #88 on the Priority List of Hazardous Substances

*Tetrachloroethylene #33 on the Priority List of Hazardous Substances

*1,1,1—Trichloroethane #117 on the Priority List of Hazardous Substances

*Trichloroethylene #16 on the Priority List of Hazardous Substances

Solvents "Materials of Concern"

There were seven solvents identified as "materials of concern," with six of the seven in the further subcategory of having potential for off-site contamination. Only one of the toxic solvents described in Figure 3.6 was not evaluated as a threat for off-site contamination, Benzene at #6. The other six with an asterisk (*) by their name will have their health risk and degree of monitoring for off-site contamination described in Figure 3.9. Several observations are in order about the contents of Figure 3.6. First, it is striking that only Benzene on this list of frequently used solvents is not identified as a threat for off-site contamination. Second, it is clear how dangerous the environment for workers was given the relatively small size of the industrial area. And every one of these solvents are on the ATSDR

Priority List of Hazardous Substances in Superfund sites ranging from Benzene at # 6 to 1,1,1—Trichloroethane at #117. They had no monitoring for off-site contamination.

Metal "Materials of Concern"

There are ten metals identified as "materials of concern" which were used at Rocky Flats with five of the ten in the further subcategory of potential for off-site contamination. Five will be described in Figure 3.7: Cadmium, Lead, Mercury, Nickel, and Chromium. These first five were not considered to have potential for off-site contamination and were not monitored. The other five with an asterisk by their name (*) will have their health risk and degree of monitoring reported in Figure 3.10 as they were considered as having potential for off-site contamination.

Figure 3.7
Ten Metals Identified As "Materials Of Concern" In the Rocky Flats Production Era and Their Listing On the Priority List Of Hazardous Substances Compiled By the Agency For Toxic Substances and Disease Registry

*Americium-241 #129 on the Priority List of Hazardous Substances

*Beryllium #43 on the Priority List of Hazardous Substances

Cadmium #7 on the Priority List of Hazardous Substances
EPA considers cadmium to be a probable carcinogen. Kidney disease can be caused by long-term inhalation exposure or by oral exposure which can lead to the buildup of cadmium in the kidneys, which in turn can cause kidney stones. According to EPA, a single acute exposure can result in long-lasting impairment of lung function, including emphysema.

Lead #2 on the Priority List of Hazardous Substances
Lead poisoning can cause numerous adverse health effects when ingested or inhaled. For adults, lead poisoning can cause damage to the sensory organs and nerves controlling the body, poor muscle coordination, and reproductive problems.

Mercury #3 on the Priority List of Hazardous Substances
High levels of exposure to mercury results in permanent nervous system and kidney damage according to EPA. It is not known what were the main exposure pathways for workers, or the methods of disposal of this toxic substance.

Nickel #57 on the Priority List of Hazardous Substances
According to the EPA, the only gaseous nickel compound is nickel carbonyl which was an essential component of glovebox work as documented in the 1957 fire. Once nickel is in surface and ground water systems, physical and chemical reactions occur which determine its fate. EPA has classified nickel refinery dust and nickel subsulfide as known human carcinogens, and nickel carbonyl as a probable human carcinogen.

Chromium (Hexavalent) #17

Chromium (Trioxide) #66 and

Chromium #78 on the Priority List of Hazardous Substances
Chromium occurs in two states in the environment according to EPA: hexavalent and trioxide. It is not clear which were sources of risk at Rocky Flats in different industrial areas. EPA has classified inhaled hexavalent chromium as a known carcinogen, especially for lung cancer. Chromium was present in "Cooling Tower Blowdown."
In their introductory section, ChemRisk (1992) stated that "although chromium is a metal, it will be evaluated in all elemental or metallic forms encountered at Rocky Flats, including salts, ionic solutions, and any other forms revealed..." indicating that there may have been other forms of chromium at Rocky Flats.

*Plutonium #124
Isotopes - 238 #122
- 239 #120
- 240 #136
- 241 no listing
- 242 no listing

*Thorium #102
Isotopes - 230 #109
- 228 #111
- 227 #227

*Uranium #97
Isotopes - 233 #239
- 234 #113
- 235 #110
- 238 no listing

Several observations are in order about Table 3.7. First, the high hazardous ranking of the metals in use at Rocky Flats reveals the great risk of exposure for workers. Second, waste disposal practices and the topography of the site at Rocky Flats were such that there was a risk that the dangerous radionuclides could migrate off-site, and in fact some of them did.

"Other" Materials

There were eight "other" substances identified as "materials of concern" with only one with an asterisk (*) to identify it in a subcategory of potential for off-site contamination, tritium.

Figure 3.8
Eight "Other" Substances Identified As "Materials of Concern" In the Rocky Flats Production Era and Their Listing On the Priority List of Hazardous Substances Compiled By the Agency For Toxic Substances & Disease Registry

Benzidine #30 on the Priority List of Hazardous Substances
Benzidine is no longer produced in the United States. It is considered by EPA "to be very toxic to humans by ingestion." It is a known carcinogen especially for bladder cancer. Production was banned in the United States in 1973.

1,3-Butadiene #154 on the Priority List of Hazardous Substances
1,3-Butadiene is classified by EPA as a known carcinogen by inhalation. It is used in producing rubber and plastics, and may be found in manufacturing plants and automobile exhaust. Neurological effects such as blurred vision, fatigue, headache, and vertigo have been reported at high exposure levels. There are no reliable medical tests to assess personal exposures.

Ethylene Oxide
According to EPA, ethylene oxide is a "chemical intermediate" in the manufacture of solvents, for example, and affects the general public through second-hand smoke. Acute exposure results in nausea, vomiting, neurological disorders, and emphysema at high concentrations.

Formaldehyde #224 on the Priority List of Hazardous Substances
In 2014 the National Research Council determined that formaldehyde is a carcinogen which is a significant departure from the conclusion of EPA in its "Hazard Summary" of 2000 A.D. which indicated that formaldehyde was a "probable carcinogen."

Hydrazine #87 on the Priority List of Hazardous Substances
Acute exposure can damage the liver, kidneys and the central nervous system. The National Research Council considers it to be a carcinogen. Short-term acute exposure may produce irritation of the eyes, nose, and throat, dizziness, headache, nausea, pulmonary edema, seizures, and coma. It is also fuel for rockets and spacecraft.

Nitric Acid
The EPA website indicates that there are no data on the toxicity of dilute forms of nitric acid following oral exposure. EPA considers nitric acid to be "a highly corrosive inorganic acid." In concentrated form it is corrosive at the point of contact but is not a systemic-toxic agent.

Propylene Oxide
Propylene Oxide is used in the production of polyurethane forms and propylene glycol. Propylene glycol is used to make anti-freeze and deicing solutions and is used in a wide range of food products including e-cigarettes. EPA has classified it as a probable carcinogen. It is among the most heavily used chemicals in the world.

*Tritium

Figure 3.9
Solvents "Materials of Concern," Used At Rocky Flats Which Have Potential Off-Site Health Impacts But Which Had No Level of Monitoring For Off-Site Contamination

There are six solvents "materials of concern" for potential off-site health impacts which in fact had absolutely no monitoring for off-site contamination. Their principal health risks in brief summary are from the EPA website. Their comparative rank as a "Priority Hazardous Substance" is by the Agency for Toxic Substances and Disease Registry (ATSDR) as of 2017. The source for the fact that there was no specific off-site monitoring is contractor ChemRisk (1992, pp. 111-133).

1,1,1 – Trichloroethane Number 117 Priority Hazardous Substance
Health Risk – effects heart and blood vessels
-toxic effects include central nervous system depression and loss of coordination and equilibrium at low concentrations
-toxic effects include anesthesia at very high concentrations and the heart may stop beating.
Monitoring Data – not routinely monitored.

Trichloroethylene (TCE) – Number 16 Priority Hazardous Substance
Health Risk – inhalation as an ingestion carcinogen
- chronic toxic effects involve kidneys, liver, nervous system, skin, the immune system, the male reproductive system, and the developing embryo.
Monitoring Data – not routinely monitored.

Carbon Tetrachloride – Number 50 Priority Hazardous Substance
Health Risk – it is an established probable carcinogen according to EPA
-inhalation or ingestion can cause serious health effects with acute exposure linked to damage to the liver, changes in enzyme levels, and damage to the kidneys (which is believed to cause in turn fluid in the lungs).
Monitoring Data – not routinely monitored (Note: This finding is very valuable for conveying the priorities for monitoring toxics in comparison to focus on supporting the production mission).

Chloroform – Number 11 Priority Hazardous Substance
Health Risk – probable carcinogen according to EPA
- acute inhalation may cause toxicity of liver and kidney, central nervous system depression.
Monitoring Data – not routinely monitored.

Methylene Chloride – Number 88 Priority Hazardous Substance
Health Risk – not classified as a carcinogen according to EPA
- acute short-term exposure can cause severe neurological effects
- inhalation studies on male rats show reproduction effects.
Monitoring Data – not routinely monitored.

Tetrachloroethylene (PCE) – Number 33 Priority Hazardous Substance
Health Risk – likely to be a carcinogen--studies of dry cleaning workers have shown association to several types of cancer according to EPA
- high chronic exposure causes liver damage and effects to kidneys, immune and hematologic systems, and reproduction according to EPA.

Monitoring Data – not routinely monitored.

Several observations are in order about Figure 3.8. The first is that three substances do not appear on the ATSDR Priority List of Hazardous Substances. But they are toxic. Acute exposure to Ethylene Oxide can cause neurological disorders and emphysema. Propylene Oxide is a probable carcinogen. Nitric Acid is highly corrosive—and there were 400,000 pounds of it used each year at Rocky Flats. Four of the eight are known carcinogens: Hydrazine at #87, Formaldehyde at #224, 1,3-Butadiene at #154, and Benzidine at #30 and are on the Priority List of Hazardous Substances. The eighth of eight, tritium, is discussed under off-site contaminants in Figure 3.10. The other seven are discussed in Table 3.11 for risk for on-site contamination.

Solvents "Materials of Concern" With Potential For Off-Site Contamination and Health Effects Which Had No Off-Site Monitoring Regime

Based on the ChemRisk research, a core finding is quite unexpected; namely, that there was no off-site monitoring for so many toxic substances. This core finding was only revealed because researchers with security clearances were given access to the Rocky Flats site archives. The previous four Figures 3.5, 3.6, 3.7, and 3.8 were presentations of the twenty-five "materials of concern." The following Figures 3.9 (six solvents) and 3.10 (five metals plus tritium) present the twelve of them judged to have potential for off-site contamination.

The sources for these characterizations are ChemRisk (1992, pp. 111-133) for monitoring data and EPA and ATSDR sites for health effects and "Priority Hazardous Substance" ranking. All of these "materials of concern" which were not monitored for off-site contamination or health effects are listed in the Priority List of Hazardous Substances.

Listing on the Priority List of Hazardous Substances is a prioritization of substances "based on a combination of their frequency, toxicity, and potential for human exposure at NPL sites" (ATSDR, 2017). "NPL sites" refers to so called "Superfund sites" under the Comprehensive Environmental Response, Compensation, and Liability Act (CERCLA) as amended by the Superfund Amendments and Reauthorization Act (SARA).

Metals Plus Tritium "Materials of Concern" Used At Rocky Flats Which Have Potential Off-Site Impacts With At Least Some Level of Monitoring For Off-Site Contamination

The following are six "materials of concern" for potential off-site health impacts as identified by ChemRisk (1992) and the level of monitoring they received for off-site contamination. Their principal health risks are from the EPA website. Their comparative rank as a "Priority Hazardous Substance" by the Agency for Toxic Substances and Disease Registry (ATSDR) is for radioactive substances with different isotopes. The report of the level of off-site monitoring is by contractor ChemRisk (1992, pp. 111-133).

Figure 3.10
Metals Plus Tritium "Materials of Concern" Used At Rocky Flats Having Potential Off-Site Health Impacts Which Have Had Some Level of Monitoring

Thorium - Numbers 102 and 109 and 111 and 227 on Priority List of Hazardous Substances
Health Risk – increased risk of cancer from irradiation of cells
– heavy metal poisoning occurs via ingestion or inhalation with exposure to large amounts of Thorium.
Monitoring Data – contributes to total alpha count in monitoring but has not been specifically analyzed routinely in effluents.

Americium – 241 #129 Priority List of Hazardous Substances
Health Risk – low dose raises cancer risk
- high dose can cause heavy metal poisoning.
Monitoring Data – airborne since 1985
- waterborne since 1971.

Beryllium – #43 Priority List of Hazardous Substances
Health Risk – probable inhalation carcinogen
- chronic exposure can cause berylliosis.
Monitoring Data – in air exhaust systems since 1963
-in waterborne emissions since 1980.

Plutonium – #120 and #122 and #124 and #136 Priority List of Hazardous Substances
Health Risk – probability of increased risk of cancer from irradiation of cells especially from particles embedded in lungs
- 0.8% of ingested amount enters the bloodstream with 80 percent of blood burden deposited in liver and skeleton.
Monitoring Data – air monitoring 1953-1973 was for total long-lived alpha, after 1973 only for plutonium
-ambient water and vegetation samples were tested for gross alpha prior to 1967, and later separations were for plutonium and uranium.

Tritium – Not Identified as a Priority Hazardous Substance
Health Risk – one of the least dangerous radionuclides since it leaves the body quickly
-tritiated water vapor or tritium gas can penetrate the skin, lungs or GI tract and can increase the risk of cancer
-tritiated water vapor is more hazardous than tritium gas.
Monitoring Data – airborne tritium was monitored prior to 1973 in anticipation of work which did not materialize but the results have not been found
-monitored for waterborne effluents since 1973 and for airborne effluents since 1974.

Uranium – Numbers #97 and #110 and #113 and #239 Priority List of Hazardous Substances
Health Risk – "can result in both chemical and radiological toxicity with the main chemical effect to the kidneys with acute exposure causing kidney failure and death"
-radioactive isotopes are carcinogens from irradiation of cells
-insoluble uranium compounds are a hazard to the lungs when inhaled, and to the bone when ingested.
Monitoring Data – in air exhaust systems since 1963
-air effluent measurements 1953-1973 were total alpha activity assumed to be uranium or plutonium
-from 1978 air effluent measurements were separate for uranium and beryllium
-water and vegetation analyses after 1961 included separations for plutonium and uranium.

Thirteen "Materials of Concern" For On-Site Health Impacts (But Not Off-Site) and Their Level of Monitoring

ChemRisk specialists determined that thirteen "materials of concern" on their original list of twenty-five were not a risk for off-site contamination. The reasoning was as follows:

> ...uses of the materials at Rocky Flats have been extremely limited in scope or duration, associated with insignificant quantities of the material, or have involved processes or forms of the material which were not expected to have significant off-site releases. These materials do not warrant further quantitative evaluation of potential off-site impacts.
>
> (ChemRisk, 1992, p. 135)

In support of this conclusion, the report allocates at least a page each to what they called a "material use profile" (ChemRisk, 1992, pp 137-161).

Figure 3.11 is derived from those pages plus the ATSDR Priority List of Hazardous Substances (if it is on the list), evidence on its amount of use, and whether or not EPA or the NRC considers it to be a carcinogen. The issue addressed in Figure 3.11 is whether the "material of concern" was potentially hazardous to workers, and whether it was monitored.

Figure 3.11
Thirteen "Materials of Concern" For On-Site Toxic Exposure But Not Off-Site, Whether Or Not They Were Monitored, Their Level of Hazard In a Superfund Site and Data On Amount of Usage

	Routinely Monitored	EPA/NRC Carcinogen	ATSDR Superfund Site Hazard Rank	EG&G Usage Estimate
Benzene	No	Yes	6	7.7 pounds/year
Benzidine	No	Yes	30	no current data
1,3-Butadiene	No	Yes	153	no current data
Cadmium	No	Probable	7	no current data
Chromium	Yes	Yes for 17, 78	17,66,78	no current data
Ethylene Oxide	No	No	No	no current data
Formaldehyde	No	Yes	224	no current data
Hydrazine	No	Yes	87	no current data
Lead	No	No	2	no current data
Mercury	No	No	3	no current data
Nickel	No	Known or Probable	57	no current data
Nitric Acid	No	No	No	400,000 pounds/yr
Propylene Oxide	No	Probable	No	no current data

As the presentation in Figure 3.11 makes clear, twelve of the thirteen "materials of concern" for on-site health impact were not monitored. While some of them were apparently used in only small quantities and early in the history of Rocky Flats, others such as Nitric Acid were basic to the plutonium production process and used in huge quantities throughout the production era.

In the absence of monitoring, it is not possible to gauge the health risk to workers at the time that they were used at the plant. A further difficulty in assessment of health risk to workers, and to the toxicity of the workplace environment in the long term, is that the amount of usage was not possible to determine based on the records available to EG&G. The ChemRisk researchers compensated for the lack of historical records available to EG&G by conducting interviews with workers from earlier periods in the history of Rocky Flats as an industrial enterprise (ChemRisk, 1991b). The limitation of historical reconstruction reflects the high degree of compartmentalization of functions and the pervasive secrecy in support of national security.

It is difficult to assess the toxic legacy of these thirteen "materials of concern" within the 1,300 acres designated as a Superfund site. Six of the thirteen are known carcinogens: Benzidine, 1,3-Butadiene, Chromium (in two states), Formaldehyde, Hydrazine and Nickel (in one form). Three are probable carcinogens: Cadmium, Nickel (in another form), and Propylene

Oxide. Only Chromium was ever monitored on-site and not at all systematically.

During the early years of Rocky Flats as an industrial enterprise, EPA had not been established, nor had these been developed elsewhere in government the extensive characterization subsequently developed by EPA and ATSDR of the health effects or identification of the priority assigned to different hazardous materials in a Superfund site.

Differing Conclusions On "Materials of Concern" In the Rocky Flats Industrial Enterprise

ChemRisk initially identified twenty-five "materials of concern." As they stated in a conclusion, "...no information was immediately available with regards to the nature of their use and potential for release" (ChemRisk, 1992, p. 257). With their exceptional access to both classified and unclassified sources, they came to three different sets of conclusions relevant to this presentation. The first set of conclusions related to four chemicals about which "no information could be found with regards to their potential use at the Plant":

	EPA Characterization	ATSDR List
Benzidine	Known carcinogen	#30
1,3-Butadiene	Known carcinogen	#154
Ethylene Oxide	Acute exposures are serious	--
Propylene Oxide	Probable carcinogen	--

In 1992, ChemRisk did not know obviously the current EPA characterization and the ATSDR characterization of 2017. As the reader will recall, after 1973 Benzidine was no longer manufactured in the United States, and is considered by EPA to be "highly toxic" to humans.

The second ChemRisk conclusion was that "...based on the nature of their use and potential for release, they do not warrant further investigation from the standpoint of potential off-site impacts":

Solvent, Chemical or Metal	EPA/NRC Characterization	ATSDR Priority Listing
Benzene	Carcinogen	#6
Cadmium Compounds	Probable carcinogen	#7
Chromium Compounds	Some carcinogens	#17, #66, #78
Formaldehyde	Carcinogen	#224
Hydrazine	Carcinogen	#87
Lead Compounds	Many adverse affects	#2
Mercury	Can lead to permanent nervous system and organ damage	#3
Nickel Compounds	Some forms carcinogenic	#57
Nitric Acid	Highly corrosive	--

There is no disputing the potential adverse health effects from these substances. But of the list, only chromium compounds were sometimes monitored. There is no data available (and probably not collected at the time) on on-site health impacts. The National Research Council recently upgraded Hydrazine to classification as a carcinogen.

The third set of conclusions related to release points for contaminants: airborne emission points, surface water emissions primarily from water retention ponds, and for groundwater which "...may have resulted from seepage in evaporation ponds as well as from various waste disposal activities or spills" (ChemRisk, 1992, p. 258). A consistent characterization of conclusions across all of the twenty-five "materials of concern" and their release points was most significantly lack of monitoring, inadequate monitoring, or gaps in the monitoring.

The overall point is that only six of the twelve "materials of concern" for off-site health impacts had any level of monitoring. And of the thirteen "materials of concern" limited to on-site risk, there was almost no monitoring at all. The findings of only an incomplete level of off-site monitoring for plutonium and tritium will be a recurring theme throughout this volume, with off-site uranium of lesser scrutiny, though the archives of Dow Chemical reveal the careless open-air incineration of uranium laced oil in the quantity of over 1,000 barrels.

Chapter Four
Failure To Anticipate Industrial Enterprise Requirements In Site Selection Criteria and Waste Management

> "Waste management tends to be a low priority project because it is desirable to do as little as possible to get rid of any waste streams."
>
> Dow Chemical Report, October 24, 1969

> "It is difficult to predict what changes will be made and this tends to develop a crisis-type situation in which effort is expended to solve a specific problem, followed by lack of activity until the next crisis."
>
> Dow Chemical Report, October 24, 1969

> "Initially hot wastes were collected in paper bags which were placed into 55 gallon drums and sealed in concrete for possible final disposal at sea. With the advent of winter and the availability of pre-mixed concrete, so plentiful during the initial construction period, these wastes were sealed in cardboard containers and stored in 91 Building pending final decision on the disposal of hot wastes."
>
> Oliver M. Windahl, Director,
> Annual Report 1952, Health Physics Division, Dow Chemical

A decision was made by General Leslie Groves (Groves, 1962) early in the undertaking to develop nuclear weapons for the United States during World War II which subsequently proved fateful to developments at Rocky Flats: to entrust private enterprise companies such as Dow Chemical and Rockwell International with managing the nuclear weapons development enterprise of the United States. At the same time, the policies of General Groves about minimum proximity of nuclear weapons production facilities to cities were disregarded with respect to the siting of Rocky Flats. General Groves had specified twenty-five miles as the minimum distance a nuclear weapons production site had to be from a city. In 1951 an expanding Denver, Colorado and Rocky Flats then at only sixteen miles away were altogether too close together for conforming to the letter and the spirit of the Groves guidelines.

Figure 4.1 General Leslie Groves, Jr.

General Groves attended the University of Washington and the Massachusetts Institute of Technology before graduating from the United States Military Academy at West Point. After graduating fourth in his class at West Point, he was commissioned into the U.S. Army Corps of Engineers. In furtherance of his training, he attended the Command and General Staff School (1935-36) and the Army War College (1938-39). In August of 1941 he was assigned to create an office complex for the 40,000 War Department staff which became the Pentagon. The next assignment in 1942 was to direct the Manhattan Project which involved most aspects of weaponizing the atom including significant engagement in choosing the sites for Oak Ridge in Tennessee, Los Alamos in New Mexico, and the Hanford Nuclear Reservation in Washington state. He remained in charge of the Manhattan Project until the creation of the. Atomic Energy Commission in 1947. General Groves retired as a Lieutenant General in 1948.

Siting Rocky Flats Upwind and Uphill From the Denver Metropolitan Area Without Understanding the Requirements of the Industrial Enterprise

The choice of the ultimate location of Rocky Flats upwind and uphill from Denver, Colorado was the product of a number of intertwining factors. The first notable factor was that only the state of Colorado had two U.S. Senators on the Congressional Joint Committee on Atomic Energy, Senators Edwin Johnson and Eugene Millikin.

Neither U.S. Senator from Colorado considered "Project Apple" (the secret code name for what became known as Rocky Flats) to constitute a danger to its surroundings. This assessment is in contrast to the known risks associated with the operation of the multiple nuclear reactors sited at Hanford and Savannah River. Both U.S. Senators had a strong pro-business constituency which valued the creation of jobs and which would provide income for the local economy. An example of the thinking which prevailed at the time is a quote from Making a Real Killing by Len Ackland: "There is one school of thought that it would be perfectly safe to put it [the Project Apple facility] on 16th Street here in Washington D.C., with which the authors do not disagree." (Len Ackland, Making a Real Killing, p. 59, note 24. The quote is attributed by Professor Ackland to two staff members of the Congressional Joint Committee on Atomic Energy of the U.S. Congress).

A second set of factors which led to the ultimate site selection was the choice of what the Atomic Energy Commission considered to be the most important site selection criteria: a dry, moderate climate; attractive community facilities and living conditions for employees; good transportation infrastructure; and well-located from the perspective of "strategic invulnerability." In retrospect, the Denver metropolitan area certainly did provide excellent community facilities and living conditions for employees at Rocky Flats. As metropolitan Denver grew, transportation infrastructure improvements continued to keep access to

Rocky Flats manageable for commuting. Denver is recognized for its "dry, moderate climate."

"Strategic invulnerability," was deficient in retrospect. As the strategic era evolved, the capability of launching ballistic missiles from halfway around the world or from submarines located just off the Pacific Coast transformed "strategic invulnerably" into "strategic vulnerably" for both Denver and Rocky Flats. This danger was exacerbated by the Soviet capability to launch multiple, independently targeted reentry vehicles each with multiple nuclear warheads (MIRVs).

In answer to the question of whether different priorities in site selection guidelines would have produced a different site location, the answer is that incorporation of broader environmental and health impact assessment criteria as is now required under the National Environmental Policy Act of 1969 (NEPA) certainly would have disqualified the site chosen. A robust risk assessment was not required under the Atomic Energy Act of 1946 and did not occur.

Reconstructing the Flawed Process of Site Selection

The availability of primary source documents from Dow Chemical archives has made it possible to reconstruct the flawed process by which Rocky Flats was chosen. The three-month site selection search was conducted by the Austin Company of Ohio from January to March of 1951. There is extensive documentation based on the contract between the Atomic Energy Commission Santa Fe Operations Office (SFOO) and the Austin Company (contract AT(29-1)-1104, Modification 1), dated February 1, 1951 and on the actual work products submitted by the Austin Company pursuant to the contract.

"Site Selection" is presented in terms of the following categories of analysis:

- The Site Selection Process
- Climate
- Prevailing Winds
- Earthquakes and Tornados
- Proximity of a Large City

This is a particularly informative presentation in light of the array of problems which developed throughout the production era (1951-1989), the genesis of environment crime, and why it was not possible to produce plutonium igniters in the quantities necessary for national security at the Rocky Flats site and still be in compliance with the Clean Water Act of 1972 and the Resource Conservation and Recovery Act of 1976.

Because the site was chosen in 1951, and relevant environmental legislation at the federal level was not passed for approximately 20 years, the absence of health and environment criteria is understandable yet regrettable.

Fifteen site selection criteria were utilized by the Austin Company:

1. General Location – Generally Nebraska, Kansas, Oklahoma, Missouri, Arkansas, Colorado, and the Texas Panhandle;
2. Site Area – a square two miles on a side preferably owned or controlled by the U.S. government (In retrospect, with Hanford originally at 670,000 square miles and operationally at 586,000 square miles, this criteria is the obvious origin for the characterization of Rocky Flats at four square miles as "very small.");
3. Plant Area – the central square mile should be reasonably level;
4. Supporting Population – the edge of the site shall not be less than five miles nor more than 25 miles from a community of at least 25,000.
   Comment: (General Leslie Groves, who administered the Manhattan Project for the United States, had mandated that nuclear weapons production sites should be more than 25 miles away from a community);
5. Isolation – There shall be a minimum displacement of homes and populations;
6. Attractive Environs – Living conditions and community facilities should be good;
7. Climate – Dry, moderate climate with evaporative cooling;
8. Rail Transportation – Within ten miles;
9. Highway Transportation – Near a good main highway;
10. Air Transportation – Airport served by major east-west and north-south airlines with an Army airfield within 50 miles;
11. Power – Dependable source of 12, 000 kilowatts;
12. Water – One million gallons per day;
13. Fuel – Heating of building only;
14. Sewage – Normal plant wastes;
15. Drainage – Site shall be above possible flood levels.

The criteria of a nearby population of 25,000 led to the consideration of cities in Arkansas (Little Rock, Fort Smith), Colorado (Denver, Pueblo, Colorado Springs), Kansas (Wichita, Topeka, Hutchinson), Texas (Amarillo), Missouri (St. Louis, Kansas City, Springfield, Joplin, Jefferson City), Nebraska (Omaha, Lincoln), Oklahoma (Oklahoma City, Tulsa, Muskogee, Enid), and New Mexico (Tyler – which is near Los Alamos, Albuquerque).

There were two levels of review of the relative importance (ranking) of these fifteen criteria. The first level of review was a discussion between the Austin Company, the AEC Santa Fe Operations Office, and Dow Chemical. This level of review had the outcome of determining the "most important" criteria for selecting a site:

> "...a dry moderate climate, adequate supporting populations, attractive environs to hold and attract workers, and accessible from Los Alamos and the Chicago and St. Louis gateway."

and

> "One of the most important criteria is that the site be within 5 to 25 miles from a community of at least 25,000 population. This is the minimum community which was believed capable of providing the 700 employees that will be hired locally and providing housing for the 200 employees that will probably be moved in from other communities."

An outcome of the meeting to establish priorities among the 15 criteria (with the addition of accessibility to the Chicago and St. Louis gateways and Los Alamos – presumably elaborations of Criteria 15-Air Transportation, of north-south and east-west air access) was the elimination of all but Colorado Springs, Pueblo and Denver.

The second level of review (conducted by the same reviewers) narrowed the finalists for site selection to only Denver:

> "...in fact, the three Colorado locations are the only ones that satisfy the climatic criteria. Pueblo is less suitable than Denver in being less attractive to workers, less accessible and in being the home of Colorado Fuel and Iron's vital industrial operations. Colorado Springs has the fundamental disadvantage of being served by a relatively small local electrical utility which has no interconnections with other utilities as emergency sources."

There are several unexplained anomalies in these explanations. It is not at all clear why being home to the vital industrial operations of Colorado Fuel and Iron was a disadvantage for Pueblo. It could be argued that having an existing industrial base was a strength related to the presence of an experienced local workforce. Without accompanying references or cited authorities, the Austin Company asserted that "Colorado provides the minimum of natural hazards..." The exhibit which accompanies that assertion refers to dust storms, tornadoes, and earthquakes.

Once the site selection had been narrowed to the general Denver area, the Austin Company declared that based on satisfactory terrain and minimum occupancy "...there are dozens of good sites in the Denver area." The Austin Company then conducted a "field survey" which reduced the "dozens of good sites" to seven sites "...which lie nearest to the City of Denver and could be economically served with power, water, railroads and highways."

The next step was a "field inspection" of all seven Denver area sites by representatives of the Atomic Energy Commission, the Dow Chemical Company, and the Austin Company. From among the seven locations, there was consensus on two of them as being superior: what was designated as Site 1 near the Rocky Mountain Arsenal, and Site 4, the ultimately chosen Rocky Flats.

The reasons for the final recommendation, presumably to the Atomic Energy Commission and to the Congressional Joint Committee on Atomic Energy, are from the Austin Company report:

> Denver is recommended because:
>
> a) It completely satisfies the Climatic criteria, permitting evaporative cooling to be used in air conditioning: this is particularly important as many areas will be supplied with 30 air changes per hour without recirculation. Low or moderate humidity is also important for the proper operation of health physics personnel protective devices.
>
> b) It is an attractive place to live and it has a popular appeal that should assist recruiting workers from other cities. It is especially important that the employees be content to remain and live in Denver, since the three months or longer security clearance period for key personnel will make any unusual turnover extremely undesirable.
>
> c) Housing in Denver is tight – but no more so than in any of the other major communities that were considered. There is considerable new housing under construction in all sections of the city and suburbs and a wide choice of desirable home sites. Denver has the advantage of being an important tourist center and has an unusual amount of transient accommodations which can house incoming employees until they have secured permanent residences.
>
> d) Labor is reported tight in Denver – as it was in all the other major communities under consideration. Despite this report there are two indications that labor should be relatively easier to hire locally; Denver has been and still is the most fruitful source of employees for the Los Alamos operations; and the Denver labor pool is still being used predominantly as a source of workers for employment outside of Colorado, rather than locally.
>
> e) Denver has a minimum of natural hazards.
>
> f) It is very accessible from Los Alamos and conveniently accessible from the St. Louis and Chicago gateways. It is a major transportation center, served by a large number of airways and railroads – providing a considerable choice of alternate routes and operating companies.

Flaws In Choosing a Site Only Five To Twenty-Five Miles From Denver, Colorado

The Lamm-Wirth Task Force (1975) among many others have concluded that given its hazardous missions, and the nature of the industrial enterprise essential to accomplishing its missions, that Rocky Flats should never have been sited where it was. The first disastrous decision in siting Rocky Flats was to disregard the nuclear weapons production site requirement of General Groves that nuclear weapons production occur further than 25 miles from a city.

In the 19th Century, Rocky Flats used to be reachable in a journey of one day from downtown Denver.

Figure 4.2 Rocky Flats Stage Coach Stop

This presentation is indebted to Charles McKay for access to archives of the Church Ranch from which this photograph was located. The rural character of the location which became Rocky Flats was already poised to be encroached upon by Denver suburbs when the siting decision was made. In the 19th Century Rocky Flats used to be a day trip from Denver and another day trip to Boulder. After crossing the plains with an Ox Team from Iowa, George Henry Church and Sarah Henderson Church homesteaded Church Ranch. When mining claims did not work out, they returned briefly to Iowa before returning with 50 cattle in May of 1862. According to Charles McKay, they noticed a lot of travelers including the Pony Express and decided to build the boarding house. The stage-stop was named Church's Crossing Stage-stop. When Wells Fargo discontinued the Pony Express stage route in Spring of 1868, their principal guests became Bull Wackers, Ox Team drivers who were hauling bales of hay to the mountains. Their most famous guest was President of the United States Ulysses S. Grant and his daughter. George Henry Church was born in 1830, one of eleven children, and lived 88 years until 1918. Sarah Henderson Church was born in 1838 and lived until 1921. Their son Frank was born in 1863, and is the father of Marcus F. Church, who was a principal litigant in the Church cluster of civil lawsuits in 1975. What became Rocky Flats in 1951 was legacy land of an enterprising pioneer family.

The second disastrous location decision was to fail to recognize the relevance of what they already knew about the nature of Denver in 1951 when they chose a site only 16 miles from Denver. This is what they knew about Denver in 1951 as part of the rationale for choosing the Denver area for what was to become Rocky Flats:

- "There is considerable new housing under construction in all sections of the city..."
- "Denver has been and still is the most fruitful source of employees..."
- "It is an attractive place to live and it has a popular appeal that should assist recruiting workers from other cities."

This is the profile of one of the most highly attractive metropolitan areas in the United States. The point is that originally siting Rocky Flats only 16 miles from an expanding Denver, Colorado was to miss the known factors driving Denver expansion. In a few short years the Rocky Flats site was surrounded by Denver suburbs and the Stage Coach stop a distant memory from a much earlier era.

Rationale For the Final Selection of the Rocky Flats Over Other Denver Sites

Site 4 on Rocky Flats is recommended because:

a) Its terrain provides a desirable combination of mesa and ravines.
b) Its deep beds of gravel provide good foundation conditions.
c) The gravel surface, its altitude above the sandy farmland to the east, and the closeness of the foothills of the Rocky Mountains provide maximum protection from widespread or local dust storms which would disrupt the important ventilation system.
d) Water from a reservoir of the City of Denver is reasonably accessible.
e) It enjoys maximum reliability of electrical power from adjacent transmission lines with multiple generating sources.
f) It is within reasonable driving distance and time of Denver, Boulder and Golden over good highways with very little commuting traffic.
g) It is about two miles from the Denver, Rio Grande and Western Railroad's Moffat Tunnel mainline. It would be feasible to extend a sidetrack to the site if this should be desirable.
h) The property is the least valuable of the seven Denver sites considered and should be obtainable for the least cost.
i) The site has the least occupancy of the seven Denver sites; only one homestead was apparent.
j) It is remote from any industrial installation or conceivable military target.
k) It is easily adaptable to any desired degree of plant security control.
l) It is ideal from the viewpoint of public relations: minimum displacement of homes, land used only for minor grazing and well removed from any residential area.

The recommended site has only one important disadvantage, which cannot be discounted; it is about 27 miles by highway from Stapleton Field, the Denver municipal airport. However, Site 4 has so many other advantages over the several sites which are closer to the airport that it was agreed that the advantages far outweigh this major disadvantage...

Flaw In Wind Direction and Velocity

The rationale for selecting Rocky Flats included the observation that "the closeness of the foothills of the Rocky Mountains provide maximum protection from wide-spread or local dust storms..." The concern was to prevent disruption of "the important ventilation system." This rationale was fatally flawed. It was based on wrong information. Evidence was available at the time which should have disqualified the Rocky Flats site:

Prevailing Winds

...a major reason for the elimination of Site 6 was that the prevailing winds...from the site are toward Denver and Marston Lake, which is a main reservoir for the city. Although this might constitute an unlikely hazard, it was considered very undesirable from the viewpoint of good public relations...The wind rose for Denver...indicates that the prevailing winds are from the south. This favors Sites 1 to 4, inclusive, as they are located north of Denver...

(Holme, 1978)

The following table is Table 4 of Lamm-Wirth Task Force Report of 1975. It is regrettable, of course, that this was prepared twenty-four years after the site was selected.

TABLE 4
AVERAGE AND EXTREME METEOROLOGICAL DATA

	Average	Extreme
Precipitation	About 15 inches/year	about 25 inches in 1969
Humidity	46%	60% in 1960
Temperature	77° F average maximum 22° F average minimum 50° F average mean	102° F, July 12, 1971 -26° F, January 12, 1963
Wind	8.1 mph average mean wind velocity	105 mph on four occasions

(Lamm-Wirth, 1975)

By stating that "Denver has a minimum of natural hazards," the report implies that the specific sites under consideration were equally safe. As the Lamm-Wirth report made clear, however winds at the Rocky Flats site could exceed 100 miles an hour as a result of the effects of the nearby (about four miles) Front Range of the Rocky Mountains.

The Austin Company's report contains many statements that are wrong. The consequences would haunt the production era and combine to create the circumstance that off-site contamination from normal operations would be impossible to avoid.

One of the original site selection criteria was the availability of one million gallons of water on a daily basis. The rationale for selecting Rocky Flats over other Denver locations was the gravel surface with "...its altitude above the sandy farm land to the east...". This rationale contained acknowledgement of the altitude differential but not its significance in terms of disposal of the byproducts of the industrial enterprise which included polluted water with an easy pathway off-site. The mention as part of the rationale of "water from a reservoir of the city of Denver is reasonably accessible" did not anticipate that only 22 years later tritium would be discovered from Rocky Flats in Great Western Reservoir.

Part of the rationale for the choice of Rocky Flats over the six other Denver area sites was that: "Its terrain provides a desirable combination of mesa and ravines." It was not explained in the particular rationale why "mesa and ravines" was seen as a positive. In retrospect, the layout chosen within the 334-acre high security area when coupled with the altitude difference from West to East exacerbated the drainage problem to the East.

Figure 4.3 Aerial View of Rocky Flats in 1951

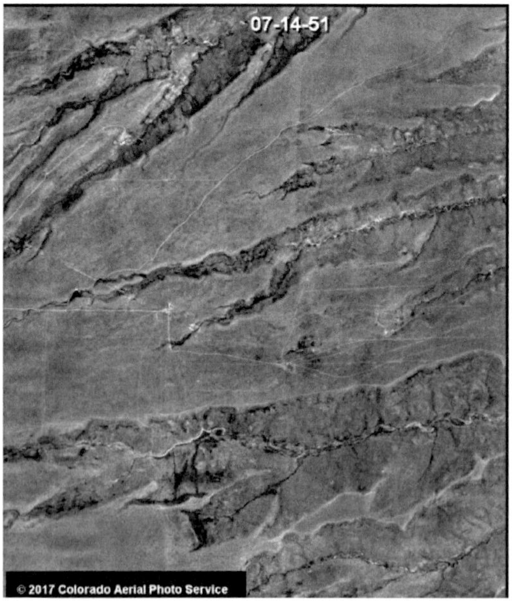

This is an aerial photograph from 1951 preserved by John Young of Colorado Aerial Photo Service. The site selection criteria and rationale for choosing Rocky Flats referred to drainage. The concern about drainage was that the site to be chosen not be vulnerable to floods. A fatal flaw in site selection was in another aspect of drainage; namely, there was a striking major altitude difference in the site chosen as Rocky Flats from West to East. The altitude on what is now the West side of Highway 93 near the old Central Avenue is 6,224 feet above sea level. The altitude on the East side near Indiana Street north of the Walnut Creek Drainage is 5696 feet above sea level, a difference of 528 feet according to U.S. Geological Survey Digital Global Map data.

The topography shaped the flow of the two creeks which traversed the site, and subsequently became highly contested terrain in the long-running Cook class action civil lawsuit of 1990.

Figure 4.4. Aerial View of Rocky Flats in 1962

This is another aerial photograph preserved by John Young of Colorado Aerial Photo Service. The production area which was chosen for development for the high security area in conjunction with the altitude difference from West to East and the ravines leading off-site, combined to make it highly unlikely that contamination could be contained within the small site and indeed it was not. The decision to cluster the major production buildings so closely together increased the likelihood that the lack of separation could lead to a circumstance that difficulties in one production facility could disrupt production in another facility.

The Austin company reported only one "important disadvantage" to the Rocky Flats site: it is 27 miles by highway distance from Stapleton Airport. At that time, Stapleton Airport was the major regional airport. The report went on to identify, as advantages of the recommended site, one asset which was true at the time, but with the advance of nuclear weapon delivery systems, soon was not: "It is remote from an industrial installation or conceivable military target." The irony is that Rocky Flats itself soon became a reachable military target, and therefore the Denver metropolitan area itself was certain to be collateral damage from a nuclear attack.

Contrasting the Siting Decisions At Rocky Flats In 1951 With the Approach Taken By the Stewards of the Siting Decisions At the Hanford Nuclear Reservation In 1942-1943

The missions of the Hanford Nuclear Reservation in southeastern Washington State and of Rocky Flats in Colorado both contributed to the national security of the United States. They have each made indelible, indispensable, and distinctive contributions to the production of nuclear weapons. Both undertook their roles in weaponizing the atom in an era that the then President of DuPont characterized in 1942-1943 as presenting "unknown hazards."

With respect to the siting at Hanford, Colonel Franklin T. Matthias had been dispatched along with representatives of the DuPont Corporation by Brigadier General Leslie R. Groves to find a suitable location for large-scale plutonium production. DuPont recommended that plutonium production be located far away from the existing uranium enrichment facility at Oak Ridge, Tennessee.

In a speech on January 14, 1967 Colonel Matthias reported (Matthias, 1967) that the Hanford location was "ideal in virtually all respects" except for the presence of the small agricultural towns of Hanford and White Bluffs. Michele Gerber of the Westinghouse Hanford Company reported the site selection criteria used to select Hanford:

- A large and remote tract of land,
- A "hazardous manufacturing area" of at least 12 by 16 miles (on a side or nearly 200 square miles),
- Space for laboratory facilities at least 8 miles from the nearest reactor or separations plant,
- No towns of more than 1,000 people closer than 20 miles from the hazardous rectangle,
- No main highway, railway, or employee village closer than 10 miles from the hazardous rectangle,
- A clean and abundant water supply,
- A large electric power supply, and
- Ground that could bear heavy loads.

(Gerber, 1992, p. 6)

The DuPont conception was predicated on the presence of ultrahazardous risk: an area of at least 12 miles by 16 miles or nearly 200 square miles must be available for a "hazardous manufacturing area."

General Groves acted under authority granted by the Second War Powers Act 56 Stat. 176 (1942). He ordered the relocation of 1,500 residents of Hanford, White Bluffs, and nearby settlements, as well as the relocation of the Wanapum people, Confederated Tribes and Bands of the Yakama Nation, the Confederated Tribes of the Umatilla Indian Reservation, and the Nez Perce Tribe. These relocations made it possible to select an uninhabited site for what became the Hanford Nuclear Reservation of the originally 670,000 square miles (not a misprint). This is an area half of the size of the state of Rhode Island. The size of Hanford was later reduced to 586,000 square miles. In contrast to the vastness at Hanford, Rocky Flats was sited on only 4 square miles.

The "hazardous manufacturing area" at Rocky Flats was selected to be 334 acres not 200 square miles. Support laboratory facilities at Hanford had to be at least eight miles from the reactors and the chemical separation facilities.

Figure 4.5 Vastness of Hanford Along the 50 Mile Border With the Columbia River

This photograph illustrates the separation between vital production areas in order that a disruption at one facility did not cause a production problem at another facility. The Columbia River provided cooling water for the nuclear reactors which produced the spent nuclear fuel containing plutonium.

At Hanford no highways, railways, or employee villages could be closer than 10 miles from the "hazardous manufacturing area." At Rocky Flats they were sited just outside the high security boundaries. In fact, ERDA in 1975 opposed the recommendation of the Lamm-Wirth Task Force to expand the "buffer zone" to reduce risk from a nuclear accident. A strength of the rationale for the current boundaries was presented by ERDA as their coherence with the natural boundaries of existing roads and community infrastructure. There was no reference by ERDA to the Seed report (1971) recommendation to the contrary. After detailed study of soil contaminated with plutonium and its spread off-site, the Seed report recommended expanding the "buffer zone."

The vastness of the Hanford Nuclear Reservation made it possible to site critical infrastructure at such great distances from each other that an accident at one constellation of facilities had very little likelihood of interfering with production of plutonium at another constellation of facilities. The high-level waste storage tanks were sited far from the chemical separation facilities located in the high plateau, while the nine plutonium creation reactors were located along the 50 mile Columbia River border of Hanford for access to cooling water.

Figure 4.6. Rocky Flats At Sunrise

This dramatic photograph of Rocky Flats at sunrise from the archives of the "Accelerated Cleanup" document Closure Legacy: From Weapons to Wildlife when juxtaposed with the Hanford site along its 50 mile border with the Columbia River in Figure 4.5 illustrates one of the fatal flaws in site selection: an original 16-mile distance from Denver soon became an ultra hazardous site almost next door to residential communities and both upwind and uphill with pronounced altitude difference causing draining of toxics off-site toward the communities.

There are of course major differences in the missions of Hanford and Rocky Flats and in their different industrial infrastructures. But the siting of Hanford was predicated on minimization of risk from the anticipated "hazardous manufacturing area." There was no such anticipation at Rocky Flats of the necessity to reduce the risk of disruption of one plutonium production facility disabling production at another facility.

Rocky Flats Confronted Challenges In Common With Other U.S. Nuclear Weapons Production Sites Which Were Not Reflected In Site Selection Criteria Or In Building Design and Construction

Rocky Flats as a site for producing plutonium igniters confronted a series of challenges encountered throughout the U. S. nuclear weapons production complex. Two of those key challenges not included in site selection criteria and infrastructure design were protecting health and safety and reducing the risk of fire.

Protecting Health and Safety

The nature of plutonium igniter fabrication is ultrahazardous: it uses toxic chemicals to separate and recover plutonium metal from manufacturing scrap, it generates enormous quantities of chemicals (and especially solvents) contaminated with traces of plutonium or uranium residues requiring environmentally safe disposal, and it creates solid materials such as containers and clothing also requiring disposal. An alternate point of view by the Atomic Energy Commission was quoted by Ackland (1999, p. 101): "The degree of radioactivity at Rocky Flats is small. However, in any work involving radioactive materials, precautions must be taken to hold exposure to a minimum. 'Radioactive sickness' is not a health factor at the plant." This comment by the AEC was in response to a newspaper inquiry; hence, it had widespread circulation. It was a comment in retrospect that has proven repeatedly to be disastrously wrong.

<u>Known Risks of New Technology: "Expect To Have a Fire"</u>

There were catastrophic plutonium fires at Rocky Flats in 1957 and 1969. <u>Making a Real Killing</u> and the <u>Bulletin of the Atomic Scientists</u> report extensively on the potentially catastrophic consequences and dangerousness of those fires. Less well known were the earlier warnings of the Atomic Energy Commission that the risk of fire should be anticipated. Experts within the Atomic Energy Commission also knew that a small amount of plutonium in the presence of water could lead to a chain reaction, a "criticality event." The release of radiation from a "criticality event" would kill anyone close to it. For that reason, a guideline for extinguishing a plutonium fire precluded the use of water. Experts within the Atomic Energy Commission also knew, based on earlier occurrences at Los Alamos and Hanford, that plutonium in small chips and the filings found in plutonium waste could spontaneously burst into flame.

Building Replacement and the Challenge of Unproven Production Processes

An illustration of the consequences of unproven production processes and the lack of fully understanding them even after nearly twenty years of operation is vividly illustrated by the decision to replace Building 771, one of the first plutonium production facilities at Rocky Flats, with a new facility, Building 371. Building 371 was to be located within the 334-acre maximum security portion of Rocky Flats. Building 771 had been opened with a projected life span of twenty-five years.

A principal advocate to acquire funding for Building 371 as a replacement for Building 771 before the U.S. Congress was Major General Edward Giller of the Atomic Energy Commission. General Giller had flown 115 combat missions in the European Theater including providing air support for the Normandy invasion. After World War II he returned to the University of Illinois and earned a Ph.D. in chemical engineering. General Giller, therefore, was a particularly credible advocate for the Atomic Energy Commission. He was responsible for the AEC programs to weaponize the atom as Assistant General Manager for Military Applications.

Figure 4.7. Major General Edward Giller

After flying 115 combat missions over Europe during World War II including cover for the Normandy Invasion, General Giller returned to the University of Illinois and earned a Ph.D. in chemical engineering. This background made him an especially credible advocate for the missions of the Atomic Energy Commission.

General Giller was asked first to explain the cost increases from the estimate in 1970 for Fiscal Year 1971 of $63 million to the estimate for Fiscal Year 1973 of $113 million. The major reasons for the cost increase were a longer projected construction period, additional demolition and removal costs ($15 million), and improvements in process design ($17 million).

Questions from Congressman Joseph Evins (Democrat) of Tennessee and the answers from General Giller are as follows:[1]

> GENERAL GILLER. The new facility is the top one. We are ready to start title II whenever we have the authority.
>
> We told the committee that we would not move into any construction or anything in title II until we had a chance to explain it and give you our decisions and, of course, get your approval.
>
> We now stand at that point in time. The procurement of long lead time items would not actually start until fiscal year 1973. Construction actually starts a little later than that. It is phased in time. We are ready to move forward when we receive approval. We have gone as far as we can go now.
>
> MR. EVINS. What unfavorable reactions have you received concerning the proposed construction of this facility?
>
> GENERAL GILLER. None
>
> ...
>
> MR. EVINS. What is the extent of dangers at the present facility? Outline the safety and other improvements being made in the existing facility pending the construction of a new facility.
>
> GENERAL GILLER. The present facility is deteriorating due to the severely corrosive atmosphere inside the glovebox lines. The corrosive atmospheres, composed principally of nitric acid fumes and hydrogen fluoride, attack the chemical process equipment and the gloveboxes which contain the process equipment. The corrosion

---

1 General Edward Giller quoted by Ackland (1999, p. 194) House Subcommittee, Public Works Appropriations, FY 1971, 1 October 1970.

> causes equipment and gloveboxes to fail, resulting in spills or leaks of contaminated materials into the working areas. These spills or leaks potentially subject personnel to radiation and to the hazard of inhaling airborne contamination. Equipment must be shut down and operating personnel are required to use respirators while the spill is cleaned up and the leak repaired. The increasing severity of corrosion with the increasing age of the equipment leads to increasing frequency of spills and the associated radiation hazard and production loss. Recent experience has shown that production interruptions due to spills and process equipment failures have increased in severity as well as frequency and are presently as long as several weeks. During production interruptions, plutonium compounds must be stored in the process areas, thereby increasing the potential radiation hazard.
>
> These safety modifications include the installation of fire doors and 2-hour firewalls to reduce the spread of a possible fire, additional corridors to guarantee the escape of occupants, improvements in the ventilation systems, additional sprinkler systems, and renovation of the heat detector system.

Following this presentation, the Congressional Hearing went on to discuss with General Giller a new facility at Los Alamos.

In subsequent testimony before a subcommittee of the U.S. Senate, General Giller characterized Building 771 as having been built in the 1950s based on old building code: "We try to keep it safe, but it has come to a point where it is time to replace that facility..."[2]

Building 371 was subsequently dedicated in the Spring of 1980. It was to undergo non-radioactive testing for one year before starting dual production processes: aqueous (the traditional chemical-based plutonium purification operation) and pyrochemical (reprocessing and refabrication of previously deployed nuclear bomb cores).

Nuclear experts associated with Rocky Flats design work did not understand completely enough the consequences of changes to technical production processes. Building 371 was designed with different operating levels: initial production processes on the ground floor, then transfer to the basement, then transfer to the subbasement, before transfer again for completion of the final processing on the ground floor (In contrast, Building 771 had all processes on one floor).

An initial discovery by Jim Weaver of Rockwell International was that in the new Building 371 twenty-five percent of the plutonium was being held up in transfer through the various piping or ducts, presumably related to moving the plutonium from one level of the building to another. With further investigation, the Department of Energy and Rockwell International could locate all but twelve percent of the plutonium. The fateful choice between competing mandates by the Department of Energy was to assign priority to meeting plutonium production goals because the

---

2 U.S. Senate Subcommittee, Public Works Appropriations, FY 1973, 26 April 1972.

military needed new nuclear warheads for its national security mission. Therefore, Building 771 was reopened even though it should have been permanently retired because, as General Giller had characterized it before the U.S. House of Representatives Subcommittee on Public Works Appropriations, Building 771 had deteriorated "due to the severely corrosive atmosphere inside the glove box lines."

Failure To Anticipate the Sheer Volume, Dangerous Volatility, and Complexity of Toxic Waste

In the earliest years of Rocky Flats there had been a failure to anticipate the sheer volume, dangerous volatility, and complexity inherent in the storage, treatment, and disposal of the waste streams associated with the production of plutonium igniters. With oversight by the Atomic Energy Commission, the years from 1951 through 1974 contained waste management choices which preordained the even more fateful choices leading to the circumstance that the production mission could not be accomplished without the commission of environmental crimes during the later years of the stewardship by Rockwell International.

Early Waste Management Was Established As a "Low Priority" Activity and Management Was Largely By Crisis

A Dow Chemical document (Dow Chemical 1969) presents waste management as a "low priority" but sometimes a "crisis-type" situation:

> Waste management tends to be a low priority project because it is desirable to do as little as possible to get rid of any waste streams. Changes in the waste program are generally made necessary by changes in shipping or governmental regulations. It is difficult to predict what changes will be made and this tends to develop a crisis-type situation in which effort is expended to solve a specific problem, followed by lack of activity until the next crisis. Some general trends in waste management can be seen and effort should be made to solve the approaching problems. These general trends are to decrease the amount of land, air, and water pollution and to improve isolation of radioactive material from man's environment.
>
> ***
>
> The waste management at Rocky Flats has been a low priority project for the past few years. Because of the recent publicity given Rocky Flats in relation to the May 11 fire, it would be highly desirable to improve our waste management program before some incident attracts attention to shipment and disposal of radioactive waste.
>
> (Dow Chemical, 1969)

Crisis-type situations occurred when it was necessary to solve specific problems. Some of the problems proved intractable. The organizing principle was that "...it is desirable to do as little as possible to get rid of any waste streams." This characterization was made 18 years after plant operations had begun, and after two major fires.

Waste Generation Was Not Predictable

Insights into early waste disposal practices at Rocky Flats is provided by Biles (1970) in a letter from Martin B. Biles, Director of the Division of Operational Safety at Rocky Flats, to Dr. Myron C. Waddell, the Chairman of the Colorado Health Planning Council dated December 20, 1970:

> ...At the time the Rocky Flats plant was placed in operation in 1951, the amount of contaminated waste that would be generated was not predictable with precision. Initially the wastes were placed in 30-gallon and 55-gallon steel drums and stored in or adjacent to the various production buildings. There were negotiations to ship contaminated oils, which comprised a large volume of those wastes, to the Idaho burial site, but because of potential leakage problems the idea of shipping the oil wastes was finally abandoned: By early 1954 about 900 drums of wastes had accumulated, making their storage a problem. Problems with shipping radioactive liquids are common. The basic Department of Transportation policy is that each primary container of radioactive liquids be surrounded with enough absorbent material to take up the entire volume of liquid if the primary container breaks. Except for small-bulk, high-value liquid shipments (for example, radiopharmaceuticals) the costs of compliance with this regulation generally make solidification before shipment preferable...
>
> ...Commencing in April 1954 and continuing until September 1958, wastes were moved to a location near the eastern boundary of the plant site and within the security fence for burial. There, in what became known as "the mound area," 1405 55-gallon drums of waste were accumulated during the period that disposal location was used. The drums of waste material were originally sent to the mound because it was uncertain at the time what the disposition of Rocky Flats waste would be. These wastes included materials such as glass and paper, oils, sand, and concreted dry materials. Except for 12 30-gallon drums of waste coming from a plutonium processing area, drums going to the mound were contaminated only with uranium (principally depleted uranium) for which there was not an acceptable disposal methods at the time....
>
> ...During the period between 1954 and 1962, a large amount of depleted uranium metal chips and scrap was accumulated from the plant operations and was disposed of in sealed drums which were buried in this trench. This disposal of the scrap uranium was necessary because the metal could not be transported. At the time that this scrap was buried, uranium shipments were not permitted unless the uranium was all in the form of uranium oxide. Some uranium scrap metal was converted to oxide and shipped, but the preponderance of the uranium was contaminated with numerous inert materials to the point of making it difficult to convert the metal to oxide. That uranium waste was buried. Probably tens of thousands of kilograms of depleted uranium were involved.
>
> (Biles, 1970)

Among the useful insights from the Biles communication are that at its onset, Rocky Flats was not able to assess the amount of contaminated waste which would be generated -- it was "not predictable with precision." It turned out early in the production era that the problem of leakage precluded shipping contaminated waste in 30-gallon or 55-gallon steel drums to the ARCO site in Idaho. This was the unanticipated beginning of what was subsequently called the "Mound" site (see Chapter Five). Biles explained that the Mound disposal decision reflected the fact that "it was uncertain at the time what the disposition of Rocky Flats waste would be." The expected opportunity to ship uranium off-site between 1954 and 1962 turned out not to be possible because contamination of the uranium with inert materials precluded it from being converted to uranium oxide, a requirement for safe shipment.

Only Three Staff Assigned To Early Waste Management

Another insight into the low priority assigned to early waste management is provided by a waste management report filed by E.S. Ryan on February 13, 1953. An important opening statement merits highlighting: "...this is a new group with little background knowledge of the problems presented by the material to be handled..."

> Since this is a new group with little background knowledge of the problems presented by the material to be handled, a series of meetings were held with other groups having a vital interest in waste disposal. These meetings were held, on the part of the waste disposal unit, with the purpose of obtaining a working knowledge of the waste disposal operation over which the unit is to have control. The group (Epp, Farrell and Ryan) began operation as a unit on January 8, 1953. On that date a joint meeting was held with supervision of 71 and 74 Bldgs., Venable, Pringle, Weidnerbaum, Ferris and Allen. During the meeting tentative procedure were outlined and discussed on how the waste disposal problem should be handled.
>
> (Ryan, 1953)

There were only three personnel assigned to the newly formed waste management group, a clear indication of the lack of magnitude of the challenge which was originally thought to be entailed by waste management.

Little Consideration Given To Disposal of Organic Waste With Trace Radionuclides

Another internal communication relevant to documenting early approaches to waste management is from K.W. Calkins to Plant Manager L.M. Joshel dated August 19, 1970:

> From the beginning of operations of the Rocky Flats Plant, certain organic liquids contaminated slightly with radioactive

> materials were generated which required disposal. Little consideration was given to this in the initial design of the plant, apparently based upon the assumption that these materials could be packaged and shipped for burial as were low-level solid wastes.
>
> Idaho Operations Office, the agency responsible for the operation of the burial site, however declined to accept these materials for burial, except for a small quantity of wastes coming from uranium operations which were shipped in 1957.
>
> These facts led to the establishment in 1954 of a "burial mound" at a location southeast of Building 991 in the eastern portion of the plant site. For the most part, liquids contaminated with uranium, either depleted or enriched, were buried there, but some drums containing plutonium contamination were also transferred to that location. Records do not indicate that they were actually buried.
>
> (Calkins, 1970)

The insight provided by Calkins, "Little consideration was given to this in the initial design of the plant...," confirms what may be inferred from the Putzier report of 1970 referring to what to do with contaminated organic liquids. This internal communication also confirms the consequences of the refusal of the Idaho Operations Office of the Atomic Energy Commission (ARCO) to authorize receiving shipments of organic liquids contaminated with radioactive waste.

The Putzier Report of 1970 Regarding On Site Radioactive Waste Disposal

This report was commissioned following the report by Edward Martell and Stewart Poet of off-site plutonium contamination which originated from Rocky Flats as well as after the negative publicity from the 1969 fire. Entitled Summary of On Site Radioactive Waste Disposal (Putzier, 1970), one surprising finding is that complete records simply had not been kept of radioactive wastes put outside of buildings at Rocky Flats. Putzier notes the limitation of his report on the first page:

> This is a compilation of extracted information from records, extrapolation, calculations and estimates from existing data along with a test of memories of a number of people in an attempt to piece together a fairly accurate account of on-site disposal of materials containing radioactivity. There are obvious omissions in records, but I believe they are complete enough to lead to a fairly accurate story.
>
> (Putzier, 1970)

Relying on memories of staff had three limitations. The first is known to us all: it is hard to remember routine transactions which occurred very recently let alone years before. The second limitation is structural: staff turnover in the nearly twenty-year time period which Putzier was attempting to reconstruct. A final limitation relying on staff memories is that access to knowledge was limited by strict compartmentalization in the service of national security.

Historical Reconstruction of Waste Disposal Practices

Putzier made distinctive contributions to the historical reconstruction of waste disposal practices despite the limitations of record keeping by Dow Chemical on waste generation and storage. The following, by way of example, is from a letter by L.R. Drake to R.R. Harrison dated January 5, 1955 noting that waste was being generated when the ultimate disposition had not yet been decided upon, and that the idea of shipping contaminated oils off-site had been dropped because of the leakage problem:

> ...concerning the disposition of D-38 oxide when the isotopic level is .29 or less. Rocky Flats Plant will make provisions to place such oxide concentration too high for disposition at ARCO, in tight metal drums and store these drums on the plant site.
>
> It is recommended that these drums be stored out-side on existing concrete slabs until some future use or disposition of this material is decided on.
>
> Some negotiations were attempted to ship contaminated oils to the Idaho burial site. Actually 10 drums, probably enriched uranium contaminated oil, were shipped. Problems relating to potential leakage eventually resulted in dropping the idea for shipping oils.
>
> (Drake, 1955)

It is clear that the planning process for the management of waste at the Rocky Flats Plant by both Dow Chemical and the Atomic Energy Commission did not anticipate the scope and depth it would encounter on problems of waste disposal.

Failure To Anticipate the Sheer Volume, Volatility, and Complexity In Waste Management

Another useful contribution to the historical record by Putzier was his preservation of the Waste Disposal Unit's Progress Reports for August of 1958 and September of 1958:

> Quote from August 1958 Progress Report: "Future Problems: No word has been received from Arco in regard to the resumption of shipping contaminated liquids to the burial grounds..."
>
> Quote from September 1958 Progress Report: "Future Problems: The rupture of a drum of 76 Building oil on the loading facility led to the conclusion that no liquid waste would be shipped to Arco, Idaho. In the following cleaning-up process, 115 drums of contaminated liquids were moved to the on-site mound after the surfaces of the contaminated drums had been decontaminated. With increased work loads resulting in an increase of contaminates [sic] wastes of oils and solvents, the problem arises as to what is to be done with them."
>
> (Dow Chemical, 1958)

The August 1958 report notes that the ARCO repository in Idaho of the Atomic Energy Commission had not replied about resuming the shipping there of contaminated oil in the necessary time-sensitive time frame. Plutonium traces in oil continued to accumulate at Rocky Flats from everyday operations. The September 1958 report notes the related problem of the leakage which was encountered from drums which had ruptured in the process of loading them for shipment to Idaho.

Owen and Steward Report 1974

A systematic reconstruction of waste disposal practices was undertaken by J.B. Owen and L.M. Steward of Dow Chemical Health Physics at Rocky Flats at the direction of the manager of the Atomic Energy Commission Area Office for Rocky Flats. It is not precisely known what precipitated ordering the report entitled <u>Environmental Inventory: A Historical Summation of Environmental Incidents Affecting Soils At or Near the USAEC Rocky Flats Plant</u> (Owen and Steward, 1974). Commissioning the report included the requirement to identify the "location of all contaminated soil at Rocky Flats." Commissioning of the report is probably linked chronologically to Dow Chemical admitting to the release of tritium into the environment off-site which was discovered in 1973.

While the purpose in commissioning the report was to focus on soil contamination, representative excerpts paint a devastating portrait of the dangerous daily workplace environment for workers:

- Radioactive infiltration of roadways
- Liquid leaking from waste drums
- High winds blew the lids off waste drums
- Laundry affected by both radioactive and chemical materials
- Uranium detected on warehouse floor
- Leaks and spills in chemical storage
- Plutonium spill outside Building 774
- Potential grows each day that drums will leak plutonium.

The report by Owen and Steward is vitally important for reconstructing specific incidents and practices with the best record available when augmented with surviving records of Dow Chemical Site Surveys (see below) and Putzier (1970, 1982). The following are illustrative excerpts from the Owen and Steward Report:

<u>Radioactive Infiltration of Roadways</u>

> Incidents involving radioactive infiltration of roadways have been noted. In May 1965, a box containing radioactive waste from Building 776 (Pu) was moved to the Storage Area near Building 663. Intermittent leakage from that box affected a strip of asphalt about 18 inches wide which was removed from the roadway. The box was stored on plastic sheeting (at Seventh and Central) until removal could be effected. No known infiltration resulted from that storage.

In June 1968, a leaking drum in transit from the 903 Area to building 774 resulted in approximately 140,000 dpm/100 cm^2 on the westbound lane of Central Avenue to and along Sixth Avenue. The entire material was spread to the ditch on the west side of Central Avenue. In August 1970, one section of the roadway between Eighth and Tenth Streets was removed and placed in the previously described asphalt dump.

(Owen and Steward, 1974)

Liquid Leaking From Waste Drums

Site Survey 9/7/61:

Site survey was called to Building 63 several times to monitor liquid leaking from waste drums from Building 44 [DU, Be], and on one occasion a drum from Building 76 [Pu]. No Contamination was found. The drums were returned to the buildings to be drained and resealed.

Owen and Steward Report at 25:

Punctured and leaking waste drums and boxes involving both solid and liquid wastes, U and Pu. Areas cleaned but residual infiltration certain. Area is now covered with gravel cap. About 280,000 square feet involved.

Owen and Steward Report at 23:

The Building 663 storage area, east of Seventh Avenue (opposite Building 444) and the adjacent storage yard west of Seventh Avenue, have both been affected several times by punctured or leaking waste boxes and drums. Both uranium and plutonium solid wastes, oils, and coolants have been involved. The areas have been scraped and the soil removed, but low-level residual plutonium and uranium are present.

Building 663, the storage and shipping facility has also been affected due to punctured drums and broken waste lines. Thus a total of about 290,000 square feet is possibly involved, to a varying [unclear in original but maybe] "...only about 160,000 square feet."

(Owen and Steward, 1974)

High Winds Blew the Lids Off Waste Drums

Owen and Steward Report at 26:

In 1953, high winds blew the lids off waste drums and uranium was released to the dock, driveway, and surrounding grounds. Direct count as high as 7,500 dm/100cm^2 and smears as high as 350 dpm were reported. The docks and sidewalks were cleaned up and the driveway areas seal-coated. A broken process waste line (June 1966) north of the building resulted in some possible infiltration as did leaking storage drums south of the building. An open ingot storage area east of Building 444 and a metal storage area south of the building have undoubtedly resulted in low-level infiltration of the soil, as has a uranium machine tool storage area west of the building.

Laundry Affected By Both Radioactive and Chemical Materials

Owen and Steward Report at 28:

> The Laundry, Building 442, is also potentially affected by both radioactive and chemical materials,...notably depleted uranium and beryllium, and in 1964 the Laundry was infiltrated by enriched uranium, impregnated in clothes from Building 883. The soil in the vicinity of this building has also been affected by instances of radioactivity release. For example, in December 1963, rag-cleaning barrels stored near the building either leaked or spilled. The liquid drained east into the ditch on the northwest side of the building. Radioactivity was detected as far east as the east end of Building 551. Cleaning efforts and subsequent runoff has reduced concentrations in that area to a low level.

Uranium Detected On Warehouse Floor

Owen and Steward Report at 29, 30:

> The Warehouse, Building 551, and adjacent grounds, have been areas of concern several times. For example, detectable uranium was discovered on the Warehouse floor in April 1953. Uranium chips and turnings were discovered in an aluminum scrap pile near the Warehouse in 1963 and again in 1964. These were removed and the grounds cleaned. In July 1963 and again in 1970, Rocky Flats received equipment and drums from off site which contained uranium above the Rocky Flats acceptable level. In 1970 the entire shipment of 55-gallon drums was returned to that vendor. These and other minor incidents lead to at least suspect areas under and around Buildings 551 and 554 (where the "hot" drums were received). A small drum storage area east of Building 552 is suspect for the same reason.

Leaks and Spills In Chemical Storage

> There have been some minor leaks and spills from drums and storage containers in the non-radioactive chemical storage containers in the non-radioactive chemical storage area east of Building 551. While a small area might have been affected, quantities involved have been so small that no impact has been noted nor expected. This area is used primarily to store drum quantities of acids, oils, soaps, and solvents.
>
> (Owen and Steward, 1974)

Plutonium Spilled Outside Building 774

Owen and Steward Report at 9:

> 1963 January: Liquid containing plutonium spilled outside Building 774 entrance. Material cleaned up.

Owen and Steward Report at 9:

> 1964 January: Coveralls containing plutonium found on west dock of Building 771. Levels to 100,000 dm/100cm^2 found on dock and in locker room. Cleaned up immediately.

Potential Grows Each Day That Drums Will Leak Plutonium

Owen and Steward Report at 10:

> 1971, Scrap drum leakage (June and July) resulted in recorded levels to 300,000 dpm/100 cm^2 on about 2,500 square feet of asphalt north of Building 771.
>
> (Owen and Steward, 1974)

J.B. Owen (Owen, 1971) wrote a letter to T.C. Jones of the Atomic Energy Commission on October 20, 1971 which provided a breakdown of the contents of 1,738 fifty-five gallon drums in temporary surface storage near Building 663:

> Rocky Flats has a backlog of 1,735 fifty-five gallon drums containing plutonium-contaminated wastes near building 663.
>
> All of these drums were generated prior to May 1971, and must be inspected prior to shipment to NRTS (Idaho). Inspection of similar wastes have resulted in the need to repackage nearly 100 percent...Continued storage of these drums outside presents the obvious potential that the drums will leak and plutonium contamination of the environment will result. This potential grows with each day.
>
> (Owen, 1971)

This portion of the Owen and Steward (1974) report, when viewed in the context of the ChemRisk (1992) report released eighteen years later of largely unmonitored "materials of concern" in the workplace, vividly combine to illustrate unique hazards at Rocky Flats.

Dow Chemical Monthly Site Surveys and Waste Disposal Group Reports

These monthly reports are invaluable for revealing what was known by Dow Chemical management and the dates they knew it from 1954 through 1960 regarding storage, treatment and disposal of radioactive and chemical waste. In the Mound area beginning in April 1954, the method of dealing with mixtures of solvents, traces of radioactivity and toxic chemicals beyond storage in 30-gallon or 55-gallon drums next to the buildings where the waste from fabricating plutonium igniters was generated was to bury it.

The excerpts below from the early years illuminate the ad-hoc nature of early waste disposal decisions: overturned, leaking and waste drums found empty; highly contaminated equipment stored inappropriately; contaminated building slabs; and internal pressure buildup in waste drums. Not noted in these terse reports is the obvious dangerous conditions routinely confronted by Rocky Flats workers. These terse reports highlight the unanticipated volatility and complexity of waste storage and disposal.

Very Temporary On-Site Storage By Temporary Buildings Before Freezing Temperatures Occur

Waste Disposal Group Report dated May 6, 1954 at page 2:

> A temporary on-site storage [area] was created by using one of the concrete slabs of the temporary buildings erected during construction of the plant. A total of 302 drums of graphite and 49 drums of liquid wastes were placed on the existing concrete slab. The drums were moved to this temporary storage site at the same time as those moved to the dirt mound for burial. Some of the liquid wastes are aqueous and while there is no impending hazard during warm weather, additional handling will be required before freezing temperatures occur next fall.

Drums of Coolant Will Be Disposed of On the Plant Site

Waste Disposal Group Report dated October 6, 1954, at page 2:

> Four drums of 44 Building [DU] waste coolant, from the concrete slab, were disposed of by blending with liquid wastes at 74 Building. In this manner the drums of coolant now in storage on the slab will be gradually disposed of on the plant site. This operation is an extremely slow method of coolant disposal.
>
> On September 20, 1954, 47 drums of graphite were moved from 44 Building [DU] to the concrete slab for off-site disposal under the supervision of the Waste Disposal Group.

Overturned Drum Containing Contaminated Waste Oil

Waste Disposal Group Report dated December 4, 1956, at page 2:

> On November 12, 1956 the contaminated oil spill on the storage slab south of 34 Building, caused by the overturning of a 20-gallon drum containing contaminated waste oil, was cleaned up and decontaminated to a low level. One hundred and twenty-four 30-gallon drums, including empties and drums containing oil, were moved from this storage slab to the "bull pen" on November 15 and 16, 1956. The area where the drums were stored was then cleaned on November 28, 1956.
>
> (Owen and Steward, 1974)

What this report identified as the "bull pen" area was immediately west of the "903" Pad. The force of the wind overturned a drum which caused it to spill. As many as 150 drums at a time were reported to have been overturned by the force of the wind at the Triangle area (see Chapter Five).

Highly Contaminated Equipment Stored On a Building Slab

Site Survey 12/5/56:

> Some of the drums containing contaminated oil from 44 Building [DU] leaked contaminating the concrete slab south of 34 Building. These drums were removed by the waste disposal group to a

fenced area for storage (probably also the "bull pen"). The leaking drums were disposed of. This group also decontaminated the area.

Subsequent monitoring of the slab disclosed the presence of highly contaminated equipment stored on the slab. This equipment was removed to a production area by the Technical Staff. The slab, however, has not been cleaned. Loose oxide was picked up and the area covered with plastic to temporarily prevent spread of the contamination. As smear up to 10,000 dpm were obtained before vacuuming, further cleaning should be done as soon as possible.

Site Survey 1/7/57:

The concrete storage slab south of 34 Building was monitored. Results showed contamination on the slab is spreading. A high of 7,245 dpm removable contamination is still present. As yet, no cleaning has been done.

Coolant Containing Plutonium Leaked Onto a Building Slab

Site Survey October 1, 1958:

SPECIAL MONITORING:
A waste drum containing coolant from Building 76 [Pu] was punctured on the east slab September 5. The contents were highly contaminated.

The contents of the drum were pumped into another barrel and the immediately area cleaned to a certain extent. Drums in the area were moved out. Those contaminated were cleaned and moved to an area southeast of 91 Building [the "Mound" area]. The last drum was moved on September 25. About 40 man-hours were required of Site Survey on this operation.

The slab is contaminated from 8,000 to greater than 100,000 c/m direct reading with up to 20,000 c/m removable contamination. This area is roped off and will be cleaned more thoroughly.

Internal Pressure Expanding Waste Drums

Site Survey 10/5/59:

Several perclene waste drums from Buildings 44 [DU, Be] and 74, [Pu] now stored south of Building 91, appear to be expanding from internal pressure. Waste Disposal has been informed and the drums will be vented.

Drums Are Rusting

Site Survey 12/3/59: Apparently as a result of this venting:

The waste drums south of building 91 still present a problem. The top of one drum from 76 Building [Pu] is covered with oil which counts about 1,000 c/m. Several drums are rusting.

Liquid Waste Drums Are Found Empty

Site Survey 11/3/60:

> Liquid waste drums stored south of 91 Building are leaking.
>
> About 550 drums from "C" plant [PU] are stored there. About 360 of these were checked. One drum is empty. One drum appears to be leaking around a vent bung on the top.
>
> About 610 drums from "A" [DU] and "B" [EU] plants were checked. Six containing "B" material and three containing "A" material are empty. (The total quantity of "A" as opposed to "B" drums there is not given.)
>
> Direct counts on pallets and the ground under the empty drums are generally greater than 100,000 dpm. The ground about twenty feet east of the "C" [Pu] drum is contaminated to 2,000 d/m, drums around this one up to 20,000 dpm.
>
> The leaking drums have been marked with red paint and the area is roped off.
>
> (Owen and Steward, 1974)

A Thoroughly Depressing Reconstruction of Early Radioactive and Toxic Chemical Waste Disposal Practices and Accidents

This overview of early Dow Chemical records of waste chronicles spills handled in a haphazard manner, the introduction of waste disposal practices that resulted in contamination throughout the site creating obvious risks to the health and safety of workers (not commented upon anywhere in available documents), the repeated occurrence of ruptured drums, drums overturned by wind, a small explosion reported by Security without documented consequences, and one drum reportedly found empty without further explanation. Particularly disturbing was the casual mention of a not casual fact; namely, that on November 15, 1962 plutonium waste had been stored on an active roadway.

As late as 1958, according to Martin B. Biles (Biles, 1970), the Director of the Division of Occupational Safety at Rocky Flats, "...it was uncertain at the time what the disposition of Rocky Flats waste would be." And by then production of plutonium igniters had been underway since at least 1953, and a minimum of at least 1,405 55-gallon drums had been buried in the Mound Area.

Assessing the Consequences of Failure To Anticipate

Site selection documentation does not include any references to a requirement of the Rocky Flats industrial enterprise for the storage and disposal of radioactive and chemical waste. This is an unexpected omission since the Atomic Energy Commission at both Los Alamos and the Hanford Nuclear Reservation had experience with radioactive waste disposal, and Dow Chemical's known preeminent experience with chemicals spanned the transformative creation of new chemicals to the safe disposal of contaminated chemicals.

It is not known why the site selection process did not include consultation with U.S. government experts with security clearances on wind, weather, and topography. This would have revealed that Denver was in fact downwind not upwind from the site selected, that winds from the Front Range of the Rocky Mountains could exceed 100 miles per hour, and that the slope of the land would send toxics into the watersheds of creeks which flowed toward drinking water reservoirs for Denver suburbs.

The testimony of Major General Giller before Congress is riveting for its portrayal of the difficult working environment inside production buildings: "The corrosive atmosphere, composed primarily of nitric acid fumes and hydrogen fluoride, attack the process equipment and glove boxes..." The failure of a replacement building for a key mission over technical design flaws illustrates the difficulty of mastering previously not fully understood production processes.

The failure to anticipate the sheer volume, dangerous volatility, and complexity of toxic wastes ultimately became the genesis of environmental crime when the new legal regimes of the 1970s were deemed applicable to Rocky Flats.

Introduction to Part II – Unbelievable Toxic Waste Disposal Fiascos and Unthinkable (and Preventable) Catastrophic Plutonium Fires and Missing and Unaccounted For Plutonium

Key Elements In the Chapter Five Narrative

- Tragic consequences became evident from the failure to anticipate toxic waste disposal in siting criteria and site selection.
- To sustain the production mission for the War Reserve stewards believed it was necessary to make unbelievably bad choices such as burying barrels of cooling oil contaminated with traces of uranium or plutonium from 1954 to 1958.
- In 1958, "all of a sudden one day" Dow Chemical stopped burying barrels with traces of uranium or plutonium. The choice was made to put them on pallets above ground on the 903 Pad exposed to the Colorado weather year-round from 1958 to 1969.
- The barrels stored above ground on the 903 Pad from 1958 to 1969 leaked an estimated 5,000 gallons of cooling oil with plutonium which migrated off-site—a fact discovered in 1970.

Key Elements In the Chapter Six Narrative

- Dow Chemical faced the fact that to keep the production mission going unabated, site selection choices had narrowed future choices available for waste disposal.
- It was acceptable to dispose of cooling oil with traces of radionuclides by open-pit burning of well over 1,000 barrels.
- To receive all manner of toxic byproducts of the production mission there were 14 trenches dug, some at least 300 feet by 50 feet.
- Environmental crime occurred in 1979 with the decision by Rockwell to use common agricultural field sprayers to operate all year-round to dispose of the astounding amount of 250,000 gallons of effluent a day or 80 million gallons a year, much of which migrated off-site.

Key Elements In the Chapter Seven Narrative

- The unthinkable at Rocky Flats is reported in depth: two preventable catastrophic plutonium fires and the revelation that 2,600 pounds of plutonium was missing or unaccounted for.

- The fire of 1957 started by spontaneous combustion of plutonium in a glovebox storage area. It was discovered by watchmen who were abused for calling the Fire Department. When the first responder arrived, he was denied entry while the fire accelerated. The fans in error were turned on high leading to an explosion which destroyed electricity and disabled radiation detection equipment.
- It is irrational [about the 1957 fire] that facilities were ever designed to permit hazards from a relatively insignificant amount of material to so strongly influence the chain of events which led to a major loss.
- First responders to the 1969 fire had not been told that water had been used on the 1957 fire to quell its initial ferocity.
- Mandated sprinklers had not been installed in the ceiling of the building where the 1969 fire occurred.
- In error the closest heat detector had been shielded so the spontaneous combustion fire went initially undetected while it accelerated.
- The roof was held on by cinder blocks. If the fire had breached the roof, it would have contaminated Denver.
- A consequence of the 1969 fire was that the capacity to produce and assemble plutonium igniters for War Reserve weapons was lost for several months.
- A Congressman's assessment on the 1969 Fire: "Nobody would permit operations to go on without firewalls and automatic sprinklers and things of this kind unless they were under the jurisdiction of the Federal Government where they can do nothing about it, I suppose."
- An informed Journalist's assessment: "It [the 1969 fire] was caused by the everyday incompetence and negligence inevitably found in any system which polices itself."
- The Department of Energy was willing to be held in Contempt of Court before a Federal District Court Judge while resisting disclosing details of the amount of missing and unaccounted for plutonium.

Part II presents the irony and tragedy which encapsulated the production era at Rocky Flats. The irony is that despite having created so much unnecessary risk and serious danger both on-site and off-site, Rocky Flats made vital and highly reliable contributions to the national security of the United States.

There are three great tragedies presented in Part II. The first great tragedy is the long-term negative consequences of failure to plan adequately for toxic waste disposal in site selection and the design of the industrial enterprise. The second great tragedy is the failure to respond to the "bad" lessons of experience by instituting constructive change of what caused the "bad" lessons in the first place. The third great tragedy is the failure of the United States government to adopt a unique hazards law to compensate nuclear workers with radiation-induced illness until 2000, eleven years after Rocky Flats closed for production.

Three unthinkable occurrences during the production era are the catastrophic fires of 1957 and 1969, and the missing and unaccounted for plutonium in the stunning amount of over 2,600 pounds between 1952 and 1993 when a nuclear bomb can be fabricated out of only 10 pounds.

Part II focuses on the fateful choices which Dow Chemical (Chapters Five, Six and Seven) and Rockwell International (Chapter Seven) made at Rocky Flats during the years of production when they were the contractors. Various fateful and subsequently disastrous choices have been evaluated in the environmental crime era, in the clash within and between U.S. institutions of constitutional government, by two juries of citizens, and by the evaluative frameworks of this book.

The first two chapters covered in Part II (Five and Six) during the stewardship of Dow Chemical have in common risky and unwise "solutions" to unanticipated waste disposal problems. A dilemma encountered by Dow Chemical across its years as contractor was captured by Dr. M.A. Thompson of the Health Physics Division when he observed in 1969 that "...many of our recent incidents have been events with small probability of occurring."

Chapter Five presents plutonium, uranium, and deadly chemicals in barrels that leaked in the Mound Area (1954-1958), the 903 Pad (1958-1969) and the Triangle Area (1966-1974). The negative consequences for contaminating the environment by storing mixed-wastes of combined radioactivity and chemicals in barrels either buried under a thin layer of soil (the Mound Area and the Trenches) or in the open air of Colorado all year around (the 903 Pad and the Triangle Area) was well understood by Dow Chemical management.

In an otherwise tense meeting of Dow Chemical management and the management of its Labor Unions on April 3, 1970, Dow Chemical representatives introduced some levity. Bill Lee observed that: "I think we have the world's barrel capitol in Rocky Flats." Jim Willging made a remark which acknowledged that the path they had embarked upon was not a wise solution for the long term: "You don't want to spend the rest of your life burying drums around here; you'll have no place to put buildings." A theme across Chapters Five and Six is that the flawed initial siting decision followed by assigning a low priority to waste management combined to put the stewards of Rocky Flats management into an untenable situation: it was simply not an alternative to fail to meet the production quotas for plutonium igniters for the War Reserve.

Previous waste management choices served to narrow the options available to subsequent stewards of Rocky Flats. Bill Lee was asked why the drums of mixed waste were buried in the Mound Area. His answer spoke volumes about the constrained choices available to management as early as 1954: "And we didn't know what else to do with them..."

Dow Chemical made two disquieting choices for radioactive waste disposal in the 1950s which were not prohibited by the controlling legal regimes of the Atomic Energy Acts of 1946 and 1954. The first disquieting choice was to make a determination in August of 1956 that, according to Putzier (1970), open pit burning of uranium laced oil was an "acceptable method" of waste disposal. There is nothing in the available record which recognizes threats from open pit burning to workers or to the community.

The open air burning of uranium based oil is discussed in Chapter Six. Putzier (1970) quoted a Dow Chemical Progress Report of February of 1961 that: "Open pit burning with air sampling of the plume indicates that these contaminated wastes from 44, 81, and 83 Buildings can be handled in this manner." Putzier indicated that Dow Chemical had disposed of over 1,000 either 30- or 55-gallon drums by open air burning of the contents, but that the records available to him contained many "omissions."

The second disquieting choice after the closure of the Mound Area was to open the 903 Pad for large scale above ground storage of barrels containing contaminated oil with traces of radioactivity. It was well known to Dow Chemical management that many of the barrels leaked into the soil. The Freiberg Report (no date) indicated that the "best estimate" was that the contents of 100 55-gallon drums had leaked into the soil totaling approximately 5,000 gallons.

When in 1970 independent scientist Ed Martell sought explanation at Rocky Flats for the plutonium he had found off-site, he was informed to his reported surprise that the source was not the 1969 fire, but most likely rather the 1957 fire and the barrels on the 903 Pad which had leaked into the soil.

The lead title of Chapter Six is "More Early Fiascos In Waste Management At Rocky Flats: From Open Air Plant Incineration To the Notorious Building 881 Hillside." It recounts an additional notorious 12 waste disposal sites of the 178 identified by ChemRisk (1992). These sites have in common providing insights into what management and workers were thinking when they made and implemented decisions which are so disastrous in retrospect. They illustrate collectively how the flawed site selection criteria narrowed the choices available for subsequent stewards, and the negative constraints which resulted from the determination to combine production and waste disposal in only 334 acres.

Chapter Seven recounts the serious plutonium fires of 1957 and 1969 which escalated quickly into catastrophic fires. Fires were a routine occurrence at Rocky Flats during both the Dow Chemical and the Rockwell International eras. Between 1954 and mid-1971, for example, there were 615 officially reported fires of which 384 were plutonium fires.

The fire of September 11-12, 1957 was reported by the Atomic Energy Commission to be the ninth most expensive accident loss in its history. The fire began by spontaneous combustion in a flammable Plexiglas plutonium storage box which was not equipped with a fire detection device. The plant protection staff who discovered the fire were reprimanded for calling the Fire Department. The first fireman who arrived was ordered not to fight the fire for ten minutes while the fire gained momentum. Then the fireman was refused permission to use water. The carbon dioxide used for the first 13 minutes was ineffective.

The fire died down when water was finally used. But in error the fans were turned up rather than off. The fire spread to the filter plenums where the electric power had been routed. An explosion in the filter plenum cut off electricity to the lights, radiation monitors, alarms and fans.

The heroic actions of first responders and Rocky Flats workers were inspiring in the face of great danger in 1957. But the use of a type of flammable Plexiglas which had been chosen by Dow Chemical management "could be ignited at its edges by one match."

The AEC Fire Investigation Committee Report of November 27, 1957 made numerous analyses and recommendations worth highlighting. But two stand out because if they had been adopted they would have been very relevant to minimizing the negative impacts of the 1969 fire.

The first excerpt was that: "...it appears (by hindsight) irrational that facilities were ever designed in such a manner as to permit hazards from a relatively insignificant amount of material to so strongly influence the chain of events which led to a major loss."

The second excerpt (and the failure to apply it to constructive change over the next twelve years) foretold why the 1969 fire did over $70 million in damage: "experience in explosives manufacture [provides]...a surprising number of analogies to guide users of radioactive material in efforts to minimize decontamination expense." Explosive manufacturers commonly reduce the maximum possible risk "at any single location by limiting both the amount of material handled and the maximum number of people permitted in one location."

As in 1957, in 1969 it was the bravery of first responders and the dedication and heroism of Rocky Flats workers throughout which prevented an even greater catastrophe as captured by the 1969 AEC Fire Investigation Committee Report:

> Only the heroic efforts of the firefighters limited this burning. Had an appreciable fraction of this shielding burned, there would have been a complete loss of Building 776-777 and its contents and a major release of plutonium to the environment.

Chapter Seven presents stunning revelations surrounding the missing and unaccounted for plutonium which were first disclosed by the Department of Energy in 1994, well after the close of production. The process of discovery in June of 1994 in the Cook class action lawsuit created a situation where the Department of Energy so resisted further disclosure that it elected to be nondisclosive until it was determined to be in Contempt of Court by a Federal District Court Judge. The struggle to force compliance with the Contempt Order was ultimately to occupy over three years, and is recounted in Chapter Fifteen.

As part of passing judgment on the combined Dow Chemical and Rockwell International thirty-eight years of production era stewardship as recounted in Chapters Five, Six and Seven, the Cook class action jury voted in 2006 to assign $200 million in punitive damages against the military-industrial contractors using the standard of "beyond a reasonable doubt." The determination of the Cook jury was that the actions of both Dow and Rockwell had been "willful and wanton."

Chapter Five
A Witches Brew of Plutonium, Uranium, and Deadly Chemicals In Barrels That Leaked: The Mound Burial Site (1954-1958) and Above Ground Storage On the 903 Pad (1958-1969) and the Triangle Area (1966-1974)

> "The rapidly increasing number of drums which contain plutonium-bearing wastes and residues, and which have been placed in outside storage is a matter of concern to Health Physics."
>
> Dow Chemical Health Physics Report, 1968

> "A serious potential fire hazard has been bought to my attention ... if one box were to catch fire, there would be a high potential for the fire to spread ..."
>
> Dr. M.A. Thompson, Dow Chemical, 1969

> "... many of our recent incidents have been events with small probability of occurring."
>
> Dr. M.A. Thompson, Dow Chemical, 1969

Risky and Unwise "Solutions" To Unanticipated Problems

If the waste disposal sites at Rocky Flats described in this chapter were introduced separately and individually, each would be considered an astounding failure. The time frame in which they operated individually and collectively without safeguards for human and ecosystem health reflected a pattern of risky and unwise solutions to waste disposal problems which were not anticipated in terms of their volume, complexity, and long-lived toxicity.

The sections which follow on the Mound Area, the 903 Pad, and the Triangle Area provide insight into one of the major evaluation questions investigated in this study; namely, "What were they thinking when they made decisions which were so disastrous in retrospect?"

The Mound Area

> "You don't want to spend the rest of your life burying drums around here; you'll have no place to put buildings."
>
> Jim Willging, Dow Chemical Rocky Flats Management, April 3, 1970

> "I think we have the world's barrel capitol in Rocky Flats."
>
> Bill Lee, Dow Chemical Rocky Flats Management, April 3, 1970

Question: "Why were they [the drums] buried?"
Answer by Bill Lee of Dow Chemical Rocky Flats Management: "And we didn't know what else to do with them..." April 3, 1970

"...the AEC licensed itself for burial grounds. So it isn't against regulations. Whether or not it's a good idea is another subject."
Bill Lee, Dow Chemical Rocky Flats Management, April 3, 1970

"You see if you start using a place for a burial ground, you can't ever use that place again for anything else."
Jim Willging, Dow Chemical Rocky Flats Management, April 3, 1970

"There's not too much that they have ever kept away from Monitors; but, by God, this is one that they got by."
Jim Kelly, Rocky Flats Steelworkers Union, April 3, 1970

"This would be a very serious thing if Dow was taking upon itself the burial of plutonium waste without going through the established procedures."
Congressman Chet Holifield, Ranking Member of the Congressional Joint Committee on Atomic Energy, April 10, 1970

Writing in their monthly report for April of 1954 dated May 6, 1954 (Dow Chemical, 1954) the Waste Management Group indicated that "to the best of our knowledge," the month of April in 1954 was the first time that sealed steel drums had been buried on-site in what came to be called the Mound Area:

> An on-site disposal of 869 drums was also made during the month of April. The method of disposal consisted of placing the sealed drums on the ground and then covering the drums with dirt. The area selected for this project is located atop the mesa to the south of 95 Building. The natural slope of the ground is to the north towards the gully which receives the effluent from the sewage treatment plant. It is felt that any water leaching the material contained in the drums will drain in the direction of the gully. Any activity not removed by the soil and carried by the water will then be diluted by the effluent from 95 Building and again by the retention ponds.
>
> (Dow Chemical, 1954)

The "direction of the gully" is presented elsewhere in this study as a tributary of what became under the law after 1972 a "navigable water of the United States." The gully was a clear pathway for toxics to migrate off-site.

Figure 5.1 The Mound Area

There is an iconic statement about the origins of the Mound Area which was made at a secret meeting of Dow Chemical management and the leadership of the Labor Unions at Rocky Flats on April 3, 1970. This was 16 years after the Mound Area had opened in April of 1954. When asked "Why were they [the drums] buried?" Bill Lee of Dow management answered, "And we didn't know what else to do with them…" The consequences of this waste disposal action were known at the time. The Waste Management Group in May of 1954 observed "the natural slope of the ground is north toward the gully which receives the effluent from the sewage treatment plant [and drains ultimately off-site]."

It is an indication of the lack of accuracy of record keeping that the Waste Management Coordinating Group itself could not attest to the validity of their own assertion that April of 1954 marked the first burial of steel drums on-site at Rocky Flats. And the consequences for off-site contamination of draining into a gully following the natural slopes of the land were not yet a concern worthy of comment and corrective action.

Ad-Hoc Nature of Waste Disposal

The ad-hoc nature of waste disposal practices, and the absence of a thoughtful management plan, is reflected in the waste disposal problem becoming so acute in 1954 that Dow Chemical placed 302 steel drums on a concrete slab associated with a temporary construction building. The comment was that this solution would work "temporarily" but would have to be replaced by the time of the on-set of cold weather in the Fall of 1954. The solution was to bury the steel drums in the Mound area where they stayed until 1970.

Included in the Putzier Report (Putzier, 1970) is a letter from E.S. Ryan to H.C. Anderson dated November 23, 1954 (Ryan, 1954):

> On November 16 and 17, 1954, the following drums of contaminated wastes were moved from 44 Building area and temporary storage (concrete slab) to the on-site mound for burial. This burial mound was formed during the movement of wastes during April 1954, at which time 862 drums of solid wastes from 44 Building were placed in the mound. The west end of the mound was reopened and the drums moved on November 16 and 17, 1954, were placed in the reopened area and recovered on November 18, 1954. Drums of liquid waste (still bottoms and oils)

placed on the concrete slab for temporary storage in April, 1954 became corroded and developed leaks. One drum of still bottoms had to be placed within a 55-gallon open top drum in order to move it. As a result of the pinhole leaks, the south end of the concrete slab used for temporary storage is now contaminated and will have to decontaminate.

There is a difference between the Waste Management Group report (Dow Chemical, 1954) that 869 drums were buried and E.S. Ryan indicating that there were 862. The astounding revelation in this letter of November 23, 1954 is that in a period of only seven months of above ground storage on a concrete slab, an unspecified number of steel drums had developed pin-hole leaks, yet they were apparently buried without re-sealing them.

Initial Efforts By Dow Chemical To Avoid Either Burial Or Large Scale Outdoor Storage

There were initial development studies undertaken in the 1950s to research alternatives to either burial or large scale outdoor storage. These efforts are referred to in a number of retrospective and more contemporaneous reports. After reporting on the establishment of the Mound Area in 1954, K.W. Calkins (1970) observed:

> These facts led to the establishment in 1954 of a "burial mound" at a location southeast of Building 991 in the eastern portion of the plant site. For the most part, liquids contaminated with uranium, either depleted or enriched, were buried there, but some drums containing plutonium contamination were also transferred to that location. Records do not indicate that they were actually buried.

<u>Problems of Permanent Disposal Were Recognized</u>

> Meanwhile, the problems of permanent disposal and of increasing quantities were recognized, and development studies were undertaken in 1956 to determine a satisfactory method of disposing of all such liquids. As a result of one study, the part IV addition to the plant included a high-speed centrifuge in Building 776 to process plutonium-contaminated organic liquids. It is not clear whether this was expected to produce a product suitable for "cold" disposal or not, but in any case the operation was disappointing and resulted in a recommendation made in 1958 that a substitute process such as distillation be developed for disposal.

Burning of Uranium-Contaminated Oil Adopted As Acceptable

> During those years, the burning of uranium-contaminated oils had been tested as a disposal method, and had been adopted as an acceptable method.
>
> Faced with the prospect of rising quantities of plutonium bearing oils the amount in the burial mound was moved south across the road to a fenced area in about July, 1958...
>
> (Calkins, 1970)

The matter of fact statement that open pit burning of uranium-contaminated oils had been adopted as "an acceptable method" of disposal is an extraordinary insight into the thinking in that era!

The immediacy of the change from burial of toxic mixed-waste in steel drums to outdoor storage is conveyed by H.E. Bowman, Manufacturing Manager in 1958 and subsequent General Manager at Rocky Flats:

> We buried them and then all of the sudden one day we struck them out--quit burying them and put them on top of the ground.
>
> (Bowman, 1970)

The first development studies to find alternative approaches to removing traces of plutonium from contaminated cooling oil were reported to have begun in 1956. The results from the use of a centrifuge to process plutonium-contaminated waste was characterized as "disappointing." The test burning of uranium-contaminated waste was "adopted as an acceptable method."

Putzier (1970) identifies additional efforts made to reduce the bulk of contaminated solvents and oils, and to recover plutonium and carbon tetrachloride:

> Future Problems: No word has been received from Arco in regard to the resumption of shipping contaminated liquids to the burial grounds. To use the limited on-site burial mound will necessitate the reduction of the number of drums to be buried. The use of the burning pit could be increased to cover additional low-level oils and combustible solvents from more buildings [i.e., the plutonium buildings]. Also stills and evaporators could be used for higher level wastes with the bottoms being buried.

No Liquid Waste Would Be Shipped To ARCO

> Future Problems: The rupture of a drum of 76 Building oil on the loading facility led to the conclusion that no liquid waste would be shipped to Arco, Idaho.
>
> Development Work: As a result of the above incident, the Waste Disposal Coordination Group started work on a program designed to reduce the bulk of the contaminated solvents and oils. The carbon tetrachloride and oil waste from 76 Building can be distilled for separation of the carbon tetrachloride from the oil. Thus far the recovered carbon tetrachloride appears to be reusable, the contamination remaining in the oil.

More Difficult Problem Is the Cleaning Up of the Contaminated Waste Oils From the Various Buildings

> The more difficult problem is the cleaning-up of the contaminated waste oils from the various buildings. Preliminary investigations show that activated carbon will remove activity, but whether the decontamination will be low enough to permit possible reuse of the oil or on-site burning, has yet to be proven.
>
> (Putzier, 1970)

The results of the initiatives reported above to reduce the bulk of contaminated solvents and oils were not reported.

Long Been a Problem of What To Do With the Contaminated Oils

Monthly Reports of the Waste Management Coordinating Group for November 2, 1954 and December 2, 1954 (Dow Chemical, 1954) refer to efforts at "sludging" which would allow for shipment of contaminated oils to the Atomic Energy Commission site in Idaho. In a November 2, 1954 report, Dow Chemical (1954c) does speak to the development of a promising approach to sludging:

> It has long been a problem of what to do with the contaminate oils and similar organics, such as perclene. From a recent study, which is being written up in detail, we found that a very simple and cheap method of sludging the oils, perclene etc., can be accomplished by using bentonite clay which has the peculiar property of swelling many times its volume when added to water or the diluted coolants, i.e., Cimcool and Texaco coolant. To the clay-water emulsion can be added the contaminated oils which help to thicken the emulsion and thereby forms a thick and very coherent sludge that can be shipped from the plant site as such.
>
> (Dow Chemical, 1954c)

The full report referred to on November 2, 1954 (Dow Chemical, 1954c) is not in available records.

> Future Problems: The Waste Management Coordinating Group completed their study on the sludging of contaminated oils using bentonite clay. A full report is being written up as to the feasibility of this process.
>
> (Dow Chemical, 1954c)

Response To an Emergency Involving a High Degree of Fire Hazard: Burial In a Trench Immediately Next To the Mound Area

The Ryan letter (Ryan, 1954) in a later passage reports the occurrence of an emergency situation and how it was handled:

> On November 17, 1954 an emergency arose which demanded immediate action. A waste of combustible nature produced in 44 Building presents a fire hazard of high degree, resulting in the spreading of contamination in the building. At the present time, 44 Building personnel are unable to safely process this waste and it must therefore be removed from the building. These wastes are presently being drummed in 30-gallon drums, which in turn are placed inside 55-gallon drums and surrounded with graphite. Upon completion of the double drumming, the waste is then moved to a trench located approximately 30 feet south of the on-site mound and parallel, east and west, to the mound. Here the drums are placed in the trench and then covered with dirt to a depth of not less than one feet. [sic] As of this date, six drums have been handled in this manner. This operation will continue until 44 Building will be able to resume processing or another method of handling is formulated.
>
> (Ryan, 1954)

The dangerous situation reported by Ryan pertained to the combustible nature of some contaminated waste which was too much of a fire hazard to be kept in Building 44. The "solution" was to place it in 30-gallon steel drums and to bury the 30-gallon drums inside 55-gallon steel drums in a special trench in close proximity to the Mound Area. The time-line was to be extended into the future "until another method of handling is formulated." Burial in the Mound Area and the negative environmental consequences of that approach to waste management will be reported in a subsequent section.

Response To the Mound Burial and the Transition To Open Pit Burning

The Putzier Report (Putzier, 1970) indicated that burial in the Mound Area continued from April of 1954 until September of 1958:

> The first mound burial occurred in April, 1954. Drums were moved to the mound area rather steadily up to March of 1957 at which time uranium contaminated oil from 90 drums was burned. In April 1957 another 79 were burned. Some negotiations were attempted to ship contaminated oils to the Idaho burial site. Problems relating to potential leakage eventually resulted in dropping the idea for shipping oils. The final actual use of the mound occurred in September 1958 at which time, 89 plutonium contaminated oil drums from Building 771 were placed in the mound. Previously 46 oil drums from Building 771 were placed in the mound.
>
> (Putzier, 1970)

The matter-of-fact statement is made that burial of uranium contaminated oil was replaced by burning it. The negative consequences of burning oil contaminated with toxic uranium upwind from a major metropolitan area and its suburbs was not deemed worthy of further comment.

Negotiations were reported as unsuccessful to truck the barrels of contaminated oil to the ARCO site in Idaho. Final use of the Mound Area as a burial site was reported to be September of 1958 with 285 total grams of plutonium being buried there from Buildings 771 and 776. The Putzier Report has an itemized count of 1606 barrels buried in the Mound Area in contrast to the Owen and Steward Report having a count of 1405 (Owen and Steward, 1974). These differences reflect the state of record-keeping, and the lack of importance assigned in the thinking of that time period of the long-lived consequences of what was being done: contributing to an unremediated ground water plume that would be part of a Superfund site with no known technological solution for remediation.

Partial Transcript of a Meeting On the Mound Burial of Toxic Waste Between Dow Chemical and Labor Unions On April 3, 1970: The Mound Area Is Rediscovered At Rocky Flats

There were regular meetings between representatives of Dow Chemical management and Labor Unions at Rocky Flats. The meeting of April 3, 1970 was chaired by V. Fruehauf from Dow Chemical. Also in attendance from Dow Chemical were H.E. Bowman, C.R. Claridge, R.D. Forest, F.J. Rindt, A.T. Scheutter, E.J. Walko, W.E. (Bill) Lee, and J.F. (Jim) Willging. There were eight in attendance from the Unions: J.D. (Jim) Kelly, R.P. Bradley, D.E. Duffy, W.L. Kennedy, V.A. Owen, J.A. Ray, P. Kelly, and I.K. Roberts. The meeting began with a dispute over whether or not to record the meeting. The reservation expressed by Chairman Fruehauf was:

> Our prime concern is we don't want information taken out of context. We (Joint Committee) are more knowledgeable about what it is all about. We want to be careful.

The resolution was to tape record the meeting and prepare a transcript. At this point Bill Lee added:

> What we do today is entirely different than what we did a year ago or ten years ago. So what we will discuss today—we probably wouldn't do now the same thing we did in 1954 or '56.

Violations of State, Federal and AEC standards

J. Kelly: …We would like to know if it's here, what is in it and why it was put there. We do think from what we have heard that it's in violation of state, federal statutes, and the AEC standards themselves. And this does bother us…

***

J. Ray: I guess what I've heard is that you've got a couple of mounds down by 91, two or three trenches inside the fence and one or two trenches outside the fence near the east guard shack…

***

J. Ray: The problem is in the mounds.

B. Lee: And whether that's a problem, I don't know, but I think politically, we should---O.K. Any other questions on these? Now there may be four over here; they only show three, but there may be four. Somebody asked me about the planking. That planking is either buried in one of these troughs or ditches or in here; I'm not sure; but the planking from the evaporation plant is indeed buried in here.

Depleted Uranium, Enriched Uranium, and Plutonium

J. Willging: Most is D38 chips.

B. Lee: Yeah, most of it was D38 chips which accounted for damned near a thousand of them, and there was 12 barrels of liquid from 71, [Pu] 89 barrels of liquid from 76, [Pu] and 40 barrels of liquid from 81 [EU]. The balance was made up of solid waste from 41, two barrels--Labs, you know the Labs. This is old 41. And the rest was 91. You know when we were assembling the nickel carbonyl stuff. It was just the wipe off and waste paper, kimwipes and things like that mixed with concrete, you know, because we thought we were going to ship them to Arco--a lot of you remember this-about 98% concrete and 2% paper.

Suppose Those Barrels Leaked

J. Willging: We are going to dig into it and see what the situation is. There's a problem. Is it better--Suppose those barrels have leaked, one of recovering some barrels or digging up a lot of contaminated dirt. Let me speak to another point right here. This is not against any regulations. It's in a federal reserve. Now if you were going to bury contaminated waste as a private individual in the state, then you run into all the regulations; but the AEC licensed itself for burial grounds. So it isn't against regulations. Whether or not it's a good idea is another subject.

J. Kelly: I thought the regulations called that they had to be buried at Arco.

J. Willging: No, I asked Putzier this morning, and he said there were no guidelines or anything like this.

No Prohibition Against Burying At Rocky Flats

B. Lee:. He has been hunting for them, Jim, and he can't find anything that would prohibit us from burying at Rocky Flats.

J. Willging: If we wanted to.

B. Lee: If we wanted to.

P. Kelly: This is why I made the statement yesterday that we better hope that everybody is pretty closed mouth because with the bag that this one private outfit is getting into in trying to establish a dump over by Lyons, plus we also know what the political climate is…

***

P. Kelly: We don't want a big razzamatazz deal, because I think we all know what would happen if this Environmental Committee (CCEI) got their fingers on this…

We Did Not Know What Else To Do With Them (Barrels)

P. Kelly: Was there any calculations made about how long it would take these barrels to rust out?

J. Kelly: Not very long.

B. Lee: It depends on the contents.

J. Kelly: I know what some of the contents are, not very long.

J. Willging: We would have to guess that most of them probably leaked; that would be my guess.

J. Ray: Why were they buried?

J. Willging: It started in 1954, is that right?

B. Lee: Correct, and finished in'58. And we didn't know what else to do with them, John...

Dow Chemical's representative Bill Lee, in response to a direct question by Jim Kelly from the Rocky Flats Steelworkers Union, downplayed the fact that there was plutonium buried in the Mound Area. The position was taken by Dow Chemical management that there was no known prohibition to burying contaminated waste at Rocky Flats. Jim Willging from Dow Chemical offered the conclusion that "most of them [the barrels] probably leaked, that would be my guess." A representative of Dow Chemical management, again Jim Willging, injected some humor into an otherwise tense meeting when he observed that "You don't want to spend the rest of your life burying drums around here; you'll have no place to put buildings."

The issue of risk surfaced at the meeting in several forms. Jim Kelly said, "My house is seven miles down the street from this place; and for the first time since I've worked here, I'm a little bit shook up, if there is a problem." Jim Willging offered his opinion that in choosing a site for a burial ground, "You want to find some place that stands no chance of ever becoming a city or a populated area for 24,000 years, at least."

Subsequent to the April 3, 1970 meeting James Stout, an attorney for the Energy Research and Development Administration (ERDA), the successor to the Atomic Energy Commission before the establishment of the Department of Energy, was quoted by H. Peter Metzger (1975) of the Rocky Mountain News as contradicting Willging on the legality of drum burial:

Burials Were Not Legal

> Last week the current ERDA attorney for the plant, James Stout, looked over the minutes of that meeting. Afterward, he sharply disagreed with Willging's 1970 view that the burials were legal. "If Willging believes that Dow could do what Arco could do, he's wrong," he said.

(H. Peter Metzger, Rocky Mountain News, March 31, 1975)

Jim Willging had remarked earlier in the 1970 meeting that: "...the AEC licensed itself for burial grounds... Whether or not it's a good idea is another subject." E.J. Walko from Dow Chemical management perhaps summarized the frustration and inconclusiveness of the meeting with the observation that "we probably don't even know who it was that made the decision to bury the God damned stuff."

Burials In the Mound Area Discussed In the Offices of the Congressional Joint Committee On Atomic Energy On April 10, 1970: The Mound Area Is Discovered By the U.S. Congress In Washington, D.C.

A significant meeting occurred in the offices of the Congressional Joint Committee on Atomic Energy on April 10, 1970 to discuss the Mound Area burials and the stream pollution problem. In attendance were:

James Kelly, President of Rocky Flats Steelworkers Union
Congressman Chet Holifield, Democrat of California and ranking majority member on the CJCAE
Congressman Craig Husmer, Republican of California and ranking minority member on the CJCAE
AEC General Manager Robert Hollingsworth, AEC
Assistant General Manager for Operations, John Erlewine, AEC
CJCAE staff member, Captain Edward J. Bauser

This is a transcript:

<u>Guess a Lot of People Didn't Know It</u>

C. Holifield: …I would like to understand what this burial ground and stream pollution problem is. Will you please tell us? This is something new to me…

J. Kelly: The Company told us a week or so ago, Mr. Chairman, that it was determined some 10 or 15 years ago to place some 55 gallon drums of hot waste, oils and what have you in trenches inside the perimeter of the plant and cover it with dirt. I didn't know this, and I guess a lot of people didn't know it…

<u>You Don't Know Whether It Is Uranium Or Plutonium</u>

C. Holifield:…I understand you don't know whether it is uranium or plutonium. [Dow Chemical knew where burials were but was uncertain what was in them]. This would be a very serious thing if Dow was taking upon itself the burial of plutonium waste without going through the established procedures. I would assume if this is low-level waste that there would be probably a prohibition against this convenient burial and that it should have been put in some permanent high-level waste burial ground like we have at Hanford.

E. Bauser: I don't know but I doubt if that site is an authorized burial site for any level of waste...

<u>It Is Inside the Fence</u>

C. Holifield: What is your understanding--and I am asking for information now--about the location of the material you think is buried out there? Is it inside the fence?

J. Kelly: Some inside and some just outside. ...

C. Holifield: Then they had better build a new fence if there is some outside the fence...You can fence it in both from the inside and outside so that nobody wanders over it. Another thing is that you should very carefully go over it with Geiger counters or other measuring devices to find out if it is buried deeply enough. The next thing is to find out if there is any seepage going into the stream or anything like that ...How long would it take you to get the facts about walking over that with a measuring device to find out if there is any seepage from the ground into the atmosphere that would be in any way unsafe or alarming?...

C. Holifield: Am I going to have to get on a plane and go out there and walk over it myself with a measuring device and try to read it as a layman?

<u>If It Were Not For the Political Situation It Would Still Be Safe To Bury There</u>

J. Kelly: [after further discussion] The company took the position as I left there last Friday that if it were not for the political situation, it would still be safe to bury there...

(Congressional Joint Committee on Atomic Energy, April 10,1970)

The most dramatic revelation was by James Kelly, President of the Steelworkers Union. Just one week before the meeting at the CJCAE headquarters, Dow Chemical had taken the position that "... if it were not for the political situation, it would still be safe to bury there..." The removal of the drums and their contents began on April 14, 1970, four days after the CJCAE headquarters meeting and concluded on April 29, 1970, in barely over two weeks time.

No Detectable Alpha Contamination Was Found In the Soil

In a letter from R.M. Vogel to E.A. Putzier on June 2, 1970 (Vogel, 1970), Vogel reported that:

> Although about 10 percent of the drums had holes, no detectable alpha contamination was found in the soil. The holes, apparently caused by rust and corrosion, were found in both the liquid and the solid waste drums. Many of the liquid drums developed leaks during handling or after exposure to the air and sun.

Absent from this surviving report on the excavation is any mention of health risk to staff or of any measurements of toxicity taken beyond a measurement of alpha contamination.

ChemRisk (1992) was able to provide more detail on the Mound Area with respect to the long-term contamination within Rocky Flats of the buried barrels, and therefore provide even more characterization of the extraordinary toxicity of the witches brew. It is not at all clear, in terms of known technology, how it was possible to achieve the "complete retrieval" as asserted by Dow Chemical. The existence of the toxic groundwater plume identified by ATSDR (2005) would belie the "complete retrieval" assertion.

Mellen (2019, p. 192) quoted both Putzier (1970) and Owen and Steward (1974) in documenting the toxicity of what was deposited in the Mound Area: between April of 1954 and September of 1958: 285 grams of plutonium which included 89 plutonium contaminated oil drums from Building 776. Owen and Steward (1974) reported that no plutonium was detected when the Mound Area was excavated in May of 1970.

Continuing On-Site Contamination In 2005: The "Mound Plume"

The "Mound Plume" is adjacent to the Solar Ponds Plume:

> The "Mound Plume"...contains many contaminants that originated from leaking drums in a waste storage area. Many VOCs have been detected in the plume, but the contaminants of greatest concern are PCE, TCE, and Vinyl Chloride.
>
> (ATSDR, 2005, p. 29)

The 903 Pad Waste Storage Area

The 903 Pad opened in late 1958 as a major location for storage of contaminated waste containing plutonium and uranium laced oil and toxic chemicals including carbon tetrachloride and trichloroethylene. This was a departure from the use of the Mound Area which from 1954 until 1958 was a location to bury (mostly) 55-gallon drums in the ground. At the 903 Pad barrels were simply placed on wooden pallets on the ground exposed to weather.

When scientists Martell and Poet (CCEI, 1970a; 1970b) first discovered plutonium off-site from Rocky Flats in 1970, Martell originally thought that the source was the fire of 1969. Rocky Flats personnel with whom he met to discuss his startling findings said that the much more likely sources were the leaking drums on the 903 Pad and the fire of 1957.

Figure 5.2 The 903 Pad
This photograph was taken in 1962. It reveals that at least the original barrels were placed on wooden pallets. In 1970 K.W Calkins wrote to Rocky Flats Plant Manager L.M. Joshel that "Little consideration was given to this [what to do with organic liquids contaminated with radioactive materials] in the initial design of the plant." When the Mound Area was closed, the 903 Pad was chosen as the site to receive 55-gallon drums of cooling oil laced with radionuclides. Three-fourths of the estimated 5,237 barrels contained plutonium. Freiberg indicated that the contents were missing from approximately 100 barrels. When Martell and Poet found plutonium off-site in 1970, Rocky Flats staff indicated that the most likely sources were the 903 Pad and the fire of 1957.

Toxic Content of Leaking Barrels On the 903 Pad

A memo from K.W. Calkins to Rocky Flats Plant Manager L.M. Joshel dated August 19, 1970 (Calkins, 1970), written twelve years after the opening of the 903 Pad, is notable for its chronicle of the toxicity of what was deposited.

Three-Fourths of the Drums Were Plutonium

> Most of the drums transferred to the field were nominal 55-gallon drums, but a significant number were completely full. Approximately three-fourths of the drums were plutonium—contaminated, while most of the balance contained uranium. Of those containing plutonium, most were lathe coolant consisting of a straight-chain hydrocarbon mineral oil (Shell Vitrea) and carbon tetrachloride in varying proportions. Other liquids were involved, however, including hydraulic oils, vacuum pump oil, trichloroethylene,

perchlorethylene, silicone oils, acetone still bottoms, etc. Originally, content of the markings were made illegible through weathering and no other good records were kept of the contents. Leakage of the oil was recognized early, and in 1959 or possibly earlier ethanolamine was added to the oil to reduce the corrosion rate of the steel drums.

(Calkins, 1970)

Mellon (2019, p. 193) quoted Owen and Steward (1974) that 5,420 drums of contaminated oil were stored at the 903 pad. Putzier (1970) indicated that 3,572 of the steel drums contained plutonium.

In addition to the well-known radionuclides plutonium and uranium, trichloroethylene is among the Agency for Toxic Substances and Disease Registry's (ATSDR) Top Twenty Priority List of Hazardous Substances at 16th. It affects the central nervous system and is a known carcinogen. Carbon tetrachloride is a known cause of liver and kidney damage. It also depresses the central nervous system. It is 50th on the ATSDR Priority List of Hazardous Substances and is a probable carcinogen (ATSDR, 2017).

The Decision To Move From Burial of Waste (1954-1958) In the Mound Area To Large Scale Outdoor Storage (the 903 Pad and the Triangle Area)

It had been the expectation at the opening of Rocky Flats that the type of contaminated waste which was ultimately buried in the Mound or placed in large scale above ground storage on the 903 Pad would be shipped to the AEC Idaho Operations Office for ultimate disposal (Calkins, 1970). According to K.W. Calkins (1970):

> From the beginning of operations of the Rocky Flats Plant, certain organic liquids contaminated slightly with radioactive materials were generated which required disposal. Little consideration was given to this in the initial design of the plant, apparently based upon the assumption that these materials could be packaged and shipped for burial as were low-level solid wastes.
>
> Idaho Operations Office, the agency responsible for the operation of the burial site, however, declined to accept these materials for burial, except for a small quantity of wastes coming from uranium operations which were shipped in 1957.

A noteworthy observation by Calkins is that "Little consideration was given to this in the initial design of the plant..."

Promising Initiatives Hampered By Design Flaws and Low Financial Priority

K.W. Calkins (Calkins, 1970) reported that promising initiatives were hampered by design flaws and other priorities for funding:

> In June 1963, information was received that the CCl_4 [carbon tetrachloride] still had been deleted from the expansion project because of funding problems. Because of this, the design capacity of the sludging process was increased to provide for processing all contaminated liquids, and funds for this project-based on a mixer-extruder system called the "jelly factory" were requested.

Major Deficiencies Required Extensive Modification

> Installation of the mixer-extruder system was completed in January, 1964, but start-up work revealed major deficiencies in the installation which required extensive modification. These modifications were not completed until late in 1965.

Most of the Drums In the Oil Field Had Deteriorated

> When this equipment was finally ready for operation, a further delay was encountered when it was found that the condition of most of the drums in the oil field had deteriorated to the point of being completely unusable for transfer to the disposal plant, even in the cart provided. As a result, FY-1966 expense funds were provided to build a temporary structure for transferring the oil into new drums, while simultaneously filtering the oil through one-micron filers for plutonium removal and recovery.
>
> After more start-up problems, the final phase of emptying the drum field began on January 23, 1967. By this time the field contained about 5,240 drums of which approximately 3,572 contained plutonium contamination. The oldest drums and those containing plutonium were processed first. To the best of our knowledge, the last of the plutonium was removed on January 25, 1968. The last of the oil was transferred to a new drum on May 28, 1968, and shipped to the disposal plant on June 5, 1968. (Calkins, 1970)

The narrative reports identify funding problems, major deficiencies in a promising approach called the "jelly factory," and that the condition of the 55-gallon drums had deteriorated, precluding their use for transfer to the disposal plant.

Dow Chemical Management Knew By 1962 That More Than Half of 903 Pad Barrels Had Leaked In Just One Year

It is clear from the archives of Dow Chemical documents that Dow Chemical, as part of routine operations, knew as early as July of 1959 that leakage from the open-air storage was occurring. By 1962 it was known to Dow Chemical management that between 50 percent and 60 percent of the barrels had leaked. The context for the construction of a rabbit-proof fence is not reported. But Jim Kelly is quoted in "Secrets of a Bomb Factory" by Frontline that some of the rabbits on the 903 Pad were "screaming ass hot."

The known high wind velocities at Rocky Flats by 1964, associated with the known contamination of the air above the 903 Pad, led to soil being placed on top of some of the "more highly contaminated areas" within the fence which had been constructed to surround the severely contaminated site. The last new drums were added to the 903 Pad in 1967. The last leaking drums were re-barreled in 1968, a year when heavy rains spread contamination further. In December of 1968 a dispute between Dow Chemical and the Atomic Energy Commission led to delay in the construction of barriers: It was finally completed in November of 1969.

Chronology of the Transfer of the Contaminated Oil Drums From the 903 Pad To Building 774

The Dow Chemical archive contains a chronology of the disposal of the contaminated oil drums from 1967 through April of 1970 prepared by K.J. Freiberg of Dow Chemical:

1. Work to remove oil from the 903 Area began January 23, 1967, under the supervision of D. M. Anderson, M.E. Maas, and R.M. Vogel.
2. From January 23, 1967, through March 10, 1967, uranium oil drums which were in good condition were transferred to Building 774 and processed.
3. Building 903 went hot on March 10, 1967, and started processing oil drums. This building was designed to prefilter the oil prior to transferring plutonium contaminated oil to Building 774 for final processing.
4. From March 10, 1967 through May 18, 1967 there were a total of 191 drums of plutonium contaminated oil filtered and shipped to Building 774.
5. On May 18, 1967 operations at Building 903 were discontinued due to the amount of time this process was taking.

(Freiberg, undated)

Points 1-5 of the 34 points in the Freiberg Report indicated how time-consuming the approach was to removing barrels from the 903 Pad which already had not leaked their contents.

The Missing Contents of 411 Barrels of Contaminated Mixed Waste Oil
The Freiberg Report (Freiberg, undated) is especially significant. Freiberg provides an explanation for the missing contents of 411 barrels of contaminated mixed-waste oil. He reports that what were presumably the first sampling wells for the 903 Pad were completed on November 11, 1969. Freiberg estimates that the content of 411 barrels could not be accounted for fully:

> 12. Taking the total number of 5,237 drums minus 4,826 drums, containing 50 gallons each, which were sent to Building 774 leaves 411 drums to be accounted for. The best explanation for the 411 drums and the volume contained within each follows:
>   A. All of the drums sent to the oil storage field originally were not completely full.
>   B. Volume taken up by the sludge which was discarded with the empty barrels.
>   C. Leakage out of the barrels and into the ground within the storage area.
>
> 13. To the best of everyone's memory and knowledge, a total of approximately 100 barrels containing 50 gallons each or 5,000 gallons of oil leaked out the of drums and was absorbed into the soil within the fenced area.
>
> ...
>
> 18. In October, 1968 weeds and vegetation were burned off the 903 contaminated barrel storage area preparatory to applying an asphalt cap over the area. No airborne contamination problems were encountered.

Their "best estimate" was that the contents of 100 barrels or 5,000 gallons had leaked into the soil beneath the 903 Pad. The comment that air contamination problems were not encountered is both unexpected and surprising. The accuracy of it cannot be verified in the absence of information about how the contents of the smoke and fire were measured.

The Nearly Three Years To "Clean Up" the 903 Pad
It took Dow Chemical from January of 1967 until November of 1969 to "clean up" the 903 Area. This extended time frame was divided as follows:

- 18 months to remove the drums from January of 1967 until June of 1968
- 5 months period of inactivity
- 12 months to pave--which was subdivided in turn into burning contaminated vegetation from the field (October of 1968), grading the field in high wind conditions (November of 1968), applying fill (July of 1969), and completing asphalt paving (November of 1969).

Continuing On-Site Contamination: The 903 Pad/Ryan's Pit Plume
The saga of the 903 Pad contamination is not over. The Agency for Toxic Substances and Disease Registry (ATSDR, 2005) has reported that there is a toxic plume in the groundwater beneath the 903 Pad. The plume contains many contaminants, including carbon tetrachloride, methylene chloride, tetrachloroethylene (PCE), trichloroethylene (TCE), and uranium isotopes (ATSDR, 2005, p. 28).

The Triangle Area (1966-1974)

A Higher Level of Plutonium Contamination Than the 903 Pad
The Triangle Area has many parallels with the 903 Pad. Some observers consider it to have had a higher contamination of plutonium than the 903 Pad based on the Barker (1974) report because all of the waste recovered from the 1969 fire was sent there. The 1969 fire destroyed inventory of $22 million in plutonium with plutonium fire ash residue apparently sent to the Triangle Area.

Figure 5.3 The Triangle Area

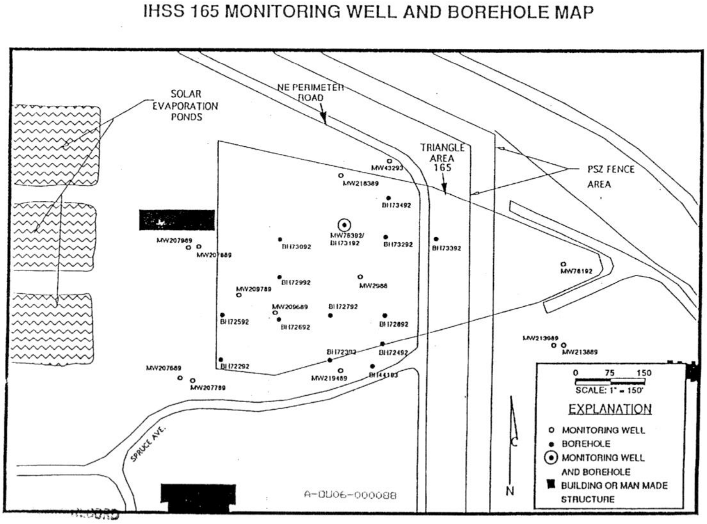

The Triangle Area was physically located between the Northeast Perimeter Road and Spruce Avenue. It opened in late 1966 out of the necessity to find a new storage site. The reason was the circumstance that construction north of Building 883 forced the removal of barrels from that area. It was the major repository for debris from the fire of 1969. But by that time 6,000 barrels of cooling oil with traces of plutonium had been deposited there. The location of test wells are marked within the Triangle Area.

<u>Excerpts From the Barker Report Characterize the Triangle Area:</u>

> The area under discussion is a triangle of land between the Northeast Perimeter Road and Spruce Avenue. This field was first used as a drum storage area in late 1966, when it became necessary to

remove a large number of drums from a field north of Building 883 due to construction of a new decontamination facility in that area.

Barker (1974)

Drums Were Placed Directly On the Ground Exposed To Colorado Weather

After being moved into the area, the drums remained stored there, sitting directly on the ground through the winter of 1966. In the spring of 1967, as weather and manpower permitted, a Chemical Operations crew categorized all drums, according to tags on them, and arranged them in an orderly manner on wooden pallets. The type of drum being used during this period of time was routinely a 55-gallon drum, bought used from off site, and re-used several times after arrival on the plant site. Liners were double polyethylene bags. This type of liner was in use until 1969 when the Chemical Operations group started cutting lids from peroxide container liners and using them as inside liners for the drums. By 1970, R&D came up with a specification for a manufactured rigid poly drum liner, which has been used since …

Barker (1974)

Fire Waste From the 1969 Fire Went To the Triangle Area

In May 1969 a fire occurred in building 776. As the cleanup from the fire began, all drums of accumulated fire waste and residues were moved into the triangle area for storage…

During the period from 1966 to the present, contamination incidents occurred in the field from time to time…

Barker (1974)

The 55-gallon drums were stored in the open air exposed to the weather until 1971 when they were placed in rail/truck type cargo containers.

The Contents Stored In the Triangle Area and Their Condition

The Barker Report (1974) quoted Waste Management Reports by R.M. Vogel (1968a; 1968b) of Dow Waste Management which characterize the contents and condition of the Triangle Area:

Environmental Contamination Is Certainly Inevitable

Health Physics Report-November 1968

The rapidly increasing number of drums which contain plutonium-bearing wastes and residues, and which have been placed in outside storage is a matter of concern to Health Physics. There are more than 6,000 such drums. Major categories are burnables and washables awaiting recovery, waste drums awaiting disposition to the burial site, and ZPPR residues. No environmental contamination has occurred to date but it is certainly inevitable if the storage becomes long-term. [Storage at the Triangle Area continued for another 6 years].

(Vogel, 1968a)

Potential For Environmental Contamination "Is of Special Concern"

Health Physics Report-December 1968

The potential for contamination to the environment is of special concern because of the storage of 5,000 drums of recoverable solids east of the nitrate ponds. Some limited soil contamination has resulted from leaking drums. Condensation inside the drums and moisture standing on the drum lids appear to be the causes of drum failures. High winds have blown over as many as 150 drums at one time.

(Vogel, 1968b)

Health Physics Report-June 1968

A leaking drum of recoverable waste stored in a cargo carrier in the 779 storage yard resulted in contamination to the immediate area. Spotty contamination to greater than one million counts per minute was found on an area of about 480 square feet. Also contaminated were three drums, the cargo carrier and two semi-trailer tires. The cargo carrier and drums were removed to 771 Building for decontamination. The incident occurred on June 21 and decontamination was completed on July 2. Effective use was made of strippable latex and foam in controlling contamination and in decontamination. No significant airborne contamination resulted.

(Quoted in Barker, 1974)

Leaking drums and exposure to strong winds were factors in common with the 903 Pad. There were reportedly over 6,000 drums stored there by November of 1968 plus subsequently an unknown number of drums from the 1969 fire.

There are differing official accounts of the Triangle Area. In Stout (1974), the Acting Area Manager for the AEC at Rocky Flats, James Stout, sent a letter to Herman Roser of the AEC Albuquerque Office which said, in part, "...storage was probably minimal until after the May 1969 fire."

The Dow Chemical Health Physics Report of November of 1968 indicated that by the November 1968 time frame there were already 6,000 drums in residence. A Dow Chemical Health Physics Report of December of 1968 reported there were 5,000 barrels of "recoverable solids" with drum failures already in evidence, 150 barrels had blown over by high winds, and the "potential for contamination of the environment is of special concern ..." (Barker, 1974).

The 903 Pad was characterized as containing more "nonrecoverable" wastes and the Triangle Area containing waste with more plutonium in it classified as "recoverable" waste, with some barrels containing more than one kilogram of plutonium. The source of the overall greater plutonium residues may have been the waste from the 1969 plutonium fire where at least 1 metric ton of plutonium was lost to production. Presumably all of it did not go up in smoke and off-site, but at least some of it remained confined within the 776/777 building as ash and therefore ended up in the Triangle Area.

Serious Potential Fire Hazard

The wastes from the 1969 fire which were stored in boxes turned out to be "A serious potential fire hazard ..." according to Dr. M.A. Thompson of Dow Chemical:

> A serious potential fire hazard has been brought to my attention by J. M. Cleveland. That is, the many wooden boxes containing residues and wastes from the Building 776 fire area that are being stored around the plant site. An example of this type of storage can be found on the north and south *sides* of Building 779. In these areas, if one box were to catch fire, there would be a high potential for the fire to spread to other boxes and to release significant amounts of contamination. Admittedly, the potential for fire to start in any one box may be small, but many of our recent incidents have been events with small probability of occurring.
>
> (Thompson, 1969)

A most memorable element of this memo (Thompson, 1969) is the recognition that "...many of our recent incidents have been events with small probability of occurring." The contemporaneous descriptions of the magnitude of risk in the previously little known Triangle Area are disturbing starting with the revelation that it contained more plutonium from the 1969 fire residue than was contained in the 903 Pad.

Anyone of the headings about the Triangle Area would have been disquieting in and of themselves for either safety of the work environment, or for long-lived toxic contamination:

- A Higher Level of Plutonium Contamination Than the 903 Pad
- Environmental Contamination Is Certainly Inevitable

- Potential For Environmental Contamination "Is of Special Concern"
- Serious Potential Fire Hazard

At the Triangle Area it had become obvious that there was not "a small probability" of an incident.

Noteworthy Decision-Making About the Disposal of Especially Toxic Waste As Reflected In the Archives of Dow Chemical Documents

The archives of Dow Chemical documents make a number of noteworthy points about decision-making:

- Despite the toxicity of both the well-known radionuclides plutonium and uranium plus the extremely toxic chemicals carbon tetrachloride and trichloroethylene, the change from burial of steel drums in the Mound Area to outdoor storage on the 903 Pad was characterized as "all of a sudden one day."
  K.E. Bowman, Manufacturing Manager, 1970 (Bowman, 1970)

- And K.W. Calkins of Dow Chemical said that, in referring to the storage, treatment and disposal of contaminated waste, "little consideration was given to this in the initial design of the plant."
  K.W. Calkins to Rocky Flats Plant Manager L.M. Joshel, 1970 (Calkins, 1970)

Reflecting On the Meaning of the Witches Brew of Toxics

There was no environmental crime subsequently directly associated with the Mound Area, the 903 Pad, or the Triangle area. Their significance is the failure to anticipate the lengths that Dow Chemical stewards would have to go to keep the vital production mission on track. It was untenable for lack of space to keep storing steel drums on the slabs of production buildings. It was obvious that burying steel drums or storing them on the ground exposed to the extremes of Colorado weather would result in exactly what did happen: barrels rusting or developing multiple pin-hole leaks and depositing their contents into the soil below the Mound area or in blowing air and deposited in the soil below the 903 Pad and Triangle Area.

There is no record in the available archives that Rocky Flats stewards thought that there would be off-site migration at the time the barrels were buried or deposited on the surface of the land. But in 1970 Ed Martell was told that off-site plutonium likely had come from either the 903 Pad or the 1957 fire.

The broader point was that site constraints at Rocky Flats were such that to keep the production mission going they obviously perceived that they had no other choice but to act as they did. The key Congressional leadership of the Congressional Joint Committee on Atomic Energy had no idea about the existence of the Mound area and took immediate action to get the buried barrels dug out. At an earlier Rocky Flats management-labor meeting, Bill Lee of management said "...the AEC licensed itself for burial grounds. So it isn't against regulations. Whether or not it's a good idea is another subject." A comment from Jim Kelly is most appropriate: "There's not too much that they have kept away from Monitors; but, by God, this is one that got by."

Since the DOE and the EPA conduct mandatory evaluations of Superfund sites, there is a mechanism in place to track change in the 1,300 acres which includes the locations of the Mound, the 903 Pad, and the Triangle Area. There is no mechanism in place for evaluating the immediate areas around the 1,300 acres where wind and water have spread toxic residue. As Jim Willging of Dow Chemical management said in 1970: "You see if you start using a place for a burial ground, you can't ever use that place again for anything else."

Chapter Six
More Early Fiascos In Waste Management At Rocky Flats: From Open Air Plant Incineration To the Notorious Building 881 Hillside

"...the incinerator that was near the west access road, I would liken to the kind of incinerator you used to see at the Safeway stores..."
E. A. Putzier, Dow Chemical Director of Health Physics, 1977

"...some of the floor drains in the manufacturing buildings were not isolated from the sewage treatment plant, and the sanitary sludge became contaminated with uranium and plutonium."
ChemRisk, 1992, p. 84

"The disposal of contaminated oil and still bottoms has been partially solved. Open pit burning with air sampling of the plume indicates that these contaminated wastes from 44, 81, and 83 Buildings can be handled in this manner."
Dow Chemical Progress Report, February 1961

"...due to the safety hazard of transporting these [uranium chips contaminated with oil coolant, dirt and other foreign materials] it is recommended they be buried on the plant site."
D.M. Bassler, Dow Chemical Management, 1961a

"Due to the fire hazard involved, the material was stored outside. Now, because of probable hydrogen pickup, it is considered a hazard to bring it back in for roasting and reduction to oxide...the drums should be buried on site, since they would not meet shipping requirements."
D.M. Bassler, Dow Chemical Management, 1961b

There were other burial sites and waste disposal practices at Rocky Flats which did not have the notoriety of the Mound Area, the 903 Pad, or the Triangle Area. On the Rocky Flats site they obviously contributed to the designation of 1,300 acres as a Superfund site. Most of these early fiascos occurred within the 334-acre high security area. The long-lived legacy of the chemical and radioactive contamination is directly relevant to future safe uses of the entire property, and to assessing the threat to surrounding communities of further off-site contamination.

The twelve early fiascos in waste management at Rocky Flats reported in this chapter share in common with the Mound Area, the 903 Pad, and the Triangle Area a distinctive feature; namely, they collectively provide invaluable insights into what management and workers were thinking when they made and implemented decisions which are so disastrous in retrospect. These twelve sites also share in common with the three sites characterized in Chapter Five in illustrating how disastrous early decisions starting with the flawed site selection criteria narrowed the choices available for subsequent stewards.

Figure 6.1 Early Fiascos In Waste Management At Rocky Flats and Their Physical Relationship To Each Other

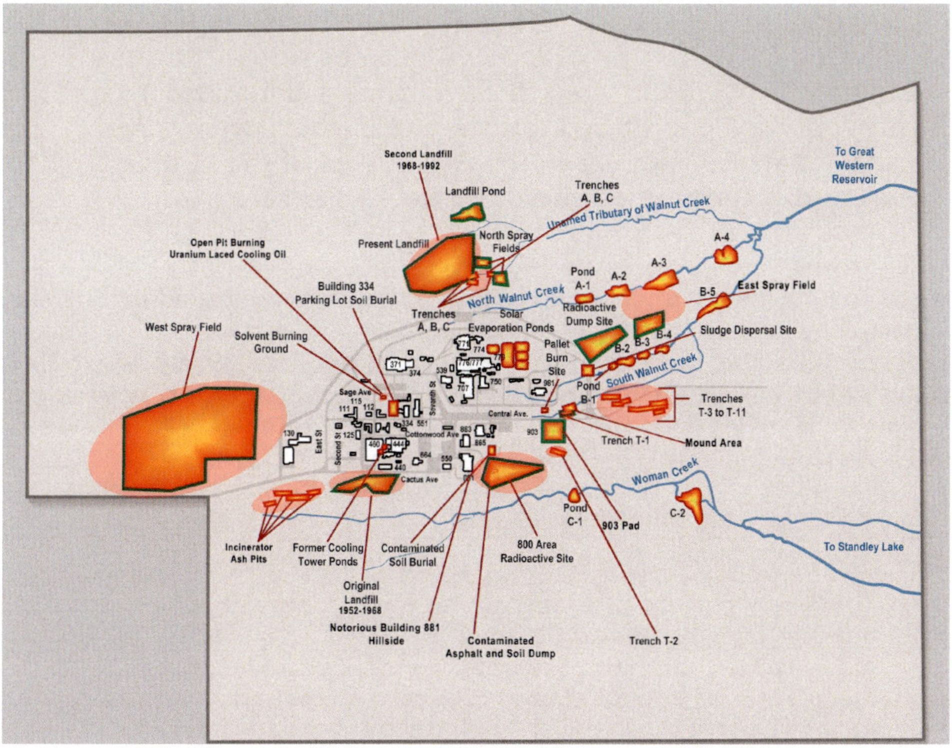

This is an exhibit introduced in the Cook class action lawsuit. It was prepared in its original form by DOE contractor ChemRisk (1992) as an important contribution to the characterizations of the Rocky Flats site after its closure for production of plutonium igniters. The labeling has been modified to highlight the locations of the waste sites covered in Chapters Five and Six and their relationship to each other.

The twelve early fiascos covered in this chapter are:

The Trenches (1954-Plan Closure)
Open Air Plant Incineration (1952-1968)
Water Retention Holding Ponds (1952-Plant Closure)
Liquid "Sanitary" Wastes (1952-Plant Closure)
Radioactive Sewage Sludge (1954-1972)
Original (Misnamed) Sanitary Landfill (1952-1968)
Second (Misnamed) Sanitary Landfill (1968-Plant Closure)
Open Pit Radioactive Laced Oil Burning (1956-1965)

Uranium Sheet Open Air Burning (1965)
Nickel Carbonyl Cylinder Burial (Dates Unknown Except for Confirmed Incidents)
Contaminated Asphalt and Soil Dump (After the 1969 Fire)
Notorious Building 881 Hillside (1952-Plant Closure)

ChemRisk researchers uncovered 178 waste disposal sites within the Rocky Flats Plant site (ChemRisk, 1992). Subsequent DOE contractors retained in the immediate years leading up to the "Accelerated Cleanup" provided evidence of soil and groundwater contamination by many known hazardous "materials of concern" which were not monitored at the time they were used. But hazardous materials appear in the early years of the 21st Century in toxic groundwater plumes under the Rocky Flats site.

The Trenches
Threats to long-term human and ecosystem health were posed by burial of radioactive waste in what are known as the "Trenches." The Putzier Report of 1970 and the Owen and Steward Report of 1974 are excellent sources about the origins of Trenches #1-8. ChemRisk (1992) was able to document Trenches #9, #10, #11, and three additional ones known as Trench A, Trench B, and Trench C.

<u>Trench #1: December 1954 to December 1962</u>
A letter from D.M. Bassler to E.S Ryan of April 12, 1961 stated:

<u>Due To the Safety Hazard of Transporting These Chips, It Is Recommended They Be Buried On the Plant Site</u>

> This letter authorizes you to dispose of fourteen 55-gallon drums of tuballoy saw chips contaminated with oil coolant, dirt and other foreign materials. The total net weight of these items is 3390 kgs…Due to the condition of the chips it is impractical to attempt to reuse the tuballoy in any of the processing equipment in 44 building. Also due to the safety hazard of transporting these chips, it is recommended they be buried on the plant site.
>
> (Bassler, 1961a)

Tuballoy refers to normal uranium not depleted uranium or enriched uranium.

Location of Authorized Burial As a Window Into the Priorities and Operating Imperatives of Dow Chemical To Deliver Plutonium Igniters
The exact location of the burial authorized by Dow Chemical in 1961 is not identified in the letter from D.M. Bassler to E.S. Ryan of April 12, 1961. It may or may not be Trench #1 (The Mound Area had been closed in 1958).

The Drums Should Be Buried On Site Since They Do Not Meet Shipping Requirements

> Please arrange to remove twenty-seven, 55-gallon drums and sixteen, 30-gallon drums from the west yard of Building 47. These drums contain tuballoy in the form of saw chips and floor sweepings, plus foreign matter and coolant. The total gross weight of the drums is 5,472 kgs. The drums should be buried on site, since they would not meet shipping requirements.
>
> This material is normally disposed of by roasting in the ore roaster, but was accumulated during the eight-month period required of engineering a suitable exhaust system. Due to the fire hazard involved, the material was stored outside. Now, because of probable hydrogen pickup, it is considered a hazard to bring it back in for roasting and reduction to oxide.
>
> (Bassler, 1961b)

This is a very instructive letter for indicating the pressures Dow Chemical was under to maintain its schedule for delivery of plutonium igniters, and at the same time to dispose on-site the mounting volumes of toxic waste. For example, since there was not the possibility of shipping mixed-waste with traces of radioactivity in oil to Idaho, the twenty-seven 55-gallon drums and thirty 30-gallon drums had to remain at Rocky Flats. The ore roaster had obviously been operating without a suitable exhaust system. For whatever undisclosed technical reason, it could no longer operate without a "suitable" system (undefined). Therefore, given the known fire hazard, and the probable build up of hydrogen gas, it was no longer suitable for treatment in the ore roaster. The direction conveyed by the Bassler letters of April 12, 1961 was to bury it on-site.

This letter is a window into the priorities, operating restrictions and imperatives of the Dow Chemical era. It is the Rocky Flats version of "The show must go on." There is no mention of whether or not it was safe to bury it, or whether it would pose a long-term threat to human or ecosystem health. The message was "get rid of it now." The location was left up to the individual instructed to "make it go away." The individual instructed was in the Waste Management Group. The available record on Trench #1 is augmented by additional information that the source of the 25,000 kilograms of depleted uranium deposited in Trench #1 was Building 444 (Putzier, 1970).

Radioactive Sewage Sludge Burial In Trenches #2 - #8

The Putzier Report (Putzier, 1970) chronicles the first burial of sewage sludge in the Trenches as occurring on July 29, 1954. After August 14, 1968 sludge was buried in the misnamed second "sanitary" landfill up until November of 1972.

The First Sewage Sludge Burial Was In 1954 In the Trenches

> The first sludge buried on the plant site dates back to July 29, 1954. There appears to be 7 locations used for sludge burial up to August 1, 1968. These are designated...as T-2 through T-8. On August 14, 1968 the [second] sanitary landfill operation was started. All sludge generated since that time have been placed in the landfill. Original authority for burial of sludge is indicated in an attached set of letters... Concentration of radioactivity in the dried sludge has not varied greatly over the years. Maximum reported was 7.9 x 10^6 dpm/kg in June 1960. Minimum was 8.4 x 10 dpm/kg in August 1964. While the information is rather sketchy, it appears that initially the activity was primarily uranium with probable increasing plutonium fraction and more recently could be principally plutonium. Something in excess of 100,000 kilograms of sanitary...[illegible]
>
> (Putzier, 1970)

There is a letter from the then Director of the Dow Chemical Health Physics and Medical Section, T.S. Chapman, to J.G. Epp (Chapman, 1954) which indicates that sludge from Building 95 should be buried in a shallow trench. Presumably that was the method employed for a brief period until the use of Trenches #2 - #8 began. It was a known violation of AEC regulations on disposal of radioactive wastes. Nonetheless, despite being 30% "above the tolerance," direction was given to put the sludge into a shallow trench.

The directions on worker safety when dealing with sewage sludge included an "isolation shower." While this constituted a recognition of the known threat to human health, the work had to be performed. It was not until November of 1972 that sewage sludge was re-classified as "non-retrievable waste" and shipped off-site to NRTS in steel barrels. The transition from putting sludge into shallow trenches, Trenches #2 - #8, or into the Sanitary Landfill, is provided by a Putzier Report dated October 22, 1970 (Putzier, 1970):

> Between August 14, 1968 and February 1970, sanitary sludge was disposed of in the sanitary landfill used for general plant disposal of solid waste (nonradioactive). Justification for this is the same as that discussed for areas T-2 through T-8.
>
> (Putzier, 1970)

It is clear that the use of the Trenches was widespread in the early years of stewardship by Dow Chemical. Some of the contaminated waste put into Trenches exceeded permissible standards. Acknowledgment of this circumstance was the reclassification of sludge to such a level of contamination that it required placement in barrels and shipment to Idaho.

ChemRisk (1992) again makes a contribution to reconstruction of the historical record based on their access to the complete Owen and Steward (1974) report:

> The contents of Trenches T-2 through T-3 contain 100,000 kilograms of "sanitary sewage sludge" and "about" 275 flattened empty drums contaminated with uranium between 1954 and 1964.

The absence of quantitative data on level of radioactive contamination reflects the fact that only gross alpha contamination data was collected during the time that these Trenches were operating.

Trenches #9-#11

The existence of Trenches #9-#11 was uncovered by ChemRisk:

> Trenches #4 - #11 were operated between 1954 and 1968, apparently mostly for disposing of flattened drums contaminated with uranium and plutonium. Asphalt planking from the Solar Evaporation Ponds containing "some" uranium and plutonium were deposited in Trenches #3-#11.
>
> (DOE, 1986b, quoted by ChemRisk, 1992, p. 191)

The source of these disclosures is identified as DOE (1986b) "Comprehensive Environmental Assessment and Response Program (CEARP) – Phase 1 Installation Assessment at Rocky Flats." It is further identified as Rocky Flats Repository Document RE-33. The importance of the disclosure is two-fold. First, the Trenches were very large in size: approximately 100 yards long (300 feet by 50 feet). Portions of the territory of the Trenches were massively spray irrigated over well past the saturation point with excess water going off-site starting in 1979 and continuing until plant closure.

Trenches A, B, C

ChemRisk (1992) supplemented the number of Trenches identified from archives of Dow Chemical documents. The ChemRisk addition of Trenches A, B, and C cited as their source DOE (1991b).

> Trenches A and B received uranium and plutonium contaminated sludge from the sewage treatment plant. The contents of Trench C are not known. Trench B began operation in 1959 with no known date of closure. Trench A operated between 1964 "and about 1974". Trench C operated from 1964 until 1974. TCE and several metals have been detected in the groundwater below Trench A but DOE (1991b) indicated that radionuclides may be at "background."
>
> (ChemRisk, 1992, p. 189)

The principal reason for supplementing the Trenches record from Dow Chemical archives is the deliberate choice which was made by Rockwell International in the 1980s to spray irrigate over ground occupied by some of the Trenches, specifically the area known as the East Trenches which drain into the watershed of Woman Creek, a "navigable water of the United States" under the legal definition of the Clean Water Act of 1972.

First Plant Open Air Incinerator

The First Plant Incinerator was located in the southwest quadrant of the Rocky Flats site. It operated from 1952 through 1968. ChemRisk (1992) drew on their access to Owen and Steward (1974) and DOE (1986b) to create a portrait of the First Plant Incinerator:

> All noncombustible noncontaminated trash and ashes from the incinerator were dumped adjacent to the incinerator and covered with dirt (Seastone, 1973). Small quantities of depleted uranium contaminated combustibles were burned along with the general combustible plant refuse over the years 1952 to 1968. It is estimated that less than 100 grams of depleted uranium would be involved (Piltingsrud, 1973). The incinerator burned office-type wastes and some depleted uranium chips. Ashes were put into pits located adjacent to the incinerator or were pushed over the side of the hill into the Woman creek drainage.
>
> (ChemRisk, 1992, p. 195)

E.A. Putzier, by then Director of Health Physics for Rocky Flats, gave a formal statement on April 26, 1977 (Putzier, 1977) which covered the description and operation of the Plant Incinerator:

```
Putzier: A.  Well, the incinerator that was near
     the west access road I would liken to the kind
     of incinerator you used to see at the Safeway
     Stores, and its intent was to burn basically
     the kind of waste, cardboard boxes and other
     burnable that were accumulated at the plant.

 Q.   And it was not filtered?

 Putzier: A.  No, there were no filters on it, and
      it was abandoned at the time, about the same
      time that there was a movement round the
      country to eliminate incinerators because of
      pollution and go to landfill type operations.

 Q.   Was your group in charge of monitoring or
      testing the smoke or ash from those
      incinerators at the west end?

 Putzier: A.  Well, the incinerator at the west
      access road, the air was not monitored to my
      recollection.  The laboratories periodically
      monitored the ash accumulation.
```

(Putzier, 1977)

The Reason For Closure of the Plant Open Air Incinerator Had Nothing To Do With Pollution

It does not take a specialist in radiation measurement to express concern about the cumulative threat of emissions from the Plant Open Air Incinerator. But the reason that the facility was closed in 1968 had nothing to do with potential health or ecosystem risks. On March 22, 1967 the Rocky Flats Industrial Relations Manager, C.M. Love, wrote to Seth R. Woodruff, the Rocky Flats Area Manager for the Atomic Energy Commission (Love, 1967) that:

> The plant incinerator, which burns approximately 100 cubic yards per day of loose combustibles containing some contamination, does not meet the emission visibility standards. A fiscal year 1969 budget request has been submitted for a gas-fired incinerator. Studies are also being made of several feasible alternate methods of disposal which might be used.
>
> (Love, 1967)

The Plant Incinerator was closed for failing to meet "emission visibility standards." It was located at the perimeter of the site. There are no known studies of off-site contamination from the Plant Open Air Incinerator which was compared by Putzier to one which might have been operated by a Safeway Store (except for the contents it burned).

Plant Incinerator Ash Disposed In Unlined Trenches

Based on DOE (1986b) and their access to the complete Owen and Steward (1974) and Putzier (1970) reports, ChemRisk (1992) documents that there were four burial trenches constructed for incinerator ash known as the Incinerator Ash Pits. Owen and Steward (1974) identify these as Trenches #1-#4:

> As estimated 100 grams of depleted uranium was burned with general combustible waste in the nearby incinerator from 1952 through 1968. Ashes from the incinerator were buried in these trenches (Owen and Steward, 1974).
>
> Some unknown quantity of depleted uranium contaminated incinerator ashes were dumped in an area south of West Road and within a few hundred feet southeast and southwest of incinerator (Putzier, 1970).
>
> Ashes from operation of the incinerator were put into pits located adjacent to the incinerator or were pushed over the side of the hill into the Woman Creek Drainage...The ash pits were covered with fill (DOE, 1986b).
>
> (ChemRisk, 1992, p. 195)

An ash sample taken by S.E. Hammond in 1957 contained "excessive amounts of enriched uranium" at a location just a few hundred yards from Woman Creek into which it drains (Hammond, 1957).

The only available analyses of the plant incinerator ash are extremely troubling. Relevant to assessing the potential for human and ecosystem harm are: Site Survey Monthly Reports (Dow Chemical 1954-1959) for September 2, 1954; February 14, 1955; January 7, 1957; April 3, 1958; May 5, 1958; July 1, 1958; August 1, 1958; February 9, 1959; April 2, 1959; May 5, 1959; and Waste Disposal Reports for November 2, 1954 and August 3, 1958. Their principal findings are:

September 2, 1954 reported the result:

> Five samples of ashes resulting from the burning of 91 building wastes were collected. The average value for the 5 results was 4.5 x 107 (sic 10^7) d/m/kg of dry ashes. These ashes were about 100 times higher in alpha activity than the usual ashes sample taken from the incinerator.

The February 14, 1955 Site Survey Report says:

> Routine sampling results from ashes collected at the general incinerator are shown in Graph No. 5. The high monthly average in August is the result of a trial burn of 91 Building wastes on August 17. This left general contamination in the incinerator. Ashes from the incinerator were buried on-site during the year.

Plan To Burn Contaminated Waste From the Beryllium Production Building

Site Survey, July 1, 1958:

> SPECIAL MONITORING:
>
> Special air sampling has been initiated on the plant incinerator stack to determine the feasibility of burning 44 building contaminated waste. Velocity readings have been obtained and samples are being taken to determine background prior to burning contaminated waste.

Contaminated Waste Actually Burned From Beryllium Production Building

Site Survey, August 1, 1958:

> Special air sampling of the plant incinerator continues. Contaminated waste from 44 and 47 Buildings is being burned. Results to date are inconclusive.

Incineration Ash From Plant Incinerator Buried In Trenches

Site Survey, May 5, 1959:

> The ash dump south of the plant was monitored. Up to 4000 c/m alpha and 30 mr/hr beta were noted. These ashes have been buried. In the future, ashes from the incinerator will be buried in a trench.
>
> (Dow Chemical Site Survey Reports, 1954-1959)

The most disquieting portion (but there are many disquieting elements) of these Plant Incinerator reports is that the location was at the edge of the site and the incinerator itself was characterized as a "Safeway Store" type.

Water Retention Holding Ponds (1952-Plant Closure)

There were four sets of water retention holding ponds at Rocky Flats:

- A-series ponds draining off-site into North Walnut Creek and then into Great Western Reservoir;
- B-series ponds draining off-site into South Walnut Creek and then into Great Western Reservoir;
- C-series ponds draining off-site into Woman Creek and then into Standley Lake; and
- Landfill Pond drains into No Name Creek, which is a tributary of North Walnut Creek.

There is a history of off-site contamination associated with these water retention-holding ponds according to the reconstruction by ChemRisk (1992) with its access to classified documents:

> 1952-1953 Direct discharge into North Walnut Creek of contaminated waste containing nitrates and radioactive substances.
>
> (ChemRisk, 1992, p. 177)

Pond A-1

- From 1953 to 1957 low-level waste was discharged into Pond A-1 for "eventual" discharge into North Walnut Creek.
- "The early discharges of low-level contaminated waste to North Walnut Creek..." (ChemRisk, 1992, p. 179).
- Reconstruction of...Pond A-1 between 1971 and 1973 revealed increased plutonium concentrations in the surface water and it "...is likely that resuspension of the plutonium allowed it to migrate downstream to Great Western Reservoir" (ChemRisk, 1992, p. 179).

Pond A-2

- Completed in the mid-1970s, Pond A-2 received water from A-1 containing chromates and algaecides.

- Pond A-2 became a source of contaminated water for the Spray Irrigation of 80 million gallons a year according to Rockwell International (1987). The assessment of ChemRisk (1992) was that "Any water that collects in the ponds is spray evaporated." The comment in Rockwell International (1987) was that this practice began in 1979 which is why it is not referenced in the available Dow Chemical archives.

Ponds A-3 and A-4

- Pond A-3 dates from 1971 and A-4 from 1980. A-3 receives water from a by-pass pipe containing surface runoff from the north area of the Plant. Water then goes to Pond A-4 before discharge into Walnut Creek.
- Pond A-4 also receives water from Pond B-5 (see below).

B- series Ponds

The B-series ponds are interconnected with the C-series Ponds. In 1991 a pipeline allowed water from C-2 to flow into B-5:

- Between 1971 and 1973 reconstruction of the B-series ponds resulted in the release of plutonium from their sediments (ChemRisk, 1992, p. 179) which supposedly then migrated downstream.
- Rockwell International (1987) reported that the dam for Pond B-5 was "destabilized" in the early 1980s requiring "costly dam repair." This situation apparently did not appear in the documents available to the ChemRisk researchers.
- Between 1952 and 1973 "decontaminated process wastewater, sewage treatment plant effluent, and laundry wastewater after 1957" were released into the B-series ponds and then into South Walnut Creek (ChemRisk, 1992, p. 179).

C-series Ponds

The intent of the C-series Ponds was to capture and control surface water runoff and release it into Woman Creek. Pond C-1 dates from 1952:

- Cooling tower blowdown water was discharged into Pond C-1 until late in 1974.
- Pond C-2 was constructed in 1980 and for undisclosed reasons was "monitored monthly and discharged periodically."
- Not reported by ChemRisk (1992) but by Rockwell International (1987, p. 3), the "East Sprayfields has excessive runoff into Pond C-2 and Standley Lake." The water from the East Sprayfields drained over the East Trenches.

Liquid "Sanitary" Wastes (1952-Plant Closure)

Liquid "sanitary" wastes describes sewage which came from restrooms, showers and sinks, food service areas, and cooling tower blowdown.

ChemRisk (1992, p. 84) described two design safeguards to prevent toxic contamination of the liquid "sanitary" waste streams:

1. The liquid sanitary waste operations are kept separate from the liquid process waste operations to prevent contamination of the sanitary waste streams.
2. In addition, the sanitary wastes which originate from plutonium areas are kept separate from those from other areas until they reach two holding tanks upstream from the treatment plant.

The purpose of the holding tank step was so that the liquid waste could be "retained and sampled to check for contamination" (ChemRisk, 1992, p. 84).

There were three approaches to disposal of liquid "sanitary" waste. The first approach from 1954 through part of 1968 was to dispose of it in Trenches T-2 through T-8. During that time frame, Trenches T-2 through T-8 "received approximately 100 tons of sewage sludge" (ChemRisk, 1992, p. 84).

The second approach was to deposit the sludge in the Second "Sanitary" Landfill when it first opened in 1968. The reason for the change to the second approach to disposal is not in available documents and it lasted only during 1968 and 1969.

Radioactive Sewage Sludge (1954-1972)

According to ChemRisk (1992) the third approach to disposal began in 1972 when "sludges were declared to be low-level radioactive waste and have since been shipped off-site for disposal at federally approved disposal sites." ChemRisk (1992, p. 84) cited a series of Repository Documents which have yet to become available: Facer (1970), Hazle (1985), and Steward (1973).

ChemRisk (1992, p. 84) revealed the reason that the precautions taken to prevent contamination were not successful:

> ...some of the floor drains in the manufacturing buildings were not isolated from the sewage treatment plant, and the sanitary sludge became contaminated with uranium and plutonium.
> (ChemRisk, 1992, p. 84)

An additional source of contamination of the sanitary sewage sludge was the inclusion of cooling tower blowdown. Cooling tower blowdown contains the unmonitored ChemRisk "material of concern" chromium in its form of hexavalent chromium.

Original (Misnamed) Sanitary Landfill (1952-1968)

The original and misnamed First Sanitary Landfill operated in the southwestern quadrant of Rocky Flats from 1952 through 1968. The purpose as described by ChemRisk (1992) based on DOE (1986b) was to dispose of "general plant waste":

> An estimated 20 kg of depleted uranium ash is buried along with normal plant waste, including small quantities of various chemicals. The 20 kg of depleted uranium resulted when 60 kg was inadvertently burned and only 40 kg were recovered. (Owen and Steward, 1974). The landfill may have received nonradioactive hazardous chemical wastes generated at the plant, including solvents. A reported old graphite dump located south of Building 44 that might have received beryllium and uranium was actually the original plant landfill.
>
> (DOE, 1986b)

Owen and Steward (1974) provided further documentation of the "inadvertently burned" 60 kilograms of depleted uranium of which only 40 kilograms were recovered.

Second (Misnamed) Sanitary Landfill (1968-Plant Closure)

The Second (Misnamed) "Sanitary" Landfill opened in 1968 and was in operation until at least 1992. According to the Owen and Steward (1974) report, between August of 1968 and February of 1970, sanitary sewage sludge containing between 800 and 8,000 dpm/g of alpha activity was buried in the Second Sanitary Landfill.

An environmental contamination problem associated with the second landfill is that it was sited in the drainage area of No Name Creek which is a tributary of North Walnut Creek. This was the location of the burial for waste sent from Lawrence-Livermore National Laboratory in California which was the source of the tritium release into Walnut Creek and Great Western Reservoir in 1973.

Open Pit Radioactive Laced Oil Burning (1956-1965)

The continuing problem faced by Dow Chemical management of disposal of contaminated cooling oil with traces of radioactivity is reflected in a letter dated June 12, 1956 from the Dow Chemical Manager at Rocky Flats, F. H. Langell, to Gilbert C. Hoover, the Manager for the Rocky Flats Area Office for the Atomic Energy Commission:

> On September 14, 1955, The Dow Chemical Company received 125 barrels of depleted uranium chips immersed in oil from UCRL for the purpose of recovery and re-use at this installation. For several months [Oct. - July = 9 months] these 125 drums of oil from which the chips were removed have been stored on a slab at Rocky Flats.
>
> Recently one or two of the drums started to leak, constituting a possible source of contamination to the Rocky Flats area. In addition, the space occupied by these drums is needed for other purposes. At the present time there are no methods at Rocky Flats for decontamination of the oil. Because of the contamination problem and the fact

> that the storage space is needed, we would appreciate your advice regarding proper disposition of these residues. Local burial is one possibility.
>
> (Langell, 1956)

The Dow Chemical offer was to bury the waste.

Fire Hazard Associated With Uranium Chips In Oil

There is a fire hazard associated with uranium chips in oil, as is indicated by a December 6, 1962 letter from L.D. Hazleton of Dow Chemical to Norman H. MacKay of the AEC Area Office:

> We have accumulated 3,390 kg of D-38 saw chips contaminated with oil coolant, dirt and other foreign materials. The contamination on these chips makes it impractical to process them in Building 44. Because of the fire hazard, these chips have been sealed in fourteen 55-gallon drums and will be buried at Rocky Flats pending approval by the Area Office.
>
> (Hazelton, 1962)

Burning Contaminated Oil Solves the Storage Problem

The fire hazard posed by uranium chips in oil, and the limited space available for on-site burial and drum storage, led Dow Chemical to begin burning uranium contaminated oil in April of 1956. Putzier (1970) quotes a Progress Report of February 1961 as follows:

> The disposal of contaminated oil and stillbottoms has been partially solved. Open pit burning with air sampling of the plume indicates that these contaminated wastes from 44, 81, and 83 Buildings can be handled in this manner.
>
> (Dow Chemical, 1961)

There were no further citations found to the Dow Chemical assessment of the (lack of) human or ecosystem health consequences from open pit burning in the available Dow Chemical archives.

According to Putzier (1970), Dow Chemical disposed of over one thousand 30 or 55 gallon drums by burning the contents. Putzier reported that burning of the uranium contaminated oil occurred "frequently." He also said that the records available to him contained "many omissions."

Collision of the Production Imperative With Safety and the Limitations of the Site For Storage and Disposal

By 1956, when open pit burning commenced of depleted uranium chips in cooling oil, space for storage of toxic waste was already at a premium. Normal operations, with the production imperative, were continuing to generate significant amounts of highly toxic waste on a daily basis.

Methods for the safe disposal of decontaminated coolant oil with trace amounts of plutonium or uranium were not available. The obvious dilemma for Dow Chemical management was more complicated than the "rational bad behavior" element of the Diamond framework of "good for Dow Chemical and bad for everyone else." The element driving Dow Chemical's dilemma was that previous flawed choices had narrowed their options. Burning contaminated oil in open pits for nine years was, however, a most regrettable response to the production imperative.

Uranium Sheet Open Air Burning (1965)

The reason or reasons why three sheets of depleted uranium totaling 60 kg were burned in the open air at Rocky Flats on May 1, 1965 await further documentation. There was an estimated 40 kg recovered from soil and shipped to the AEC repository in Idaho. The Owen and Steward Report (Owen and Steward, 1974) indicates that an estimated 20 kg of uranium ash were buried in the original Sanitary Landfill.

Nickel Carbonyl Cylinder Burial (Dates Unknown Except For Confirmed Incidents)

Nickel Carbonyl is both flammable and volatile. It is a liquid which becomes a gas at only 43° C with high vapor pressure. Acute toxicity can come from either inhalation or contact with the skin. While its role in constructing plutonium igniters is not disclosed, it is known that Nickel Carbonyl has applications for vapor coating in the metallurgical and electronics industries. Because of its flammability and tendency to explode, it is recommended that Nickel Carbonyl should be stored in tightly closed containers in a cool area.

<u>Nickel Carbonyl Cylinder Disposal Areas</u>

One of the most gripping parts of the narrative of first responders to the 1957 plutonium fire was the discovery of nickel carbonyl cylinders in the anteroom to where the fire started. There were even more cylinders within the space where the fire had spread. The first responders carted out those they could retrieve, a crane assisted, and they were buried within the high security area after being soaked with water.

ChemRisk (1992) drew on Hobbs (1972) and Smith (1975) in supplementing Owen and Steward (1974) in documenting the special care which had been taken in the disposal of what they called "Nickel Carbonyl Bottles." ChemRisk (1992, p. 190) cites the Hobbs report that disposals between March and August of 1972 involved seven 25-pound cylinders, two 5-pound cylinders, and one "lecture bottle" (not further identified). The burial site was described (but not further identified) as "a remote area of the plant site."

The process of disposal was to drill a "dry well" three feet in diameter and fifteen feet deep, then:

> ...the cylinders were opened by individuals wearing supplied air packs and suspended in the hole until they were drained. In

some cases, the chemical ignited immediately after release to the well. In other cases, the well remained silent for long periods before a muffled ignition occurred. Samples at the lip of the hole indicated concentrations around 10 parts per million during the disposal.

(ChemRisk, 1992, p. 190)

Owen and Steward (1974) cite a different location for the disposal of the nickel carbonyl cylinders removed at great risk during the 1957 fire or "when ready for discard." This location was about 200 feet east of the Solar Evaporation Ponds.

Another disposal site is identified in a letter by R.E. Smith (Smith, 1975) to D.J. Cockeram of September 2, 1975 as "one-half mile north and west of the current sanitary landfill." This description could fit the vague description of Hobbs (1972) as "a remote area of the plant site." These reports are three years apart.

The glaring omission in the historical record of disposal of this very volatile, dangerous, and toxic substance was what was done with the nickel carbonyl cylinders before 1957, between 1957 and 1972, and (presumably) after 1975 until the closure of the production era in 1989.

Contaminated Asphalt and Soil Dump

The already contaminated Asphalt and Soil Dump was the location in 1969 for the dumping of debris from the 1969 fire in Building 771/776 according to the Putzier Report (Putzier, 1970):

```
About 321 tons of asphalt and soil contaminated
with plutonium at the time of the Building 776
fire on May 11, 1969, were dumped in an area
designated as asphalt dump ... Average plutonium
concentration for the total amount dumped is
estimated at 7.4 dpm/gm.  Total contained
plutonium is estimated at 0.97 mCi or about 14 mg.
```

(Putzier, 1970)

Another presentation about the Asphalt and Soil Dump is provided by ChemRisk (1992):

> Approximately 320 tons of plutonium-infiltrated asphalt and soil from the 1969 Building 776 fire were buried in 1969 under 1 to 2 feet of fill dirt. Less than 1 mCi of plutonium is estimated to be dispersed in about 250 cubic yards of material, with an estimated alpha activity of about 7 dpm/g. About 60 cubic yards of plutonium contaminated soil from the Building 774 waste storage tank area was placed on top of the asphalt disposal area in 1972, and covered with 3 feet of fill dirt. Estimated activity of the soil was less than 250 dpm/g total long-lived alpha (Owen and Steward, 1974). Total contained plutonium is estimated at 0.97 mCi or about 14 milligrams (Putzier, 1970).
>
> (ChemRisk, 1992, p. 194)

Building 881 and the Notorious Building 881 Hillside (1952-Plant Closure)

During the early years of operation, the AEC had a policy of having more than one nuclear weapons production site perform the same production function. Highly enriched uranium was primarily a specialty of the Oak Ridge site in Tennessee. As part of redundancy policy, Building 881 at Rocky Flats processed highly enriched uranium. Prior to the transfer back to the Oak Ridge Reservation, employees of Rocky Flats who worked in Building 881 between 1952 and 1966 are covered under the unique hazards law because of exposure to neutrons from process work with enriched uranium.

Building 881 Construction

Construction began in 1951 for Building 881, originally known as B Plant. The workload was described by ORAU as "heavy":

> The basic operations involved casting and machining. The components were solid pieces of uranium, machined to particular shapes, which were then assembled with plutonium, stainless steel, and depleted uranium in D Plant [Building 991 - the center of final assembly operations].
>
> (ORAU, 2004, p. 8)

Building 881 was idled in the early 1960s "while the enriched uranium areas were decontaminated and decommissioned" (p. 8). It was converted subsequently to stainless-steel operations.

881 Hillside Designated As a "High Priority" Site For Remediation By EPA

DOE (1991a) indicates that the 881 Hillside was listed as a "High Priority" site for remediation because it included eleven individual hazardous substance locations and the presence of volatile organic compounds, radionuclides, and metals in the groundwater. Among the toxics of concern are: trichloroethane, carbon tetrachloride, dichloroethane, benzene, and uranium.

EPA (1990) characterized the 881 Hillside as posing "a potential long-term threat to the public health and environment." One reason for this characterization was due to "the presence of contaminated groundwater and its proximity to Woman Creek." The EPA analysis was based on twelve sites (only eleven confirmed by evidence) comprising the 881 Hillside.

The twelve areas which were described originally by the EPA are as follows:

1. Oil Sludge Pit--a small pond located south of Building 881 was used for disposal of oil sludges in the late 1950s.
2. Chemical Burial Site--a small pit was used for disposal of liquid waste southeast of Building 881 in the early 1960s.
3. Liquid Dumping--an area east of Building 881 was reportedly used for disposal of unknown liquids prior to 1969. This was

reportedly not substantiated by drilling in 1987. Therefore, this site may not exist.

4. and 5. No. 6 Fuel Oil Tanks--two fuel oil tanks are located south of Building 881; they are out of service and filled with concrete.
6. Outflow Site--an overflow line from the sanitary sewer sump south of Building 881 daylights on the slope below the building.
7. Hillside Oil Leak--oil was discovered flowing from a Building 881 footing drain in early 1973. The source of the oil was never positively identified but the oil was collected in a skimming pond and transported off site. There is an on-going discharge of water from the footing drain.
8. and 9. Multiple Solvent Spills--two areas east of Building 881 were used for barrel storage between 1969 and 1972.
10. Radioactive Site--soil contaminated with low levels of radionuclides were placed on the hillside east of Building 881 and covered with soil between 1969 and 1972.
11. Sanitary Sewer Line Leak--the sanitary sewer line leaked on the hillside southwest of Building 881 in early 1981.
12. Drum Storage Area--Building 881 is currently used for satellite collection and 90-day accumulation of RCRA-regulated wastes.

(EPA, 1990, p. 2-3)

According to ORAU (2004), only gross alpha measurements of water borne emissions were made at Rocky Flats before 1970:

> In general, there was no monitoring of plutonium, beryllium, or organic chemicals during this time, even though laundry wastes from Building 771 and effluents from the sanitary sewer system were discharged directly into Walnut Creek until 1974.
>
> (ORAU, 2004, p. 17)

ChemRisk (1992) enhanced the historical record on the 881 Hillside by their scholarship drawing on Helmstedt (1988):

> From 1951 until 1972, portions of the "Hillside Area" near Building 881 were used as oil sludge pits, chemical burial sites, liquid disposal sites, solvent drum storage sites, and fire damage refuse disposal sites. As a result soil and groundwater have been contaminated with volatile organic compounds including carbon tetrachloride, TCE, and PCE. Alluvial groundwater contains 1,1,1-TCA, and chloroform. Uranium was the only radionuclide occurring above estimated background concentrations (USDOE, 1990). One of the Solid Waste Management Units (SWMUs) in the 881 Hillside area, SWMU 130, is sometimes called the 800 Area Radioactive Site #1 (Helmstedt, 1988).
>
> (ChemRisk, 1992, p. 194)

These contributions are distinctive for documenting the extent of different sources of contamination, and the presence of dangerous volatile organic compounds.

The DOE Characterization of the 881 Hillside In 1992

There were characterizations of the human and ecosystem health risk associated with the 881 Hillside after the closure of the production era with different emphases: DOE (1992) and ATSDR (2005). DOE (1992) is an extensive report reflecting the fact that:

> The 881 Hillside was originally identified as a high priority area due to high concentrations of volatile organic compounds (VOCs) detected in ground water and proximity of the 881 Hillside area to Woman Creek.
>
> (DOE, 1992, p. xviii)

The extensiveness of the report reflects the thoroughness of the characterization of the physical subsurface features of the site, and provides a baseline risk assessment that considers contaminant fate and transport.

Given the thoroughness of the investigation, there were numerous findings. The principal findings from the point-of-view of threats to human and ecosystem health are:

- In general, the contamination at OU1 is relatively minor in terms of contaminant concentrations...
- Surface soils in the eastern portion of OU1 are contaminated with wind-blown plutonium -239, -240, and americium -241 transported from the 903 Pad.

(DOE, 1992, p. xix)

The principal DOE conclusions based on their investigation were:

- ...there also does not appear to be any predicable future ecological risk based on long-term exposure to the contaminants of concern at the site.
- The public health evaluation at OU1 has shown no measurable impact upon human health from contamination present at the site.

(DOE, 1992, p. xx)

Special interest was focused on three sites within the Building 881 Hillside: the eastern portion with plutonium -239 and -240 and americium -241 from the 903 Pad, and two other sites. One of the sites labeled IHSS 130 contained solid semi-volatile organic compound asphalt materials. The other site labeled IHSS 119.1 was contaminated by volatile organic compounds. IHSS 119.1 and IHSS 119.2 were "multiple solvent spill sites." They were drum and scrap metal storage areas:

> It is speculated that elevated concentrations of solvents in the ground water may be from residual solvents or oils on the scrap metal or from the adjacent liquid disposal trench at IHSS 109 (Trench T-2) in OU2.
> (DOE, 1992, pp. 1-11)

IHSS 130 was the location for dumping approximately 320 tons of contaminated asphalt from the 1969 fire and the 400 feet of roadway removed when an unexpected leak during transit contaminated the roadway. This location is presented earlier in this chapter as the Contaminated Asphalt and Soil Dump.

Continuing On-Site Contamination: The "881 Hillside Plume"

The "881 Hillside Plume" is located on the southern edge of the Industrial Area, but north of Woman Creek. Past waste disposal practices in this area caused several contaminants to enter the groundwater. These contaminants include carbon tetrachloride, methylene chloride, TCE, PCE, selenium and uranium isotopes (ATSDR, 2005, p. 29).

Reflections on Disastrous Decisions Made Out of Necessity To Maintain the Production Mission and the Consequences For Off-Site Contamination

This chapter is notable for disastrous decisions in waste management. The pattern is clear across all of the toxic waste disposal sites chronicled in these less well known disposal sites in comparison to the more notorious Mound Area or 903 Pad. But the open pit burning of radioactive laced oil from 1956 through 1965 certainly rivals the Mound Area and the 903 Pad for notoriety. A Dow Chemical Manager revealed this practice to the AEC Manager at Rocky Flats in 1956. Drums had started to leak and "the space occupied by these drums is needed for other purposes." Further, "there are no methods at Rocky Flats for decontamination of the oil." The solution offered by Dow Chemical management was "local burial." Instead it was burned. The broader issue revealed by open pit burning was the overwhelming challenge posed by the production imperative which clashed repeatedly with the absence of a technology to recover traces of plutonium and uranium from cooling oils exacerbated by the limitations of the small site for hazardous waste disposal.

The recent availability in 2018 of the Seed Committee report of 1971 and the partial availability of the results of the soil sampling ordered by the Federal District Court in the Church cluster of civil lawsuits of 1977 and 1978 makes it possible to document more fully the relationship between the waste disposal fiascos presented in Chapters Five and Six with the extent of off-site contamination. The consequences of these early fiascos for plutonium in soil are presented in Chapter Thirteen.

Under the general auspices of the Centers for Disease Control (CDC) researchers at the Oak Ridge Associated Universities (ORAU, 2004, p. 17), with access to classified records, reported that Dow Chemical had only conducted gross alpha measurements of water bourn emissions before

1970. There had been no independent monitoring for plutonium, beryllium, or organic chemicals during that time period. Further, and more inexplicable from the point-of-view of preventing off-site contamination, "effluents from the sanitary sewer system were discharged directly into Walnut Creek until 1974" (ORAU, 2004, p. 17).

What makes this such an astounding revelation is that it was well known (in secret) that the sanitary sewer system was not sanitary! Instead, some plutonium process lines flowed into the sanitary system lines and were spray irrigated off of the Rocky Flats Site. Because management failed to reduce the risk of encountering sewage sludge contaminated with low levels of radioactivity, workers were required to take an "isolation shower" after a work shift.

Chapter Seven
The Unthinkable At Rocky Flats: The Preventable Catastrophic Plutonium Fires of 1957 and 1969, and Missing and Unaccounted For Plutonium 1952-1993

"...it appears (by hindsight) irrational that facilities were ever designed in such a manner as to permit hazards from a relatively insignificant amount of material to so strongly influence the chain of events which led to a major loss..."

AEC Fire Investigation Committee Report, November 27, 1957

"Experience in explosives manufacture affords a surprising number of analogies to guide users of radioactive material in efforts to minimize decontamination expenses. Explosive manufacturers commonly limit the maximum possible risk at any single location by limiting both the amount of material handled and the maximum number of people permitted at one location."

AEC Fire Investigation Committee Report, November 27, 1957

"Only the heroic efforts of the firefighters limited this burning. Had an appreciable fraction of this shielding burned, there would have been a complete loss of Building 776-777 and its contents and a major release of plutonium to the environment."

AEC 1969 Fire Investigation Committee Report

"The damage to Building 776-777 and its equipment was very extensive...the building was grossly contaminated with plutonium...The capacity to produce and assemble plutonium pits for War Reserve weapons was lost for a period of several months and production of special devices for the weapons laboratories was interrupted."

AEC 1969 Fire Investigation Committee Report

Senator Gordon Allott: "What stupid engineer did that?"

Mr. Erlewine (AEC Assistant General Manager): "It was just about that case. Someone, in making a change in the glovebox, had failed to put back the heat detection device where it should have been."

Hearing Before the U.S. Senate on the 1969 Fire

"Admittedly, the potential for fire to start in any one box may be small, but many of our recent incidents have been events with small probability of occurring."

Dr. M.A. Thompson, November 11, 1969 (Dow Chemical, 1969).

"It [the 1969 fire] was caused by the everyday incompetence and negligence inevitably found in any system which polices itself."

M. Peter Metzger, The Atomic Establishment, 1972

The archives of Dow Chemical documents contain riveting accounts of both the 1957 and 1969 fires. The most stark element across both reports is that the brave first responders to the 1969 fire were at greater personal risk than they already were by responding promptly because they had not been informed of the obvious lessons from fighting the 1957 fire. In addition, and compounding their risk, building modifications to Building 776/777 made it even less safe than it already had been based on its original construction.

The unthinkable was an all too regular an occurrence at Rocky Flats. This chapter begins with the explosions and fires which occurred during the early years and ends with the astounding revelation that there are over 2,600 pounds of missing and unaccounted for plutonium when a bomb reportedly can be constructed with 10 pounds. In between these sections will be presented narratives on the 1957 and 1969 fires.

Many disturbing details and commonalities arise from study of the 1957 and 1969 fires. Surprisingly, after the 1957 fire, no meaningful building modifications were made to Building 776/777 to reduce the likelihood of a criticality event in case of another fire in what had become a major plutonium production facility.

Explosions and Fires At Rocky Flats Were a Routine Occurrence: 1954-1971

Fires at Rocky Flats were a routine occurrence. A letter from the Manager of the AEC Rocky Flats Area Office to the Manager of the AEC Albuquerque Operations Office of August 15, 1972 reported that between 1954 and mid-1971 there had been 615 officially reported fires of which 384 were plutonium fires.

There were numerous explosions reported to the Rocky Flats Executive Safety Committee, and memorialized in their minutes. The following list reveals the great variety of causes underlying the reported explosions and fires [Note: it has been asserted that many fires and explosions were not reported]:

October 1954	Perchloric acid explosion
June 16, 1955	Violent reaction between hydrogen peroxide and sodium hydroxide
October 1955	Explosions and production of steam
June 14, 1957	"Simmons-Holman" lost time injury
September 8, 1959	Explosion in peroxide make-up tank
November, 1960	Explosion of unsuspected accumulation of combustible gases in dry box
October 25, 1961	Explosion in boiler
June 16, 1965	Cyclohexane explosion resulted in broken glovebox window and contamination of small portion of room
June 28, 1965	Explosion in lathe box resulted in window cracked and glovebox blown out
September 29, 1965	Plutonium filings exploded and contaminated room

November 9, 1965	Glovebox explosion and fire when flammable solvent (acetone) contacted ignition source
October 24, 1966	Incinerator explosion, cause undetermined, window failed
April 3, 1968	Explosion--leak permitted acetylene gas accumulation in glovebox. Sample take-out was blocking exhaust filter
September, 1968	Acetylene escaping from regulator on tank exploded and burned
November 6, 1968	Chemical explosion
October 7, 1970	Explosion--exact cause unknown
October 14, 1970	Explosion and fire after heater placed in a dissolution vessel
October 22, 1970	Explosion and fire during metal dissolution
August 22, 1971	Explosion in can of sample chips. Reaction between plutonium chips and residual carbon tetrachloride "Since we do not know how to preclude such a reaction, it can be concluded that we do not know how to store or ship the chips with an adequate margin of safety."

This list demonstrates the danger involved in operating a nuclear weapons production plant when where 615 fires were reported to have occurred during only 17 years. The list of explosions and fires from the minutes of the Rocky Flats Executive Safety Committee does do not include almost all of the significant fires and explosions reported by Putzier (1982) in The Past 30 Years At Rocky Flats.

Edward A. Putzier began his career at Rocky Flats as a Radiation Engineer on June 11, 1952, and retired several months after finishing his report in November of 1982. At the time of his retirement he was Health Sciences Manager for Rockwell International at Rocky Flats. While the bulk of this chapter focuses on the catastrophic 1957 and 1969 fires, the explosions and fires documented below by Putzier (1982) were all serious, and could have been even more serious.

Chemical Explosion June 14, 1957 (also in records of the Executive Safety Committee)

This explosion occurred during the processing of plutonium nitrate. The operator and the radiation monitor knew something wrong was going on and tried to get away. The glass vessel containing the plutonium nitrate was destroyed. The force of the explosion sent projectiles flying. The two victims were not protected by respirators. Three employees who aided the victims were highly contaminated.

Fire At the Filtrate Recovery Box – Building 771, July 1963

According to Putzier (1982) "it was a relatively serious fire":

> There must have been relatively high levels of contamination released locally and quite a number of the people who actually

did some of the initial work in coping with the situation were protected only by half-mask respirators...with survey meters on the filters of the respirators [it] is indeed scary if we think about our concern over proper respiratory protection today.

(Putzier, 1982, p. 50)

The classified version of the Filtrate Recovery Box Fire has not been made available.

Fire In Aero-Tech Unit, Beryllium Shop In February 1964

The focus of Putzier's (1982, p. 51) report on the Beryllium shop fire is the question of whether or not beryllium will burn. This fire was in a ventilation –filtration-type unit in Building 444. There had been a buildup of both beryllium chips and "oily residues"—it "was questionable that the beryllium had anything to do with the fire or that the beryllium actually burned."

Putzier cites an AEC Bulletin of January 10, 1963 which discusses five fires involving metallic beryllium which suggest that "beryllium can burn under various conditions, especially very fine powdered beryllium." He cites a caution contained in an internal letter in the file from 1962 which suggests that using CO_2 on beryllium fires can cause the fires to become much more violent.

Degreasing Explosion, Building 776 On June 12, 1964

One of the main lessons learned from this explosion as cited by Putzier (1982, p. 51) was "the incompatibility of burning plutonium in carbon tetrachloride could not have been better demonstrated." Another general lesson Putzier derived from this explosion was that "curiosity can lead to trouble," and that "whatever kind of response...might be involved needs to be done keeping in mind the jeopardy an employee responding may put himself into." These general lessons were derived when a worker was seriously injured and exposed to radiation by an explosion. The security guards from a contractor who aided the injured employee were themselves unknowingly contaminated. The injured worker subsequently had fingers and a thumb amputated.

Americium Explosion, Building 771 On September 25, 1964

The cause of this explosion was distinctive. The root causes were to be found in a shipment to Rocky Flats from another DOE entity, in this case by the Lawrence-Livermore National Laboratory (LLNL) in California. Rocky Flats had first supplied americium to LLNL in solution form in 1958. In 1964, LLNL returned the americium solution along with some americium in oxide form. Upon arrival at Rocky Flats, the americium had been stored in an unoccupied room (Room 180A in Building 771). The cause of the explosion was the failure to vent the solution form of americium before storage. Putzier concluded that the solution form of americium should have been converted to oxide before its shipment from LLNL.

What he referred to as "radiolytic decomposition" under pressure can release oxygen and hydrogen gas. The ignition source for the gases was thought to be static charges which can occur with plastics [Note: not further explained]. The formal accident report has not been declassified.

Oil Leak From Wastebox On May 4, 1965

This was an oil leak, not an explosion or a fire. A strip of roadway became contaminated with 45 grams of plutonium which had not been known to exist in cooling oil contained within the confines of a glovebox. The glovebox was being prepared for shipment to long-term storage at the Idaho repository:

> In this case the oil did not show up until the box was transported to an outside area in preparation for loading and shipping to Idaho. The box leaked and contaminated a narrow path of road some 400 feet long. The contaminated asphalt was put in drums and was buried at a sludge pit near Building 995. The box also leaked at the storage area immediately east of Building 444 and there was an area of soil contamination at that location that was picked up and committed to contaminated waste and sent off to Idaho.
>
> (Putzier, 1982, p. 50)

This matter-of-fact recital by Putzier of an oil leakage incident reveals the danger of contamination which can be undetected in a routine disposal task. It also reveals the ad-hoc nature of on-site contaminated waste disposal. It is for that reason that the recounting by Putzier of the leak of oil contaminated with 45 grams of plutonium is reported here. In this case the incident did not lead either to an explosion or to a fire. But it reveals hidden dangers in an industrial enterprise where there was not continuity in knowledge of risk.

Fire In Lathe Coolant System, Building 776/777 On October 15, 1965

Putzier (1982, p. 54) characterized this incident as both "very serious and probably unique in nature" and that "it wound up as probably the most serious incident in numbers of people affected that Rocky Flats has had." The fire was "triggered by the pyrophoricity of plutonium but beyond that was again a demonstration of the incompatibility of burning or sparking plutonium and carbon tetrachloride" (p. 55). The task which led to the fire was cleaning out of a blocked coolant line which was thought to contain plutonium in chip form. Carbon tetrachloride was known to be in the coolant line. The carbon tetrachloride had been used earlier to try to clear out a blockage in the line.

In trying to use mechanical force to clear the line by inserting a brass rod, there had been some sparking:

> This did not impede their further attempts, however, and they went into the line with a center punch creating more force in the line, more friction, and once they had created enough energy to

> start the plutonium chips on fire that were in the line, the carbon tetrachloride exothermic reaction took over and in my estimation, vaporized plutonium that got out into the room.
>
> (Putzier, 1982, p. 50)

Employees working on clearing the line were contaminated. There were no continuous air monitors present to warn people to put on respirators. According to Putzier, the small size of the plutonium particles (9.32 micron) could lead to lung burdens.

This recounting of explosions and fires (at least some of them) by Putzier (1982) and the Rocky Flats Executive Safety Committee provides a context for considering the catastrophic fires of 1957 and 1969. The narrative in this chapter will now shift to the catastrophic plutonium fires of 1957 and 1969.

1957 Plutonium Fire

Introduction To the 1957 Plutonium Fire In Building 771

The fire of September 11-12, 1957 was reported by the Atomic Energy Commission to be the ninth most expensive accident loss in its history. The lack of accurate record keeping of the amount of plutonium in Room 180 where the fire broke out, and in the filters at the time of the fire, makes it impossible to reconcile the amounts of plutonium lost with the available internal Dow Chemical reports.

Building 771 was characterized at the time of the 1957 fire as the main plutonium building at Rocky Flats with numerous design defects, few fire breaks, and overcrowded. Dow Chemical was reported to have deliberately replaced the original gloveboxes made of fire resistant steel and laminated safety glass with more flammable Plexiglas. The reason for this action, reportedly, was to increase the effective shield for glovebox workers from exposure to neutrons.

The fire began by spontaneous combustion in a flammable Plexiglas plutonium storage glovebox which was not equipped with a fire detection device. The plant protection worker who discovered the fire was reprimanded for calling the Fire Department. The first fireman who arrived at the scene was ordered not to fight the fire for ten minutes while the fire gained momentum.

- The fireman initially asked to use water on the rapidly growing fire but was refused permission.

- The result was that only carbon dioxide was used for the first 13 minutes while the fire gained greater momentum.

- When water was subsequently used for only one or at the most a few minutes, the fire died down.

- But in error, the fans were ordered to be turned up rather than off. The effect of this error was to spread the fire.

- An explosion followed which also cut off electricity to the lights, radiation monitors, alarms, and fans since electric power had been routed through the filter plenum.

The 1957 fire burned from 10:10 p.m. on September 11, 1957 until the next day until declared out "officially" at 11:28 a.m. on September 12, 1957.

The following narrative details the general construction of the 771 Building which made the fire worse, what happened during the 1957 fire itself and in its immediate aftermath, the environmental contamination which was known to have occurred, and the lessons from the 1957 which were known but not subsequently used either to prevent or to contain the 1969 fire.

Plutonium Lost - "This record is the best that can be garnered..."
A report to the Fire Investigation Committee in a letter by J.F. Quinn, Jr. of September 24, 1957 (Quinn, 1957) stated:

<u>Amount of Plutonium Lost Is the Best Estimate Which Can Be Obtained</u>

> From records kept by the Accountability Group, there was 42.3 kg of material in the area. However, some late records are not available yet and this figure will be changed. This record is the best that can be garnered by a combination of recollection, talks with the operators working in the area, and the September 3, 1957, inventory report to the Accountability Group...
>
> The total of the listing [of plutonium removed from Room 180 after the fire] is 28.0 kg. The difference of 14.3 kg is material transferred out and not recorded, and material whose exact amount and location are not known."
>
> (Quinn, 1957)

The most noteworthy element is the imprecision of the letter. The estimate of 14.3 kilograms of plutonium apparently lost does not include "material whose exact amount and location are not known." In a later acknowledgment, it is reported that the filters which contained plutonium had not been changed in four years. Among the unknowns are the amounts of plutonium in the ductwork and in the filters.

The Fateful Choice of Flammable Plexiglas For Glove Box Construction
An AEC Fire Investigation Committee Report identified the use of flammable Plexiglas as one of the major factors in contributing to the spread of the fire and to its severity:

Plexiglas Contributed To the Spread of the Fire

> The use of Plexiglas in glovebox construction without metal separators between Plexiglas panels contributed to the spread of the fire. Those units most completely destroyed were those composed of 100% Plexiglas construction without metal frames.
> (AEC, 1957)

Original construction of glove boxes at Rocky Flats used stainless steel frames and laminated safety glass. This choice of construction materials constituted better protection against the spread of fire. This original choice was discontinued with the cryptic comment that the flammability of the type of Plexiglas chosen by Dow Chemical over less flammable forms of Plexiglas as the substitute for laminated safety glass was known but not "appreciated" or "understood."

Elsewhere it is revealed that only after the 1957 fire was it "appreciated" that the form of Plexiglas chosen could be ignited at its edges by one match. The reason for the fateful choice is given as:

> The original dry box equipment was stainless steel-laminated safety glass; however, in some areas, Plexiglas was used due to the difficulties of fabricating enclosures and rigidly framing them so that glass could be used to give a contamination-tight joint without breaking. Where gloves are required, transparent panel glass has not been satisfactory from a fabrication-tightness point of view, and Plexiglas has been used … (Classified Deletion)
>
> This facility used considerable amounts of Plexiglas; and, in some cases, it was used not only as window material but also as the sole structural material (no metal framing).
> (AEC, 1957)

There were safer materials studied. The reasons for their rejection were not in the available documents.

Storage of Nickel Carbonyl Cylinders

Nickel carbonyl is a poisonous substance which is both flammable and volatile. At least 21-nickel carbonyl cylinders were stored near the filter plenum and the room where the fire started by spontaneous combustion. According to the testimony of John Pringle reported in an Atomic Energy Commission document of October 7, 1957 entitled "The Report of Investigation of Serious Incident in Building 71 on September 11, 1957," (AEC, 1957) he was concerned sufficiently about the substantial hazard posed by the nickel carbonyl cylinders that he tried to remove them as the fire accelerated:

As the firemen started down the plenum with the hose, I noticed a nickel carbonyl cylinder lying on the floor of the anteroom. By this time, my Chemox canister was out of time and I went back to the access shaft and obtained a new canister. As I returned through the shaft, I met Bob Vandegrift who told me that he had seen Floyd Parker in front of the building. Bob Vandegrift and I went immediately to the anteroom of the filter plenum, and we moved seven nickel carbonyl cylinders to a storage position in the old Pipe Shop area. We then went to the East entrance of the filter plenum to try to determine the extent to which the fire had spread.

(AEC, 1957)

Nickel Carbonyl Cylinders Were Burning Around the Top

Much later (sometime between 4:30p.m. and 8:00 a.m. [on September 12, 1957]), Pringle reports:

"…a fireman called and reported that he had found several small cylinders in the exhaust filter plenum which were burning around the top. We immediately concluded that these were additional nickel carbonyl cylinders…

Mr. Venable called Will Wright to determine the number and location of carbonyl cylinders still in storage in Building 71. I went immediately to the exhaust filter plenum in a Chemox mask and found 14 nickel carbonyl cylinders along the south wall of the downstream side of the filters and about twenty feet from the west end of the plenum. Some of these cylinders were still burning…

While Mr. Venable got a burial pit dug and a crane with a drum lift, lined up outside the access shaft, Jesse Cleveland, Vic Ryan, and myself, went back in the plenum and put the remaining cylinders in barrels. We then filled the barrels with vermiculite and carted them to the access shaft. They were then lifted from the access shaft by the crane and placed in the pit which had been dug, and the drums filled with water before the lids were replaced.

(AEC, 1957)

Another report indicates that those nickel carbonyl cylinders were buried subsequently east of the Solar Evaporation Ponds. The bravery of those who responded to this extreme threat is not even acknowledged in their matter-of-fact recounting of a dangerous part of a harrowing night. It is recognized for what it is: a tribute to first responders everywhere.

Fire Detection and Alarms

A disturbing finding in the AEC (1957) report is that there were no working fire or heat detectors where the fire in the plutonium area began:

> <u>Heat Detecting Equipment Did Not Function</u>
>
> Heat detecting equipment installed in the [main building] filter [plenum] did not function. It was later found that this equipment had been blocked out to eliminate operating difficulties which were being experienced. This detection equipment was to have sounded an alarm in the Building Superintendent's office and in the Boiler Room and simultaneously to have shut down the fan system.
>
> (AEC, 1957)

The fire detectors which had been installed in the main filter plenum had been "blocked out" (were inoperable). Therefore, the fire was discovered by plant protection staff and there was no automatic alarm ring within the Fire Department. AEC (1957) reports that a supervisor had ordered that: "No one should turn in an alarm for Building 71 except Building 71 supervision..."

Timeline of the Start of the Fire

September 11, 1957

10:10 p.m. Spontaneous combustion of plutonium casting residues in flammable Plexiglas glovebox line discovered by plant protection staff.

10:10 p.m. Order given not to use carbon dioxide fire extinguishers, and plant protection staff ordered out of Room 180. One plant protection staff member was reprimanded for notifying Fire Department.

10:12 p.m. Lieutenant Eminger of the Fire Department arrives in protective gear but ordered not to enter Room 180.

10:12 pm. - 10:24 p.m. Fire gains momentum.

10:24 p.m. Lieutenant Eminger and Shift Supervisor Vandegrift enter Room 180 and start applying carbon dioxide to the fire with no effect.

10:25 p.m. Against standing orders and design of the fire detection system, the exhaust fans were turned on high. This carried hot gases into the main building filter plenum and fanned the fire.

10:28 p.m. The boiler operator saw smoke from the exhaust system.

10:37 p.m. Use of water began on the fire.

10:39 p.m. Explosion occurred which blew up the filters and caused a column of smoke 80 to 100 feet high.

10:40 p.m. Fans went off.

10:40 p.m. All activity after 10:40 p.m. was anticlimactic.

10:58 p.m. Second fire truck was called.

11:05 p.m. Second fire truck arrives.

September 12, 1957
11:28 a.m. The fire is declared extinguished after the use of 30,000 gallons of water in a building with no criticality drains.

(AEC, 1957)

The dry, matter-of-fact tone of the AEC report on the fire does not reflect the extent of the bravery in what was an ordeal at great personal risk with an accelerating dangerous fire, and with an uncertain outcome. This was the first experience at Rocky Flats with a major plutonium fire.

The early moments of the fire were described by one of the Plant Protection staff, Patrolman Anderson:

Irvin Said There Was Something Wrong

As Irvin and I entered the room #179, Irvin said there was something wrong, he said "Do you smell anything?" I said it smells like rubber burning. We went into room #180 and one of the hoods was on fire, the gloves were almost gone and the glass was crumbling. Irvin went to turn on alarm and I went for a CO_2 fire extinguisher and noticed in the office off of room 180, Parker on telephone. I told him about fire and I started to gather up CO_2 fire extinguisher. Parker ordered me not to use CO_2 on fire. I was ordered out of the room by Bruce Owens.

(AEC, 1957)

The Fire Gained Great Momentum

V.F. Eminger, the Lieutenant of the Fire Department who was the first responder stated:

On arrival, I told Fireman Hoar to bring Chemox masks to air lock and stand by, then proceeded to room 180 at double doors in hallway 196, and at the doors I was stopped and told to not go in although I was properly dressed (bunker coat, asbestos gloves, booties and chemox mask) and as I stood at the door, to my left R. A. Vandegrift was putting on a chemox mask making preparation to go into room 180. As I stood there, the fire gained great momentum in the right corner of the room near the lathe.

(AEC, 1957)

Robert Vandergrift, the Shift Supervisor, who assumed responsibility with no apparent regard for his own safety, reported:

The First Lieutenant Suggested Water and We Said No

As we were completing putting on our masks outside 180, the fire Lt. suggested using water and both Owens and I said no. I was ready first so I entered 180 with a standard size CO_2 extinguisher from the hall. I was able to get within 4 feet of fire and emptied the extinguisher I had. I returned to the door and picked up a second CO_2 extinguisher. At this time the Lt. returned with me. I was unable to get within 6 feet of the fire. I believe we used two more standard CO_2 extinguishers without appreciable effect. However, we may have only used the three smaller extinguishers. At this point, George Stapleton met us at the door with the large CO_2 extinguisher on wheels from Fab Line. I took the nozzle and the Lt. took the cart, and we were able to get to a point between the small mixer and the centrifuge before the heat stopped us. At this point we emptied the large CO_2 extinguisher into the fire without any effect.

We Could Feel the Temperature Drop After Water Was Used

We returned to door of 180 and held a hurried council. The fire Lt. again wanted to use water. Owens was not in favor of using water in the area. As the large CO_2 extinguisher had done no good, I said we could use the hose as a fog or spray nozzle. The firemen had already strung the hose up the west hall to the door of 180. We entered the room, the fire Lt. holding the hose. We were unable to get past the door of the small room to the right of the main door. We opened up the hose with only the spray and shot it over the top of the dry box in front of us onto the fire. We could feel the temperature drop and after a few minutes of spraying the fire, we stopped the water to determine what effect there was, if any. While standing discussing the situation, the first explosion occurred.

(AEC, 1957)

Building 771 Not Designed For Water Use

The AEC Fire Investigation Report indicated that when Building 771 was designed, the necessity to use at least some water in fighting plutonium fires had not been anticipated:

Automatic Sprinklers Were Deliberately Excluded In Design and Construction

Because of contamination, accountability and criticality problems, as well as the difficulty of extinguishing plutonium fires with water, automatic sprinklers, water standpipes and floor drains were deliberately excluded at the time of design and construction of the building. The building was equipped with 15% CO_2 extinguishers and 100% CO_2 carts. There are fire hydrants outside the entrances to the building and near the roof hatch to the boiler ventilating area.

(AEC, 1957)

It turned out that existing water pressure was inadequate for aggressively fighting a fire. Moreover, there was not enough length in the available fire hoses to reach the fire. More hoses had to be obtained which led to further delay.

The AEC (1957) Fire Report indicated that Building 771 had not been designed to use water to fight plutonium fires. There had not been the safety features built into the building to prevent criticality events or explosions caused by water.

The Decision To Turn the Fans On High

The decision to turn the fans on high was one of the most fateful decisions. The order to turn the fans on high was conveyed according to AEC (1957) by a Shift Captain to the boiler room operator who actually controlled the operation of the fans:

Putting the Fans On High Was Against Orders

…if the fans should be on high and he said yes. So I called the boiler room and asked Lieker to put the fans on high. He said he could not do this without an order directly from Vandegrift. I told him Vandegrift was already fighting the fire and that we had to have the fans on high. He agreed to this but said it was against his orders.

(AEC, 1957)

A post-fire assessment indicated that in normal operations if the heat detectors had not been blocked, the fans would have been shut down automatically.

A post-fire analysis by I.B. Venable, stated that:

Assuming the fire was first detected at 10:10 p.m., the explosion occurred at 10:39 p.m., and the fans went off at 10:40 p.m., the largest portion of the damage resulting in loss of production occurred in that … period. As far as building operation is concerned, all activity after 10:40 p.m. was anti-climactic…

(AEC, 1957)

The central point in the Venable report is that activity after 10:40 p.m. was "anti-climatic" after the explosion occurred and the fans went off because the damage already had been done when the fans ran at high speed for the

15 minutes between 10:25 p.m. and 10:40 p.m., during which time the fire accelerated.

Power of the 10:39 p.m. Explosion
Chilling details reported in AEC (1957) are according to Shift Captain T. Eckert:

A Column of Smoke 80 To 100 Feet High Was "Very Dark in Color"

> The time was 10:39 p.m. I then heard Owens shout that it had blown up and for everyone to get out.
>
> I went on out the west air lock to the outside where the men were with the fire truck and they said it had blown up the filters. I looked up at the stack and a column of smoke 80 to 100 feet high was coming from it and was very dark in color.

(AEC, 1957)

A report from Lieutenant V.F. Eminger, a fireman, stated:

An Explosion Went Off Knocking Mr. Vandegrift To the Door

> …[A]n explosion went off knocking Mr. Vandegrift to the door and I was pushed through the doors into the hall. Neither was hurt, the fire, after the explosion was out (blew itself out). We came out of the room again to go up to the filters and as I started up the stairs, someone advised the fire had started again in room 180, so we again went to 180, this time through the office 180B.
>
> The fire was small in two or three different places--used two 15% CO_2 and was out.

(AEC, 1957)

A tribute to the courage of all of those at risk from the explosion is that they kept fighting throughout the night and into the later morning in the most dangerous conditions.

Filter Plenum Fire "Smoke Was So Intense…"
The magnitude of the fire was affected by flawed design and construction of Building 771. There was no backup generator. When the 10:40 p.m. explosion occurred it knocked out electric power because the electric power lines had been routed through the filter plenum in an area which had been destroyed by the explosion.

Battery powered lights ran out of battery life (charge) during the long night. One of the design and construction failures exposed by the fire was that there was no sprinkler system. Fire detectors had been blocked.

The report of Section Assistant Superintendent John Pringle in AEC (1957) provides a first hand account:

Large Fire In the Filters

> I entered the downstream side of the exhaust filter plenum from the east end and observed a large fire in the filters approximately two-thirds the distance from the east entrance to the west entrance. The exhaust filter plenum was not particularly smoky, and I could see the entire length of the plenum to the fire clearly…I was informed by somebody that Mr. Venable had already gone to the access shaft with the fire truck.

The Smoke Was So Intense

> I then went immediately to the access shaft and into the second floor area where I encountered firemen stringing hose to the east door of the filter plenum. I led the way to the east door and opened it for the firemen to bring in the hoses. It was my impression that one of the doors was actually open and the other was closed. As we entered the anteroom outside the filter plenum, the smoke was so intense that we could not find the doors to the plenum with lamps. We therefore sprayed down the anteroom with a fog nozzle, which also reduced the intensity of the heat to the point where we could get the hoses into the door of the plenum.

We Could Not See Beyond the Door

> …We could not see beyond the door of the plenum due to the smoke and intense heat. We then went back to the west end of the plenum where firemen were beginning to string new hose to the center hatch of the plenum in order to fight the fire from that point…

(AEC, 1957)

A classic understatement by Section Assistant Superintendent John Pringle was the closing of his dramatic report: "…the firemen reported that all fire had been extinguished. This was approximately 10:00 a.m., and I left for home at approximately 10:30 a.m." The obvious risks to their very lives which had been confronted by all of the Dow Chemical staff and the first responders did not merit even an acknowledgment that their extraordinary service throughout a harrowing night was more than a routine "I left for home at approximately 10:30 a.m."

Environmental Contamination From the 1957 Fire

The explosion at 10:40 p.m. destroyed the power to the smokestack sensors. This eliminated the source of measurement of radioactivity released up the smokestacks. The lack of a precise inventory of plutonium in the filters and ducts or even in Room 180 makes it impossible to reconstruct plutonium release into the environment.

AEC (1957) does report the recollections of employees who saw large quantities of smoke leaving all night from the 150-foot Building 771 smokestack.

Section Superintendent I.B. Venable:

The Exhaust Filters Were On Fire

> ...I then proceeded to the plant and arrived there at 10:55 p.m. coincident with the arrival of J. W. Pringle, Supt. of 71 Building.
>
> It was immediately apparent that the exhaust filters were on fire judging from the smoke coming out of the stack. I instructed Pringle to locate the fire lieutenant and to get water into the filter plenum as soon as possible. I further instructed Springer and C. W. Stone to survey the 76 Building and 91 Building areas for contamination spread. I determined that a site survey team had already been dispatched for the purpose of determining the environmental contamination, if any, in the area surrounding the Rocky Flats.

(AEC, 1957)

Misrepresentation of Radioactivity Released

Dow Chemical subsequently reported that: "Early estimates of the total known and unknown quantities released was given as one gram (0.07 curie)" (Dow Chemical, 1957c). But Dow Chemical did not speak with one voice of denial. T.S. Chapman, Director of the Health Physics and Medical Section, indicated that the releases of smoke up the Building 771 smokestack continued at least until dawn (Chapman, 1957). It was not credible to estimate that only one gram of plutonium was released because the smokestack monitoring of radiation release failed at 10:40 p.m. when the electrical system ceased to function after the explosion.

1957 Fire Report By the AEC Fire Investigation Committee and Its Supplement

There was a formal fire report compiled by the AEC Fire Investigation Committee. The full report is not yet available. But the excerpts which are available are consistent in their recommendations to separate physically related units. The following are examples:

> 1. Design and build future plutonium facilities to provide positive separation of related units in such fashion that fire, explosion, ventilation failure, etc., will not shut down the whole facility. Consider wing-type buildings with separate ventilation for each wing...

Irrational That Facilities Were Ever Designed In Such a Manner

> "When these facts are viewed in retrospect, it appears (by hindsight) irrational that facilities

were ever designed in such a manner as to permit hazards from a relatively insignificant amount of material to so strongly influence the chain of events which led to a major loss...The same facts suggest the desirability of approaching design of dry boxes used for radioisotope handling by applying the same "containment" philosophy applied to reactors; namely, to design facilities to withstand and to contain all radiological hazards in the event of the worst creditable incident. The separation of facilities is also suggested for reduction of the total risk."

(AEC Fire Investigation Committee Report, November 27, 1957)

With respect to the analogy to the experience of the manufacture of explosives, the authors of the report viewed that realm of experience to contain valuable lessons:

Explosives Manufacture Offers Analogies For Radioactive Material Handling

Design Simplicity

Experience in explosives manufacture affords a surprising number of analogies to guide users of radioactive material in efforts to minimize decontamination expenses. Explosive manufacturers commonly limit the maximum possible risk at any single location by limiting both the amount of material handled and the maximum number of people permitted at one location. Explosives handling areas are kept as free as possible of extraneous materials and facilities. Exposed surfaces are minimized... Extreme care is of course given to avoidance of fire risks. While few radioactive material will involve any explosion risk, the extreme importance of minimizing radioactive decontamination expense in the event of fires suggest that radioactive material users may well profit by the hazard limitation and housekeeping technique that have been found effective and economically justifiable in the explosives manufacturing field.

(AEC Fire Investigation Committee Report, November 27, 1957)

The approach recommended for design of a plutonium fabrication area was to limit the "maximum possible risk" at any single location. This could be achieved by a two-fold approach:

1. Limit the amount of material (plutonium) handled; and
2. Limit the maximum number of people permitted at one location.

This sensible approach at Rocky Flats was not reported to have been followed at least in available documents in the twelve years leading up to the 1969 fire.

H.A. Spavin (Spavin, undated) wrote a two-page memorandum to John Epp of the 1957 AEC Fire Investigation Committee in which he advocated design choices which would minimize the risk of a broad Plant shutdown from a more local fire or contamination:

> Isolate Component Operations In Construction of Plutonium Processing Facilities
>
> It is recommended for all future construction of plutonium processing plants that the various component operations be independent and isolated by separation or physical cutoff so that a fire or serious contamination incident will not cause a shutdown of operations other than the component immediately involved.
>
> (AEC Fire Investigation Committee Report, November 27, 1957)

Plutonium Storage

The 1957 AEC Fire Report contained numerous recommendations concerning plutonium storage and handling. For example:

> 1. Provide plutonium metal storage areas separate from production areas, if possible, for skulls, chips…sludge, etc.
> 2. Provide the storage areas with fire detection systems.
> 3. Design storage areas with the possibility of fire in mind, and so space them that appropriate extinguishing means can be used safely (G-1 powder, graphite, CO_2, water spray). Consider use of powder extinguishers above each storage pot so that fire would automatically dump extinguisher on fire.
> 4. Provide humidity control to decrease water vapor content of air in areas handling plutonium. (Water vapor accelerates rate of oxidation of plutonium.)
> 5. Provide automatic CO_2 extinguishing systems in lines where oils, Plexiglas, etc., are used in conjunction with plutonium metal.
>
> (AEC Fire Investigation Committee Report, November 27, 1957)

This recommendation reiterated the necessity to assure functioning fire detection systems and to separate storage areas from actual daily-use production areas. One of the recommendations was to store plutonium in Number 10 cans where the atmosphere had been purged with helium and then the cans sealed. There is no evidence that this was the approach taken in Building 776/777 before the 1969 fire. One commentator concluded that helium purging of storage cans of plutonium "probably also would have prevented the 1969 fire if it had been used at the time."

The following excerpt is from the 1957 AEC Fire Investigation Committee Report section containing commitments which were made by Dow Chemical:

> Dow Chemical Made Commitments It Did Not Carry Out
>
> It is, of course, most desirable to minimize the quantity of fabrication residues intermediately stored. To do this several techniques have been developed which have:
>
> 1. markedly decreased the quantity of residue created per unit weight of parts produced, and
> 2. considerably increased the capacity to recover plutonium from these residues and thereby convert it back to useful metal.
>
> Extraordinary administrative effort is being devoted to assuring ourselves that no excessive quantities of plutonium, as skull, turnings, or other highly pyrophoric residues are being accumulated in any given area. Additionally, a continuous fire watch is being provided in Building 76, our major fabrication facility, and a fire watch will be established in Building 71 to cover those times when production personnel are not operating in the building.
>
> (AEC, 1957)

This "extraordinary administrative effort" envisioned by Dow Chemical in the 1957 Fire Report was not in evidence at the time of the 1969 fire, at least in the available archives of Dow Chemical. Furthermore, there is no evidence in the archives of "a continuous fire watch" being a matter of routine in either Building 776 or Building 771.

Plexiglas Construction

Construction with plexiglas is one of the most critical areas of the 1957 AEC Fire Committee Report. It reveals why at the highest level of Rocky Flats management of both Dow Chemical and the AEC there was not an effort made to reduce the flammability of dry box and glovebox construction. With respect to dry box construction, the following was the recommendation:

> Dry Box Construction
>
> Eliminate or minimize the use of flammable materials of construction. If flammable plastic windows must be used, build the boxes so that flammable panels are separated by metal framing of suitable width to act as a firebreak.

There was an additional 1957 Fire Supplementary Report presenting alternatives to flammable Plexiglas G, including the less flammable one known as Plexiglas 5900 which was $3.36 per square foot compared to $2.21 and $2.77 per square foot for more flammable varieties of Plexiglas.

Why Were Sensible Recommendations Not Adopted By Top Management of Dow Chemical and the AEC Area Office At Rocky Flats?

Quite fortuitously, the answer to this question is provided by a letter dated December 20, 1957 from a Dow Chemical Manager for Rocky Flats, to Seth H. Woodruff, Jr., Manager of the Rocky Flats Area Office for the AEC:

> Operations have been reviewed since the fire with a view to eliminating hazards due to the flammability of Plexiglas, where possible. In those areas where an undue hazard was deemed to exist, substitution of glass or metal panels for Plexiglas has been made...

Dow Chemical Plant Manager: Plexiglas Does Not Present an Undue Fire Risk In the Majority of Areas

> ...The Plexiglas in the majority of the areas in Buildings 76 and 77 [776/777] does not present an undue fire hazard, and it does not appear feasible from the standpoint of interruption of operations nor of cost, since there are roughly 29,000 square feet of plastic window to replace all Plexiglas at this time. However, a 24-hour fire patrol has been instituted in the affected areas to provide early detection until such time as adequate detecting devices are installed.
>
> Since the fire, pyrophoric metal has not been stored closer than twelve inches to the Plexiglas panels. In addition, chip-handling facilities provide metal tunnels for shielding plastic from oily chips collected in cans in the machining operations. The degreased chips are canned in tin cans to minimize fire hazards.
>
> (Langell, 1957)

Incredibly and erroneously, of the three reasons offered for not installing in retrofit safer Plexiglas, one was to state that for the majority of areas in Buildings 776/777, Plexiglas "does not present an undue fire hazard."

With respect to the issue of cost of safer Plexiglas, there needed to be 29,000 square feet replaced at a cost for the "most promising" at $3.36 per square foot. It is not known from available archives if the $3.36 per square foot cost included installation. The loss from the 1969 fire was over $48 million in damage to the Rocky Flats facility, and $22 million in lost plutonium.

Even if the labor costs to install the new Plexiglas were not included, the assessment that the current Plexiglas did not present "an undue fire hazard" was to underestimate the role which flammable Plexiglas was to have in accelerating the 1969 fire. The Rocky Flats Manager's bottom line that "redesigning and rebuilding these gloveboxes would be prohibitive" is not supported in the available Dow Chemical archives.

The third argument appears to be the controlling one for the Rocky Flats Manager given his choice of phrasing: "...does not appear feasible from the standpoint of operations...." 'Operations' is used as a synonym for not interfering with production of plutonium igniters.

The institution of a "24-four-hour patrol" in the "affected areas" was not implemented at least in available Dow Chemical Archives.

Sprinklers and Water Availability

It had been clear during the 1957 fire that water pressure had been too low, and that fire hoses had not been placed where they could be quickly extended to where water was most needed.

The issue of protection of the integrity of roof structures came up in the 1957 fire reports. If the integrity of the roof structure became breached during a plutonium fire, dangerous particles would be released into the wind and off-site without the impediment of the roof.

A Dow Chemical letter of December 20, 1957 (Dow Chemical, 1957b) addresses roof construction:

> <u>Protecting the Integrity of Roof Construction Did Not Happen</u>
>
> Due to the problems of roof construction in the new buildings, sprinkling and other protection for roofs is being planned as follows:
>
> Building 76 roof will be sprinkled (Note: this important pledge was not fulfilled).
>
> Building 77 roof will be sprinkled [This important pledge was not fulfilled].
>
> Building 81 roof will be protected with an asbestos cement board ceiling.
>
> Building 83 will be sprinkled over the press area only.
>
> We do not plan external box line sprinkling systems; however we are in the process of providing fire hoses distributed at 50 to 75 feet intervals throughout Building 76 and 77 so that, in the event of a Class A fire, adequate extinguishing means will be readily available.
>
> Dow Chemical (1957b)

Dow Chemical (1957b) is notable for pointing out the roof construction planned for Buildings 776/777 included the pledge to install sprinklers. This did not happen.

Failures To Pass On Lessons From the 1957 Fire
The following points stand out concerning the failures to pass on lessons of the 1957 fire to the stewards for Dow Chemical responsible for the safety of Buildings 776/777:

- Fire Breaks were not provided in Buildings 776/777;
- Storage for plutonium in work areas was actually made more dangerous;
- Flammable Plexiglas was not changed for known less flammable versions; and flammable Benelex was not excluded; and
- Sprinklers were not installed on the roofs of Building 776/777 which were known to be vulnerable to fire, and were mandated by AEC policy.

And as already has been reported, the use of water successfully on the 1957 fire was not known to those risking their lives to fight the 1969 fire.

Lessons of 1957 Fatefully Not Followed
From the perspective of fire prevention, the assessment by a Dow Chemical Rocky Flats Manager on December 20, 1957 proved to be incorrect when he stated that the Plexiglas in the majority of the areas in Building 776/777 "does not present an undue fire hazard." He went on to prioritize not interrupting plutonium production operations. He also identified the cost involved in installing safer Plexiglas. As noted earlier, it is possible to quantify this judgment: there were 29, 000 square feet at $3.36 per square foot for one form of more fire resistant Plexiglas. The fire in 1969 did $70 million in damage.

Putzier indicated that discussions after the 1957 fire had concerned the need for partitioning and modularizing future buildings. That had not occurred with respect to Building 776-777. A letter from the Manufacturing Manager to the Head of Engineering before this fire had not been translated into action. The Manufacturing Manager had urged that partitions be installed "to prevent spread away from or to confine spread to a more localized situation." (Putzier, 1982, p. 56)

<u>Why Building 776/777 Was Not Made More Fire Safe?</u>

Why were the recommendations of the Manufacturing Manager not followed? This is how Putzier in 1982 characterized the reason for this decision <u>not</u> to act to make Building 776/777 more fire safe:

> The argument that this was never done relates to (1) the expense that would have been involved, and (2) someone expressing concern over employees getting trapped within some of these areas and not getting out.
>
> (Putzier, 1982, p. 56)

What was Putzier's assessment of the decision not to put partitions into Building 776/777 and the consequences of the failure to act?

> In retrospect, I think it was a rather lame excuse at the time that compartmentalization was not accomplished. Even the partition between 777 and 776 had been removed for convenience of space. This in itself [if the partition between the buildings had been kept] would have prevented the majority of the spread [of the 1969 fire].
>
> (Putzier, 1982, p. 56)

1969 Fire In Building 776/777

Introduction To the 1969 Fire In Building 776/777

Four elements of the 1969 fire stand out immediately. The first was the magnitude of the loss as reported by the Atomic Energy Commission: $22 million in plutonium lost and $48 million in repair costs for Building 776/777. These amounts make up half of the losses which were admitted to by the Atomic Energy Commission in its entire history.

Figure 7.1 Destroyed Main Filter Plenum in Building 771 From the Fire of 1957

There was a double catastrophe associated with the Main Filter Plenum. Electric power for lights, fans and the vital radiation release sensors had been routed through this Filter Plenum. Its destruction made it impossible to document the amount of plutonium released up the smoke stack into the environment.

The second element which stands out is that instead of being prepared to prevent a catastrophic fire in 1969, it was as though the 1957 fire never had happened, and that there were no lessons to be learned from it. Building 776/777 was the largest plutonium producing facility at Rocky Flats after Building 771. There were clear parallels in risk. Yet there had been no apparent effort to improve it structurally to make it more fire safe, starting with remedying its original construction flaws. Specific commitments which had been made, such as installing a sprinkler system, were inexplicably not honored despite an AEC requirement for sprinklers in all new construction.

The third element which stands out is how the fire started, the combustible surroundings which had been allowed to occur, the inadequacy of heat detectors or firebreaks, and the obstacles put in the way of first responders.

The 1969 fire is reportedly (like the 1957 fire) to be a product of spontaneous combustion. A problem encountered immediately after the start of the 1969 fire was that the ventilation systems were not segregated from one another. It had to have been incredulous to a student of the 1957 fire that carbon dioxide was used first on the 1969 fire while it accelerated in intensity. The Dow Chemical documents reported two further matters which are also incredulous:

- The first responders to the 1969 fire had not been told that water had been used successfully on the 1957 plutonium fire.
- The 776/777 building had not been retrofitted to reduce the likelihood of a criticality event.

Dow Chemical had not even hired a fire protection engineer until 1967. That individual is reported to have had no direct fire engineering experience.

The fourth element which stands out is that the safety practices at the Atomic Energy Commission's Los Alamos National Laboratory, predating the Dow Chemical era at Rocky Flats, provided a model for handling plutonium which presumably was transferable in whole or part to Buildings 771 and 776/777. This did not occur.

There is a strong sense of "having been there before" in reading reports on the 1969 fire. The heroism of Dow Chemical workers and first responders again is inspiring, as it was during the 1957 fire. The power and deadly nature of the plutonium fire in 1969 is, in essence, a second encounter with the terrible nature of what Kai Erikson called "a new species of trouble" for humanity. In 1969, as in 1957, this deadly form of the "new species of trouble" was set off by spontaneous combustion.

There had been an expectation that the lessons from the 1957 fire in Building 771 and the extensive AEC Fire Investigation Committee Report would form the core of efforts to improve the safety of Building 776/777 as a facility. This did not occur. There had been an expectation that the hard fought lessons from fighting the 1957 fire itself such as the known ineffectiveness of carbon dioxide extinguishers, the prior planning and installation necessary to allow the rapid deployment of water hoses, and the quick application of water would be translated into standard operating procedures. Not only did this not occur, but there had not been physical modifications to Building 776/777 in the form of drains to reduce the risk to workers of a deadly criticality incident from the application of water.

Design and Construction of Building 776-777

After the 1957 fire in Building 771, Building 776-777 became the primary plutonium production facility at Rocky Flats. That circumstance makes it imperative to examine the design and construction of Building 776/777:

Building 776/777 Was Not Designed and Constructed To Meet AEC Fire Safety Standards

> Building 776-777 complex was not designed or constructed to meet AEC fire safety standards – nor does it appear that any concerted efforts over the years have been made to achieve these standards.
>
> 2. Since original construction, decisions were made to add new equipment and structures on ad hoc bases. This "Topsy-like" growth led to overcrowded conditions that made compliance with fire safety regulations even more difficult if not impossible.
>
> ...Building 776-777, a vital weapons production facility, did not meet even the minimum AEC fire safety standards. This increased the fire loss and unnecessarily exposed firemen to fire, plutonium, and nuclear criticality hazards.
>
> (Atomic Energy Commission, 1969)

The AEC Fire Committee Report characterized the shoddy construction of the roof which nearly burned through during the fire:

Shoddy Construction of the Roof of Buildings 776/777

> The roof consists of metal pan deck, 1/4 inch styrofoam insulation nailed to this deck, covered with asphalt-saturated felt hot mopped to the Styrofoam, 1/2 inch fiber hardboard nailed through the preceding layers and a final ply of butyl membrane applied to the hardboard with butyl mastic. As a result of recurrent wind damage, the roof is presently being held in place by concrete blocks randomly place on the butyl sheet approximately 4 to 5 feet on centers.
>
> (AEC, 1969)

"Nearly Catastrophic"-Testimony of General Edward Giller of the AEC Before Congress 1970

On October 1, 1970, Major General Edward Giller, the Atomic Energy Commission Assistant General Manager for the Division of Military Applications, testified before the Subcommittee on Public Works of the House Committee on Appropriations. One part of his testimony is riveting about how close the 1969 fire came to being truly catastrophic:

How Close the Fire Came To Being Truly Catastrophic

> GENERAL GILLER. They have sufficient material in them during a day's operations and heat generating flammables that are part of the

process so that if a major fire were to break out and break through the building, that is breach the roof, then hundreds of square miles could be involved in radiation exposure and involve cleanup at an astronomical cost as well as creating a very intense reaction by the general public exposed to this. Our problem is first to prevent the fire from starting, and then if it starts to keep it within the building.

In the fire we had last year we kept it in the building. If the fire had been a little bigger it is questionable whether it could have been contained.

CONGRESSMAN JOHN RHODES. There is no danger of an explosion?

GENERAL GILLER. The possibility of explosion is not the major problem. It is a dumping of radioactivity in the homes, streets, cars, and clothes of people which would require a cleanup such as we did in Palomares or Thule where it was relatively easy, where you pick up every 2 inches of soil, wash people's houses down, buy their clothes, and so on. In addition, there is possibility of lawsuits by people involved. No one would perhaps become medically sick, but this is the problem.

(Giller, 1970)

General Giller elaborated before Congress: "If the fire had been a little bigger, it is questionable whether it could have been contained…hundreds of square miles could be involved in radiation…"

"Thick Black Smoke" and Like 1957, the Loss of Vital Utilities

In the 1957 fire the explosion at 10:40 p.m. destroyed the electric power for Building 771 which had been routed through the filter plenum which was the site of the explosion. In the circumstances of the 1969 fire it is less clear why the construction of Building 776/777 did not protect either the radiation monitoring equipment or the building utilities which went out at 4:00 p.m., about one-and-a-half hours after the fire began:

Aside from temporary decontamination partitions and other non-fire rated walls, Bldg. 776-777 is one fire area. This includes the second floor where ductwork and piping pass through the floor to reduce the effectiveness of the cutoff…

Almost from the beginning it was virtually impossible to see because of the thick black smoke and the loss of lights in the main fire area.

(AEC, 1969)

"The Fire Condition Was So Bad" - Excess Crowding
Constructive recommendations from the 1957 fire had not been implemented concerning reducing crowding and isolating as much as possible radioactive materials which were not in use from glovebox exposure:

Crowded Conditions In the Fire Areas

> The crowded conditions in the fire areas made firefighting very difficult. In some areas the firefighters could not get in between gloveboxes because of the air packet on their backs...The fire condition was so bad in the building that it was virtually impossible to size up the situation and provide effective direction of the firefighting efforts.
>
> (AEC, 1969)

One report stated that changes in the building had not been reflected in the floor plans available to the firefighters and other first responders.

Changes Within Building 776/777 After the 1957 Fire
There were at least two major changes within Building 776/777 after the 1957 fire. Contrary to the precautionary principle of keeping the structure divided into small sections to contain a fire, four years after the 1957 fire the cinderblock wall was removed which had separated Buildings 776 and 777 into two internal buildings. The effect of this was to create one large area where the fire could spread without structural impediment:

> Originally Building 776 was separated from Building 777 by a common concrete block wall penetrated only by a single glovebox conveyor section and a single personnel door that was normally locked...
>
> In 1961, the concrete block wall between the buildings 776 and 777 was removed in order to accommodate additional equipment.
>
> (Factory Insurance Association, 1970, Vol. I, p. 14)

The second change after the 1957 fire was to increase the amount of Benelex shielding. There are two reported reasons for this. Dow Chemical explanations were protection for workers and cost:

Increase In Benelex To Improve Neutron Shielding

> The use of Benelex...as a neutron shield in place of methacrylate plastic has been proposed. This proposal was prompted by a cost savings that would result from lower cost of Benelex over the methacrylate plastics [such as Plexiglas]...This is a report of the work conducted to determine the relative burning properties of these materials. Benelex is a product of Masonite Corporation and is a dense, dark brown, filled organic material...The

most common type of methacrylate plastic used at Rocky Flats is Plexiglas SE-3, which is a self-extinguishing type. Methacrylate plastics that do not contain a flame retarding additive burns [sic] quite vigorously.

(Factory Insurance Association, 1970)

Another explanation for the increase in the use of Benelex shielding was to reduce the exposure of workers to penetrating radiation:

Reducing Worker Exposure Levels To Penetrating Radiation

Concern about the increasing levels of penetrating radiation to which employees in Building 776 were exposed led to actions in late 1967 and early 1968 to reduce exposure levels. In addition to various administrative actions, a decision was made to install additional shielding on the glovebox systems in Building 776-777 in order to reduce the exposure of employees. The shielding work called for installing lead, leaded glass, and both Benelex and Plexiglas of various thicknesses on the gloveboxes and conveyor lines.

(Factory Insurance Association, 1970)

Additional Shielding Authorized By the AEC

The additional shielding work was authorized by the AEC in January 1968, and started immediately. Although limited initially by availability of materials, lack of firm criteria, and manpower, a maximum effort was underway by March 1968.

The design work and some of the installation was performed by Dow; the other installation work was performed by Swinerton and Walberg (under its cost-type contract with the AEC). The total cost of the Building 776-777 shielding project was estimated at $3.5 million.

As of the date of the fire, most of the shielding work in Building 776 had been completed. Approximately 1,170,000 pounds of Benelex and Plexiglas were used…

(Factory Insurance Association, 1970)

"Only the Heroic Efforts of Firefighters" Prevented Consumption of Benelex and Plexiglas In the 1969 Fire

At two points in the AEC Investigation Report there is explicit mention of how close the fire came to being truly catastrophic:

Less Than One Percent of Shielding Burned

The Plexiglas windows contributed heavily to the spread of the fire and the extent of the loss. These windows, a major structural part of the containment system, provided a fuel surface on the inside of the glovebox conveyor systems. Continued operation of the glovebox ventilation systems provided a copious supply of air to support the combustion. Under these conditions the burning of the windows and plutonium would have resulted essentially in the same loss as was experienced even if no other combustible materials had been present.

(AEC, 1969)

<u>Only the Heroic Efforts of Firefighters Limited This Burning</u>

Less than 1 percent of the total of almost 600 tons of Benelex and Plexiglas shielding was consumed in the fire. Only the heroic efforts of the firefighters limited this burning. Had an appreciable fraction of this shielding burned, there would have been a complete loss of Building 776-777 and its contents and major release of plutonium to the environment.

(AEC, 1969)

These obvious sources of danger were exacerbated by the previously reported overcrowding and the failure to remove plutonium from the work areas when it was not in use.

Viewpoint of a Dow Chemical Fire Protection Engineer

A Dow Fire Protection Engineer was asked for his views on the installation of Benelex and Plexiglas. Since the AEC had notified Dow Chemical that Benelex was combustible, the recommendation was to install sprinklers:

<u>No Review of the Benelex and Plexiglas Decision By the Fire Protection Engineer</u>

No drawings for installation of Benelex and Plexiglas were submitted to or reviewed by the Dow Fire Protection Engineer or the Fire Department. However, the Dow Project Engineer verbally requested the views of the Dow Fire Protection Engineer. The latter was not familiar with the material and talked to the Fire Chief and the AEC/AL Fire Protection Engineers. Subsequently, he advised the Dow Project Engineer that AEC had pointed out Benelex was a combustible material and consideration should be given to installing sprinklers in the operating area of Building 776.

(AEC, 1969)

A choice was made not to install sprinklers even though they were mandated for new construction under AEC rules, and the Dow Fire Protection Engineer and Dow Fire Department were not consulted formally.

Concept For How the 1969 Fire Started

In the absence of eyewitness accounts of how the fire started, and what the Investigation Board called the "nullifying" of the "heat sensing system" in Glovebox 134-24, the following is the reconstruction of how the fire started and what made it so dangerous:

> The Investigation Board has examined all the available evidence and proposes a concept that is consistent with the known facts as to how the fire might have started and spread.

A Plutonium Briquette Spontaneously Ignited

> A plutonium briquette in an open can located at the south end of the Benelex-Plexiglas storage container in Glovebox 134-24 in the North Foundry Line ignited spontaneously. The heat of the burning plutonium metal caused the Benelex and Plexiglas of the storage container to char and out-gas, thus generating a flammable gas which was ignited by the burning plutonium.

Heat Detectors Were Shielded and Did Not Function

> The Benelex-Plexiglas storage container continued to char and smolder for some time, releasing both smoke and heat. The smoke was pulled into the Booster System No. 2 exhaust system and gradually clogged the filters. The heat generated caused other briquettes to ignite and initiated a slow burning of the storage container, particularly in the cracks between the joined sections. After a period of smoldering, flames erupted on the outer surfaces of the container and spread to the combustible gloves and Plexiglas windows on Glovebox 134-24. At this point in time, the fire probably was still undetected because no one was in the area and the heat detectors were located outside and under the glovebox.

Large Quantities of Particularly Pyrophoric Forms of Plutonium Were Present

> Glovebox 134-24…in Building 776 contained large quantities of particularly pyrophoric forms of plutonium…The Investigation Board made…conclusions concerning this plutonium and glovebox…as follows:
>
> G. The recycling of scrap, chips and skull materials has been the source of almost all the reportable plutonium fires in Building 776-777.
>
> H. The storage of plutonium briquettes in cans without lids exposed potential ignition sources.

I. Without the large cellulosic fuel source (the Benelex-Plexiglas storage containers), it is unlikely that a plutonium briquette burning in an open metal container in Glovebox 134-24 would have ignited the Plexiglas windows.

J. Nullifying the heat sensing system in Glovebox 134-24 by the addition of the Benelex-Plexiglas storage container prevented an earlier warning of fire which probably would have greatly reduced the size of the fire and consequent damage.

(AEC, 1969)

"Absolute Negligence" U.S. Senate Hearing

It turned out that the "nullifying" of the heat sensor system had not been a deliberate plan of management. Rather, as part of the reconstruction of the glovebox where the fire began, the installed Benelex served as a de-facto heat shield of the heat sensor system which had not been replaced where it was supposed to be located after routine maintenance by an engineer.

On November 21, 1970, there was a hearing before the U.S. Senate Subcommittee on Deficiencies and Supplementals of the Committee on Appropriations. The excerpts below are from the testimony of:

- Wilfrid E. Johnson, Acting Chairman of the Atomic Energy Commission;
- John Erlewine, Assistant General Manager of Operations; and
- General E. B. Giller, USAF, Assistant General Manager for Military Applications

May 1969 FIRE AT ROCKY FLATS, COLO

Senator Gordon Allott. I do, Mr. Chairman. This is very critical to my State.

I would like to go into it to at least clear the record in my mind. The fire that you speak of occurred in May of the last year. Has the cause of this fire been ascertained?

The Cause Was One of Self-Ignition of Plutonium

Mr. Erlewine. Yes, sir. We believe that the cause was one of self-ignition of plutonium. It can do this. The fire started in a storage cabinet where cans of plutonium were stored. These were briquettes that were made up of scrap in the process of recycling. When you divide plutonium into fine pieces you create more surface area, and the prospect of catching fire by self-ignition increases. The fire clearly started in this cabinet during the work hours and it reached serious proportions before it was discovered and it spread through the gloveboxes that General Giller described.

Senator Allott. I am asking this for my information and the information of my constituents. This characteristic of plutonium has been known for some time?

Mr. Erlewine. Yes, sir

Senator Allott. Why had not adequate steps been taken before this fire to protect against such a fire?

Mr. Erlewine. Many steps had been taken in this particular instance. Warning systems in that glovebox did not convey the warning at an early enough time. There are plutonium fires in these gloveboxes. There are measures ordinarily taken by the operators to extinguish the fires. The warning system to call the fire department in this particular instance came too late.

Burning Plutonium Is a Fairly Common Thing

Mr. Johnson. Senator, it might be well to point out that burning plutonium is a fairly common thing. It is done purposely in many cases to produce plutonium oxide that is used as fuel. These pressings of chips and what not are a little bit difficult to predict. Normally they are pressed tight enough so that you don't have the risk of a spontaneous combustion.

As General Giller has pointed out, the normal reaction should a fire start is to put it out and there are plenty of ways of doing it.

What Stupid Engineer Did That?

Senator Allott. In other words, there was no backup for a detection system in the storage area?

Mr. Erlewine. In this particular case, Senator, when they revised the chamber in which these were stored the heat detecting devices that are normally inside had been placed elsewhere, and were insulated so that the did not give the warning in time.

Senator Allott. What stupid engineer did that?

Mr. Erlewine. It was just about that case. Someone, in making the change in the glovebox, had failed to put back the heat detection device where it should have been.

Senator Allott. Who in the AEC is responsible for the supervision of such engineering?

Mr. Erlewine. We have on-site an area office that does make safety inspections and our Albuquerque office does too. But this is an enormous plant. These inspections are made about every 6 months to a year. This is the kind of thing that would have to be caught in most instances by people operating the plant and the supervisor of the line.

Absolute Negligence On the Part of the Government

Senator Allott. As you describe it, it seems to me like this is a situation of absolute negligence of the part of the Government, on the part of the AEC people specifically. In other words, you are storing pyrophoric material, you stored it in a place where you did not have adequate detection. Your detection was inadequate. Now somebody was responsible for this. You could not expect me or Senator Byrd to know what steps to take in a situation like this, but certainly in the AEC the people should have been competent to do this.

Responsibility For Operation Is Broadly Delegated To Our Operating Contractor

Mr. Johnson. Senator, of course we can't get out from under the responsibility we have. However, as you know, the responsibility for operation is broadly delegated to our operating contractor. Normally it is the operating contractor's responsibility to include this in the provision to supervise this kind of thing. We don't have, as far as I know, the specific detailed inspection of everything that is done. We do have safety regulations.

Serious Deficiencies In Organizational Management Practices

General Giller. It is certainly true in reviewing the operating practices at our Rocky Flats plant, which was done by our committee after the fire, we identified what we considered to be serious deficiencies in organizational management practices. They have changed the safety operation by giving it an independent status and charging it with a much more visible responsibility.

There Is No Excuse For the Particular Situation That Was Allowed To Generate the Fire

> We have learned as much as we can from the fire, both from the management standpoint as well as from other technical aspects such as engineering things that one should do. It is certainly true that there is no excuse for the particular situation that was allowed to generate the fire.

(U.S. Senate Subcommittee on Deficiencies and Supplementals of the Committee on Appropriations, November 21, 1970)

Human Error Nullified the Heat Detector

This testimony by Mr. Erlewine revealed that it was human error which nullified the heat detector:

> Someone, in making the change in the glovebox, had failed to put back the heat detection device where it should have been... This is the kind of thing that would have to be caught in most instances by people operating the plant and the supervisor of the line.

Senator Gordon L. Allott, Republican of Colorado, characterized this human error as "...absolute negligence on the part of the Government, on the part of the AEC people specifically." General Edward Giller's testimony was that Dow Chemical had "...serious deficiencies in organizational management practice." Further, he testified that "...there is no excuse for the particular situation that was allowed to generate the fire."

"Two Fireballs" Led to the Decision On Initial Use of Water

The decision to use water was made by Captain Jesser of the Rocky Flats Fire Department. He and the other firemen began fighting the fire with hand-carried carbon dioxide extinguishers. The 1969 firefighters had not been informed of the failure to contain the 1957 fire with carbon dioxide:

The CO_2 Extinguisher Was Discharged Without Effect

> Captain Jesser directed the fireman with him to attack the fire at this point with a hand-carried CO_2 extinguisher. He then proceeded to the east end of the North Foundry Line to attack the fire with a 50-pound CO_2 extinguisher mounted on a cart. When he opened a Benelex door to insert the nozzle in the conveyor line, the gloves were already burning and the fire on the inside of the conveyor line was moving east at a rapid rate. He discharged the CO_2 extinguisher without effect and returned to the location where he first observed the fire. The CO_2 extinguishers being used by the firemen on the North Foundry Line also had no effect.

Two Fireballs Went To the Ceiling

> One of the firemen heard two loud reports (like rifle shots) and saw two fireballs (about basketball size) go to the ceiling in the area of the North Foundry Line. This occurred while the firemen were laying out a fire hose, and before any water had been used on the fire.
>
> At approximately 2:34 p.m., Captain Jesser concluded it would be necessary to attack the fire with water.
>
> (AEC, 1969)

It is not clear from this report if the loss of time caused by using ineffective carbon dioxide had the same negative consequences of allowing the fire to accelerate as had occurred in 1957. General Giller before the U.S. Senate on November 21, 1970 identified a key challenge faced by Captain Jesser in fighting the fire and how difficult it was:

The Primary Requirement For Fighting a Plutonium Fire Is To Keep It In the Building

> ...That is the primary requirement of fighting a plutonium fire, keep it in the building, and keep the smoke in the building. When you design a plutonium building you design it to have a fire going inside, to fight the fire, to put it out, have the ventilators run all the time and filters run all the time and never get touched by the fire. You don't want to have a wall switch burn through and turn off the fans, then your material gets out and you have a severe problem.
>
> (transcript of U.S. Senate Testimony by General Giller, p. 14)

"No Built-In Fire Protection" Revealed In Visit By AEC

Two-months before the 1969 fire, Rocky Flats had been visited by a member of the Atomic Energy Commission Headquarters Industrial Safety and Fire Protection Branch (between March 11 and March 15, 1969). He reported as follows on his visit to Building 776-777:

Fateful AEC Visit Two Months Before the 1969 Fire: "No Built In Fire Protection"

> On March 14 we visited a Manufacturing Building No. 776-777. This is a very large undivided complex with glovebox lines. This facility is one of the most important to the program, but had practically no built-in fire protection or suppression equipment. There were a number of lesser problems which were discussed with local management people and will not be discussed here due to the nature of operations.
>
> (AEC, 1969)

This informative report underscores the importance to national security of Building 776/777, and its lack of fire suppression or built-in fire protection equipment.

"Serious Fire Hazard" Recognized By Both Dow Chemical and the Atomic Energy Commission

The AEC Fire Investigation Report contains a damning recognition by Dow Chemical and the AEC about the seriousness of the fire hazard in Building 776/777:

Building 776-777 Represented a Serious Fire Hazard

> In March 1969, AL [Albuquerque] safety personnel had discussions with Dow officials, who acknowledged that Building 776-777 represented a serious fire hazard and inquired when AL was going to require that something be done about it. However, no report or recommendations were made to AL management.
>
> (AEC, 1969)

The devastating part of this recognition made in secret is that "no report or recommendations were made."

Consequences of Storage Practices Placed Both Workers and Plutonium In Jeopardy

Given that the fire in 1969 caused so much damage and loss, and that the extreme fire risk was known before hand, it is useful to reconstruct the reasoning behind the decision-making on storage practices. The AEC Fire Investigation Report provides an answer:

Relatively High Radiation Exposures to Workers

> The decision to store relatively large quantities of plutonium in the conveyor lines and gloveboxes may have provided greater ease of operations and for cost reduction, but this also contributed to relatively high radiation exposures to the worker and, in the event of an accident, placed in jeopardy a large quantity of plutonium.
>
> (AEC, 1969)

An example of the amount of stored plutonium in the congested building is provided by the AEC Fire Investigation Report about what is known as the "4 High Mill Pit":

> The 4 High Mill Pit, which was cleaned between November 1968 and March 1969 for the first time, contained approximately 74 Kg of sludge from which 13 Kg of plutonium was recovered.
>
> (AEC, 1969)

The "4 High Mill Pit" [not further explained], beyond its accumulation of 13 kilograms of plutonium between only November of 1968 and March of 1969 (and was not cleaned in that time frame), posed an obvious fire risk.

"A Gradual Deteriorating Condition" Reported By General Giller At the House of Representatives October 1, 1970

The question posed to General Giller was why the Atomic Energy Commission had not come to the U.S. Congress sooner with the recommendations for site improvements at Rocky Flats:

<u>It Is a Gradual Deteriorating Condition</u>

CONGRESSMAN EDWARD BOLAND. I am surprised this recommendation did not come long before today. If these facilities are in this state, and I applaud you for making this request, why was this not done before? You can see with the naked eye, I presume, the inadequacy of the buildings, equipment and many of the fire and safety hazards, you have described here. Why was there not some recommendation made on this before?

GENERAL GILLER. It is a gradual deteriorating condition.

CONGRESSMAN BOLAND. It did not just suddenly arrive. You do not have to have a catastrophe, do you, to be alert to the problem?

GENERAL GILLER. You will recall a year ago we had a near catastrophe.

<u>Do You Have To Wait For a Catastrophe</u>?

CONGRESSMAN BOLAND. Yes; but do you have to wait for a catastrophe before you actually know that the facilities and the equipment are in the bad shape that the Commission now finds them?

GENERAL GILLER. Certainly as managers of a program like this one, we should be able to replace outmoded equipment and buildings on a regular basis so that your plant does not get into such a condition. I think as a result of our management practices, worrying mostly about radiation perhaps, rather than fire and safety, and as the result of several years of very stringent budgets, we did sort of have the subject rather forcefully brought to our attention by the fire.

The AEC requested $25 million as the first payment on $118 million worth of upgrading that needed to be done at AEC facilities.

(Subcommittee on Public Works of the House Committee on Appropriations, October 1, 1970)

The AEC was proposing that $78 million of their total upgrading requests of $118 million, should be allocated to Rocky Flats as follows:

New plutonium recovery facility, Rocky Flats;
Procurement and installation of firewalls, doors, heat detection systems, etc.; filter plenum improvement; and extensions of utilities.
Total estimated cost, $63,000,000.

Filter plenum improvements, manufacturing buildings, Rocky Flats;
Additions and improvements to existing filter plenums, including additional stages of filtration, and additional filter plenums.
Total estimate cost, $8,000,000.

Liquid plutonium waste treatment facility improvements, Rocky Flats;
Procurement and installation of 2-stage filter plenums, waste holding tanks, process equipment, exhaust monitors, filters, etc., and associated modifications.
Total estimated cost, $2,000,000.

Inerting of hazardous gloveboxes, various manufacturing buildings, Rocky Flats;
Procurement and installation of equipment for converting gloveboxes to a closed system and reroute and/or add new ducting, controls, cooling coils and analyzers, other related equipment, and modify plenums.
Total estimated cost, $5,000,000
(AEC 1970)

Urgency of Upgrading the Facilities At Rocky Flats Known In 1970

The urgency of upgrading the facilities at Rocky Flats, in the viewpoint of the Atomic Energy Commission, is reflected in their determination to spend $15 million of the first $25 million of additional allocation and $78 million of the total of $118 million of additional allocation on only Rocky Flats.

As part of the October 1, 1970 Congressional Hearing at which General Giller testified, two members of the House of Representatives reflected on what they had heard and the conclusions they had drawn:

The Risk of Radiation Damage

Congressman Joe Evins: And the committee can reasonably conclude from all this testimony that some of these facilities were built over 20 years ago and are not meeting current code standards. We also have the risk of radiation damage for which there certainly is concern by the public.

Are We Subject To the Charge of Negligence?

> Congressman Glenn Davis: Are we fairly subject to the charge of negligence in failing to have some of these basic things [fixed] at these installations? Nobody would permit operations to go on without firewalls and automatic sprinklers and things of this kind unless they were under the jurisdiction of the Federal Government where they can do nothing about it, I suppose.

The Atomic Energy Commission, based on the record as presented by General Giller, was aware of "a gradual deteriorating situation." Congressman Edward Boland posed an important question to General Giller:

> You do not have to have a catastrophe do you, to be alert to the problem?

Worried More About Radiation Exposure Than Fire

The answer from General Giller was instructive as it contained a number of elements:

- The task is to replace outmoded equipment and buildings on a regular basis: "So that your plant does not get into such a condition."
- The management focus was to reduce radiation exposure, "worrying mostly about radiation perhaps, rather than fire and safety…"
- The lack of results was from "several years of very stringent budgets…"

Congressman Glenn Davis posed the question about the issue of negligence:

> Are we fairly subject to the charge of negligence in failing to have some of these basic things [fixed] at these installations?

The answer by Congressman Davis to his own question was that "nobody would permit operations" the way Dow Chemical was running Building 776/777 "without firewalls and automatic sprinklers." An interesting next response by Congressman Davis seemed to mitigate the responsibility of Dow Chemical since "they were under the jurisdiction of the Federal Government where they can do nothing about it, I suppose."

The Fire Was Caused By Everyday Incompetence and Negligence

An investigative journalist H. Peter Metzger, who followed Rocky Flats actively, had a harsher, more structural assessment of responsibility as he wrote in the 1972 The Atomic Establishment:

> "Clearly then, this fire was not simply an accident that could happen anywhere, as Dr. Lloyd M. Joshel, then Dow's general manager at Rocky Flats, implied at a meeting in Denver...
>
> It was caused by the everyday incompetence and negligence inevitably found in any system which polices itself.

The Metzger analysis focused on the consequences of a faulty governance system where a nuclear weapons complex policed itself behind a cloak of secrecy.

"This Is a Situation of Absolute Negligence"

Senator Gordon Allott from Colorado, at the U.S. Senate Subcommittee on Deficiencies and Supplementals of the Senate Appropriations Committee, assigned several levels of responsibility. At the level of the individual, he asked, "What stupid engineer did that?" referring to the person who so incorrectly reinstalled the heat detection device so it did not report the start of the 1969 glovebox fire.

Senator Allott assigned another level of responsibility to the Government and to the Atomic Energy Commission directly: "...this is a situation of absolute negligence..." The reason was that fire detection was inadequate.

The response to Senator Allott by Wilfrid E. Johnson, the Acting Chairman of the Atomic Energy Commission was to accept that "...of course we can't get out from under the responsibility we have." But he the went on the make a governance argument:

> However, as you know, the responsibility for operation is broadly delegated to our operating contractor.

It is the role of the AEC to establish safety regulations, not to conduct "specific detailed inspection of everything that is done."

AEC Regulations For Emergency Planning Failed On Preventing To the 1957 and 1969 Fires

The issue of fire protection planning is a useful lens to evaluate emergency planning. Though both the Atomic Energy Commission and Dow Chemical apparently aspired to maintaining all facilities at a high level, both the 1957 and 1969 fires revealed that this aspiration had not been realized with respect to fire prevention.

The Atomic Energy Commission had regulations governing emergency planning, and regulations very relevant to fire prevention. The record of the condition of both Building 771 in 1957 and Building 776/777 in 1969 is clearly that both facilities were not fire-safe.

The strength of emergency planning provides a prism for determining the level of concern attached to the need for emergency preparedness in the area of fire prevention. Quite astounding is the Dow Chemical record on off-site emergency planning as revealed by the AEC Fire Investigation Report:

No Need to Have Plans For Off-Site Damage

> The Dow official in charge of nuclear safety emergency planning advised the Board that there was no need to have plans for possible off-site damage or personal injuries, since it was not possible for serious off-site contamination to occur, and expressed the view that if such contamination were possible the plant should not be located where it is. He seemed singularly unswayed by the fire on May 11. The possibility that it might have resulted in a significant off-site release of plutonium was countered by the observation that it did not happen. The same general attitude and approach was reflected by the Dow General Manager's comments when this subject was discussed with the Operating Board members...

No Need to Alert the Community of Risk

> From the information furnished to the Investigation Board it appears that any evaluation of a particular situation and any actions initiated would depend on the on-the-spot judgment of the personnel in charge. On the occasion of the May 11 fire, the senior management officials made inquiries about possible off-site contamination after arrival at the plant, and concluded there was no necessity to establish roadblocks on access or other roads in the vicinity, or to initiate other alerts with area police or communication media.
>
> (AEC, 1969)

A key absurdity is that the Dow official responsible for the Dow Chemical Off-Site Emergency Plans did not envision any "need to have plans for possible damage or personal injuries..." The logic was astounding: "...if such contamination were possible the plant should not be located where it is."

The AEC Fire Investigation Report Findings and Conclusions

It turns out there are two versions of the AEC Fire Investigation Report. One version is unclassified known as "Report by United States Atomic Energy Commission on the May 11, 1969 Fire at the Rocky Flats Plant near Boulder, Colorado," presented in Hearings Before the Sub-Committee on Public Works of the Committee on Appropriations, House of Representatives, Ninety-First Congress Second Session, Part 4, 418 et seq., 1971. Using the legal authority provided by the Atomic Energy Act of 1954 as amended, elements of the AEC Fire Report were classified as "Restricted Data," the release of which would compromise the national security of the United States. The commentary provided to the public contrasted greatly with the classified version.

Issue: Standard For Fire Protection
Public Congress Version

> 1. Facilities such as building 776-777 require a higher standard of fire protection than previously provided.

The Classified Version said:

> C. Building 776-777, a vital weapons production facility did not meet even the minimum AEC fire safety standards. This increased the fire loss and unnecessarily exposed firemen to fire, plutonium, and nuclear criticality hazards.

Issue: Combustible Shielding
The Public Congress Version said:

> 3. Less than one percent of the total of almost 600 tons of combustible radiation shielding was consumed in the fire.

The Classified Report stated:

> F. Less than one percent of the total of almost 600 tons of Benelex and Plexiglas shielding was consumed in the fire. Only the heroic efforts of the firefighters limited this burning. Had an appreciable fraction of this shielding burned, there would have been a complete loss of Building 776-777 and it contents and a major release of plutonium to the environment.

Issue: Significant Omissions To Congress
The Classified Report also stated:

> G. The recycling of scrap, chips and skull materials has been the source of almost all the reportable plutonium fires in Building 776-777.
>
> K. The installment of Benelex and Plexiglas shielding made access through glove ports more awkward and interior cleanup more difficult. Further, this shielding substantially reduced visibility in some parts of the system. Both conditions added to the potential for excess combustible debris to accumulate inside the gloveboxes and conveyor lines.
>
> M. The fire caused no lost time injuries but one employee received a significant lung burden of plutonium.

These conclusions (G,K,M,) were omitted from the Public Report which was submitted to Congress. Their release was seen as a threat to national security.

Issue: Omissions From the AEC Fire Investigation Report Submitted To Congress Related To Management Conclusions

In the Classified Report, but not in the report submitted to Congress, the following "Management Conclusions" were stated:

A. The failure of both AEC and Dow (Rocky Flats) management to evaluate fully the consequences of certain decisions had an adverse effect on fire safety in Building 776-777.

Not Constructed To AEC Fire Safety Standards

1. Building 776-777 complex was not designed or constructed to meet AEC fire safety standards – nor does it appear that any concerted efforts over the years have been made to achieve these standards.

New Equipment and Structures Added On Ad Hoc Basis

2. Since original construction, decisions were made to add new equipment and structures on an ad hoc basis. This "Topsy-like" growth led to overcrowded conditions that made compliance with fire safety regulations even more difficult, if not impossible.

Plutonium Storage Practices Increased Risks To Workers

3. The decision to store relatively large quantities of plutonium in the conveyor lines and gloveboxes may have provided greater ease of operations and for [achieving] cost reductions, but this also contributed to relatively high radiation exposures to the workers and, in the event of an accident, placed in jeopardy a large quantity of plutonium. When relatively high radiation exposures were encountered, combustible shielding materials were installed without full fire safety evaluations. These actions moved the facility further away from AEC fire safety standards.

Policy Not Followed To Mandate Sprinklers In New Construction

4. Although a policy was established in 1967 by Albuquerque Operations Office to include sprinkler systems in all new construction,

including Building 707, no action was taken to install sprinklers in Building 776-777.

Granting Waivers Circumvented Fires Safety Evaluations

5. The decisions by AEC management in 1967 to grant certain waivers from the requirements of AEC Manual Chapter 6101 – regarding submission of preliminary construction proposals – circumvented that part of the normal process of making fire safety evaluations with respect to the additional shielding material added in Building 776-777.

B. The AEC and Dow (Rocky Flats) organizations and procedures were not adequate to ensure safe conditions in Building 776-777.

Divided and Unclear Understandings of Safety Responsibilities

1. There are divided and unclear understandings of safety responsibilities at the various levels of management at Dow (Rocky Flats). For example, construction modification to structures and operating equipment could be made without assurance of full safety evaluations. The significant modifications made in Glovebox 134-24 are a relevant case. According to the Dow (Rocky Flats) organization and safety philosophy, there were several individuals and groups who shared the responsibilities for these actions, yet there was no single overall evaluation made of the safety implications of the modifications; and none made with respect to fire safety.

2. The Dow (Rocky Flats) organization and procedures do not assure an adequate understanding of, and compliance with, the established fire safety policies at all levels of supervision. Some supervisors did not interpret the Dow (Rocky Flats) policy, that safety is a line responsibility, as including responsibilities for evaluation of fire protection of buildings and equipment furnished them; rather, they believed these responsibilities resided with "experts."

No Centralized Safety Organization

3. While Dow (Rocky Flats) does have such "experts" and they provide advice to line management when called upon, there is no one centralized safety organization with

appropriate responsibility to assure that overall safety evaluations are performed. Without detracting from obvious responsibilities of line management, a competent and centralized safety organization conducting independent inspections could have been a major influence in correcting or avoiding fire safety deficiencies that played a significant role in the fire on May 11, 1969.

4. AL did not understand the purpose of the safety analysis procedure required by AECM 6101 – Administration of the Construction Program – and as a result did not apply it properly at Rocky Flats.

5. Fire Inspections of facilities at Rocky Flats were conducted by AL, but they did not perform in-depth evaluations of the Dow (Rocky Flats) Organization and procedures for assuring fire safety.

6. No formal fire safety appraisals of AL or Dow (Rocky Flats) have been made by Headquarters' Division of Operational Safety, although there have been informal communications and contacts.

<u>AEC Management System Failed To Assure Compliance With AEC Fire Safety Policy</u>

C. A major loss to a vital weapons production facility occurred because of the failure of the AEC management system to assure compliance with AEC fire safety policy.

(Marcus Church, Freedom of Information Act Filing, 1975)

The management conclusions offer candid assessments of where the breakdowns occurred. For example, "major loss to a vital weapons production facility occurred because of the failure of the AEC management system to assure compliance with the AEC fire safety policy." Even in the Cold War context of the early years of the 1970s it is not obvious why release of this assessment to the Congress and to the American people would have been a threat to national security.

Reflections On the Catastrophic Plutonium Fires

There was no collective memory from the lessons of the 1957 fire which led either to preventing the 1969 fire or to fighting the 1969 fire itself. The lessons from 1957 were not transmitted to first responders in 1969. An unacceptable omission in transmitting lessons from 1957 to first responders in 1969 was that water had been used successfully on the 1957 fire to quell its initial ferocity. The extensive design recommendations and needed operational approaches from the AEC 1957 report were not implemented.

Perhaps it was the routineness of daily operations plus the turnover of staff and secrecy-based compartmentalization in the production area which contributed to the amnesia about the warning from M.A. Thompson that: "Admittedly, the potential for fire to start in any one box may be small, but many of our recent incidents have been events with small probability of occurring." A routine safety inspection could have detected the fact that a fire safety sensor had been shielded.

Perhaps it was the magnitude of the $70 million in damages in 1969, or the realization that production of plutonium igniters had been lost for several months, or General Giller's "nearly catastrophic" warnings to policy makers which served to alert all involved to how close the 1969 fire had come to contaminating parts of the Denver metropolitan area which led the Rocky Flats enterprise to embed fire safety more completely in the operating culture for the last twenty years of the production era.

There are some anomalies. There was an original "culture of safety" at Los Alamos during World War II. Which was not carried forward to Rocky Flats. Dow Chemical had agreed after the 1957 fire to make significant fire safety changes but did not. One promised change had to do with the precautionary principle of keeping the Building 776/777 divided into small sections to block a small fire from spreading. Putzier said the reason they did not implement that was concern that workers might get trapped. (Putzier did not concur with the reasoning of Dow management). Another change was to substitute Benelux which was more flammable in the glove boxes because it was both a better neutron shield for workers and cheaper.

The AEC had notified Dow Chemical that Benelux was combustible with a recommendation that sprinklers be installed. No drawings were submitted to the Dow Fire Protection Engineer. The Dow Project Engineer did ask the Fire Protection Engineer for his opinion. That led to an inquiry to the Rocky Flats Plant Fire Chief and to the Fire Protection Engineers at AEC Albuquerque. The answer back was that consideration "should be given to installing sprinklers in the operating area of Building 776." There is no evidence in the available record that this happened.

Two months before the 1969 fire, Rocky Flats had been visited by a representative of the AEC Headquarters Industrial Safety and Fire Protection Branch. He reported on his visit to Building 776/777 that: "This facility is one of the most important to the program, but had practically no built-in fire protection or suppression equipment." Before the U.S. Senate, General Giller said that, "It is certainly true that there is no excuse for the particular situation that was allowed to generate the fire."

The Continuation of Plutonium Fires During the Rockwell International Era

The Atomic Energy Commission had warned from the beginning to "Expect to have a fire." The reports available to the Atomic Energy Commission after the catastrophic fires of 1957 and 1969 provided a roadmap for the Department of Energy and Rockwell International to follow to avoid catastrophic fires. The record is clear that none occurred. It is not at all clear what the reasons were for this positive accomplishment.

DOE (1986a) and Rockwell International (1987) had focused in secret but separately on the record of environmental contamination and compliance with the new suite of environmental laws. Neither the FBI search warrant, the two known reports of the Special Federal Grand Jury, the focus of the Wolpe subcommittee of the U.S. Congress, nor the Department of Justice rebuttal addressed the problem of reducing the number of recurrent fires.

<u>The Insights Into Fires During the Rockwell Era From the Church Cluster of Lawsuits</u>

Howard Holme near the end of his 520-page submission to the Federal District Court on behalf of the plaintiffs in the Church cluster of lawsuits described the known circumstances of a beryllium fire which occurred during the Winter of 1978. The Holme report was supplemented as a source of information by an article in <u>The Denver Post</u> and by a letter from an official of Rockwell International to an official in the Office of the General Counsel of the Department of Energy in March of 1978.

Holme's observation about the fire on February 23, 1978 was that: "neither Rockwell nor Rocky Flats is immune from further fires and negligence..." At least one of the causes of the severity of the fire, according to Holme, was "welding in a filter plenum while the ventilation system is running" (1978, p. 471).

Building 776/777 lacked sprinklers. A letter dated March 1, 1978 from Robert E. Yoder, the Director of Health, Safety and Environment of Rockwell International to J.J. Chavez in the Office of the General Counsel of DOE reported that all 16 roughing filters in the plenums were destroyed and 6 HEPA filters evidenced signs of fire and water.

<u>Insight Into Fires During the Rockwell Era From the Cook Class Action Lawsuit</u>

The Cook class action lawsuit submitted Figure 7.2 to the Federal District Court and the jury. The factual basis for the submission was DOE (1991e) "Advisory Committee on Nuclear Facility Safety, Final Report" of November 12, 1991, p. 156.

Figure 7.2 Nuclear Facility Safety Committee Report On Fires During the Rockwell Era

This presentation is important for two reasons. The first is that plutonium fabrication is dangerous to those doing it. The second is that the Rockwell era did not experience a catastrophic fire. Given the dangerous nature of plutonium this was an important accomplishment. The reason why the frequency of fires increased during the last years of the Rockwell stewardship is not yet known. Cook Class Action Exhibit.

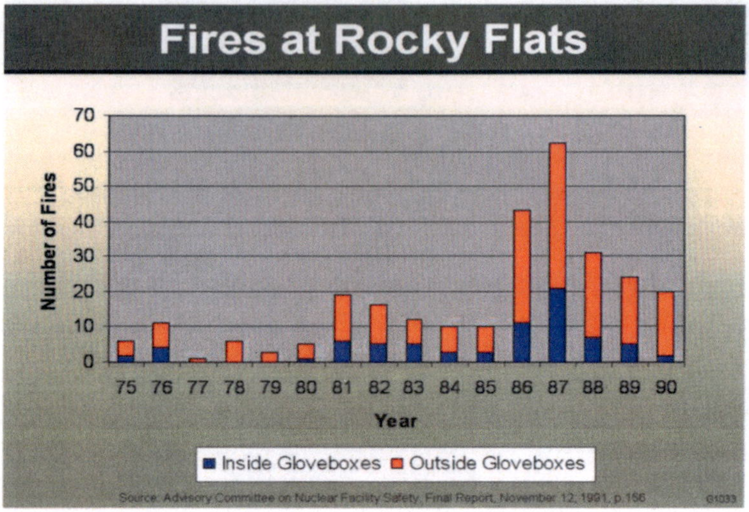

The narrative which is available currently about occurrences of fires during the Rockwell era does not explain why the number of fires increased during the closing years of their stewardship.

Reflections Across the Dow Chemical and Rockwell International Eras On Waste Management Choices and the Causes of Fires and Explosions

Howard Holme reflected on what he had learned across the Dow Chemical era about why crises kept occurring. This is what he said toward the end of his 520-page report that is relevant to both waste fiascos chronicled in Chapters Five and Six and the fire and explosions documented in Chapter Seven:

> As was the case in the areas of fire protection and waste management, it appears that earlier practices at the plant were in many respects better than later ones. Some practices which were done in the earlier years were cut back and monitoring was characterized by increasing laxity until a major incident, the discovery of contamination by outside agencies, or a request by AEC headquarters officials caused an expansion of monitoring.
>
> (Holme, 1978, p. 478)

The observation that increasing laxity characterized the Dow Chemical era over the years was not expected. The role of crises in spurring short-term action had been a recurrent theme in the available Dow Chemical Archives.

The Unthinkable Across the Production Era: Missing and Unaccounted For Plutonium

The Department of Energy so resisted disclosure of the amount of missing and unaccounted for plutonium that it was found in Contempt of Court by a Federal District Court Judge. Figure 7.3 reveals the submission to the Court once the information was finally provided.

Figure 7.3. Missing and Unaccounted For Plutonium At Rocky Flats As Finally Revealed By Department of Energy Documents After It Had Been Found In Contempt of Court By a Federal District Court Judge
This meets the definition of the unthinkable. Cook Class Action Exhibit

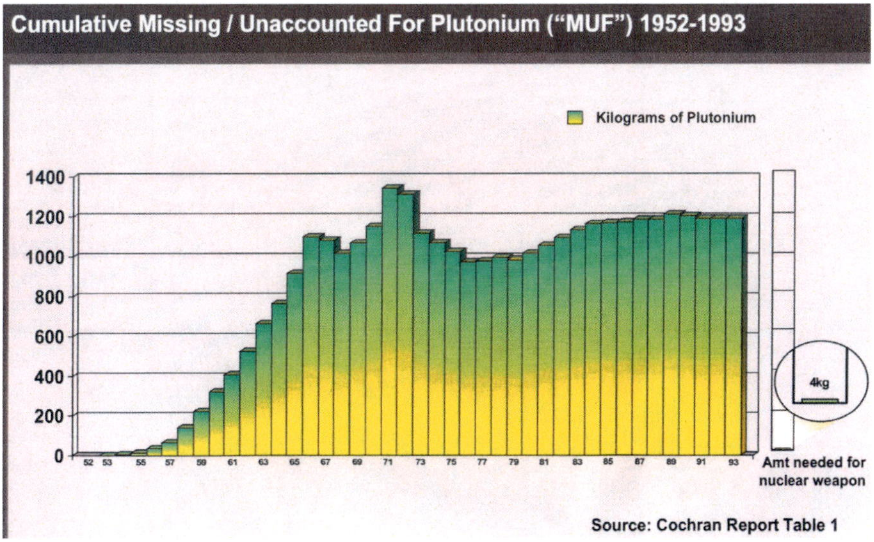

This revelation is unthinkable. There is over 2,600 pounds of missing and unaccounted for plutonium from Rocky Flats. This is especially noteworthy because of the years of assurances by contractors that plutonium had not been released on-site or off-site during routine operations, or mixed in smoke during the fires of 1957 and 1969 and escaped into the atmosphere.

EPOCH TWO

DISCOVERY OF ENVIRONMENTAL CRIME, AND THE TRANSFORMATION OF ROCKY FLATS BY THE CRIMINAL JUSTICE SYSTEM AND BY THE "SEPARATE BUT EQUAL" CONSTITUTIONAL "CHECKS AND BALANCES" BETWEEN THE EXECUTIVE, THE LEGISLATIVE AND THE JUDICIAL BRANCHES OF THE U.S. GOVERNMENT: 1986-1994

Introduction To Part III: The Environmental Crime Era, the Clash Within and Between the "Separate But Equal' Branches of U.S. Government and the Findings of the Special Federal Grand Jury

Key Elements of the Chapter Eight Narrative

- Not constructing the first Solar Evaporation Pond in the 1950s with an impervious liner;
  - In the 1960s and 1970s the next four Solar Evaporation Ponds were constructed with only semi-permeable liners; and
- Failure to anticipate the sheer volume of contaminated water which would be a byproduct of the fundamental nature of the industrial enterprise.
  - Disposing of contaminated water after 1979 by Spray Irrigation off-site year around in all weather conditions.
  - The Atomic Energy Commission knew in the 1960s that the Solar Evaporation Ponds leaked into the groundwater.
  - The Department of Energy knew in the 1980s that the monitoring wells were deficient, and that the Solar Evaporation Ponds leaked into the groundwater.
- The EPA National Enforcement Investigation Center in 1985 knew that Rocky Flats was not in compliance with RCRA groundwater monitoring requirements when the top Rockwell and DOE officials certified under penalty of law that it was.

Key Elements of the Chapter Nine Narrative

- An FBI Search Warrant Affidavit in June of 1989 said that the illegal storage of hazardous wastes without a permit was a felony under RCRA.
- The first FBI raid on another facility of the U.S. government began at Rocky Flats on June 6, 1989 and lasted 18 days seizing over 1 million pages of records.
- Secretary of Energy James Watkins dispatched a "Tiger Team" in August of 1989 to Rocky Flats whose scathing report said that information provided to EPA was false.
- Secretary Watkins established that there should be equal weight of the production mission with environmental protection, health, and safety. DOE awarded Rockwell with only 40 percent of the possible performance bonus award in 1989.
- The Rockwell International era ended on December 31, 1989.
- Rocky Flats closed forever to production of plutonium igniters.

Key Elements of the Chapter Ten Narrative

- On June 22, 1989 the Department of Justice asked Judge Sherman G. Finesilver to empanel a Special Federal Grand Jury which received over 2000 exhibits, heard over 200 witnesses, and identified individuals it wanted indicted for environmental crime from both DOE and Rockwell.
- The Special Federal Grand Jury was disbanded and discouraged from writing a report.
- The U.S. government chose to allow Rockwell to plead guilty to ten felony and misdemeanor violations of RCRA and the Clean Water Act, pay a fine of $18.5 million, not have any Rockwell people indicted, and keep its eligibility for federal contracts.
- Judge Finesilver released a heavily redacted version of the Report of the Special Federal Grand Jury because the original did not meet "necessary standards." A person or persons unknown leaked the first Grand Jury Report to Westword. A principal conclusion of the Grand Jurors was that Rocky Flats was "an on-going criminal enterprise."
- The Wolpe subcommittee of the U.S. House of Representatives began months of hearings and concluded that the Department of Justice had "bargained away" the right of the American people "to know the truth..."
- The Department of Justice spent two years crafting a rebuttal to the U.S. Congress.
- Attorney General Janet Reno concluded that the case had been professionally handled.

Key Elements of the Chapter Eleven Narrative

- "The Second Citizen Special Federal Grand Jury Report" was discovered in 2013 in the archives of the Wolpe Congressional Subcommittee. It had not been identified because it lacked a title page.
- The Second Citizen Special Grand Jury Report is different from the one leaked to Westword in that it cites specific exhibits relevant to environmental crime and used the notation [Indict] to convey where that was appropriate in their judgment.
- Among the central conclusions in the Second Report are:
  - operators have placed themselves "above the law" and hid behind "the cloak of 'national security' ",
  - operators have committed the "most serious crime that can ever be committed in a democracy...breached the public trust."

- Rockwell manipulated DOE to endorse "Rockwell's illegal conduct in pursuit of exorbitant bonus and contract fee awards."
- The FBI/EPA raid found "compelling evidence" in the opinion of the Grand Jurors that:
    - operators engaged in a pattern of behavior "which had converted the DOE and its contractors into a racketeering influenced corrupt organization…",
    - operators embodied a "culture of greed" and "criminal misconduct" which insisted that "the end of corporate bonuses always justified the illegal means to achieve those bonuses."
- Federal, Colorado and local regulators were found to have demonstrated "an unblemished record of ineptitude, failure and bureaucratic incompetence,"
- Employees have defended themselves with four core arguments:
    - they are immune from prosecution because of the sovereign immunity of the U.S. government,
    - since they did not direct illegal activities, they are not accountable,
    - incompetent and negligent management is not a crime, and
    - they were ignorant of the law.
- The position of the Grand Jurors was that there is no excuse for "directly or indirectly permitting the violation of the law by others."

Epoch Two is presented in four chapters. Chapters Eight, Nine, Ten, and Eleven are distinctive between 1986 and 1994 for the significance of what became known about the extent of contamination at Rocky Flats, for the transformation of Rocky Flats by the criminal justice system, and for the spirited public clash between "separate but equal" branches of U.S. constitutional government. The spirited public clash between the Legislative branch and the Executive branch Departments of Justice and Energy were highly contentious over the right of the American people to know the truth about Rocky Flats, and whether or not the prosecutors in the U.S. Attorney's Office in Denver and the Department of Justice in Washington, D.C. had used appropriate discretion in obtaining justice.

Chapter Eight begins unexpectedly with evidence that the genesis of environmental crime was not the disposal of radioactive or chemical waste but rather the disposal of contaminated water. Key evidence in support of this conclusion was assembled by Rockwell International itself in 1987. Rockwell International (1987) indicated in secret that top management of both the Department of Energy and Rockwell International at Rocky Flats had certified under penalty of law that Rocky Flats was fully in compliance with relevant provisions of the Resource Conservation and Recovery Act.

A startling revelation was that at the time the "fully in compliance" affirmation was made the National Enforcement Investigation Center of the Environmental Protection Agency had determined that the facility was not in compliance with RCRA groundwater monitoring requirements. While not a violation of the then controlling legal regimes of the Atomic Energy Acts of 1946 and 1954, the Atomic Energy Commission had known since 1960 that the first Solar Evaporation Pond was leaking into the groundwater. The reason for the leaking from the first Solar Evaporation Pond was its lack of an impervious liner. When the next four Solar Evaporation Ponds were constructed, the liners were only semi-impervious.

The Solar Evaporation Ponds were used to receive in 1962 unmonitored beryllium oxide and enriched uranium (73 shipments totaling 179,700 gallons) from the Coors Porcelain Company. Elevated tritium levels were also found in the Solar Evaporation Ponds with several different explanations offered for why this happened.

Chapter Nine recounts the unbelievable commission of environmental crime not merely its genesis. The criminal investigation began unexpectedly in secret in the Department of Energy itself in 1986 almost in parallel with the Rockwell International assessments in 1987 also in secret. The candid DOE assessment had in common with that by Rockwell International that they were not intended to be publically disclosed. The DOE document was prepared by an attorney within DOE for the purpose of a high-level briefing for department officials. It was released routinely to the U.S. Congress and then to an investigator for the Environmental Protection Agency. He in turn asked a Special Agent of the FBI based in Denver, Colorado to join him in what became a criminal investigation under the Resource Conservation Recovery Act and the Clean Water Act.

Special Agent Jon Lipsky prepared a Search Warrant Affidavit for a Magistrate Judge in June of 1989. The Affidavit said in part that Rockwell had been storing hazardous waste without a permit which was "an apparent RCRA felony, involving the illegal storage of hazardous wastes without a permit." The first ever FBI raid on a facility of the U.S. government followed on June 6, 1989 and lasted 18 days.

One outcome of the FBI raid was the start of a chain of events leading to the departure from Rocky Flats of Rockwell International on December 31, 1989. The dilemma for Rockwell was summed up by S.I. Iacobellis when he wrote to a DOE top manager that they were troubled by the prospect that Rockwell and Rockwell people "may be subject to legal liability for actions taken in good faith..."

There are three rationales offered for why the Rockwell era ended at Rocky Flats the way it did. One rationale was that of S.I. Iacobellis that vulnerability to legal liability was a major concern: "We are reluctant to continue to expose our people and our company to potential criminal liability for actions taken under DOE's direction."

Another rationale was offered by FBI Special Agent Lipsky: Secretary of Energy James Watkins in 1989 was intent on honoring the pledge in his confirmation hearings to change the DOE culture. Implementation of this pledge was the reduction in the annual bonus to Rockwell of only 40 percent of the possible award over issues of health, environment and safety.

A third (possible) rationale was that in this time frame Rockwell was subsequently found to have engaged in violations of federal laws and regulations. Administrative Law Judge David W. James, Jr. determined in a finding under the Major Fraud Act of 1988 that Rockwell had been charged with or implicated in "criminal, fraudulent or improper acts." Secretary of Energy Watkins had dispatched "Tiger Teams" in August of 1989 to Rocky Flats whose reports were scathing.

The broader point is that the plutonium igniters fabricated by Rockwell International were a mainstay of the national security of the United States in a world where nuclear weapons were the weapon of choice to defend America's way of life.

The legacy which Rockwell inherited in 1975 of a 334-acre high security production site for fabricating plutonium igniters was a mismatch with the requirements of the industrial enterprise. When the site constraints were coupled with the new suite of environmental laws, it made the commission of environmental crime inevitable.

Chapter Nine and Ten are notable for the transformation of Rocky Flats by the criminal justice system of the United States, and ultimately the clash between constitutionally "separate but equal" branches of the government. The transformation of Rocky Flats by the criminal justice system began with a request on June 22, 1989 by the Department of Justice to Judge Sherman Finesilver to empanel a Special Federal Grand Jury. In December of 1989 at its swearing in ceremony, several elements of the instructions stand out: "You are free to exercise your own judgment without fear or favor...you will diligently inquire into and true presentment make, of all public offenses against the United States...of which you shall have or can obtain legal evidence."

After hearing over 200 witnesses and receiving over 2,000 exhibits, the jury was disbanded. Judge Finesilver released a heavily redacted version of their report. Subsequently, a person or persons unknown "leaked" this first Grand Jury report to Westword magazine. Judge Finesilver accepted a plea agreement by Rockwell calling for a fine of $18.5 million with no criminal indictments of officials or employees of Rockwell.

After the controversial settlement between the Department of Justice and Rockwell International, a subcommittee of the House of Representatives began months of hearings. When the Departments of Justice and Energy were judged not to be cooperating sufficiently with the investigation, the subcommittee exercised its power of subpoena. The judgments of the Wolpe subcommittee were harsh: the Department of Justice had "bargained away" the right of American people to "know the truth…" The Department of Justice spent two years concluding that: "No basis existed for the sweeping criticisms by the Wolpe report." Attorney General Reno concluded in 1994 that the DOJ report was accurate in its conclusion that the criminal investigation had been "properly and professionally handled by the attorneys who conducted the lengthy investigation."

Chapter Eleven concludes the presentation of the transformation of Rocky Flats by the criminal justice system. In 2013 a second and more detailed version of the report of the Special Federal Grand Jury was discovered in the archives of the Wolpe subcommittee. It had lacked a cover page and its significance not recognized at the time. The Second Citizen Special Federal Grand Jury Report, as it is identified in this study, is notable for the specific exhibits cited in support of its notations {Indict}. It was especially critical of the Office of the U.S. Attorney in Denver for offering immunity to individuals in return for their testimony—some of whom in their judgment should have been indicted.

In 1993 an anonymous Grand Juror summed up the transformation of Rocky Flats by the criminal justice system by focusing on what was perceived as a violation of the public trust "as the greatest crime of all."

Chapter Eight
The Genesis of Environmental Crime Was Failed Management of Contaminated Water, Not Plutonium Or Other Radioactive Substances

"On November 8, 1985, A. E. Whiteman and J.E. Dorr certified to the U.S. Environmental Protection Agency (EPA) that the solar ponds... were in compliance with all groundwater and financial requirements ..."

Rockwell International, 1987, Internal Report Briefing Memo

"I certify under penalty of law that I have personally examined and am familiar with the information submitted in this document..."

Rockwell International, 1987, Internal Report Briefing Memo

"The review [by the EPA National Enforcement Investigation Center] indicated that the facility was not in compliance with the RCRA groundwater monitoring requirements when the certification was made."

Rockwell International, 1987, Internal Report Briefing Memo

"DOE has known since at least 1960 that the ponds leaked..."

Rockwell International, 1987, Internal Report Briefing Memo

"In 1962, 73 shipments of liquid wastes were received from CPC [Coors Porcelain Company] for a total of 179,700 gallons. Liquid wastes containing beryllium oxide and enriched uranium...were transported to the Rocky Flats solar evacuation ponds..."

ChemRisk, 1992, p. 204

"Rockwell used the Sewage Treatment Plant to treat hazardous waste from the Chromic Acid Spill...The discharge into the "B" series ponds was therefore partially untreated sewage, and sewage along with the toxic chromic acid was spray irrigated [off the Rocky Flats site].

Second Special Federal Grand Jury Report, 1992 with Notation [Indict]

Three flawed decisions and a fateful omission led directly to the genesis of environmental crime related to contaminated water. The first flawed decision by Dow Chemical was not to construct an impervious liner for the first Solar Evaporation Pond in the late 1950s. When subsequently the next four Solar Evaporation Ponds were created in the 1960s and 1970s, the second flawed decision by Dow Chemical was to construct those next four ponds with only semi-permeable liners consisting of timbers covered primarily with asphalt and/or cement.

The third flawed decision in 1979, this one by Rockwell International, was to utilize spray irrigation year around in all weather conditions for disposing of contaminated water. The error of this decision was compounded by saturating the soil with water over known locations of buried radionuclides and toxic chemicals. The East Spray Fields flow of water was over the unlined East Trenches before the contaminated water migrated into the Woman Creek watershed and off of the Rocky Flats site into the surrounding community.

An omission with the known topography of the site was the failure to anticipate the sheer volume of contaminated water which was a byproduct of the fundamental nature of the industrial enterprise. It will be recalled from the siting criteria that the industrial enterprise would require at least 1 million gallons of water a day. The volume of contaminated water requiring disposal reached the staggering amount of between 70 million and 80 million gallons a year by the middle of the 1980s. This omission led directly to the commission of environmental crime.

Genesis of Environmental Crime

The Atomic Energy Commission in the 1960s that the Solar Evaporation Ponds were leaking into the groundwater. The Department of Energy knew in the 1980s that the monitoring wells either were deficient or not correctly placed, and that the required assessment monitoring was not being conducted in order to comply with the requirements of RCRA.

In parallel with the secret 1986 "patently 'illegal' " Mary L. Walker Briefing Memo (DOE, 1986a) prepared by DOE (Chapter Nine), Rockwell International had prepared a secret briefing document in 1987 which included their confirmation of the following:

> On November 8, 1985, A.E. Whiteman and J.E. Dorr certified to the U.S. Environmental Protection Agency (EPA) that the solar ponds at the U.S. Department of Energy (DOE) - Rocky Flats Plant near Golden, Colorado were in compliance with all applicable groundwater monitoring and financial requirements stated or referred to in the Hazardous and Solid Waste Disposal Act, as amended. Mr. Whiteman is the Acting Area Manager of the Rocky Flats Plant and certified on behalf of the owner, the DOE. Mr. Dorr is the vice president and general manager of Rockwell International and certified on behalf of the operator.
>
> and
>
> The review [by the EPA National Enforcement Investigation Center] indicated that the facility was not in compliance with the RCRA groundwater monitoring requirements when the certification was made.

The certification signed by A.E. Whiteman, J.E. Dorr, and Project Director Hatch from a consultant contained the following text:

> I certify under penalty of law that I have personally examined and am familiar with the information submitted in this document and all documents and that, based on my inquiry of those individuals

> immediately responsible for obtaining the information, I believe that the information is true, accurate, and complete. I am aware that there are significant penalties for submitting false information, including the possibility of fine and imprisonment.
>
> (Rockwell International, 1987)

A principal outcome of the false statements was that the production mission continued without interruption.

Solar Evaporation Ponds (Late 1950s-1989)

According to the Government Accountability Office (GAO, 1991):

> One of the highest priority environmental remediation efforts at the plant has been cleaning up the five solar evacuation ponds. These ponds are surface evacuations for storing and evaporating various liquid wastes. Some of these ponds have leaked waste into the ground. On-site monitoring has shown that the ground water has been contaminated with radioactive material, nitrates and heavy metals.
>
> (GAO, 1991, p. 3)

At the time of the original construction of the Solar Evaporation Ponds, an impermeable lining was not required. Instead, they were lined with clay and other semi-permeable materials such as asphalt planking. Putzier (1982) reported that there were contaminated groundwater plumes consisting of nitrates; a variety of radionuclides including strontium and plutonium; metals such as chromium, arsenic, beryllium, cadmium, nickel; and volatile organic compounds such as phenols, byphenols, and phthalates. During 2002 and 2003, DOE initiated remediation of the Solar Evacuation Ponds including removing all pond lining and contaminated soils. ATSDR (2005) concluded that the groundwater plume was still in existence under the Solar Evaporation Ponds.

According to DOE (2002a), the purpose of the Solar Evaporation Ponds was:

> ...to store and evaporate low-level radioactive process waste containing high concentrations of nitrates and treated acidic wastes. The sludge and sediments resulting from the process were periodically disposed of at the Nevada Test Site.
>
> (DOE, 2002a, p. 1)

The remediation process to close the Solar Evacuation Ponds permanently required initially that "approximately eight million gallons of excess liquid in the ponds must be removed" (DOE, 2002a, p. 1). Evaporators were used to accomplish this with vapors sent to a condenser. The remaining sludge was converted into Pondcrete and sent to the Nevada Test Site for burial.

Tracing the Genesis of Environmental Crime

The background of the genesis of environmental crime is traced by Rockwell International (1987) back to the known facts concerning Solar Evaporation Pond construction. In the 1950s, Pond 207A was constructed with planking and asphalt. The bottom buckled which caused leaking. In the 1960's the three B Ponds were constructed with asphalt and cement over planking but "on hot days the asphalt cement would slide, crack, and leak." In 1970 Pond 207C was constructed over an original clay-lined pond.

Installation of a Groundwater Interceptor System

As early as 1972 it was known that samples collected from the 60, 66, and 71 series wells indicated that the Solar Evaporation Ponds had been leaking. In 1982 samples of intercepted water revealed that trichloroethylene had been found at 3 parts per billion, and chloroform had been found at 2 parts per billion.

Solar Evaporation Pond Disposal of RCRA Hazardous Waste was captured by EPA Hazardous Waste Code (Rockwell International, 1987):

EPA Hazardous Waste Code	Solar Evaporation Pond Waste Description
D001	Ignitable (nitrate salts)
D002	Corrosive (acid waters)
D003	Reactive (reactive chemicals & metals)
D004	EP toxic (barium chloride)
D007	Spent cyanide plating bath solutions from electroplating operations

Rockwell International (1987) identified RCRA Groundwater Monitoring program deficiencies:

1. A monitoring program was conveyed to DOE on November 13, 1981 and by DOE to EPA on November 19, 1981.
2. The monitoring program was for detection monitoring despite the fact the "Rockwell knew impoundments were leaking..." and that more rigorous assessment monitoring was required.

(Rockwell International, 1987)

Conclusions Made By Rockwell International (1987) Regarding RCRA Compliance In November 1985

Rockwell International (1987) went into extensive detail about compliance deficiencies with RCRA. The required monitoring under RCRA should have begun in 1983. The assessment monitoring program required by the Compliance Agreement of July of 1986 between EPA and the Colorado Department of Health had not been implemented.

The major program deficiency at the time of certification was that detection monitoring (See 40 CFR Part 265.92) was being conducted instead of assessment monitoring (See 40 CFR Part 265.93). Assessment monitoring is conducted to determine:

1. whether hazardous waste or constituents have entered the groundwater,
2. the rate and extent of migration of hazardous waste or constituents in groundwater, and
3. the concentrations of hazardous waste or constituents in groundwater.

DOE has known since at least 1960 that the ponds leaked, yet they initiated a detection monitoring system under RCRA. RCRA regulations [40 CFR Part 265.90(d)] allow an owner/operator to decide whether to implement initially a detection or an assessment monitoring program...The procedures include confirmation testing and notification of the EPA Region VIII Regional Administrator. Subsequently, an assessment monitoring program should have been initiated. These steps were simply not done. As a result, an assessment monitoring program was required as part of the Compliance Agreement signed by EPA, DOE and the Colorado Department of Health (CDH) in July 1986.

Furthermore, the first time that DOE notified EPA and CDH of leakage from the impoundments was in the initial Part B which was submitted in November of 1985.

A further statement included in Rockwell International (1987) was that the data from the monitoring which was done "were judged to have limited validity because of poor wells, sampling equipment and monitoring equipment." DOE itself concluded that "...it is doubtful that statistical analysis of the RCRA groundwater data will provide any meaningful information." Nonetheless, DOE had used that data in the November 1985 Part B application.

RCRA Groundwater Monitoring Wells Deficiencies: History of Deficiency Determination

This important section from Rockwell International (1987) documents the historical record of prior determinations that the Solar Evaporation Ponds were not in RCRA compliance:

- The report on Nonradioactive Groundwater Data Summary 1975-1985, dated 9/84 by N.D. Hoffman [RF-14] states on page 13:

  "many of the wells are over 10 years old: it may be necessary to drill new wells following current improved drilling practices and recent regulations."
  This statement applies to all wells not just RCRA wells.

- The memo on History and Status for RCRA Ground Water Monitoring Program, dated 11/9/84 by T.C. Greengard [RF-15] summarizes samples collected for the RCRA program:

  states that wells were contaminated by salt poured into them in April 1983.
  The final section on Current Status indicates that:

a comprehensive review of the RCRA groundwater monitoring program was being conducted and that a consultant was going to assist in the project beginning in Dec. 1984.
The section concludes that major revisions in the existing program were expected.

- The memo on RCRA Groundwater Monitoring Data, dated 2/7/85 by G.L. Potter and T.C. Greengard [RF-16] reiterates much of material presented in
RF-15 by T.C. Greengard:

states that RFP evaluated groundwater monitoring program in 1984. Deficiencies and desired improvements to be examined in more detail by a contractor beginning in about March 1985 include:

1. Well placement and construction,
2. Sampling methods...

(Rockwell International, 1987)

Rockwell International (1987) Conclusions About Compliance of the Monitoring Well Network In November 1985

This section conveys Rockwell International's own internal conclusions about RCRA compliance as of November of 1985:

> The monitoring well network did not comply with RCRA regulations when the certification was made because:
> 1. Construction details were not and are not available for determining whether criteria stated in 40 CFR part 265.919(c) were met.
> 2. The wells were not close enough to the ponds to satisfy the requirements of 40 CFR part 265.91(a) and (b)
> 3. The wells were contaminated by rock salt in April 1983: whether the wells had been adequately rehabilitated was not determined.
>
> (Rockwell International, 1987)

Certification of RCRA Compliance of the Solar Evaporation Ponds

Rockwell International's secret (1987) conclusions about whether the Solar Evaporation Ponds were in compliance with RCRA directly parallel the previously secret DOE (1986a) Mary L. Walker Briefing Memo which had concluded that some of the operations at Rocky Flats were "patently 'illegal.' "

> The facility was inspected in April 1987 by the EPA Hazardous Waste Groundwater Task Force. The inspection was coordinated by personnel from the EPA National Enforcement Investigations Center (NEIC) and the objective was to determine if the facility was in compliance with groundwater monitoring requirements of RCRA regulations.
>
> An initial review of information collected by the Task Force suggested that the facility was not in compliance with the applicable groundwater monitoring requirements when the

certification was made. At the request of the Director of NEIC, the information obtained during the Task Force inspection was thoroughly reviewed to determine if the facility was in compliance with the RCRA groundwater monitoring requirements in November 1985.

The review revealed that the facility was not in compliance with the RCRA groundwater monitoring program and the wells composing the monitoring network were grossly deficient. The deficiencies were noted, in writing, by DOE, Rockwell and the subcontractor personnel both before and after the certification was made....

(Rockwell International, 1987)

The "Zero Discharge" Policy of 1973

Prior to 1973 during the Dow Chemical era, Rocky Flats would discharge process waste into off-site surface waters. It is not known if the change in policy to "zero discharge" was related to the public becoming aware of off-site plutonium contamination in 1970, or if its origins were in the revelation in 1973 that tritium had migrated off-site into Walnut Creek and Great Western Reservoir.

> Prior to establishment in 1973 of the policy that Rocky Flats would attempt to discharge no process waste to off-site surface waters (a "zero discharge" policy), the aqueous wastes from this process [creation of a slurry in Building 774 after solids containing radionuclides were removed] went to either the solar evaporation ponds or to the "B" series of holding ponds depending upon the concentrations of radioactivity present. The water in the "B" ponds went on to Great Western Reservoir.
>
> (ChemRisk, 1992, p. 81)

There were several releases of waterborne radioactivity to the Solar Evaporation Ponds from spills at the Critical Mass Laboratory of uranyl nitrate solution [a water soluble uranium salt made from depleted uranium from which radioactive isotopes have been essentially removed]. It was the wastewater from mopping up the spills which was sent to the Solar Evaporation Ponds. The uranium content had to be less than one gram per liter (ChemRisk, 1992, p. 93).

Waste Dumped Into the Solar Evaporation Ponds Which Had Been Received From Off-Site Sources: Coors Porcelain Company

There are two references in ChemRisk (1992) to disposal of contaminated wastes from the Coors Porcelain Company which were dumped into the Solar Evaporation Ponds. The Coors Porcelain Company was one of a number of local off-site institutions and government agencies which were allowed to send contaminated wastes to Rocky Flats. Among others granted such permission were the U.S. Bureau of Reclamation, the Colorado School of Mines, Martin Aircraft Company, The Rocky Mountain Arsenal, and the Veterans Administration Hospital.

The Coors Porcelain Company (CPC) had entered into a contract with the U.S. government in September of 1960 to manufacture 756 beryllium and beryllium-uranium fuel elements. The fuel elements were intended for use in a low-altitude supersonic missile system powered by a nuclear propelled air breathing "ramjet" engine. The solid wastes from CPC were for transshipment to federally approved radioactive waste burial grounds. According to ChemRisk (1992):

> In 1962, 73 shipments of liquid wastes were received from CPC, for a total of 179,700 gallons. Liquid wastes containing beryllium oxide and enriched uranium from the CPC fuel element fabrication process were transported to the Rocky Flats solar evaporation ponds in a 2,500-gallon tanker truck.
>
> (ChemRisk, 1992, p. 204)

ChemRisk researchers reported that in order for uranium waste to be eligible for disposal in the solar ponds, the uranium concentration "was required to be less than 20 parts per million." ChemRisk cited a CPC final operations report which concluded that the amount of uranium deposited in the Solar Evaporation Ponds was 962 grams along with at least 631.4 kilograms of beryllium, 1.0 kilogram of yttrium [a rare earth metal] and .06 kilogram of zirconium [a metal used to house fuel elements which is very resistant to corrosion].

The final CPC report indicated that while they thought that 962 grams of uranium had been sent to the Solar Evaporation Ponds, 5, 276 grams "had not been located," and 3,016 grams were "unaccounted" for in their process of material control. ChemRisk (1992, p. 204) indicated that Rocky Flats during the Dow Chemical era had not monitored what was in the CPC wastes which were discharged into the Solar Evaporation Ponds. In reporting that the CPC wastes were "reportedly removed" in 1970 and 1971, there was no mention that the Solar Evaporation Ponds were known to have leaked into the groundwater.

Tritium Release Into the Solar Evaporation Ponds

ChemRisk (1992) researchers were given access to a report by the Atomic Energy Commission (AEC, 1974) and quoted it as follows:

> There are five known sources of tritium effluent release at Rocky Flats. They are Building 779, Building 561, Building 771, which released tritium in the 1973 incident...Building 774 where tritium contaminated water is evaporated, and the four solar evaporation ponds adjacent to Building 779. The solar ponds are the source of water fed to the Building 774 evaporator.
>
> (AEC, 1974, p. 24 quoted in ChemRisk, 1992, p. 242).

The 1973 Tritium Release

Another explanation is offered for the elevated tritium levels which were found in the Solar Evaporation Ponds. This second explanation is from the AEC (1974) official tritium release report. According to the ChemRisk report (1992, p. 246) quoting AEC (1974, pp. 20-27), there was a complex process in Building 779A which involved converting plutonium scrap from its plutonium metal form into plutonium hydride and then into plutonium oxide. The resulting plutonium oxide was then reconverted into plutonium metal in Building 771. Tritium was vented in this process:

> It appears likely that curie quantities of tritium oxide were transferred to Building 771 with the plutonium oxide, thus providing an explanation for the elevated tritium levels in the process waste streams of Building 771. Process wastes from Building 771 subsequently go to Building 774 for further waste treatment, or to the sanitary sewers or the solar evaporation ponds, providing a plausible explanation for the elevated tritium levels found in Building 774 process wastes and in the solar evaporation ponds.
>
> (AEC, 1974, pp. 20-27 quoted in ChemRisk, 1992, p. 246)

Five Different Sites For Environmental Crime By Spray Irrigation

Rockwell management used at least five different sprayfields and had two more potential sites being analyzed for suitability as a sprayfield when production ceased.

ChemRisk concluded based on DOE (1991c) that the initial three sprayfields were:

> The South Area sprayfield was used first, until runoff was found to be draining into North Walnut Creek. Use of that field was discontinued, and use of the North Area Spray Field was also found to flow into North Walnut Creek. Spraying was then moved to the Pond Area Spray Field and drainage flowed back into the existing landfill pond.
>
> (ChemRisk, 1992, p. 189)

The fourth spray field at Rocky Flats was the West Spray Field which operated between 1982 and 1985:

> ...the West Spray Field was spray irrigated with water from solar evaporation ponds that contained elevated levels of nitrates and other wastes. The practice may have contaminated the ground water and the water in the soil lying just above the ground water.
>
> (ChemRisk, 1992, p. 196)

The DOE (1991d) study is referenced in relationship to contaminated groundwater.

The fifth operational spray field identified by ChemRisk (1992) was the East Spray Fields which were notorious for a dubious distinction. The dubious distinction was spraying contaminated water over the buried toxic contamination in the East Trenches. This practice was memorialized in a marked photograph by Rockwell International which is contained in the archives of the Congressional Wolpe subcommittee.

Figure 8.1 Rockwell International's Submission To the Environmental Protection Agency In the Process of Obtaining the Renewal of Their National Pollutant Discharge Elimination System Permit

This was part of the submission to the EPA circa 1989 to obtain a renewal of their NPDES Permit. This image was provided to EPA to demarcate the no-spray areas at Rocky Flats. Spray irrigation over the marked areas (East Trenches, Mound Area etc.) could possibly recharge the waste burial sites and contribute to contaminant migration. Rockwell International ignored their own caution and the constraint on their renewed NPDES Permit and spray irrigated over the marked areas. This is one of the environmental crimes they subsequently pled guilty to as part of their plea bargain in 1992.

Federal Civil Action 89-M-1154 USA ex rel James S. Stone Versus Rockwell International

In Judge Matsch's court Jack Kolar, an attorney for the U.S. government, focused his opening statement on a series of major issues which had emerged in the USA v. Rockwell International plea agreement of 1992 related to the mishandling of waste:

- Pondcrete - sludge from Solar Evaporation Ponds was supposed to harden into 1,500-pound blocks which could then be shipped off-site.

- Saltcrete - nitrate crystalline wastes from a plutonium recovery process were supposed to be encapsulated into hardened concrete for shipment off-site.

- Radiotoxic water from the Sewage Treatment Plant was spray irrigated on a year round basis on top of East Trenches land east of the 903 Pad. The down gradient drainage was to Woman Creek and then off-site toward Standley Lake.

The essence of the presentation on spray irrigation was that the contaminated water was clearly not evaporating or being absorbed by the vegetation or the soil. It was clearly visible that water known to be radiotoxic was running off into waters leading off-site. Key testimony was provided by William Rask, a DOE Operations Branch Chief at Rocky Flats.

Rockwell Violated RCRA and the Terms of Their NPDES Permit By Spray Irrigating Chromic Acid and Raw Sewage Onto the South Sprayfields

The Chromic Acid spill occurred on February 22, 1989 when a workman inadvertently left a hose running which was the cause of the spill. The Grand Jury singled out six actions taken by Rockwell International for special criticism related to the Chromic Acid Spill:

1. Rockwell used the Sewage Treatment Plant to treat hazardous waste from the Chromic Acid spill. There was not a permit to treat chromic acid as it is a RCRA listed hazardous waste.
2. The chromic acid destroyed the microbiological bacteria in the Sewage Treatment Plant, and thereby its capacity to treat sewage. The discharge into the "B" series ponds was therefore partially untreated sewage, and sewage along with the toxic chromic acid was spray irrigated.
3. Rockwell International spray irrigated the chromic acid onto the South Spray Field to avoid publicity. The South Spray Field was frozen at the time and covered with snow and ice.
4. Rockwell International acted too slowly to identify the unknown liquid that contained hexavalent chromium, a known carcinogen.
5. Rockwell International failed to tell DOE or Colorado regulatory agencies in a timely way that the chromic acid spill had occurred, or to tell the surrounding communities, including that spray irrigation of untreated sewage had occurred.
6. Rockwell International violated both RCRA and the Clean Water Act. The level of chromium exceeded by five times the permitted level in drinking water.

The Chromic Acid spill was considered by the Grand Jury to be an environmental crime. It is considered here as another illustration of the linkage of effluent from the Sewage Treatment Plant to spray irrigation itself, and of spray irrigation to environmental crime (see Chapter Eleven).

Spray Irrigation During the Rockwell International Era

Rockwell International (1987) reported that water usage and spray irrigation volumes were both increasing at Rocky Flats. The spray irrigation practice had begun in 1979. Among the problems with spray irrigation were that the B-5 Pond had become destabilized in the early

1980s. This led to costly dam repairs and was the source of the last two NPDES permit violations by the EPA. It was known to Rockwell that the spray site as of 1987, the East Trenches, was proximate to RCRA sites. Further, the East Spray Fields had excessive runoff into Pond C-2 and Standley Lake, the raw water supply for Westminster/Thornton/Northglenn.

In the Rockwell International (1987) report, an unsolved challenge was the "ineffective mechanism for eliminating water during the Winter months." The solution recommended was to relocate the East Sprayfields "to an alternative buffer zone site." The siting criteria recommended were: "distance from RCRA sites, upstream from the RFP holding ponds, and the installation of runoff controls to protect offsite drainages."

The need was expressed to assess both groundwater and surface water impacts of spray irrigation. Coming eight years after spray irrigation started in 1979, and only two years before Rocky Flats closed for production, there is no evidence in the available records that this recommendation was ever acted upon by Rockwell management.

Continuing Legacy of the Genesis of Environmental Crime: The "Solar Evaporation Ponds Plume"

The "Solar Evaporation Ponds Plume" is located in the northeastern part of the Industrial Area. Contamination in this area originated from selected liquid wastes at the Rocky Flats Plant that were previously discharged to temporary holding ponds. The contaminants of concern for this groundwater plume include nitrates and uranium isotopes (ATSDR, 2005, p. 29).

Reflections On the Genesis of Environmental Crime

It is highly ironic that it was contaminated water which was the genesis of environmental crime under the new suite of environmental laws dating from the 1970s.

At the time the Solar Evacuation Ponds were constructed early in the production era, there had been no requirement for impervious linings. This lack of impervious linings led to contamination of the groundwater.

The Atomic Energy Commission had known about the leaking to groundwater since 1960. The Department of Energy had known since the 1980s. Rockwell International had been fully briefed in 1987 on the legal vulnerability stemming from both DOE and Rockwell management certifying in 1985 that Rocky Flats was in full compliance with "all applicable groundwater monitoring..."

There was simply no way that the volume of 80 million gallons of contaminated water a year by the late 1980s could be evaporated or otherwise dispersed onto a comparatively small site with two creeks running through it. And when the creeks were reclassified as "navigable waters of the United States" under the Clean Water Act of 1972, standards for compliance with the law changed dramatically.

By Spray Irrigating over the East Trenches as reflected in Figure 8.1, Rockwell International blatantly violated its National Pollution Discharge Elimination System permit. By Spray Irrigating chromic acid mixed with raw sewage over the South Spray Field to escape detection, Rockwell International not only violated the public trust, in the judgment of the Special Federal Grand Jurors, they committed a serious environmental crime.

Chapter Nine
The Commission of Environmental Crime, the First FBI Raid Ever On Another U.S. Government Facility, the End of the Rockwell International Era, and the Closure of Rocky Flats for Production

"We have basically no RCRA groundwater monitoring wells, our permit applications are grossly deficient (some of the waste facilities are patently 'illegal')"...

DOE, 1986a, Mary L. Walker Briefing Memo

"...approximately twenty 55-gallon drums of hazardous wastes have been stored there [at Rocky Flats] ...without a permit...This conduct was an apparent RCRA felony, ... involving the illegal storage of hazardous wastes without a permit."

FBI Search Warrant Affidavit, 1989

"We are very troubled by the prospect that Rockwell people, as well as Rockwell, may be subject to legal liability for actions taken in good faith under our contract to manage and operate the Rocky Flats Plant."

S.I. Iacobellis, Rockwell International, 1989

The future of Rocky Flats has been transformed by the commission of environmental crime, the first FBI raid ever on another U.S. government facility, the dispatch of an investigative "Tiger Team" to Rocky Flats by Secretary of Energy James Watkins with its scathing report, the circumstances surrounding the end of the Rockwell International era, the closure of the site for production, and the subsequent ruling by an Administrative Judge of the Armed Services Court of Appeals that in the same general period of time as its management of Rocky Flats, Rockwell International had been charged with or implicated in "criminal, fraudulent, or improper acts" related to Federal contracts.

Secret Department of Energy Briefing Memo: "Patently 'Illegal' ": 1986
A "Briefing Memo" was prepared in 1986 for a meeting within the Department of Energy with Assistant Secretary of Energy Mary L. Walker. Among the assessments in the briefing memo was that part of the operations at Rocky Flats under the authority of the Department of Energy were "patently 'illegal' ". The following is a key excerpt from the Mary L. Walker Briefing Memo:

> We have basically no RCRA groundwater monitoring wells, our permit applications are grossly deficient (some of the waste facilities there are patently 'illegal')". We have serious contamination, and we have extremely limited environmental and waste characterization data for a site of this complexity. There are CERCLA problems here which are within EPA's exclusive jurisdiction.
>
> (DOE, 1986a)

That memo certainly would have remained secret had it not been released routinely to the U.S. Congress, and then to a staff member of the Environmental Protection Agency who gave it to an official in the area of environmental crime. It was used as a rationale for opening what EPA calls a "case investigation." The agent in charge on behalf of the EPA was William Smith. William Smith contacted Special Agent Jon Lipsky of the FBI about the memo. William Smith indicated he would only work on the case if Special Agent Lipsky agreed to investigate Rocky Flats as well (Lipsky, 2016).

The subsequent raid by the Federal Bureau of Investigation had been authorized by Acting Associate Attorney General Roger Mazzulla in January of 1989. In August of 1989 Secretary of Energy Admiral James Watkins dispatched a "Tiger Team" of investigators to Rocky Flats after the FBI raid. The "Tiger Team" findings are scathing.

Lipsky (2016) indicated that he personally thinks Admiral Watkins approved the FBI raid "to make an example of Rockwell and turn the tide against the evil 'DOE culture'..." which he was "...hell-bent to change, per his confirmation hearing in February of 1989." Lipsky (2016) went on to comment that "Admiral Watkins could have scuttled the raid if he had chosen to do so. It was that political."

FBI Search Warrant and the Search of Rocky Flats For Evidence of Environmental Crimes

The next events which were transformative occurred in early June of 1989 when Special Agent Jon Lipsky's application for a Search Warrant was granted by a U.S. Magistrate. The FBI/EPA raid on June 6, 1989 lasted for eighteen days, interviewed 200 witnesses and involved seizing over a million pages of documents.

The "synopsis of probable cause," the heart of the affidavit to the U.S. Magistrate, was that during the management era of Rockwell International, both Rockwell and DOE officials respeatedly had stated to the public and certified to agencies of the U.S. government that "Rocky Flats was in compliance with federal and state environmental laws." The Mary L. Walker Briefing Memo (DOE, 1986a) and a 1988 Goverment Accountability Office (GAO, 1988) study were cited in the Search Warrant application that this was not the case.

Attorney General Orders Search Warrant Unsealed At Start of Raid
It is unclear in available documents why Attorney General Richard Thornburgh ordered that the Search Warrant Affidavit be unsealed on only the second day of the FBI raid. McKinley and Balkany (2004) uncovered a very interesting undated memorandum from Martin C. Carlson, a senior legal advisor in the Department of Justice, to Lawrence Lippe, Chief, General Litigation and Legal Advice Section of the Criminal Division of the Justice Department. In it, attorney Carlson stated:

> If the contents of the [search warrant] affidavit were to be publically disclosed, potential defendants would be provided with a virtual roadmap of the areas under investigation by the government.
>
> (McKinley and Balkany, 2004, p. 249)

It is clear that the Carlson formulation was standard policy in the Department of Justice.

McKinley and Balkany (2004) assembled a collection of local and national newspaper accounts of the FBI raid. There was a consistent theme that the FBI raid was intended to stop illegal activities by the contractor which were, or could be, imperiling the community and which had to be stopped:

> "Flats Burned Waste Secretly" (Lindsay and Day, 1989)
> "Rocky Flats Conspiracy Alleged: FBI Used Spy Plane" (Pankratz, 1989)
> "Flats Charges Include Illegal Waste Burning: Document Claims Chemicals Put in Creeks Feeding Water Supplies" (Pierce, 1989)
> "Investigation at Rocky Flats: A Cause for Contract Cancellation." (Steeth, 1989)
> "Rockwell Has History of Run-Ins with Regulators Everywhere" (The Denver Post, 1989)
> "Energy Department Freezes Millions in Bonuses to Rockwell" (Graf, 1989)
> "Rocky Flats Illegally Burned, Dumped Waste, U.S. Claims" (Finley and Graf, 1989)

Perhaps the Rocky Mountain News three years later captured the essence of newspaper reporting with their headline of March 27, 1992 which said, "Safety Fears Escalated with FBI Raid."

The Beginning of the End of the Rockwell Era

Rockwell International Initiatives After the FBI/EPA Raid

Rockwell International made initiatives to the Department of Energy in the immediate months after the FBI/EPA raid: a letter (Beall, 1989) from the Rockwell Chairman on August 14, 1989 to the Secretary of Energy proposing what Rockwell was willing to do as the Rocky Flats contractor, and a letter (Iacobellis, 1989) from Rockwell dated August 17, 1989 indicating that what the Department of Energy required of its Rocky Flats contractor put both the company and its employees at legal risk. The Secretary of Energy responded to the Rockwell Chairman on August 25, 1989 (Watkins, 1989a). These letters are supplemented by the contemporaneous notes of a conversation between the Rockwell Chairman and Secretary Watkins on September 18, 1989 (Wakefield, 1989) and a letter by Secretary Watkins (Watkins, 1989b) dated September 20, 1989.

Rockwell Letter of August 17, 1989 – "Serious concern relating to the continued operation"

S. I. Iacobellis, the President of Aerospace Operations for Rockwell International (Iacobellis, 1989) wrote to Deputy Secretary of Energy W. Henson Moore on August 17, 1989 indicating that Rockwell International could not continue as a contractor at Rocky Flats with its company and employees at criminal risk for doing what the Department of Energy required them to do. The bottom line objection is to criminal prosecution and "that Rockwell people, as well as Rockwell, may be subject to legal liability..." There is no known reply by the addressee of this letter, Deputy Secretary W. Henson Moore.

Lipsky (2016) provided a context for understanding the Iacobellis communication to Deputy Secretary Henson:

> Regulators had at their disposal the ability to enforce civilly the environmental laws of the United States. The Colorado Department of Health (later re-named the Colorado Department of Public Health and Environment) had received authority in 1984 delegated from the Environmental Protection Agency to enforce the Resource Conservation and Recovery Act at Rocky Flats.
>
> (Lipsky, 2016)

An assessment by Lipsky (2016) was that this approach "is not how to enforce the law. The entire civil regulatory framework was beginning to change for Rockwell in the six months leading up to the end of the production era.

Secretary Watkins Sent a "Tiger Team" To Rocky Flats In August of 1989

The Tiger Team (1989) report was scathing about the environmental conditions and monitoring at Rocky Flats including equipment more than 15 years old and astoundingly not measuring for americium 241 or that there was equipment which had not been calibrated since the 1970s. The following is a representative excerpt:

> The assessment of the RFP contribution to public radiation...does not fully address all potential exposure pathways and radionuclides. The quality assurance and quality control practices for radiochemical analyses...cannot adequately verify

the validity of analytical results. Chemistry Standards lab also non-complying. General lab for radiochemistry analysis non-complying. Information provided to the EPA is false.

Memo to the File About the Conversation Between the Rockwell Chairman and the Secretary of Energy On September 18, 1989 and the "Bonus Fee" Letter of September 20, 1989

This "memo to the file" (Wakefield, 1989) is important for both what was reported to have been said directly as well as what was implied such as the reported assertion by Chairman Beall that "his lawyers were concerned about the politics of the situation" related to criminal prosecution of Rockwell executives. As a business matter, Rockwell wanted to "continue as operator under the contract." Admiral Watkins expressed his concern that "Rockwell had not acknowledged properly his directives to accept a new culture which would place environment, safety and health on a level equivalent to production." Without elaboration, Admiral Watkins is quoted as saying that "assurance would not be forthcoming" from the Department of Justice regarding criminal prosecution.

<u>The Award Fee For Rockwell: Correspondence From the Secretary of Energy To the Chairman of Rockwell On September 20, 1989</u>

This letter is dated two days after the telephone conversation between Chairman Beall and Admiral Watkins. Admiral Watkins (Watkins, 1989b) indicated that "special care" had been given to determining the proper bonus award fee for the period between October 1, 1988 and March 31, 1989. Rockwell was to receive 40 percent of the "total possible award fee." Subsequent to this communication Rockwell filed an unsuccessful lawsuit on the grounds that under their DOE contract the Rockwell bonus was to be determined by the Albuquerque Office of DOE, not by the top management in Washington, D.C.

Rockwell Charged With Or Implicated In Criminal, Fraudulent, Or Improper Acts

During the period of time roughly paralleling the period of its contract to manage Rocky Flats, Rockwell International had become embroiled in a series of Federal Contract disputes. These disputes culminated in 2000 in the courtroom of Administrative Judge David W. James Jr. of the Armed Services Court of Appeals.

The case before Judge James in 2000 was ASBCA No. 49994, an appeal by Boeing North America, Inc. Boeing had earlier acquired Rockwell International. The appeal was over the disallowances of legal costs incurred by Rockwell in litigating and settling a shareholders lawsuit against the directors of Rockwell.

According to Judge James, "the suit alleged that the directors failed to institute and enforce adequate internal controls, and fostered a corporate climate that encouraged employee misconduct under federal contracts and result in criminal and civil penalties and fines."

In his opinion denying Boeing's appeal, Judge James cited the following criminal, fraudulent, or improper acts with federal contracts by Rockwell:

1. Settlement of allegations of mischarging costs incurred from 1975 to 1977, Rockwell in 1982 paid a fine of $500,000 and entered into a consent degree "enjoining it from submitting false claims in the future."
2. In October of 1985, Rockwell agreed to pay $1 million in fines and restitution as part of a guilty plea over six employees submitting false timecards under a 1979 Air Force contract.
3. In February of 1987, Rockwell pled guilty to violating the 1982 consent decree and paid a $5.5 million fine. The Truth in Negotiations Act had required Rockwell to disclose fully to the U.S. government the price terms of a subcontract under its May 1983 Air Force prime contract.
4. Rockwell terminated a facility manager at Rocky Flats who had required employees to perform personal work which was then charged to the Rocky Flats DOE contract. Rockwell was not prosecuted for its employee's acts.
5. In May of 1988 and June of 1989, the Colorado Department of Health [and Environment] discovered 32 violations of the Colorado Hazardous Waste Act. The $18.5 million fine and guilty plea to ten misdemeanor and felony violations of the Clean Water Act and RCRA in March of 1992 followed from the FBI raid and the work of the Grand Jury.

Among the reasons cited by Judge James for denying the appeal was a report to implement the Major Fraud Act of 1988, Public Law 100-700, which constituted a "guiding principle":

> Government should not pay for wrongdoing, the defense of wrongdoing, or the results or consequences of wrongdoing by contractors.

Judge James could "discern no benefit to the Government in a contractor's defense... in which the contractor's prior violations of federal laws and regulations were an integral element..."

Reflections On the Commission of Environmental Crime, the Unprecedented FBI Raid and the End of the Rockwell Era and Closure of the Production Era

The extent of the commission of environmental crime was well known to the top management of both the Department of Energy (DOE, 1986a) and to the top management of Rockwell International (Rockwell, 1987).

Secretary Watkins had commissioned an in-depth "Tiger Team" on-site investigation. Chairman Beall had the internal Rockwell investigation concerning RCRA monitoring violations and of alleged false statements under penalty of law. From the unsealed Search Warrant Chairman Beall had to be fully informed about the purposes of the FBI raid.

The timing of the unsealing of the Search Warrant Affidavit, under the authority of the head of the Department of Justice, Attorney General Richard Thornburgh, had the effect of forewarning Rockwell of the statements of "probable cause" and what proof of criminal behavior was being sought by the FBI while they were in the process of interviewing witnesses and seizing over a million pages of documents. To make an obvious understatement, the targets of a mafia investigation would never had been told what the FBI was looking for in the middle of a criminal raid.

The point is that the Rockwell/DOE formal correspondence masked some of the real issues which both organizations knew existed. The exception is the steadfast opposition of Rockwell to the criminal indictment of its employees, some of whom had received “target letters” from the Office of the U.S. Attorney in Denver, Colorado. That issue was not to be resolved until the final plea agreement.

Chapter Ten
Transformation of the Case of Rocky Flats By the Criminal Justice System and the Clash Between "Separate But Equal" Institutions of U.S. Constitutional Government

"...You are free to exercise your own judgment without fear or favor and shall not be deterred or influenced by the criticism of the public..."

Charge to the Special Federal Grand Jury by Judge Sherman G. Finesilver, 1989

"To the extent to which DOE may have authorized Rockwell to break the law, DOE acted more often than not at Rockwell's direction and after Rockwell had independently formed an intent to break the law."

Findings of the Special Federal Grand Jury

"As the situation before the FBI raided the Plant in June of 1989, DOE continues to direct and endorse this course of illegal activity in violation of applicable environmental laws and in the name of political expediency. It is primarily for this reason that this Grand Jury has recommended that the Rocky Flats Plant be closed!"

Findings of the Special Federal Grand Jury

"...the actions of Rockwell do not come close to meeting the prosecutorial threshold suggested by the Attorney General [rogue acts]...[the period of the 1980s was one of] ...extraordinary regulatory confusion that surrounded the application of environmental laws to DOE facilities..."

Williams and Connolly, Criminal Defense Attorneys for Rockwell International

"The crimes were attributed to a "culture" at DOE--not the actions of responsible individuals. This is the white collar equivalent of blaming an armed robbery on "society"--not the individual holding the gun."

Wolpe Subcommittee, 1993

"Second, it should be noted that the most important thing that Federal prosecutors bargained away in negotiations with Rockwell was the truth. By entering into this plea agreement, the prosecutors bargained away the right to fully and accurately inform the American people and the Congress about the conditions, activities, and crimes at the Rocky Flats facility."

Wolpe Subcommittee, 1993

"Rockwell's conviction of multiple felony and misdemeanor offenses and payment of $18.5 million in fines without the possibility of indemnification by federal taxpayers provided substantial benefits to the government and constituted a proper exercise of prosecutorial discretion."

Department of Justice, 1994

"The big story is that they all knowingly violated environmental laws. Congress passes laws it doesn't intend to see enforced. The government has a lack of will to enforce the law. The government is lying to us. It's a breach of public trust. That is the biggest crime of all."

An Anonymous Grand Juror, Los Angeles Times, 1993

Appointment of a Special Federal Grand Jury

Following closely after the FBI raid, on June 22, 1989, the Department of Justice asked Judge Sherman G. Finesilver to empanel a Special Federal Grand Jury. Subsequent instructions to the Special Federal Grand Jury by Judge Finesilver in December of 1989 have a number of noteworthy elements in light of the subsequent choice made by the Department of Justice to disband them and the determination by Judge Finesilver not to release their report as failing to meet "necessary standards." The following are excerpts from Judge Finesilver's charge to the Grand Jury:

> ... You are free to exercise your own judgment without fear or favor and shall not be deterred or influenced by the criticism of the public ...You are the defenders of the innocent as well as the accusers of the guilty, and in both respects you vindicate the integrity of the law. ...Your proceedings are secret and must remain secret permanently unless and until the Court determines that the proceedings or a portion of them should be revealed in the interests of justice...
>
> (Finesilver, 1989)

The oath made it obligatory to identify "all public offenses against the United States...."

The Criminal Defense Case By Rockwell Attorneys: "Not Rogue Acts"

Williams and Connolly is a premier Washington, D.C. law firm specializing in criminal defense. It is also very well connected politically. Lawyers from Williams and Connolly made the following points:

- Rockwell International had ceased being the contractor at Rocky Flats on December 31, 1989. It was not until 1990 that Judge Lewis Babcock ruled in the Sierra Club lawsuit that the Resource Conservation and Recovery Act applied to Rocky Flats.
- Attorney General Richard Thornburgh described the nature of environmental prosecutions brought by the Department of Justices as follows: "Generally these crimes were either rogue acts in direct violation of environmental regulations–so called midnight dumping—or surreptitious maneuvers that undermine the regulatory system through concealment or fraud. Sometimes both. In addition, there was also criminal negligence—in which careless disregard of environmental regulation caused potentially disastrous results."

An essential point of the Williams and Connolly lawyers was that "the actions of Rockwell do not come close to meeting the prosecutorial threshold suggested by the Attorney General..." Williams and Connolly lawyers characterized the period of the 1980s as one of "...extraordinary regulatory confusion that surrounded the application of environmental laws to DOE facilities...." During the 1980s when Rockwell was the contractor at Rocky Flats, the Department of Justice had joined with the Department of Energy in defending RCRA exemptions for Rocky Flats.

Report of the Special Federal Grand Jury: Its Suppression and Then Government Omissions

The Special Federal Grand Jury was disbanded after having heard testimony for two and a half years. They had prepared a long Grand Jury Report which called for indicting individuals whom they held responsible for committing environmental crimes from both Rockwell International and the Department of Energy. Judge Finesilver refused to release the Grand Jury Report (First Special Federal Grand Jury Report, 1992) indicating that it did not meet "necessary standards." The foreman of the Grand Jury made a public statement requesting that President-elect Bill Clinton appoint a Special Prosecutor to investigate the conduct of the Department of Justice (Frontline, 1993). The Department of Justice shredded the proposed indictments of individuals. Judge Finesilver subsequently approved the release of an edited version of the Grand Jury Report. A "leaked" version of the Grand Jury Report was published by Westword magazine. The website of the Constitution Society published both versions highlighting in blue what Judge Finesilver and the Department of Justice had edited out.

Among the Leaked First Grand Jury's Principal Conclusions In 1992: "An Ongoing Criminal Enterprise"

The following presents a summary of the principal conclusions of the First Special Federal Grand Jury Report based on their report leaked by person or persons unknown:

Ongoing Criminal Enterprise

> The Department of Energy ("DOE"), its contractors - Rockwell International, Inc. ("Rockwell"), EG&G, Inc. ("EG&G"), and many of their respective employees have engaged in an on-going criminal enterprise at the Rocky Flats Plant ("the Plant"), which has violated Federal environmental laws. This criminal enterprise continues to operate today at the Rocky Flats Plant, and it promises to continue operating into the future unless our Government, its contractors, and their respective employees are made subject to the law.

Compelling Evidence

> When agents of the Federal Bureau of Investigation ("FBI") and the Environmental Protection Agency ("EPA") raided the Plant on June 6,

1989, they found compelling evidence that hazardous wastes and radioactive mixed wastes had been illegally stored, treated, and disposed ("STD") of at the Plant in violation of the Resource Conservation and Recovery Act ("RCRA").

Failure To Obey the Law

For 40 years, Federal, Colorado, and local regulators and elected officials have been unable to make DOE and the corporate operators of the Plant obey the law. Indeed, the Plant has been and continues to be operated by government and corporate employees, who have placed themselves above the law and who have hidden their illegal conduct behind the cloak of national security. Government and corporate employees have breached the public's trust by engaging in a continuing campaign of distraction, deception, and dishonesty.

Grand Jury Actions About the Criminal Enterprise

The ongoing nature of the criminal enterprise at the Rocky Flats Plant has prompted this Special Grand Jury to take three actions. First, the Special Grand Jury has approved indictments against certain current and former Federal employees, corporate employees, and corporations. Second, the Special Grand Jury has made presentments to this Court of evidence of criminal conduct by certain corporations and persons. Third, this Special Grand Jury strongly recommends that the Rocky Flats Plant be closed as the only means to stop the continuing nature of these criminal acts.

Why Illegal Conduct Has Continued At Rocky Flats

(a) DOE managed the Plant with an attitude of indifference toward environmental laws;
(b) DOE actively participated in and directed a conscious and ongoing effort to evade the application of certain environmental laws to the Plant;
(c) the Colorado Department of Health ("CDH") and EPA were lax and ineffective in attempting to enforce environmental laws and regulations at the Plant; and
(d) Rockwell and its successor (EG&G) - with the apparent knowledge of DOE - have excused their illegal conduct by asserting that they cannot operate the Plant without violating one or more environmental laws.

The Deletions From the First Grand Jury Report By the Department of Justice and Judge Finesilver: 1992

There are now three versions available of the first "leaked" Grand Jury Report. The first version is the double-spaced typewritten draft of the Grand Jury report which was leaked by person or persons unknown to Westword without the section on indicting the individuals held responsible by the Grand Jury. The second version is the one released to the public by Judge Finesilver. The third version is on the web by the Constitution Society with the deletions by the Judge, the Department of Justice, and / or the Office of the U.S. Attorney highlighted in italics.

The general intent is obvious from these examples of what twenty-three citizens concluded:

1. DOE continues to tolerate violations of environmental laws;

2. DOE and the new contractor, EG&G, continue to violate the law and cannot restart Rocky Flats without violating the law;

3. The Grand Jury did not investigate the issue of serious health risks to workers or area residents, or violations of the Clean Air Act;

4. The DOE Plant Manager has made false written statements;

5. EG&G has perpetuated many of the RCRA violations which prompted the FBI raid;

6. EG&G operates Rocky Flats in violation of the groundwater monitoring requirements of the Clean Water Act; and

7. Rockwell violated RCRA and the Clean Water Act with inappropriate spray irrigation with water that migrated off-site and into Woman and Walnut Creeks.

Department of Justice Actions After Disbanding the Grand Jury

In addition to aborting the legal process underway which was leading to the indictment of specific individuals from DOE and Rockwell held responsible for deliberately violating environmental laws, the Department of Justice entered into protracted negotiations with Rockwell International on a settlement.

The Congressional Investigation of 1992

The U.S. Congress (Wolpe, 1993) began an investigation after the settlement between the Department of Justice and Rockwell International. After months of hearings it issued a report including transcripts of testimony and evidence with the ultimate charge that the Department of Justice had "bargained away" the right of the American people "to know the truth..." to fully and accurately inform the American people and the Congress about the conditions, activities and crimes at the Rocky Flats facility...The prosecutors forfeited the opportunity...to create the political momentum for political reform throughout the DOE complex."

Some of the key conclusions of the bipartisan Congressional subcommittee are as follows.

Legitimate Criminal Charges Not Pursued

Even before the federal district court in Colorado accepted the settlement agreement, the Subcommittee received allegations that the legitimate charges against both the company and individual Rockwell and DOE employees had not been pursued. Allegations were also made that the fine, although large, served no punitive purpose as it did not even represent the profits and fees paid to Rockwell in excess of its costs for operating the plant during the three years for which environmental violations had been charged.

Lack of Cooperation By the Department of Justice

The Subcommittee's investigation began slowly due to an extreme lack of cooperation by the Department of Justice. The Subcommittee was forced to subpoena the FBI and the Justice Department in order to conduct interviews and examine requested documents.

"Systemic" Defiance of Environmental Laws and Regulations

However, there was evidence of troubling behavior and attitudes in the manner with which Justice Department headquarters viewed environmental crimes cases, particularly those which result from systemic defiance of environmental laws and regulations...

Department of Energy "Culture"

1. The "culture" at DOE has been cited as a critical factor in this case, but the "culture" at DOJ was just as significant. The extreme conservatism and lack of aggressiveness of Justice headquarters was a major--perhaps the overriding--factor which influenced both the strategies adopted throughout the course of the investigation and ultimately, the outcome of the case.

Ignoring Off-Site Releases of Pollutants

7. Several areas in which the prosecutors had collected indictable evidence showed that there had been off-site releases of pollutants. None of those off-site releases were included in the plea. This coincided with Rockwell's demand that such charges be excluded from the plea or rewritten as permit violations...

Government Lack of Public Accountability

First, our investigation revealed a very troubling lack of public accountability in the Federal Government. Serious environmental crimes were committed, but no individuals were held responsible. The crimes were attributed to a "culture" at DOE--not the actions of responsible individuals. This is the white collar equivalent of blaming an armed robbery on "society"--not the individual holding the gun.

Federal Prosecutors Bargained Away the Truth

Second, it should be noted that the most important thing that Federal prosecutors bargained away in negotiations with Rockwell was the truth. By entering into this plea agreement, the prosecutors bargained away the right to fully and accurately inform the American people and the Congress about the conditions, activities, and crimes at the Rocky Flats facility. Conditions that I might add, continue to this day even through Rockwell has been replaced as the facility's contractor. And these conditions are not unique to Rocky Flats; they can be found throughout the DOE complex. By papering over the situation at Rocky Flats, the prosecutors forfeited the opportunity to focus National attention on the crimes committed at Rocky Flats; attention that is necessary to create the political momentum for meaningful reform throughout the DOE complex.

Department of Justice 1994 Rejoinder To the Congressional Report of 1993

The Justice Department (Department of Justice, 1994) took two years to compile a report which Attorney General Janet Reno (Reno, 1994) declared had demonstrated the integrity of the actions of the prosecutors. The essence of the Department of Justice Rejoinder is as follows:

Department of Justice Did Not Improperly Influence Prosecutors

As detailed below, no basis existed for the sweeping criticisms by the Wolpe Report. The evidence demonstrated that neither DOJ nor anyone else improperly influenced the prosecution. Although DOJ monitored the case closely, almost all major decisions regarding the investigation and plea negotiations were made by the prosecutors in Colorado. The plea agreement between Rockwell and the government was the product of extensive negotiation between the parties and was reviewed and approved by a wide range of experienced and senior officials not only within DOJ, but also at the Environmental Protection Agency (EPA], and the Colorado Department of Health (CDH).

Rockwell's Conviction a Proper Exercise of "Prosecutorial Discretion"

Rockwell's conviction of multiple felony and misdemeanor offenses and payment of $18.5 million in fines without the possibility of indemnification by federal taxpayers, provided substantial benefits to the government and constituted a proper exercise of prosecutorial discretion.

Opposing Release of Grand Jury Report Proper Decision

DOJ's decision to oppose issuance of a special grand jury report was the only notable example of Main Justice overruling the line prosecutors. Decisions on the issuance of special grand jury reports, however, are required by the United States Attorneys' Manual to be made by the Criminal Division. There was nothing improper about the process or the substance of the decision, which was made by a senior career official based on a reasonable and longstanding Criminal Division interpretation of the special grand jury statute.

****

Statement of The Attorney General Concerning the Internal Report on the Rocky Flats Prosecution
Attorney General Janet Reno

Proper and Professional

"I am extremely pleased that the internal report on the Rocky Flats case concluded that it was properly and professionally handled by the attorneys who conducted the lengthy investigation and obtained a 10-count guilty plea and record penalty. In light of the detailed analysis and conclusions in the report, there should be no remaining doubt that the prosecutors acted energetically and consistently with the highest professional standards."

(Reno, 1994)

Unreconciled Elements In the Transformation of the Case of Rocky Flats

Two of the three "separate but equal" branches of American constitutional government engaged in a fundamental clash over the environmental crime era at Rocky Flats. The clash revealed a recurrent theme: citizen concerns not addressed fully in one historical era surely will re-emerge in the next era and subsequent eras.

There are compelling differences in the specificity of evidence cited by the U.S. Congress and the Department of Justice. The U.S. Congress version of the evidence was published in three volumes containing transcripts of testimony and extended documentation. The Department of Justice did not supply an Appendix with evidence to support their main conclusions. The Department of Justice was simply inaccurate in its claim that there were no "political or personal influences" in the actions of the Department of Justice despite the known fact that even the White House

Counsel had a role in approving the 1991 Federal Facilities Compliance Agreement without any enforcement penalties for noncompliance.

Another example is that when the Regional EPA official refused to approve a compliance agreement with no penalties for noncompliance, a political superior in Washington, D.C. stripped him of authority for the decision and made the issue of compliance meaningless.

Not even discussed by the Department of Justice was the travesty of a Federal District Court Judge lending the credibility of his office to the release of a false version of the First Special Grand Jury Report with the preposterous claim that the true version lacked "necessary standards." This was a breach of the public trust.

The U.S. Congress could have been describing a totally different case with their assertions that the Department of Justice had chosen not to pursue "legitimate charges" and that the fine paid by Rockwell served "no punitive purpose." The U.S. Congress published its detailed findings in 3 volumes of documents and transcripts of testimony under oath by numerous participants with first hand knowledge of what had transpired at Rocky Flats.

But the most uncompromising definition of the unresolved clash over Rocky Flats was by an anonymous Grand Juror who had concluded that the Congress had passed laws "it doesn't intend to see enforced," and that the Executive Branch has a "lack of will" to enforce the law and is "lying to us." But the "biggest crime of all" is what they were all doing: a "breach of public trust."

Chapter Eleven
The Second Citizen Special Federal Grand Jury Report On Environmental Crime At Rocky Flats

The Citizen Grand Jury members worked on a Second Report in 1992 which was buried for reasons unknown.

> "...the Plant has been and continues to be operated by government and corporate employees, who have placed themselves above the law and who have hidden their illegal conduct behind the cloak of 'national security.' "
>
> "...government employees and corporate employees, who have operated the Plant, have committed the most serious crime that can ever be committed in a democracy...breached the public trust through a clever public relations campaign of distraction, deception, and dishonest denial."
>
> The Special Federal Grand Jury grouped criminal conduct at Rocky Flats into three broad categories:
>
> "1. Single event or discontinued crimes like the chromic acid spill incident;
> 2. Criminal offenses which were on-going such as the failure to monitor adequately groundwater contamination or the on-going non-permitted storage of radioactive plutonium 'residues';
> 3. Crimes against the public trust."
>
> Preceded by the notation "[Indict]"
> "...no one at any time has ever sought to exempt DOE's employees, its contractors, or the employees of the contractor from prosecution for the criminal nature of this on-going course of conduct and conspiracy to violate the criminal laws of the United States."
>
> "...Rockwell displayed a callous disregard for the effects of its actions on the surrounding communities and the environment."
>
> "Rockwell and its managers were able to consistently manipulate and control DOE policy to assure DOE endorsed Rockwell's illegal conduct in pursuit of exorbitant bonus and contract fee awards."
>
> Second Citizen Special Federal Grand Jury Report, 1992

This "Second Citizen Special Federal Grand Jury Report" (Second Citizen Special Federal Grand Jury Report, 1992) on environmental crime at Rocky Flats was clearly a "work in progress." It ended up in the archives of the Congressional investigation of 1992 (Wolpe, 1993) without an identifying cover page. It has a very polished "Introduction" which reads like an overview of central conclusions from the work of the Grand Jury. The Grand Jury had been in session for more than two years and heard the testimony from 200 plus witnesses, and received 2,000 now sealed exhibits which the U.S. government in 2019 declared to be "lost". In many cases the essence of the sealed exhibits is revealed in the text; in others, the record of "best evidence" will have to come from other sources.

Organization of the Second Citizen Special Federal Grand Jury Report
The organization of the Second Citizen Special Federal Grand Jury Report has the following headings after the First Introduction:

- Scope of the Investigation—what remains for future study and what was not investigated completely (parts of this are surprising);
- The citizen drafters conceptualized criminal conduct at Rocky Flats in three broad categories: single event crimes, on-going criminal offenses like inadequate ground water monitoring; and crimes against the public trust;
- Government Agency Negligence and Malfeasance—as the citizen drafters phrased it, "how such extensive illegal conduct could have gone undetected for such a lengthy period of time";
- Chromic Acid Spill—a single occurrence environmental crime where a chromic acid spill ended up being spray irrigated off the Rocky Flats site: "Rockwell displayed a callous disregard for…the surrounding communities and the environment";
- False Statements—violations of the essence of public trust: not telling the truth under oath;
- Groundwater Monitoring—or most accurately, criminal negligence in lack of monitoring, a reason why DOE itself had said in 1986 that toxic ground water at Rocky Flats posed the greatest threat to public health in the entire U.S. multi-state nuclear weapons production complex;
- Pondcrete—serious deficiencies in attempting to produce a solid hardened monolith from solid and liquid low-level hazardous and radioactive mixed-waste. The DOE Regional Office waited two-years even to request funds to cover the expense of providing an outdoor storage cover for the "gooey" slime, "play-doh" blocks which had started to spill into the environment;

- RCRA Problems—the heart of the guilty pleas by Rockwell International over permit violations of the storage, treatment and disposal of land-banned hazardous and mixed-wastes including revelations about violations of environmental law in the operation of the Building 771 incinerator, and recounting the DOE eight-year effort to avoid RCRA compliance;
- RICO—the magnitude of environmental crime at Rocky Flats with DOE and Rockwell as "indistinguishable co-conspirators" in violating the law invoked the federal racketeering statues known as the Federal Racketeering Influenced Corrupt Organizations Act;
- Salt Brine—environmental crime committed in disposing of small quantities of radioactive materials together with corrosive and organic hazardous materials, "mixed-wastes," in a Solar Evaporation Pond without a permit (which could not have been obtained); and
- Spray Irrigation—of hazardous and mixed wastes...of up to 250,000 gallons a day over a multi-year period onto only 17 acres. "Sheet runoff" went into Woman Creek and Walnut Creek which ultimately flowed downstream into Great Western Reservoir and Standley Lake, sources of drinking water for Denver suburbs.

It seemed to this reader that this Second Citizen Special Federal Grand Jury Report ended in mid-sentence, through of course literally it did not. But after two Introductions and a carefully written scope of their investigation (and what was not included in the two-year investigation), it was noteworthy that what they obviously considered violations of the public trust were not more fully developed. And after their intense work for over two years, there was no more fully informed citizen opinion concerning the big issues of whether Rocky Flats could be safely restarted, or whether the plant could operate without violating RCRA and the Clean Water Act.

This Second Citizen Special Federal Grand Jury Report was written without the assistance of the U.S. Attorney's Office in Denver. Yet this report is notable for its specificity of what citizens considered to be environmental crimes, and the detailed evidence which they cited in support of their assessment. A challenge for the future is locating further sources of "best evidence" from other credible primary sources. This will be necessary until the time, if it ever comes, when the 2000 plus exhibits considered by the Grand Jury are unsealed by court order. A key current source of "best evidence" is coming from such civil lawsuits as the recently concluded (2016) Cook civil litigation.

Using the Notation ["Indict"] Before Some of the 165 Paragraphs In the Second Citizen Special Federal Grand Jury Report

Beginning with the section on the Chromic Acid Spill, the Grand Jurors began some paragraphs with" [Indict]" often followed by the name or names of the individual or individuals or organizations the citizen drafters wanted the Grand Jury formally to vote to indict. In fact, there has been "target letters" sent to the attorneys for a number of individuals from both the Department of Energy and Rockwell International. The Office of the U.S. Attorney for Denver had a central role in authorizing the target letters which notified individuals that they were under investigation by the Grand Jury for having themselves committed environmental crimes.

The Grand Jury intended that specific individuals should be indicted. This version of the Second Citizen Special Federal Grand Jury Report is critical of the Office of the U.S. Attorney for granting immunity from prosecution to individuals, at least some of the Grand Jury believed, should have been indicted rather than receiving immunity from prosecution in return for their testimony.

The actual list of individuals who were on the list to be indicted voted by the Special Federal Grand Jury was shredded by the Office of the U.S. Attorney and not included in the report "leaked" to Westword, nor is it known except to their attorneys and the individuals themselves. The actual names have been deleted of the individuals named to be indicted in this second version of the Citizen Special Federal Grand Jury Report.

In four instances it would be unclear in the narrative if the role of the individual was not identified. Two of the instances are the top official of the Department of Energy at Rocky Flats and the top official of Rockwell International at Rocky Flats. Otherwise, using paragraph 157 as an example, it would say "Mr. X reported in a letter to his Albuquerque supervisor..." This analysis of the Second Citizen Special Federal Grand Jury Report, therefore, will modify paragraph 157 to begin with "The DOE Area Manager at Rocky Flats reported in a letter to his Albuquerque supervisor..." The same modification will occur with identifying the Rockwell International Plant Manager only by his role. Other named individuals will only be identified as Rockwell employees or DOE employees and their roles if necessary for clarity of understanding the reasons the citizen drafters considered them for indictment.

A third instance is the individual who held the combined role of Director of Plutonium Operations and Director of Environmental Compliance for Rockwell International. It is hard to imagine a more significant conflict of responsibilities or a more important role in fulfilling the mission of Rocky Flats in furtherance of the national security of the people of the United States.

An example of the necessity of identifying roles in the fourth instance is the internal dispute over the "By-Product" Rule of May 1, 1987 as reported in paragraph 92 on RCRA. By 1987, DOE had acknowledged that RCRA applied to the hazardous portion of all mixed-wastes at DOE facilities. Subsequently, an attorney for DOE at the Rocky Flats Plant and the attorneys for Rockwell International at the Rocky Flats Plant advised both the DOE and Rockwell Plant Managers that under the 1987 court decision in *American Mining Congress*, both the Environmental Protection Agency and the Colorado Department of Health lacked authority to regulate mixed-waste residues because the residues were not being "discarded" and could therefore be stored indefinitely without being subject to RCRA regulation. The point in this fourth instance is that identifying their roles as lawyers is important to conveying the significance of the legal advice in its context.

Section-By-Section Narrative of the Second Citizen Special Federal Grand Jury Report

This narrative will now shift to a presentation section-by-section of the Second Citizen Special Federal Grand Jury Report. Judge Finesilver had rejected for cause the substance of the First Citizen Special Federal Grand Jury Report. He made no known assessment of this Second Citizen Special Federal Grand Jury Report.

One of the purposes of this volume is to examine how citizen concerns can be expressed effectively within American democracy when there are sensitive national security priorities at stake. The twenty-three citizen members of the Special Federal Grand Jury became more knowledgeable collectively about Rocky Flats than any other group of American citizens ever has been with the possible exception of the members of the jury in the Cook class action lawsuit which had exceptional factual submissions made to them. The importance of this Second Citizen Special Federal Grand Jury Report is that they are sharing insights as they alone chose to formulate their assessment, along with the evidence that they determined supported their assessment.

The First Introduction

The Rocky Flats Plant is introduced in the first paragraph as an "on-going criminal enterprise." Further, Rocky Flats is projected to continue as an "on-going criminal enterprise" unless the U.S. government, its contractors, and their respective employees "are made subject to the law."

The FBI/EPA raid of June 6, 1989 is presented as finding "compelling evidence" that DOE and Rockwell International had been:

- Illegally storing, treating, and disposing of hazardous and radioactive mixed-waste in violation of RCRA.
- Violating the Clean Water Act with illegal discharges of pollutants, hazardous materials and radioactive matter into the Platte River and the drinking water supplies for cities of Colorado.
- Embodying a culture of "greed" and "criminal misconduct" which insisted that "the end of corporate bonuses always justified the illegal means to achieve those bonuses."

- Engaging in a pattern of behavior "which had converted the DOE and its contractors into a racketeering influenced corrupt organization within the meaning and in violation of the Federal Racketeer Influenced and Corrupt Organizations Act ("RICO").

In response to what they characterized as the "insidious, on-going nature of the criminal enterprise," the Grand Jury took four actions:

1. "...the Grand Jury has indicted certain current and former Federal employees, corporate employees, and corporations."
2. "...the Grand Jury has made presentments of criminal conduct by certain persons whom the prosecutors chose to immunize from criminal prosecution in exchange for their testimony..."
3. Recommend that certain government employees be disciplined or terminated with prejudice for "noncriminal misconduct" or "malfeasance in office."
4. Recommend strongly that the Rocky Flats Plant be closed as "the only means to stop the continuing nature of these criminal acts."

Several of the elements of this first part of the Introduction go beyond the First "Leaked" Citizen Special Federal Grand Jury Report. This version echoes the characterization of "an on-going criminal enterprise" and repeats the recommendation that the Rocky Flats Plant be closed. It goes beyond the first "leaked version" by identifying a "culture of greed" tied to corporate bonuses and characterizing the "pattern of behavior" as constituting the establishment of a "racketeer influenced corrupt organization" within the meaning of the Federal RICO statute. This part of the Introduction also criticizes the prosecutors for offering immunity to certain individuals in return for their testimony.

A final note of commentary on the initial paragraphs of the Introduction is the observation that this Second Citizen Special Federal Grand Jury Report ends abruptly in terms of both style and substance on page 43. It cannot be known whether or not there are more pages to the draft unless the Grand Jury records are unsealed.

The first part of the Introduction indicated that citizens on the Special Federal Grand Jury had intended to recommend that certain government employees be sanctioned for "noncriminal misconduct" or "malfeasance in office." That section is not included in this Second Citizen Grand Jury Report, at least in the version included amongst the documents obtained from the Congressional archives of Wolpe (1993).

This Introduction section includes harsh assessment such as:

- Federal, Colorado, and local regulators and elected officials have demonstrated an unblemished record of ineptitude, failure, and bureaucratic incompetence in their "feeble efforts over the last 40 years to make the DOE and the corporate operators of the Plant obey the law."

- "...the Plant has been and continues to be operated by government and corporate employees, who have placed themselves above the law and who have hidden their illegal conduct behind the cloak of 'national security.' "
- "...government employees and corporate employees, who have operated the Plant, have committed the most serious crime that can ever be committed in a democracy...breached the public trust through a clever public relations campaign of distraction, deception, and dishonest denial."

The issue of governmental service as a public trust is fundamental to how American democracy should function. This explicit raising of the extensive violation of the public trust in the operation of Rocky Flats indicates the depth of conviction of the drafters of this section.

Defenses Encountered By Citizen Grand Jurors To Explain Government Employee Behavior Judged To Be Criminal

The Second Citizen Special Federal Grand Jury Report was especially harsh on what the authors saw as criminal behavior by employees of the U.S. government—employees they obviously held to a much higher standard of expected conduct.

Employees of DOE and other government departments and agencies had defended themselves before the Special Federal Grand Jury with versions of four core arguments, all of which were rejected by the citizen members:

Argument 1: *Employees were immune from prosecution.* This assertion was rejected because the sovereign immunity of the U.S. government does not extend to immunizing individual government employees from prosecution for criminal conduct.

Argument 2: *Since they did not direct the illegal operations, they could not be held to be accountable for criminal conduct.* The response of the citizens on the Grand Jury was that employees of the U.S. government are subject to the laws of the country.

Argument 3: *Incompetent and negligent management is incompetent and negligent behavior, not a crime.* The response of the Grand Jurors was that "Criminal conduct should never be a part of a government employee's work."

Argument 4: *They were ignorant of the law or relied on faulty legal advice, and therefore should not be held accountable for criminal acts.* The response was that "If government employees do not obey the law, we cease to be one nation under law."

Another variation of behavior by government employees before the Special Federal Grand Jury was the assertion by some of them that it was not their assigned duty to make sure that the private contractor followed applicable law in operating the Plant. The response to this argument was that government bureaucrats placed "form or process over substance," and that they ignored "repeated warnings from multiple sources that the Rocky Flats Plant was being operated in violation of numerous environmental laws."

Government employees, in the view of the Second Citizen Special Federal Grand Jury Report, must be held "accountable for their complicity with and toleration of criminal acts at the Plant." There is no excuse for officials of the U.S. government "directly and indirectly permitting the violation of the law by others." This was further characterized to be the bureaucratic approach of "hear no evil, see no evil, do no evil."

The concluding section of the first Introduction to the Second Citizen Special Federal Grand Jury Report was sharply critical of attorneys for DOE "whose sole objective was to thwart attempts to subject the Rocky Flats Plant to the rule of law." A theme is the obligation of the legal community toward bringing the rule of law to Rocky Flats.

Another recurrent theme is what the Introduction called the "root of the problem" at the Plant: "...the negligent mismanagement of money and wastes...", and DOE's "aggressive efforts to place the Plant and its operators above the environmental laws by which all other companies must abide."

The Introductory section concludes with the observation that: "The limits of time and space prevent this Grand Jury from detailing each and every example of the criminal culture at the Rocky Flats Plant." That is a declarative statement of what circumstances had limited the Grand Jury, and what they had not investigated. The first circumstance is the limitations in the scope of what the Grand Jury actually investigated. This is explained in the next section: much of it is a surprise. The second circumstance is a surprise as well—why there had not been discussions subsequently made public of seeking a "Presidential exemption" which would have authorized the Rocky Flats Plant to operate out of compliance with environmental law. Such an exemption would not have reduced the massive environmental pollution and risk to human and ecosystem health, but it would have taken the issue of criminal liability "off the table" in the interest of preserving national security. A Presidential exemption was never granted. There is no evidence that one was ever requested by either the Department of Justice or the Department of Energy even though the issues at Rocky Flats did engage the White House Counsel.

Scope Of the Investigation

The principal message of this short section of the report was what remains for future study, and what was not investigated completely.

The Grand Jury did not attempt to determine the facts on such major topics as whether or not there had been a serious risk to the health of employees or area residents from the criminal and negligent acts at the Plant.

"Specifically, the Special Grand Jury did not attempt to determine:

(a) whether the criminal and negligent acts at the Plant (including illegal STD of hazardous and radioactive mixed wastes) pose a serious health risk to employees of the Plant or area residents;
(b) if the Plant was and is illegally storing significant quantities of plutonium within certain ventilation ducts located within buildings at the Plant;
(c) whether illegal wiretapping has occurred or is occurring at the Plant;
(d) whether beryllium dust, which has collected within various ventilation ducts throughout the plant, poses a significant health risk to employees at the Plant or area residents;
(e) whether Rockwell or EG&G have filed false claims with the Federal Government because they have paid thousands of production employees at the Plant an average annual salary of more than $40,000 for each of the last two years, although industrial production stopped at the Plant in 1989;
(f) whether DOE and EG&G have been storing radioactive wastes in gloveboxes at the Plant as a means to avoid exceeding storage limits at the Plant."

(pp. 3-4)

This lack of focus on health effects in the 1989-1992 time period is understandable for two reasons. The first reason is that law at the federal level recognizing the uniquely hazardous conditions faced by workers in a nuclear weapons production complex was not even passed until 2000.

The second reason is that studying health effects is a multi-year undertaking. With the Hanford Nuclear Reservation as an example, worker health studies were multi-year undertakings. Dose reconstruction studies are labor-intensive and time-prolonged undertakings. It was clearly beyond the scope of their charge and the possible for a limited-time Special Federal Grand Jury both to attempt to set in motion comprehensive health studies, obtain the resources to fund them, and to be in office long enough to receive the results.

With respect to the issue of whether or not there was plutonium in ventilation ducts, the answer was clearly "yes." In a relatively short time frame, and with existing technology and concern for worker health and safety, it was not feasible to measure the quantity of plutonium in the ventilation ducts in relationship to environmental crime.

It is not clear why the 200 plus witnesses before the Grand Jury were not asked about wiretapping and what the relevance was of the issue.

The threat of beryllium dust was both clear and present. But the same limitations apply to documenting the health effects of beryllium as to documenting other health effects.

The issue of filing false claims is useful to note. But Rockwell International left on comparatively short notice. A problem for EG&G was that some production processes were abandoned in "midstream" by Rockwell International and had to be completed by EG&G. Also, DOE had wanted to reopen Rocky Flats to produce a different configuration of plutonium igniter for new warhead designs. Therefore, it is understandable that employee levels were maintained in the 1990-1992 time frame.

The final issue is that of storing radioactive wastes in glove boxes as a means of avoiding exceeding Plant storage levels. It is not clear why this was not investigated, especially since storage of dangerous amounts of plutonium in gloveboxes was a contributor to the ferocious nature of the start of both the 1957 and 1969 fires.

Not mentioned in this section, but in the one immediately following (the Second Introduction which starts on page 5 of the original report), is the reason for not addressing Clean Air Act violations in contrast to the intense focus on the Resource Conservation and Recovery Act and a lesser focus on the Clean Water Act. The reason given as to why the Grand Jury avoided the Clean Air Act is that it "contains virtually no criminal sanctions for violations of the Act."

With respect to violations of the Clean Water Act, this Second Citizen Special Federal Grand Jury Report disclosed that they had "heard testimony that radioactive material had been released from the Plant into the drinking water supplies of various urban users and downstream agricultural users...." But the Grand Jury had been advised specifically that the Supreme Court of the United States had ruled in *Train v. COPIRG* that the Clean Water Act (and no other federal law as well) did not prohibit dumping radioactive waste into water supplies. Therefore, this area had not been a focus of the Grand Jury.

Second Introduction

The Second Introduction is less of a contextual overview of the work of the Grand Jury than was the First Introduction. The Second Introduction began by stating that the criminal conduct which occurred at Rocky Flats was grouped by the Grand Jury into three broad categories:

1. Single event or discontinued crimes like the chromic acid spill incident;
2. Criminal offenses which were on-going such as the failure to monitor adequately groundwater contamination or the on-going non-permitted storage of radioactive plutonium "residues"; and
3. Crimes against the public trust.

The Second Introduction was used to define further what the Grand Jury did <u>not</u> do:

1. The Grand Jury "has not attempted to evaluate the safety of prior, current, or proposed activities at the Plant."

The reason given for not attempting to assess safety was that the charge to the Grand Jury was to "determine if crimes had been committed at the Plant." This was explained further in terms of the scope of their work day-to-day: "If a particular condition or activity at the Plant did not appear to violate an environmental law, the Grand Jury did not receive or review any testimony on the subject."

2. "In particular, The Grand Jury did not evaluate whether the Plant can be restarted or operated in a safe manner."

This Second Introduction distinguished between "safe operation" and the restarting decision, and whether or not the "on-going criminal enterprise" could be ended. That distinction is a recurring element in the balance of their report.

Time Allocation By the Grand Jury

The major time allocation by the Grand Jury went to determining if DOE, EG&G or Rockwell International had violated "any of the liquid or solid waste restrictions set forth in the Resource Conservation and Recovery Act...." This time allocation was not quantified in terms of percentage of time spent, but was represented as "most of the time." This time allocation is reflected in the amount of space allocated to RCRA violations in the Grand Jury report narrative.

The drafters distinguished between RCRA violations related to the storage, treatment and disposal of various hazardous and radioactive mixed-wastes, the illegal incineration of radioactive mixed-wastes in the incinerator in Building 771 which was characterized as "antiquated" and which did not have any type of RCRA permit, using the Sewage Treatment Plant to treat hazardous wastes without a RCRA permit, "excessive and ill-advised" spray irrigation of effluent from the Sewage Treatment Plant, spray irrigation of chromic acid, flushing "hazardous and radioactive materials" throughout the groundwater, and failure to monitor "adequately" the growing groundwater contamination problem.

There were a series of different issues at Rocky Flats of concern to the Grand Jury. These included:

- "...criminally false statements regarding environmental conditions and compliance at the Plant";
- Defrauding the U.S. government and taxpayers through "improper performance bonus awards...";
- The refusal of the U.S. Attorney and his staff "...to assist the Grand Jury in its effort to write this report..." and in drafting "the indictments and presentments" voted by the Grand Jury;
- The refusal by the U.S. Attorney and his staff to permit the Grand Jury "to question certain witnesses (including an agent of the Federal Bureau of Investigation) ...";

- The U.S. Attorney in December of 1991 indicated that the Grand Jury should not write a report which the drafters believe that "it is permitted to do by statute;"
  Note: The U.S. Attorney is directly quoted in the draft as saying that "it would be inadvisable for the Grand Jury to meet again." The word "inadvisable" is circled with a line going to the margin of the page with the word "inapropos" written in by hand.

- The U.S. Attorney "granted immunity to virtually ever person who testified before the Grand Jury without consulting in advance with the Grand Jury."

This last issue raised two important points. The first point about this Second Introduction is that the phrase "probable cause" is used to refer to their assessment of the evidence. It is clear that at least one Assistant U.S. Attorney "expressed with vigor" his opposition to indicting "any government employee" with only one exception.

Elsewhere, the drafters used the phrase "preponderance of evidence" which can refer to the standard for deciding a civil case, as opposed to the "beyond a reasonable doubt" standard for a criminal case, which they never used.

Government Agency Negligence and Malfeasance

A central focus of the opening section of this portion of the report is "how such extensive illegal conduct could have gone undetected for such a lengthy period of time." As they framed it, there are four factors which contributed to this circumstance:

- DOE and its operators "acted with haughty indifference to environmental laws...." The reason was "the firm conviction that they were pursuing a goal that was more important than compliance with the law..."

Note: There is a handwritten change in the text from "a goal that was" to "goals that were." Whatever the final version would have been, this factor states an element of what Secretary of Energy James Watkins and many others considered to be a hallmark of the DOE "culture."

- DOE directed "a conscious and ongoing effort to evade the application of environmental laws...." DOE was also a participant.

- The Colorado Department of Health and the U.S. Environmental Protection Agency were "lax and inept in attempting to enforce environmental laws and regulations..."

Note: There are significant instances where "higher ups" took authority away from appropriate line managers who had intended to enforce compliance with environmental law. A concrete example was provided on negotiations in 1991 on the Federal Facility Compliance Agreement. James Scherer, the Denver Regional Administrator for EPA, refused to sign the agreement because there were no enforcement penalties for DOE in continuing its RCRA noncompliance. An EPA (unidentified) Deputy Administrator relieved Mr. Scherer of responsibility for approving the agreement. According to the draft report, a two-year agreement was subsequently signed without civil penalties or enforcement mechanisms with the concurrence of the White House Counsel and other high-ranking political appointees.

- The assertion that Rockwell and EG&G--"with the full knowledge and blessing of DOE--have excused their illegal conduct by asserting that they cannot operate the Plant without violating one or more environmental laws."

There is a context for considering this last factor. According to the citizen drafters, "DOE did not review nor monitor Rockwell's budget for environmental compliance." One reason cited was that the DOE Manager at Rocky Flats in 1985 had 55 total employees. Only two of them had even part-time assignments for all of the issues of safety and the environment. Therefore, DOE was dependent on Rockwell for compliance information. DOE was in essence a "dependent endorser" of Rockwell's actions.

The citizen drafters cited sealed Exhibit 1478 as providing concrete illustrations of the "overwhelming number and complexity of the environmental areas for which the DOE was supposed to monitor Rockwell in 1987 and failed to do so effectively."

Instead of "independent oversight," DOE acted "more often than not as the 'puppet' of Rockwell." More than being a "puppet," the DOE Plant Manager "frequently made false written statements with knowledge of the falsity of his statements or with a reckless disregard for knowing whether his statements were false."

Note: It was this DOE Plant Manager that an Assistant U.S. Attorney reportedly did not speak against indicting. Whether or not the Grand Jury did indeed vote to indict this individual, any indictments that they might have voted were shredded by the U.S. Attorney's Office.

Failure To Budget Environmental Compliance and "Bureaucratic Games"

Paragraphs 26-36 contain fundamental conclusions about the flawed interactions, to the detriment of the environment, between Rockwell International at Rocky Flats, the DOE Oversight Office at Rocky Flats, the DOE Regional Office in Albuquerque, and the headquarters offices of both DOE and EPA in Washington, D.C.

The critique has the following elements:

- DOE announced policy is that "obeying the law is our No. 1 priority."
- Regional DOE "...instructed Rockwell to obey the law, they failed to budget the amounts necessary to comply with RCRA

and other laws, when Rockwell advised them specifically of many of these needs."

- DOE's regional and headquarters offices..." explicitly discouraged Rockwell from complying with environmental laws by omitting environmental compliance from the list of criteria for the award of large bonus fees."
- DOE officials "dangled huge production incentives in front of Rockwell managers that could be earned most easily only if environmental compliance was ignored."

Three final major points made in paragraphs 26-36 get to the heart of conclusions reached by at least some of the Grand Jurors:

- In the absence of a Presidential exemption, the headquarters offices of DOE and EPA approved Federal Facility Compliance Agreements which allowed Rocky Flats to continue violating the groundwater monitoring requirement of the Clean Water Act, store residues in violation of RCRA, generate mixed-wastes for which there is no permanent storage site, and continue to commit "environmental crimes."
- DOE would not sign a compliance agreement "which contained any enforcement penalties." DOE employees knew that "one Federal agency (EPA) could not impose sanctions against another Federal agency (DOE) for violating environmental laws."
- DOE has become "a self-regulating agency, which is above the law and without accountability, except to this Grand Jury."

The Second Grand Jury report states that they cannot indict a Federal agency for violating the law. Therefore, "DOE is identified as an unindicted coconspirator with Rockwell, EG&G, and individuals in an ongoing conspiracy to violate the environmental laws of the United States."

Harsh Assessment of Action (Seldom Taken) and Inaction (Mostly) By the Environmental Protection Agency and the Colorado Department of Health

A harsh assessment by the Grand Jury is of the inaction by the Colorado Department of Health. The Citizen Grand Jury does not cite evidence in support of their assertion that before the FBI raid in June of 1989, CDH knew of criminal violations of both RCRA and the Clean Water Act, including groundwater monitoring and did "nothing about it." CDH also did "nothing about it" over the false elements in the RCRA Part B application which subsequently proved to be such a cornerstone of environmental crime allegations. CDH also "responded in an inappropriate, ineffective bureaucratic method" to spray irrigating "massive volumes" of Sewage Treatment Plant effluent onto the East Trenches.

EPA also failed to respond to the environmental crime committed by spray irrigating onto the East Trenches which was not permitted by Rockwell's NPDES Permit. EPA only wrote to DOE about this in January of 1990. While CDH and EPA focused on one sealed bag of beryllium dust stored in a closet which did not have a storage permit, there was no attempt to force DOE and its contractors to remove the substantial quantity of beryllium dust which they knew resided in the air ventilation ducts throughout the 400-series and 800-series buildings at the Plant.

The drafters of this section of the Second Citizen Special Federal Grand Jury Report chose to characterize the (lack of) response by CDH and EPA as "a classic illustration of bureaucratic form over substance." In this illustration:

> ...CDH and EPA have elected not to argue with DOE's position on the beryllium, which remains in the air ducts.

The DOE argument is presented by Grand Jurors to be that since the beryllium dust had not "been generated" as a "waste" within the meaning of RCRA, despite its extremely hazardous nature, it is not subject to RCRA. The drafters of the Second Citizen Special Federal Grand Jury Report concluded that workers within the Rocky Flats Plant could be breathing the dust, and beryllium dust could be escaping into the atmosphere outside of the Plant. The EPA's December 1988 position on the regulation of beryllium is conveyed in sealed Exhibit 1473.

A final example cited in this portion of the Second Citizen Special Federal Grand Jury Report is their negative assessment of the granting of immunity by the U.S. Attorney to an attorney for DOE assigned to Rocky Flats. The first comment was that the DOE attorney would have been indicted if it had not been for the grant of immunity. The context for this harsh judgment was the decision by DOE's Deputy Plant Manager in 1987 to classify all mixed-wastes as subject to being managed as RCRA mixed-wastes under DOE's "By-Product" Rule. The DOE Plant Manager was persuaded not to act on the Deputy Plant Manager's position by the DOE resident lawyer. The DOE Plant Manager did not include mixed-waste residues among the wastes listed in DOE's 1987 RCRA, Part A Permit application.

Rockwell's private attorneys, Williams and Connolly, later cited the *Sierra Club* decision of 1990 as finally subjecting Rocky Flats to RCRA. The point that Williams and Connolly lawyers made was that as a matter of settled law, the decade of the 1980s was a period of "regulatory confusion" as to whether the new environmental laws applied to Rocky Flats and to other government sites for the production of nuclear weapons. The drafters drew attention to sealed Exhibit 849 as presenting the position of the DOE Albuquerque office.

Earlier in the Grand Jury report in paragraph 33, the drafters described what they understood as a pattern of conduct by administrators and attorneys:

> On those occasions when Rockwell advised DOE officials that it might be violating an environmental law or that its conduct could be challenged by regulatory agencies, DOE officials either ignored such notices from Rockwell, joined with Rockwell in concocting a rationalization for such conduct, or actively participated in plans to shield Rockwell from attack and conceal potentially damaging information from being disclosed to the public or regulatory agencies.

It was not until 1999 in the *Stone* False Claims Act lawsuit that the extent of notification by Rockwell to DOE was disclosed: there were 165 instances cited by Rockwell where notification had been provided to DOE.

Chromic Acid Spill

The Chromic Acid Spill was a one-time occurrence which had been described for the Grand Jury in the now sealed Hobbs report as Exhibit 1480. This section of the draft report was written without access to the specifics as presented by Exhibit 1480.

Chromic Acid, presented in the draft report as a bright green fluid, should have been diverted from the Sewage Treatment Plant, and the Rocky Flats Plant shut down. The reason this should have been done was to prevent what apparently occurred: spray irrigation of untreated wastewater with chromic acid in it. Rockwell purportedly decided to spray irrigate onto the South Spray Field to avoid "discharging the colored effluent directly downstream." When a Rockwell employee expressed concern to a DOE counterpart, Rockwell was requested to discontinue the spray irrigation until the "foreign substance" had been identified. This request was presented by the Grand Jury Report as ignored.

The conclusion drawn by the citizen drafters of this section was that:

> ...Rockwell displayed a callous disregard for the effects of its actions on the surrounding communities and the environment.

In the judgment of the citizen drafters, Rockwell should have advised "Bloomfield, Westminster, CDH, EPA of the toxic spill, the spray irrigation of a toxic substance in an area supplying Broomfield's drinking water, or the discharge downstream from the spray irrigation fields of partially treated or untreated wastewater effluent." The notation "[Indict]" appeared for the first time in this citizen report.

Note: Subsequent fact-finding, not apparently available to the Grand Jurors, revealed that a worker had left a faucet running which caused the spill. Chromic Acid contains sulfuric acid among other toxic chemicals.

False Statements
This section focused on violations of the public trust: not telling the truth under oath. Apart from the content, the most striking feature of this element of the Second Citizen Special Federal Grand Jury Report is the level of specificity of the charges supported by sealed Exhibits and accompanied by the identifying number of the Exhibit.

False Statements To a Colorado Department of Health Inspector

By way of example; DOE and Rockwell employees lied to a CDH inspector regarding the use of the Solar Evaporation Ponds which is a felony under the False Statements Law for making material misrepresentations. Rockwell had in fact deposited salt brine in Solar Evaporation Pond 207-C on a regular basis "during 1986, 1987, and 1988." The notation "[Indict]" is used.

False Statements About Waste Streams At Rocky Flats

It is a crime to underreport the number of waste steams at Rocky Flats. FBI Agent Jon Lipsky had prepared a table:

> ...to illustrate the dramatic jump between 1987 and 1988 in the number of waste streams at the Plant which Rockwell reported or acknowledged to exist. The reported number jumped in this short period from 2,000 to more than 4,000 waste streams.

The citizen drafters cited now sealed Exhibits 2025 and 2017 and invited comparison with the number of waste streams reported in 1984 for RCRA and NPDES permit applications with the comment that:

> This information can also be used as an illustration in the report of the Plant's practice of lying to the public regarding the scope and nature of environmental problems at the Plant.

Broader False Statements About the Status of the Solar Evaporation Ponds

This section begins with the notation "[Indict]." Now sealed Exhibit 454 contains a letter from a top Rockwell administrator at Rocky Flats to the DOE Plant Manager which stated that "no waste had been deposited in any of the solar ponds during the previous 12 months." The DOE Rocky Flats Manager subsequently made the same material misrepresentation to the Colorado Department of Health. In a written report, Rockwell represented to DOE that Solar Evaporation Pond 207-C had been closed (sealed Exhibit 1677). This was a material misrepresentation (see above).

Deceiving Regulators By Repackaging Containers of Hazardous and Mixed-Waste

This section began with the notation "[Indict]." Deceiving regulators by repackaging containers of hazardous and mixed-waste was a "shell game" which had the effect of establishing a new "accumulation date." Establishing a new "accumulation date" was a means of bypassing and avoiding "certain prohibitions against storage of such wastes in a non-permitted facility for longer than a specific period of time, typically 90 days or one year."

Rockwell Graded Itself Falsely For Purposes of Generating Bonus Fees
This section began with the notation "[Indict]." During the last five years that Rockwell International operated the Plant, according to the Second Citizen Special Federal Grand Jury Report, "DOE relied almost exclusively on Rockwell to supply the information and suggested grades for each category on which the amount of the bonus would be determined." Once initial recommendations by Rockwell for the Rockwell bonus for Rocky Flats had received whatever modification by DOE and been incorporated, "DOE's Plant Manager would submit the revised bonus fee report to DOE's Albuquerque Office on DOE letterhead, as his recommended bonus fee for Rockwell."

Two comments on this practice are in order. The first comment is on the citizen drafters' observation that "many of these reports contained false and misleading statements concerning management and environmental practices at the Plant." These false statements are a violation of the public trust as well as what the citizen drafters saw as an indictable crime. The second comment is that it is not correct to state that this practice continued for all of the last five years during which Rockwell operated the Plant. The process became a matter of discussion between DOE Secretary James Watkins and Rockwell Chairman Donald Beall in the Fall of 1989. Rockwell International's position was that the amount of their bonus was to be determined under their contract by the DOE Regional Office in Albuquerque, New Mexico.

The actual bonus amount was not determined in Albuquerque but in Washington, D.C. According to Secretary Watkins (Watkins, 1989b), Rockwell was determined to have earned only 40 percent of the possible bonus award. Rockwell subsequently filed a lawsuit which was one more instance of the deteriorating and increasingly contentious relationship between DOE and Rockwell.

Groundwater Monitoring
Groundwater contamination is one of the most contentious and least understood legacies of the production era at Rocky Flats. One reason for the contentiousness is the danger posed by contaminated plumes of groundwater. Another and related reason is the lack of adequate groundwater monitoring in both the Dow Chemical (1952-1974) and Rockwell International (1975-1989) production eras. The Mary L. Walker memo from 1986 (DOE, 1986a) specifically cited the inadequacy of groundwater monitoring. The Mary L. Walker memo is sealed Exhibit 1629.

Citizen drafters of this section of the report concluded that DOE never invoked the authority from the DOE national office to direct the DOE regional or DOE Rocky Flats offices to bring the Plant into compliance on groundwater monitoring. This section on not taking corrective action includes the notation "[Indict]." EG&G is criticized for not bringing Rocky Flats into compliance with the groundwater monitoring requirements of the Clean Water Act. This observation about EG&G begins with the notation "[Indict]."

The citizen drafters noted that DOE itself had concluded that toxic groundwater threatened the water supplies of the downhill cities of Broomfield and Westminster, and had determined that toxic groundwater was the number one threat to public health throughout the entire nuclear weapons multi-state production complex.

According to the citizen drafters, EPA had known as early as 1985 that DOE and Rockwell were violating their earlier certificate of compliance with requirements for groundwater monitoring. Sealed Exhibit 1058 is cited as establishing unequivocally that DOE knew in July of 1985 that the groundwater monitoring system at Rocky Flats was deficient.

This section of the report is not well developed in relation to documenting the severity of the groundwater problem.

Figure 11.1 Slumping Pondcrete Presented In the Cook Class Action Lawsuit

This visual presentation of Pondcrete captures the magnitude of the failure. It is not known without access to the Exhibits themselves what was actually available to the Grand Jury about the formula change in the ratio of concrete to low-level mixed-waste from the Solar Evaporation Ponds which caused the failure to solidify. The original formula contained a greater amount of cement in relationship to low level mixed waste under the initial supervisor.

Pondcrete

Despite its brief length of only eleven paragraphs, the section on Pondcrete is notable for not including the notation to "[Indict]" anyone, and for the number of exhibits which are referred to (sealed exhibits 1486, 1610, 819,1618, 1617, 1619, 1620, 1621, 1624, 2047). Some of the "best evidence" on the background of the Pondcrete problem and its significance is in the FBI interviews in the Congressional archives. According to Exhibit 2047 cited as the source by the Grand Jury, it was in August of 1987 that Rockwell employees first learned that "there were serious deficiencies" with the manufacturing process by which Pondcrete was being fabricated.

The background is reported to be that, according to sealed Exhibit 1486, DOE officials in both Washington, D.C. and the Albuquerque regional office were involved in the decision to produce and store

Pondcrete and Saltcrete at Rocky Flats. The following is the chronology on Pondcrete as constructed by the citizen drafters:

- 1986—The intent of DOE and Rockwell was to make Pondcrete and store it for six months before shipping it to the Nevada Test Site for permanent underground storage. The Pondcrete was intended to be a solid form of concrete containing a mixture of solid and liquid low-level mixed-wastes from the Solar Evaporation Ponds.
- Date Not Reported in Grand Jury Report—A Nevada Test Site official questioned whether Pondcrete could be stored at their location because it contained some land-banned substances.
- Date Not Reported in Grand Jury Report—The Rocky Flats Plant Manager for Rockwell made a decision without consulting DOE to store the Pondcrete on the uncovered asphalt 750 Pad.
- August of 1987—Rockwell employees first learned that "there were serious deficiencies" in that the Pondcrete did not form into the expected "hardened" monolithic blocks. Pondcrete reportedly had the consistency of "gooey" slime.
- August of 1987 for "almost" two years—Rockwell and DOE continued to manufacture "play-doh" Pondcrete blocks without substantially changing the formula with the knowledge that "the Pondcrete mixture was not sealing the hazardous and radioactive constituents into concrete blocks."
- May of 1988—Rockwell first informed DOE Plant Management of the Pondcrete spill of mixed-waste from a container.
- April of 1989—the Albuquerque DOE office made their first request for funds to pay for covering the outdoor storage of the cardboard boxes containing the Pondcrete, as cited in the sealed Exhibit 1618.

This is the essence of the citizen jury report on Pondcrete. There are sealed Exhibits reporting the situation. There were no notations of "[Indict]."

The Understanding By the Second Citizen Special Federal Grand Jury of RCRA and Its Requirements

RCRA problems featured prominently in the charges against Rockwell International after the Grand Jury was dismissed. RCRA violations were a main element of the felonies admitted to by Rockwell International in the final plea agreement settlement. The citizen drafters of the Second Citizen Special Federal Grand Jury Report devoted their most extensive coverage to reporting conclusions on criminal violations of RCRA. It is useful, therefore, to begin by conveying their understanding of the requirements of RCRA. The citizen drafters also provided their understanding of the reasons why DOE made the decisions they did from the inception of the RCRA legal obligation through 1989.

As part of the Wolpe subcommittee Congressional investigation (Wolpe, 1993), it was revealed that there had been large purchases of oxygen in November of 1988. That in itself is not proof of operation of the Incinerator in December of 1988. But the Building 771 Incinerator is reported to be the only user of oxygen at Rocky Flats. Since the issue never came to trial, and was not included in the settlement agreements, the opinion formed by citizens on a Grand Jury which heard testimony under oath for over two years is a distinctive perspective on the Building 771 incinerator dispute and the requirements of RCRA.

The citizen drafters of this section of the Second Citizen Special Federal Grand Jury Report highlighted their understanding of three elements of RCRA as they apply to Rocky Flats. The first two elements are:

- RCRA generally prohibits the storage, treatment, or disposal of hazardous waste at private and government facilities without a permit issued by EPA or an authorized state government such as Colorado.
- RCRA permits the temporary storage for less than one year of land-banned hazardous and mixed-wastes, if the waste stored at that site is in the process of being treated before it is stored permanently elsewhere.

The assertion by the citizen drafters is that there are no "permitted methods or techniques...for the treatment of the liquid, low level, mixed-waste." The commission of environmental crime stems from continuing to store liquid, low-level mixed-waste on the Rocky Flats site beyond the time allowed for temporary storage for treatment. The third element of RCRA considered relevant in Rocky Flats is that RCRA requires a permit for all STD operations at the Rocky Flats Plant which would include the Building 771 incinerator:

- From November of 1980, DOE has been obligated through its operator of the Plant to apply for a RCRA Permit for all hazardous waste STD operations at the Plant.

The assertion is that DOE initially failed to apply for a permit for the Building 771 Incinerator, claiming that either all hazardous waste was exempt from RCRA, or that RCRA does not apply to mixed radioactive and hazardous waste at nuclear weapons production facilities covered by the Atomic Energy Acts as amended.

Chronology of Arguments Advanced By DOE Regarding RCRA Permits As Compiled By Citizen Grand Jurors

The Second Citizen Special Federal Grand Jury Report provided the following chronology:

- 1982 DOE Order No. 54.880.2 established an internal DOE program for hazardous waste which was a parallel to but exempt from EPA oversight, but which conformed to RCRA.

- 1983 According to the DOE General Counsel, RCRA did not apply to DOE nuclear weapons facilities because they were expressly exempted, including Rocky Flats, by provisions of the Atomic Energy Act.

- 1984 In *LEAF v. Hodel*, a Federal District Court Judge in Tennessee held that purely hazardous wastes at DOE facilities was subject to regulation by EPA under RCRA.

- 1984-1985 In response to *LEAF v. Hodel*, DOE and Rockwell claimed that all mixed-wastes were exempt from RCRA because they contained traces or greater amounts of radioactivity.

- 1985 In a DOE memorandum from July of 1985, the opinion was expressed that all mixed-wastes at DOE facilities were subject to EPA's regulatory authority. (The original DOE source for this citation is not further identified).

- 1985 In November of 1985, Rockwell and DOE claimed that the Colorado Department of Health did not have authority over the mixed-wastes which were stored at Rocky Flats. In the DOE RCRA Part B application to CDH of November of 1985, DOE did not disclose the storage of mixed-wastes.

Note: Part B is a site-specific narrative.

- 1985 In December of 1985 CDH issued a Notice of Deficiency because DOE did not acknowledge storing low-level mixed wastes.

- 1986 In July of 1986, a compliance agreement was reached between DOE, CDH, EPA and Rockwell. DOE did not submit to the jurisdiction of CDH and EPA over the storage, treatment or disposal of transuranic mixed-wastes (such as those containing traces of plutonium residues).

- 1986 In November of 1986, EPA delegated to CDH the authority to regulate low-level, mixed-wastes at Rocky Flats.

- 1987 In announcing compliance with what was called the "By Product" Rule of May 1, 1987, DOE acknowledged that RCRA applied to the hazardous portion of all mixed-wastes at DOE facilities, including transuranic mixed-wastes.

- 1987 Attorneys for DOE at Rocky Flats, and Rockwell International at Rocky Flats, citing *American Mining Congress v EPA* of July 31, 1987, advised the DOE and Rockwell International Rocky Flats Plant Managers that residues were not being "discarded" and therefore were not subject to RCRA. Residues could be stored for an indefinite period of time at Rocky Flats without being subject to RCRA regulations as part of a "continuous manufacturing process."

- 1987 On October 3, 1987 the DOE Regional Deputy Manager at Albuquerque adopted the position taken by both sets of attorneys at Rocky Flats as documented in sealed Exhibit 849; namely, that mixed-waste residues were not being "discarded." While "residues" are hazardous wastes, they contain some plutonium along with some hazardous constituents. The "residues were not being immediately reused" but were part of a "continuous manufacturing process" with an ultimate goal of extracting plutonium for "reuse." Subsequently, DOE submitted a revised RCRA Part A Permit application without disclosing the stored plutonium-laced residues.

Rationale For the Evolving DOE and Rockwell Position On RCRA As Understood By the Second Citizen Special Federal Grand Jury Report

Citizen drafters had obviously given careful thought to what they considered to be the faulty and deceptive reasoning behind the evolving positions of DOE and Rockwell International on the mixed-waste residues issue, and the DOE/Rockwell corollary claim that plutonium residues in mixed-waste were an essential part of a "continuous manufacturing process."

The RCRA problems section began in paragraph 94 by characterizing the *American Mining Congress* interpretation by Regional and Rocky Flats DOE and Rockwell as "strained" and choosing to "ignore the controlling facts in the court's written opinions."

> Unlike the situation in *American Mining Congress* the residues were not being immediately reused at the plant in a continuous manufacturing process to extract plutonium for reuse.

An argument in the Second Citizen Special Federal Grand Jury Report was that at Rocky Flats, some of the steel drums containing residues had been stored for over 15 years. And none of the residues "were being immediately processed as part of an on-going manufacturing cycle." DOE and Rockwell were characterized as "so obsessed with secrecy and avoiding any regulation of operations at the Plant by CDH or EPA."

That was the background explanation offered for the "strained" interpretation of *American Mining Congress* by DOE's Regional Deputy Manager, and the failure to disclose "any information about the residues, which Rockwell was storing at the Plant." This portion of paragraph 94 contains the designation "[Indict]" but it does not further identify the specific target for the indictment. The citation for the position of the DOE Deputy Regional Manager is identified as now sealed Exhibit 849, and the immediate consequence was the submission of another false revised RCRA Part A Permit.

Explanations For the Choices Made By DOE and Rockwell International Over the "Residues" Issue and the "Continuing Manufacturing Process" Claim

Both EPA and CDH are represented as "having backed down" in 1987 when "DOE and Rockwell refused to provide this information to them." "This information" refers to the number of barrels of residue stored, how long they have been stored, and "what hazardous materials (if any) were mixed with the radioactive component of the residues."

The apparent "best evidence" for understanding the motivations of Rockwell International and Rocky Flats DOE for "wanting to avoid RCRA regulation" are set forth in a letter of December 9, 1987 to the DOE Regional Manager at Albuquerque. These reasons are memorialized in the now sealed Exhibit 1056. The drafters noted in paragraph 96 that the reasons should be inserted "into the text" (Note: This did not occur as part of the drafting process).

These were different reasons offered for the choices which were made by Rockwell International and by DOE. Paragraph 97 with the notation "[Indict]" offers the citizen drafters' explanation for the Rockwell choices:

> [Indict] Rockwell had a tremendous incentive to avoid disclosing the status of residue storage at the Plant. Rockwell management knew that the residues were being stored and labeled at the Plant in a manner which violated several provisions of RCRA. In addition, Rockwell management placed a high priority on excluding CDH and EPA inspectors from all areas of the Plant, whenever and wherever possible, in part, because they feared that CDH or EPA might discover, during the course of their inspections of the storage of the residue drums, other RCRA operating violations at the Plant, such as the usage of the incinerator in Building 771 as a non-permitted process for treating mixed-wastes.

The issues surrounding the Building 771 Incinerator will be addressed under a separate heading below.

The reason why DOE adopted the "strained application" of the *American Mining Congress* opinion was presented in paragraph 98:

> The cost of RCRA compliance was a major factor in DOE's decision to advance the strained application of the *American Mining Congress* opinion to residue storage at the Plant. DOE and Rockwell officials did not want to spend the money, which would have been required to comply with RCRA in storing the residue drums at the Plant, e.g. the construction of the residue drum storage facility.

In the judgment of the Citizen Grand Jurors the cost of RCRA compliance was presented as "a major factor" in DOE's decision.

Paragraphs 97 and 98 contain the assessment of the citizen drafters as to the motivations of DOE and Rockwell for not complying with the RCRA storage requirements. Paragraph 99 begins with a call for the indictment of Rockwell International "[Indict]" and addresses the "40-year inventory" of radioactive residues:

> [Indict] The Rocky Flats Plant has a "40-year inventory of radioactive residues." Some of these residues have been "temporarily" stored at the Plant for plutonium "recovery" for more than 20 years, and the most optimistic projections suggest that it will take at least 20 years to process all of these residues and recover the plutonium in them. These residues include radioactive wastes, which were collected after a massive industrial fire destroyed significant portions of the Plant in 1969. DOE has never applied for nor received a RCRA Part A Permit or RCRA interim status to STD residues at the Plant.

Elsewhere in the draft (paragraphs 121, 122, and 123) it is reported that in 1988 and 1989 DOE local management instructed Rockwell to "recover plutonium from those residues from which the plutonium could be most easily extracted." According to the citizen drafters, this raised two insurmountable problems: cost which is easily understood, and the assertion that: "there was no known procedure by which the plutonium could be extracted from the other waste substances."

Paragraph 122 begins with the "[Indict]" notation in the context of calculating an "economic discard level" as a strategy for avoiding RCRA regulation:

> [Indict] In an attempt to avoid RCRA regulation DOE officials calculated an economic discard level (herein "EDL") for certain residue materials, which have been collected and stored at the Plant for several years. However, the calculation of the EDL for these residues was fraudulent because no method exists by which the plutonium can be extracted from these mixed-wastes. Similarly, other mixed–wastes (in addition to incinerator ash) have been and are being stored at the Plant without a RCRA Permit, and they are labeled as "residues" although it has been many years since anyone has attempted to recover the plutonium from these wastes and it is unknown whether the extraction process will work.

Two observations by the citizen drafters are worth highlighting: the first is that all of the residues could have been sent to the DOE Idaho National Experiment Laboratory (INEL) and as the drafters phrased it, "could be shipped there today." This first observation meant that shipment to INEL [also referred to in Dow Chemical Archives as Arco], would have allowed Rocky Flats residues to be placed in retrievable storage for recovering the plutonium "...if a process is developed for recovering the plutonium from the residue" and if it is "economically wise to attempt to recover the plutonium from such residues."

The second observation is unelaborated: "...DOE chose for political reasons to leave the residues in storage at the Plant without a RCRA Permit." While the DOE choice is characterized as "political," paragraph 120 characterizes Rockwell's choice "to terminate its contract" as one where continued operation "...would expose the company to criminal prosecution under RCRA:"

> [Indict] Rockwell decided in the fall of 1989 to terminate its contract to operate the Plant when its National Director of Environmental Affairs advised the Chairman of Rockwell that any operation of the plant would expose the company to criminal prosecution under RCRA. The Director of Environmental Affairs recommended this action because RCRA makes it a crime for any person or company to generate mixed-wastes for which there is no permitted, permanent storage facility. Since no facility in the United States has a permit to permanently store low-level, mixed-waste, Rockwell violated RCRA, prior to the termination of its contract to operate the Plant, each and every time that it generated any low-level, mixed-waste at the Plant.

Paragraph 120 is accompanied by the notation "[Indict]." The drafters noted that "since no facility in the United States has a permit to permanently store low-level, mixed-waste..." they are not saying that DOE as a total organization had no such facilities under its authority (like INEL), only that the military-industrial contractors operating nuclear weapons production facilities did not.

As a comment on the explanations presented immediately above, paragraph 120 indicates "Rockwell decided in the fall of 1989 to terminate its contract ..." The exchange of correspondence between Donald Beall, the Chairman of Rockwell (Beall, 1989), and Energy Secretary James Watkins (Watkins, 1989a, 1989b) is instructive for identifying causal factors behind the end of the Rockwell era. This correspondence, however, did not reach the level of disclosure known to each of them. For example, Chairman Beall knew the extent of the problem of compliance from Rockwell International (1987). Secretary Watkins had to have known of the "Tiger Team" findings.

The Building 771 Incinerator In the Second Citizen Special Grand Jury Report

In the previously excerpted paragraph 97, the citizen drafters identified the "usage of the incinerator in Building 771 as a non-permitted process for treating mixed-wastes..." as a RCRA violation at Rocky Flats which Rockwell did not want CDH or EPA to discover. Paragraph 97 is accompanied by the "[Indict]" notation.

Paragraph 102 introduces the issue of the "incinerator ash for which there is no process to separate the plutonium from the ash":

> A substantial portion of the residue, which is stored at the Plant, consists of incinerator ash for which there is no process to separate the plutonium from the ash. This residue has been collected at the Plant for decades as the by-product of burning mixed-wastes in the 771 incinerator. In 1971, Dow (the prior operator of the Plant) expended $50 million attempting to develop a process by which the plutonium could be separated from the ash that was produced when mixed-waste was burned in the incinerator. However, this experiment was

abandoned as a failure and no one has attempted subsequently to extract plutonium from the large volume of incinerator ash residues, which the Plant has generated and which are stored at the Plant.

The RCRA Part A permit signed by the DOE Plant Manager on November 1, 1985 "contained a substantial lie." It said explicitly that the incinerator in Building 771 "was not being used to burn or treat hazardous waste." The direct lie was that in fact the incinerator was being used for that purpose. The historical misrepresentation was that in fact the "...incinerator had been used for an extended period of time before that date [November 1, 1985] to reduce the volume of hazardous waste..."

In commenting on the validity of the 1985 RCRA Part A Permit with respect to what CDH knew at the time, the citizen drafters did not reach a conclusion. One possibility was that CDH genuinely did not know. The concern for the citizen drafters was phrased as follows:

> ...it is important to know whether CDH was asleep at the switch again or if DOE was actively concealing from CDH the real use of the 771 incinerator.

CDH, according to someone who had testified before the Grand Jury, had issued previously a Clean Air Act Permit for the Building 771 incinerator. Since the records are sealed, this cannot be independently verified.

Paragraph 107 begins with the "[Indict]" notation, and discloses that:

> [Indict] Rockwell used the incinerator in building 771 on a regular basis during the 1980s (until ???1990???) as a RCRA hazardous and mixed-waste treatment facility. By burning various hazardous and mixed-wastes in this incinerator, Rockwell substantially reduced the volume of such wastes, which DOE and it had to store at the Plant.

The reason that no RCRA permit had been applied for to operate the Building 771 incinerator are conveyed in the paragraph 108.

> [Indict] DOE and Rockwell knew when Rockwell was using this incinerator for this purpose that DOE had never applied for a RCRA permit to operate the 771 incinerator as a hazardous and mixed-waste treatment facility. Moreover, they knew that DOE had not applied for such a permit because the 771 incinerator could not meet the environmental standards and safeguards which RCRA imposes on incinerators that are used to burn hazardous materials. They also knew that the 771 incinerator could not be modified to bring it into compliance with the requirements of RCRA. Instead, they chose to ignore the RCRA requirements and they proceeded to use the 771 incinerator as a treatment method to reduce the volume of hazardous materials, which were stored at the Plant. In the process, they consistently released hazardous materials into the air of metropolitan Denver at levels substantially greater than those amounts which RCRA allows for incinerators.

A bottom-line for the citizen drafters in Paragraph 109 is that "Rockwell and DOE management also knew that no amount of modifications to the 771 incinerator could bring it up to the operating standards, which RCRA requires for an incinerator that treats RCRA hazardous and mixed-wastes."

Returning to the issues raised in the *American Mining Congress* case, and in *Sierra Club v. DOE and Rockwell International,* residuals generated at Rocky Flats were not exempt from regulation under RCRA. This paragraph 110 is accompanied by the notation "[Indict]."

Several issues remain regarding the Building 771 incinerator on which this Second Citizen Special Federal Grand Jury Report sheds light. The first issue is about cost of compliance with environmental law as reported in Paragraph 114:

> DOE and Rockwell chose not to upgrade the 771 incinerator because it would have been cheaper to build a new incinerator to treat RCRA wastes at the Plant. The 771 incinerator relied on outdated, inefficient technology.

The Second Citizen Special Federal Grand Jury Report does not provide a chronology for the use of the Building 771 incinerator in the 1988-1989 time frame. The Building 771 Incinerator became a contentious issue during the Wolpe subcommittee findings of fact and the Department of Justice rebuttal.

Officials from both DOE and Rockwell including the DOE Rocky Flats Plant Manager were reported to have been advised in October of 1988 at a time when the Building 771 incinerator was restarted that "no radioactive or mixed-wastes would be burned in the incinerator." That assertion (without attribution) in paragraph 112 is contradicted by information in paragraph 113.

Paragraph 113 begins with the "[Indict]" notation:

> ...Although DOE managers at the Plant and in the Albuquerque office knew that DOE did not have a RCRA permit to operate the 771 incinerator, they instructed Rockwell to restart the 771 incinerator in October of 1988 and to continue operating it without seeking a RCRA Permit or interim status to treat hazardous and mixed-wastes in the 771 incinerator. Despite its awareness of the potentially illegality of burning hazardous and mixed-wastes in the 771 incinerator, the [DOE Rocky Flat Plant Manager] and Rockwell's management directed Rockwell's employees to restart the 771 incinerator and to burn hazardous and mixed-wastes in it.

Paragraph 115, again with the "[Indict]" notation, "Rockwell employees continued to burn hazardous and mixed-wastes in the incinerator in Building 771 until Judge Babcock ruled that such conduct was illegal and a violation of RCRA." This statement as drafted is actually incorrect. Judge Babcock made his ruling on April 12, 1990. Rockwell ceased being the contractor on December 31, 1989.

Issues Related To Environmental Compliance: What Rockwell Knew and When They Knew It–and Why They Did Nothing About It As Presented In the Second Citizen Special Federal Grand Jury Report

An insight into the Rockwell approach to environmental compliance is provided by Rockwell's Director of Plutonium Operations (who was also tasked with supervising Rockwell's Environmental Compliance Programs). The reasons for the approach he advocated are presented in paragraph 133 from an argument he made on December 10, 1987 that DOE should not include residues in the Plant's RCRA, Part A Permit Application:

(a) it would keep CDH and EPA out of Rockwell's production areas and, thereby, avoid inspections which might uncover RCRA and other environmental operating violations;
(b) DOE would have to request an exemption to continue storing transuranic mixed-waste residues at the Plant for which there was no plutonium recovery process;
(c) by avoiding the RCRA permitting process, Rockwell would still be able to conceal from CDH and EPA that Rockwell was incinerating some of the residues in the 771 incinerator although it did not have RCRA permit: (????check on this one in the exhibit) and
(d) the Plant lacked sufficient floor space to store the residues in compliance with RCRA's requirements for separation between storage containers.

The Exhibit cited as the source of the views of the Director of Plutonium Operations is the now sealed 1471. The central points, however, are unmistakably clear, especially when it is recalled that he had the separate (and perhaps unreconcilable) task of overseeing both plutonium operations and environmental compliance. First, there was concern that CDH and EPA could uncover RCRA and other operational violations of environmental law. It would serve a preventive purpose if inspectors could not enter the production areas. Second, while this is the first explicit mention this reader has seen of a Presidential exemption, it is clear from the context that this is not something that DOE wanted to request. Third, and right at the heart of the dispute over operation of the Building 771 incinerator, it was an admission that mixed-waste residues were being burned there without a RCRA permit—one that Rocky Flats could not possibly have obtained if it had applied for one. And it is clear that to someone responsible for both producing plutonium igniters and for environmental compliance at Rocky Flats that the "continuous production cycle" for recovering plutonium from mixed-waste residues was a fiction of the first order.

Now sealed Exhibit 850 from November 14, 1988 is the source which the citizen drafters cited as indicating that DOE and Rockwell "admitted in an internal memorandum that, to the extent to which below economic discard levels were burned in the incinerator in building 771, such burning constituted an illegal, unpermitted treatment of hazardous waste." This paragraph, number 137 was preceded by the "[Indict]" notation followed by "DOE (???)" and Rockwell (???)."

> [Indict] DOE (???) and Rockwell (????) admitted in an internal memorandum on November 14, 1988 [pull Exhibit 850, p. 30] that, to the extent to which below economic discard level materials were burned in the incinerator in building 771, such burning constituted an illegal, unpermitted treatment of hazardous waste.

The final paragraph of the RCRA section, paragraph 138, is especially important for illuminating the contentious nature of the relationship between Rockwell and DOE:

> Rockwell concealed from DOE that Rockwell was using the incinerator in Building 771 as a means to treat mixed-waste by reducing its volume. Rockwell always represented to DOE that the 771 incinerator was being used only as a "process" by which plutonium was being recovered from mixed-waste residues. Rockwell did not disclose to DOE or EPA that Rockwell was using the 771 incinerator as a means to reduce the volume of mixed-waste, which would have to be stored at the Plant or be shipped elsewhere for storage.

RICO

Preceded by the notation "[Indict]" before the sixth paragraph, this short six paragraph section concludes with a short quote from the citizen drafters which sums up their attitude based on the entire section and the proceeding five paragraphs:

> ...no one at any time has ever sought to exempt DOE's employees, its contractors, or the employees of the contractor from prosecution for the criminal nature of this ongoing course of conduct and conspiracy to violate the criminal laws of the United States.

The controlling "no one at any time..." phrase written by the citizen drafters is their impassioned declaration that the United States is a country of laws. The preceding five paragraphs describe the lengths Rockwell (and DOE) went to for the purpose of invoking the sovereign immunity of the country for the purposes of shielding Rockwell, DOE, and their employees from civil and criminal liability.

Paragraph 139 describes the attempt by Rockwell to immunize itself, its employees, and DOE's employees from prosecution for criminal acts:

> Rockwell skillfully and in complicity with certain DOE officials conspired over a period of years to hide its criminal acts and the criminal acts of it employees behind the "skirts" of the sovereign immunity for an agency (DOE) of the Federal Government. Certain Rockwell employees conspired to persuade the DOE Plant Manager and other DOE employees to endorse their illegal acts as the official DOE policy. Further, they drafted correspondence and reports for DOE's approval, which approved of many of Rockwell's criminal acts

> and, thereby, set in motion for Rockwell to subsequently argue that the defense of estoppel by entrapment bars prosecution of Rockwell, Rockwell's employees, and DOEs employees. In the process, Rockwell cleverly attempted to immunize itself, its employees and DOE's employees from subsequent prosecution.

Rockwell employees testified before the Special Federal Grand Jury that they should not be prosecuted for doing what the DOE Plant Manager and other DOE employees told them to do. The citizen drafters had a key phrase: "DOE more often than not acted at Rockwell's direction in authorizing Rockwell to break the law."

DOE employees, in testifying before the Special Federal Grand Jury, sought to become "...a *law unto themselves* and attempted to immunize themselves from prosecution by hiding behind the blanket of the sovereign immunity of the U.S. Government." It clearly was the conclusion of the citizen Grand Jurors that: Rockwell and DOE were indistinguishable co-conspirators to violate RCRA. Rockwell used DOE "as a shield against environmental reporting, regulation, compliance, and enforcement."

The citizen drafters of the RICO section were especially harsh on Rockwell in Paragraph 140 linking failing to disclose criminal activity and manipulating DOE policy "in pursuit of exorbitant bonus and contract fee awards."

> ...Rockwell controlled all of the material information, data, and analysis regarding environmental matters at the Plant. DOE more often than not acted at Rockwell's direction in authorizing Rockwell to break the law. Since Rockwell often failed to disclose all of the relevant facts to DOE's employees, Rockwell and its managers were able to consistently manipulate and control DOE policy to assure that DOE endorsed Rockwell's illegal conduct in pursuit of exorbitant bonus and contract fee awards.

Salt Brine

Salt Brine is a mixed-waste from the evaporator process in building 374. It is a "mixed-waste" because it "contained small quantities of radioactive materials, together with corrosive and organic hazardous materials." The linkage of salt brine to the commission of environmental crime is described in paragraph 147 preceded by the notation "[Indict]":

> [Indict] [Pull Exhibits 1425 (p. 12), 1426 (p. 2), and subsequently numbered exhibits for evidence citing specific dates and quantities.] On November 15, 1987, Rockwell employees transferred 7,800 gallons of salt brine "concentrate" (a hazardous substance within the meaning of RCRA) from Building 374 to Building 774 and ultimately into Solar Evaporation Pond 207(C), which was a closed RCRA unit. [Insert statements regarding specific, multiple other occasions through 1988 of transfer from the exhibits for other days within the statute of limitation.] Use this information also to support the False Statement indictment and

> report section. [See Testimony of Rockwell employee of March 18, 1991.] At the time at which these transfers occurred, the Solar Evaporation Ponds were marked with signs, which stated clearly that the Ponds were "under closure" and that the Ponds should not be used for any purpose. At the time at which each of these transfers into the Ponds occurred, the Plant was not operating under a "state of emergency."

Other paragraphs underscored the fact that the "[Indict]" notation came from the specific approvals which had to have been obtained before the "salt brine concentrate" could be transferred into Solar Evaporation Pond 207-C, an illegal transfer because Pond 207-C was not a "RCRA permitted operating unit during this period...when they stored, treated, and disposed of salt brine and other hazardous substances in Pond 207-C."

The use of Pond 207-C would have been permitted in an "emergency situation" as conveyed in paragraph 145:

> The 1985 RCRA Part A Permit for the Plant permitted DOE and Rockwell to use three solar evaporation ponds for the storage of liquid hazardous wastes in an "emergency" situation, such as a major spill, ruptured storage tank, or other event that posed a serious threat to human health or the environment. However, this Permit did not allow DOE and Rockwell to use the solar evaporation ponds as a regular STD area for process wastes containing hazardous material.

No "emergency situation" had been invoked for what was a routine operation, not a major spill, not a ruptured storage tank, nor other event "that posed a serious threat to human health and the environment," hence the rationale for indictment.

Spray Irrigation

In its Second Introduction, the Grand Jurors made two interrelated observations about violations of the public trust in relationship to offsite contamination from Rocky Flats especially in relationship to downstream water supplies. The first observation was that the Grand Jury had heard testimony that:

> ...radioactive material had been released from the Plant into the drinking water supplies of various urban users and downstream agricultural users, the Grand Jury was advised that it could not indict Rockwell or DOE officials for endangering the public in this manner.

The reason for this observation was that an Assistant U.S. Attorney had advised that the U.S. Supreme Court had determined in <u>*Train v. COPIRG*</u> that the Clean Water Act (and for that matter any other federal law) did not prohibit DOE, Rockwell, or EG&G from:

> ...dumping radioactive wastes or other radioactive material directly into Standley Lake, the Great Western Reservoir, the South Platte, or any other river or tributary of the United States.

The second observation is that in the area of RCRA liquid wastes, the Grand Jury found by a "preponderance of evidence" that various persons had violated provisions of RCRA through the improper storage, treatment, and disposal of hazardous waste. The East Spray Fields represent the improper disposal of hazardous waste as well as the "excessive and ill-advised spray irrigation of treated wastewater effluent" from the Sewage Treatment Plant. This latter activity was a violation of RCRA and NPDES. The use of the Sewage Treatment Plant to treat hazardous waste was an environmental crime because the Sewage Treatment Plant did not have either a RCRA Permit to do this nor interim status under RCRA to treat such wastes, let alone spray irrigate effluent.

In the Second Introduction, the drafters also reported finding additional environmental crime in that:

> ...bypassing permitted wastewater discharge points by spray irrigating the related wastewater effluent in excessive quantities onto severely sloped hillsides at the Plant.

The definition of "excessive" is not provided in the Second Introduction.

In the section of the report on Spray Irrigation, citizen drafters mention the mind-boggling figure that Rockwell spray irrigated "an average of 80 million gallons of treated wastewater effluent onto only 17 acres of land during 1987, 1988, and 1989, which was in excess of the amount of fluid which the ground could absorb or which could be expected to evaporate into the air without discharging off-site into the waters of the United States." Elsewhere in the Second Citizen Special Federal Grand Jury Report, the citizen drafters characterized the spray irrigation in the equally mind-boggling "...massive volumes of water (as much as 250,000 gallons of treated wastewater effluent per day)..."

The citizen drafters report indicates that "sheet-run-off of the sprayed effluent" went into Woman Creek and Walnut Creek which ultimately flowed downstream into either Great Western Reservoir or Standley Lake. Testimony before the Grand Jury revealed different consequences of spray irrigation during the summer and winter months:

> ...During the driest months of the summer, no more than 35% of the water, which was spray irrigated, was absorbed into the ground on the Plant site. The remaining volume of the treated wastewater effluent continued downstream to Standley Lake and the Great Western Reservoir. During the winter months, Rockwell continued the spray irrigation even during blizzards, when the ground was covered with snow and ice, and it was impossible for any of the treated wastewater effluent to penetrate the surface of the ground. Massive "ice castles" were formed around these agricultural spray irrigation devices at those times, but the flow of irrigation water never stopped.

Figure 11.2 Rockwell Spray Irrigation Device Illustrating an "Ice Castle"

The spray irrigator device was of the type used in agricultural irrigation. According the Grand Jury, the environmental crime came from discharging effluent over steeply pitched hillsides and beyond approved discharged points, causing contaminated groundwater to enter waters of the United States, violating discharge standards, violating "good engineering practice" under the NPDES Permit, and also under the NPDES Permit, recharging a solid waste disposal site with additional water.

There are no sealed exhibits cited to substantiate the estimates of run-off differences between the summer and the winter months. However, the "Tiger Team" sent to Rocky Flats in 1989 by Secretary of Energy James Watkins devoted a section in their report to the East Spray Fields, as did whistleblower Jacque Brever. The most unexpected confirmation of the magnitude of use of the East Spray Fields is by Rockwell itself during the discovery phase of the Stone litigation before it went to the U.S. Supreme Court. The Rockwell confirmations are in their paragraphs 163, 164, and 165 of their rejoinder to the Department of Energy.

Environmental Crime At the East Spray Fields

Beginning with their notation "[Indict]," the drafters laid out their five-point case substantiating the commission of environmental crime in the East Spray Fields as follows:

> [Indict] Rockwell's practice of spray irrigating its treated wastewater effluent violated the conditions of its NPDES Permit and RCRA in the following respects. First, the large volume of water, which ran directly off of the steeply pitched hillsides and beyond the designated outfalls at the B-5 Pond and C-2 Pond constituted unauthorized discharges into the waters of the United States. Second, some unknown volume of the water, which was spray irrigated, entered into the ground and resurfaced as "seeps" discharging into the waters of the United States. Third, spray irrigation moved some of the effluent from one collection pond to another collection pond, where the discharge standards were substantially more lenient and, thereby, illegally bypassed a discharge point. [Compare Exhibit 2026, p. 1827 for discharge standards in B-3 and B-5 ponds] Fourth, the massive volume of spray irrigation on a relatively small parcel of land violated a condition of the NPDES permit that spray irrigation could occur if and only if it was "consistent with good engineering practices." Since

> it was impossible for the ground to absorb the large volume of treated wastewater effluent, which was spray irrigated onto it, the practice of spray irrigation at the Plant violated "good engineering practices." Fifth, the practice of spray irrigating over the "east trenches" (i.e. the western portion of the south spray field) was inconsistent with good engineering practices and, therefore, a violation of the NPDES permit, because it recharged a solid waste disposal site with additional water.

Now sealed Exhibit 2026 contains evidence of bypassing what are called "discharge points." The allegation of "unauthorized discharges into waters of the United States" is straightforward. The discharge of spray irrigation into groundwater is among the least developed of assertions by the drafters of the report.

The critical relationship between the East Spray Fields and what are called the "East Trenches" is not developed in this citizen report. In the Second Introduction it is revealed that the drafters had found a "preponderance of evidence" that environmental crime had been committed in violation of provisions of RCRA through "flushing hazardous and radioactive materials throughout the groundwater in an area known as the east trenches..."

The explanation for the environmental sensitivity of the "east trenches" is provided by paragraph 151 which began with the notation to "[Indict]":

> ... Rockwell followed a continuous practice from 198? through 1989 of spray irrigating massive quantities of treated wastewater effluent from the Plant's Sewage Treatment Plant (herein the "STP") onto an area known as the "east trenches" where massive quantities of radioactive and hazardous material had been buried during the 1950's, 1960's, and 1970's[?]. By following this practice with the knowledge and implicit consent of DOE, Rockwell flushed unknown quantities of radioactive and hazardous materials into the underlying groundwater and greatly expanded the plume of contaminated groundwater, which lies under the ground in that area of the Plant. This practice also accelerated the migration of the contaminated groundwater downstream in the direction of Standley Lake.

The reference to "massive quantities" of radioactive and hazardous materials having been buried in the "east trenches" is authenticated by the records in archives of Dow Chemical documents. The Dow Chemical archives are augmented by the findings by ChemRisk researchers as reported in ChemRisk (1992) on the toxic content of the Trenches.

The Debate Over Good Engineering Practices Regarding the East Spray Fields

The importance of this issue is framed by what the citizen drafters called EPA's use of "vague and ambiguous" terms regarding the NPDES permit for Rocky Flats:

Rockwell chose to exploit the EPA's use of vague and ambiguous terms, which the EPA chose not to define in the NPDES permit for the Plant. For example, EPA directed DOE and Rockwell to engage in spray irrigation of the STP's wastewater effluent in a manner "consistent with good engineering practices." However, EPA never defined what it meant by this phrase.

The lack of more precise definition of key terms by EPA provided latitude for Rockwell and DOE to offer their own definitions.

Dual Voices Within Rockwell On Spray Irrigation

The Second Citizen Special Federal Grand Jury Report reveals a dispute within Rockwell itself on Spray Irrigation:

> The Rockwell manager, who was responsible for supervising operation of the spray irrigation system, recommended to Rockwell's management that Rockwell should stop its practice of spray irrigating the wastewater effluent when the spray fields were frozen. However, Rockwell management ignored his recommendation and directed him to continue spray irrigation regardless of the meteorological conditions which might be present.

The reason Rockwell maintained the practice of spray irrigation was explained by the citizen drafters in paragraph 152:

> Rockwell was totally committed to the goal of avoiding any direct discharge of water effluent downstream. Consequently, Rockwell spray irrigated the effluent under any circumstance--including blizzards--if the irrigation equipment was working.

Lack of Diligence By EPA On the East Spray Fields

Citizen drafters of this Second Citizen Special Federal Grand Jury Report were very critical of how EPA exercised its role as revealed in paragraph 156:

> EPA never inspected the spray irrigation system at the Plant nor any records for operation of the spray irrigation system. If EPA had inspected the records or if EPA had inspected the spray fields, EPA would have discovered what would have been obvious to even the most ill-informed observer. Rockwell was engaging in spray irrigation as a means to avoid reporting discharges of its treated wastewater effluent downstream. Likewise, EPA would have realized that Rockwell was spraying ???times more water onto the spray fields than they could absorb under ideal circumstances. EPA would have also learned from a casual inspection of the records of the spray fields that Rockwell was spray irrigating at times when it would have been impossible for any of the sprayed water to have evaporated or infiltrated into the ground before it travelled off of the Plant site.

There is clear representation by Rockwell in the *Stone* litigation in 1999 that what they were doing in the East Spray Fields was well known to DOE. The now sealed Exhibit 1035 is a Rockwell internal report explaining why spray irrigation "had to result in an unpermitted downstream surface water run-off" to quote the summary by the citizen drafters. Not explained are reasons:

1. As reported in paragraph 161, why DOE's NPDES Permit "allowed direct discharge downstream from the Plant and into Woman Creek only from the outfall at the B-5 Pond," and
2. As reported in paragraph 160, why Rockwell "spray irrigated the treated wastewater effluent without prior testing to determine if the effluent contained unhealthy concentrations of any substance."

The End of the Second Draft Grand Jury Report

It seemed as though this citizen draft report ended in mid-sentence, though in the literal sense it obviously did not. The reason for this impression is that having written two Introductions and the Scope of the Investigation at the beginning, it ended without fully developing "crimes against the public trust" in the same depth the way they had "short term, single event crimes or discontinued crimes" and "continuing criminal offenses."

A mystifying omission is what they said in the Second Introduction; namely, that "the Grand Jury did not evaluate whether the Plant can be restarted or operated in a safe manner." Having spent over two years of their lives in examining over 2000 Exhibits and hearing from over 200 witnesses, it is not possible to imagine any group of twenty-three citizens who could speak more authoritatively to this constellation of fundamental issues relating to safe future operation without further harm to the clean water of Denver, Colorado or risk to human or ecosystem health.

Reflections On the Second Citizen Special Federal Grand Jury Report

The Second Citizen Special Federal Grand Jury Report is a remarkable document. It identifies the content of some of the 2,000 exhibits which provided evidence on RCRA related criminal offenses as well as other crimes related to the Clean Water Act.

Public Trust

In contrast to the various versions of the first report, the Second Citizen Special Federal Grand Jury Report was eloquent about the importance of the public trust. Drafters were particularly critical of government employee behavior judged to be criminal. Arguments of government employees which were rejected included that as government employees they were immune from prosecution, that since they did not direct the illegal operations they were not accountable for criminal conduct, that negligent management is negligent management but not a crime, and that they were ignorant of the law or received faulty legal advice.

Related to violations of the public trust was what Grand Jurors called a "culture of greed" and the negligence and malfeasance by a government agency--"how such extreme conduct could have gone undetected for such a lengthy period of time". The jurors were caustic about individuals who did not testify honestly under oath, thereby violating the essence of public trust.

Response to the National Security Defense

Response of the citizen jurors was particularly compelling on this issue. They were critical of Federal, Colorado and local regulators and elected officials for their "feeble efforts over the last 40 years..." The Rocky Flats Plant is operated by government and corporate employees who have "placed themselves above the law..." The public trust has been breached, the most serious crime in a democracy, through a "clever public relations campaign of distraction, deception, and dishonest denial."

EPOCH THREE

ENVIRONMENTAL JUSTICE DELAYED AND THE CONTINUING STRUGGLE OVER THE FUTURE OF THE ROCKY FLATS SITE: 1990-CONTINUING

Introduction to Part IV: Continuing Clashes Over the Future of the Rocky Flats Site

Key Elements of the Chapter Twelve Narrative

- Comprehensive Environmental Response, Compensation, Liability Act (CERCLA) of 1980 provided the legal foundation of the contamination standards for the "Accelerated Cleanup" based on future uses of the site.
- There are three organizational frameworks:
  - A transformational model of Executive branch and Legislative branch collaboration was conceived for the "Accelerated Cleanup" to overcome the existing model of collaboration which neither worked during the Cold War production era, nor was working as evidenced by the lack of cleanup progress at the Hanford Nuclear Reservation.
  - Closure Legacy: From Weapons to Wildlife (2006) by Frazer Lockhart provides a detailed account of the "Accelerated Cleanup"
  - Three Governmental Accountability Office reports undertaken at Congressional request in 1999, 2001 and 2006 provide substantive appraisals of the "Accelerated Cleanup,"
  - The Citizen "Precautionary Principle Cleanup," which did not occur remains highly critical of the U.S. government "Accelerated Cleanup."
- The U.S. government in 1995 had estimated that the Rocky Flats cleanup would take 65 years and cost $37 billion.
- The importance of the GAO studies beyond Rocky Flats is that they address the reasons for the early completion, what work remains, the level of protection of public health, and the lessons for other DOE sites.

Key Elements of the Chapter Thirteen Narrative

- There are three overarching features of the contentious debate over the future uses of the lands of Rocky Flats now a Wildlife Refuge:
  - irreconcilable conceptions,
  - continuing clashes, and
  - the meaning of the NEPA law as it applies to Rocky Flats land over fifty years after the law was passed in 1969 remains to be litigated in Federal District Court for the District of Colorado.
- There have been five flashpoints (and counting):
  - over the Comprehensive Conservation Plan of 2004-2005,
  - over the Fish and Wildlife Service plan to conduct a "prescribed burn" on over 701 acres in 2015-2016 to counter excess underbrush and invasive species,

- over the attempt of some civil society groups to obtain Preliminary Injunctions to keep the FWS from opening trails to the public across the Wildlife Refuge (these were unsuccessful),
- a sample of plutonium five times higher than that of the U.S. government standard for the top three feet of soil was found at the Eastern edge of the Wildlife Refuge in September of 2019, and
- there are two NEPA lawsuits and one "legal petition" before the Federal District Court for the District of Colorado as of the end of 2019.

There are irreconcilable futures for the Rocky Flats site which are presented in Chapter Twelve and Thirteen. These irreconcilable futures are driven by different worldviews and models of risk, and different assessments of the history of contamination by toxics across the entire 6,500 acres. There does not appear to be either a political constituency, the financial resources, or the motivation to try to bridge the chasms separating them, if bridging is even possible now that there is so much invested on all sides, and it is far too late to start over.

There are diametrically opposed conceptions about what would have been the best outcome which was feasible to reducing the long term toxicity of the site itself. For some informed members of local civil society, the closer the site came to its original pristine state the better. For the respected DOE on-site manager in 1996, the site would only be cleaned up to the level prescribed by law, which by then was CERCLA. For the U.S. Congress, the nation had higher priorities and more pressing obligations than to invest in new technology to remediate pollution which sociologist Kai Erikson called a "new species of trouble" for humanity. It was beyond the realm of the possible to spend 65 years and $37 billion dollars on stabilization and site restoration. This was a lesson which had already been learned from the Hanford Nuclear Reservation where there were more people working on cleanup with limited results than had ever worked on producing plutonium.

The unlikely conversion of 4,000 acres of what was mostly the "buffer zone" into the Rocky Flats National Wildlife Refuge under the jurisdiction of the U.S. Fish and Wildlife Service is now a flashpoint between civil society and the U.S. Fish and Wildlife Service. The Fish and Wildlife Services manages a part of the former Hanford Nuclear Reservation but not as a Wildlife Refuge open to the public in a metropolitan area. From the start, there were irreconcilable differences memorialized in a clash in 2004-2005 over the obligations imposed on the Wildlife Refuge by the National Environmental Policy Act. These differences have never been bridged. A plan of the Fish and Wildlife Service to proceed with a "prescribed burn" over 701 acres turned into a major skirmish in 2015 and 2016. The specter of fire on the former "buffer zone" united segments of the Denver community which usually do not make common cause. The incident is just one example of the different priorities for the future of the site.

There are irreconcilable differences between the Departments of Interior and Transportation, the Federal Highway Administration and the US Fish and Wildlife Service on the one hand, and an empowered group of civil society on the other. Their intractable conceptions for the future of the Wildlife Refuge in 2017 and 2018 have ended up in the Judicial branch of government in Federal District Court for the District of Colorado.

The organizational framework for Chapter Twelve is to begin with the difficult experience which Rocky Flats encountered during the Cold War with even receiving cooperation within the Executive branch of government on transporting toxic materials to either ARCO/INEL in Idaho or to the Nevada Test Site. These repositories were controlled by the Atomic Energy Commission and subsequently, the Department of Energy, under the legal authorities of the Atomic Energy Acts of 1946 and 1954. As Chapters Five and Six made clear, insufficient attention had been given during the founding and early years of Rocky Flats to the storage, treatment and disposal of the toxic byproducts of the operation of the industrial enterprise associated with the fabrication of plutonium igniters.

The 178 known waste disposal sites at Rocky Flats, and the fifteen fiascos characterized in Chapters Five and Six, reveal the urgency of creating a different model of cooperation between Executive and Legislative branches of government and regulatory agencies if there was to be meaningful and affordable cleanup and environmental restoration of the Rocky Flats site.

In order to increase understanding of the different conceptions of cleanup and environmental restoration between the U.S. government and some informed members of civil society, Chapter Twelve uses the device that there have been two cleanups at Rocky Flats in the post-production era: the U.S. government "Accelerated Cleanup" which actually occurred, and the civil society driven "Precautionary Principle Cleanup" which was rejected as a model.

Chapter Thirteen presents one of the most intractable and as yet unresolved conflicts about the nature of Rocky Flats as an "ultrahazardous" enterprise where the contractors from the military-industrial complex were found by a civil jury to have behaved in a manner which was "willful and wanton" (Chapter Fifteen), and what should be in the public interest for the long term future use of the site given the toxic history.

While there are many issues on both sides of the increasing divisive dispute, they come down to answers to two fundamental questions:

- What are the obligations under the rule of law which are posed by the requirements of the National Environmental Policy Act?

- What are the special vulnerabilities of children to low dose exposure to ionizing radiation, and is there a threshold below which exposure to plutonium will not cause the release of free electrons which can cause cell damage and potentially cancer?

There have been two fundamental clashes over the requirements of the National Environmental Policy Act as they pertain to the Wildlife Refuge in 2004-2005 and in 2017-2018. The answer to the question of obligations under law will be delivered in Federal District Court. The role of legal precedent will be a defining consideration in the ultimate decisions of Federal District Court Judges.

The answer to the second question from a scientific point-of-view will be based on definitive U.S. government associated science. The prestigious National Research Council of the National Academies of Sciences, Engineering and Medicine in 2006 delivered a comprehensive report which is highly relevant to understanding the cancer risk which is associated with low-dose exposure to ionizing radiation. The Agency for Toxic Substances and Disease Registry reported in 2010 on the effects of exposure to plutonium with a special section on the vulnerability of children.

Chapter Twelve
A Tale of Two Cleanups At Rocky Flats: The U.S. Government "Accelerated Cleanup" Which Was, and the Citizen "Precautionary Principle Cleanup" Which Was Not

> There is no formula that can be applied to every site, since each site is different geographically, in terms of cleanup scope and future missions and with different cultural and political issues.
>
> "Why was the project successful?"
> ...the regulators were cooperative...the contractors incentivized and motivated...the DOE delivered most of the government furnished services and equipment on time...the budget appropriations were consistent and reliable at $650 million per year...closure was managed as a finite project using project management principles...stakeholders were involved in project planning...workers were involved in work planning.
>
> ...no single factor was responsible for achieving accelerated closure, but that in some measure all of these factors and more were necessary for success.
>
> Frazer Lockhart, Closure Legacy: From Weapons to Wildlife

There have been two cleanups in the post-production era at Rocky Flats: the U.S. government "Accelerated Cleanup" which actually occurred, and the civil society driven "Precautionary Principle Cleanup" which was rejected as a model. The first cleanup was conducted by contractor Kaiser-Hill for the Department of Energy on behalf of the U.S. government. The Department of Energy worked closely with the Environmental Protection Agency and the state of Colorado through the Colorado Department of Public Health and Environment in establishing the technical standards for the "Accelerated Cleanup" which conformed with the legal requirements of the Comprehensive Environment Response, Compensation, and Liability Act (CERCLA) of 1980.

Prior to the "Accelerated Cleanup" there had been extensive studies of the site conducted by agencies of the U.S. government and private contractors. There have been a series of reports assessing aspects of the government cleanup by the Government Accountability Office (1999, 2001, 2006). As mandated by the CERCLA, there are five-year reviews of the 1,300 acres of the portion of the Rocky Flats site classified as a "Superfund site." The U.S. government sponsored and funded "Accelerated Cleanup" was completed by contractor Kaiser-Hill in 2005.

Formal consultations for the citizen "Precautionary Principle Cleanup" were under the auspices of a series of formal DOE sanctioned working groups which extended from 1993 through 2002. These working groups did not use the phrase "precautionary principle" but the discourse by civil society centered around the "exercise of caution," and "avoiding actions which were risky" (Moore, 2016). The phrase to describe the civil society vision of the cleanup "which was not" is a device of the author to capture the spirit of the civil society approach as containing central elements of "better safe than sorry."

Several continuing organized citizen groups including the Rocky Mountain Peace and Justice Center have an extensive critique of the "Accelerated Cleanup" which begins with the basic fact as they see it that what the U.S. government conducted at Rocky Flats is misnamed as a "cleanup."

Organizational Frame for Chapter Twelve

There are three main sections to Chapter Twelve. The first section is based on Closure Legacy: From Weapons to Wildlife by Frazer Lockhart (2006). This definitive presentation of the "Accelerated Cleanup and Concept" presents a transformative model of Executive branch and Legislative branch collaboration which was conceived to overcome the existing model of collaboration which neither worked during the Cold War production era, nor was working after the Cold War as evidenced by the lack of cleanup progress at the Hanford Nuclear Reservation.

The second section draws on reports by the Government Accountability Office undertaken at Congressional request in 1999, 2001, and 2006. The GAO report in 2006 addressed four fundamental questions:

Question 1 - What factors contributed to the cleanup's early completion?

Question 2 - What work is remaining and at what cost?

Question 3 - What measures will assess whether the cleanup had achieved a protection of public health and environment consistent with the cleanup agreement?

Question 4 - What specific lessons may be drawn from the Rocky Flats Cleanup for other DOE sites?

The third section is a presentation of the citizen "Precautionary Principle Cleanup" which was not (Moore, 2016). It is premised on assumptions that are in diametric opposition to the assumptions underlying the U.S. government "Accelerated Cleanup." This section is based on lectures and writings by Dr. LeRoy Moore, a co-founder in 1983 of the Rocky Mountain Peace and Justice Center.

Significant Challenge Confronting the U.S. Government

The significant challenge confronting the U.S. government after the close of production was an imperative that decisive action be taken at Rocky Flats to eliminate as much as possible the known toxic burden at the site.

There were well-known fiscal and technology constraints. The Environmental Crime era from 1986 until 1994 had definitely documented in-depth multi-faceted risk. DOE itself had declared in 1994 that Rocky Flats contained the two most dangerous buildings in the entire multi-state nuclear weapons production complex. The astounding magnitude of nuclear materials at Rocky Flats was declassified in 1994. The time it would take, and the cost, were estimated in 1995 to be 65 years and $37 billion. A final and compelling element of challenge was a combination of the known consequences of site selection deficiencies with an urban area now encroaching in close proximity to Rocky Flats itself.

New Strategic Plan For Rocky Flats

The U.S. government funded "Accelerated Cleanup" is especially well documented by Lockhart (2006). Closure Legacy is over 400 pages plus extensive appendices. The approach taken by the U.S. government is straightforward. Silverman (1994) presented a stakeholder-inclusive new strategic approach. The Clinton administration "reinventing our government" led to an incentive and performance-based new federal contracting system. Contractor Kaiser-Hill applied the new approach to Rocky Flats. Lockhart's (2006) comprehensive presentation is highlighted in this volume by covering the handing of special nuclear materials and by the use of new applications of technology.

Silverman indicated that the primary focus of the Rocky Flats Strategic Plan is to:

> ...reduce, eliminate or mitigate the significant environmental liabilities which exist as a result of past years of operating a nuclear weapons production facility...
>
> (Silverman, 1994, p. 1)

What is fundamentally different from initiatives at Rocky Flats during the Cold War production era is the engagement of stakeholders in broad definition of the term:

> We believe strongly that the people directly affected by our decisions, especially decisions affecting human health, the environment, and economic interests, should be involved early in the planning process.
>
> (Silverman 1994, p. 1)

As this broadened definition of stakeholders was implemented as reported in Lockhart (2006), it included high level members of both the Executive and Legislative branches of government, and groups which never before were included in-depth in the process such as regulators.

Making Contracting Work Better

The overall structure of Making Contracting Work Better and Cost Less was an initiative of the Clinton administration to "reinvent our government". The translating of "reinvention" by the Department of

Defense was: "As the largest civilian contracting agency, the Department must significantly reform its contracting practices to meet this critical commitment" (Silverman, p. 1).

The essence of this contracting reform as it applied to the Kaiser-Hill contract is the following principle:

> The Department must clearly state what we expect from our contractors and develop meaningful ways to measure whether they are meeting our expectations. We must rapidly inject into our contracting system performance criteria and measurement mechanisms..."
>
> (Silverman, 1994, p. 1)

Translated into the Kaiser-Hill contract, there were "meaningful" ways to measure performance, and the establishment of new "reward systems that better distinguish between superior and inferior contractor performance..."

Insights of "Accelerated Cleanup" By Leader Frazer Lockhart
The Rocky Flats Closure Legacy report (Lockhart, 2006) chronicles the approximately decade of engagement from initial planning in the 1992-1996 time frame to when it was considered to have been completed in 2005. The Closure Legacy report itself was characterized by its author as a "lessons learned" document spanning technical and scientific matters as well as policy and programmatic issues which were encountered during the decade.

Structural Framework "Pillars" In Lockhart's Model

There were four "pillars" of the transformative concept which formed a structural framework:

- Gaining Congressional and Executive Administration Support,
- Creating a New Model for Engaging Regulators Actively and From the Onset of Planning,
- Implementing the New 1994 DOE Performance Based Contracting Approach, and
- Implementing Fifteen Elements of the Closure Project.

Strategic Planning Process and the People Who Mattered

A strategic planning process for what became the "Accelerated Cleanup Concept" began, according to Lockhart (2006), with a realization that: "any progress would require alignment of interests of the Site, [DOE] headquarters, regulators, contractors, Congress, and multiple stakeholders." Lockhart credited a critical subset of individuals who were "passionately and energetically committed to accomplishing the vision" of a different approach:

- Rocky Flats Plant Managers,
- Kaiser-Hill Senior Managers,

- DOE Assistant Secretaries for Environmental Management,
- Secretaries of Energy, and
- Office of the Governor and Lt. Governor of Colorado.

(Lockhart, 2006, I-9)

Finally, the Comprehensive Environmental Response, Compensation, and Liability Act (CERCLA) provided a framework in law for determining that the required level of restoration was directly related to future uses of the site.

Implementing the Closure Legacy Project

Once the new innovative framework was in place, there were ten elements of the building blocks of the implementation. They were:

- Safety Integration
- Special Nuclear Material Removal
- Decommissioning
- Waste Disposition
- Environmental Restoration
- Security Reconfiguration
- Technology Deployment
- Future Land Use End State and Stewardship
- Federal Workforce
- Stakeholder Involvement

(Lockhart, 2006, I-9)

While it is beyond the scope of the possible in this study to review all of the elements of this innovative framework, "Special Nuclear Materials" and "Technology Deployment" will introduce readers to the distinctive substance.

Special Nuclear Material Removal

The information in this section on "Special Nuclear Material Removal" was highly classified until the Department of Energy declassified it in 1994. Rocky Flats had the largest plutonium inventory of any facility of the Department of Energy (12.9 metric tons) as well as an inventory of 6.7 metric tons of highly enriched uranium.

The Various States of Plutonium

An initial problem to address was the various states that the plutonium was in:

> 6.6 metric tons of relatively pure plutonium metals, 3.2 metric tons of plutonium in approximately 6 bulk tons of plutonium oxides; and 3.1 metric tons of plutonium in approximately 106 bulk tons of plutonium residues. The enriched uranium was in various forms. Additionally, the Site had numerous "Special Items" that

requires special handling due to weapons classification and nuclear safety concern.

(Lockhart, 2006, 7-1)

According to Lockhart, these all required "characterization, stabilization, packaging for long-term storage, consolidation, repackaging/over packing into approved shipping containers and removal from the site..." These tasks had to be accomplished before "decontamination and decommissioning" could begin (Lockhart, 2006, 7-1).

A Concluding Challenge For Special Nuclear Material Removal

There were technical challenges at almost every turn. But the only legal challenge reported was a lawsuit filed against the Department of Energy by the Governor of South Carolina to prevent shipments from Rocky Flats to Savannah River until DOE "identified a disposition path" for the plutonium already in storage there. The Governor did not prevail.

Major Destinations For Special Nuclear Materials

Once the shipping could begin, the primary receiver for the highly enriched uranium was the Y-12 Plant in Oak Ridge, Tennessee. For the plutonium pits, the primary receiver was the Pantex Plant, in Texas. The plutonium metals oxides were primarily sent to the Savannah River site. It was an assessment of safety related issues which revealed the vulnerabilities in Buildings 771 and 776 because of the presence of "large quantities of plutonium-containing solutions" and of plutonium packages "that were improperly packaged" (Lockhart, 2006, 7-5):

> The Pu and HEU solutions were identified as the most significant hazards at Rocky Flats. These acid solutions had been sitting idle in tanks for years and presented a leaking, corrosive, explosive, radioactive spill hazard. Also, the criticality safety risk is higher with solutions, exacerbating the risks involved with venting, draining, and processing these solutions.
>
> (Lockhart, 2006, 7-6)

Principles of Technology Deployment

There were three guiding principles used in the deployment of technology:

1. To achieve improvements in worker safety,
2. To shorten the time to completion, and
3. To streamline the disposition of waste.

Lockhart, 2006, p. 12.5)

Surprising for this writer, Lockhart indicated that: "Choosing which technologies would provide the most improvement at the beginning of the Closure Project was a speculative process" (Lockhart, 2006, p. 12.5). The reason for the surprise was that the organizers turned out to be seeking incremental change from the use of technology: evolution not revolution.

An example of the evolution not revolution was in the application of non-expansive foam to diverse cargo containers which prevented the shifting of cargo during transit. During the production era Department of Transportation regulations had been an obstacle that could not be overcome.

Another example of evolution not revolution was with regard to the size reduction of large pieces of equipment. Size reduction of equipment such as uranium metal forming equipment which came as massive single units would have required for example, diamond wire cutting. The process which evolved was to build custom pallets, shrink-wrap the equipment and spray it with "InstaCote" coating. This coating was like a truck bed liner. The result was a "strong-tight" container, a Department of Transportation approved approach. Lockhart's comment was that "Easily thousands of man-hours of difficult and dangerous size reduction in anti-contamination clothing were avoided by use of InstaCote" (Lockhart, 2006).

Visualizing Applications of Significant Technologies

Each of the technologies in operation were applied to significant challenges facing the "Accelerated Cleanup." Five examples are provided.

The first example is directed to the challenge of packaging radioactive process equipment. This particular challenge was met by the use of polyurethane foam rather than utilize carpenters to stabilize the equipment for shipment. This is illustrated in Figure 12.1.

Figure 12.1 Polyurethane Foam Utilized to Block and Brace

The disposal of equipment used at Rocky Flats which was contaminated with low-level radioactive waste required meeting a number of regulatory requirements. One requirement was that the contaminated equipment had to be "blocked and braced" in order to prevent damage to the Waste Package during shipment. The figure demonstrates how polyurethane foam was used to meet the regulatory requirement without carpenters having to crawl inside of the Waste Package risking exposure to low-level radioactive waste while "blocking and bracing" with lumber and hammers.

The second example is directed to the challenge of minimizing manual activity by workers to increase their safety in the process of cutting up large storage tanks for shipment. These large tanks are contaminated with both radioactivity and chemicals. This is illustrated in Figure 12.2.

Figure 12.2 Ultra-High Pressure Water Jet Used to Remotely Cut B774 Tank
Hundreds of tanks for storage and treatment of plutonium and its processing were used at Rocky Flats. Some of the tanks were so large that they had been installed inside of buildings before construction had been completed. Some tanks were also contaminated with polychlorinated biphenyls (PCBs). Most of the tanks were too large to be removed from the buildings. This 7,300-gallon tank in Building 774 is illustrative. There would have been great risk to workers even using personal protective equipment if they had been required to manually cut the tanks. The solution was to use the technology Jet Edge from a firm in St. Michael, Minnesota. Jet Edge is an ultra-high pressure, abrasive water-jet cutting system. It uses a 50-horsepower intensifier pump to generate 55,000 psi output pressure. According to Lockhart (2006) "the pump supplies water to an abrasive cutting head that is configured on a 14-foot ladder to guide the cutting head..."

The third example is the use of chemical decontamination to both improve worker safety and reduce transuranic waste. This is illustrated in Figure 12. 3.

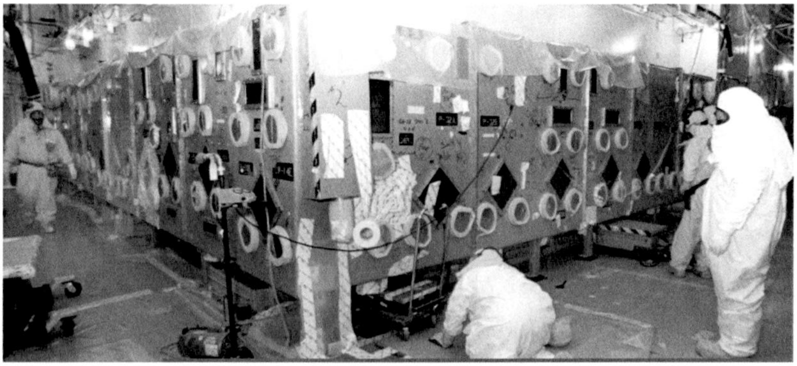

Figure 12.3 Chemical Decontamination of Gloveboxes and Tanks Improved Safety and Reduces Transuranic Waste
The gloveboxes during the production era were used to protect workers from the hazards of processing plutonium, americium, uranium and other radioactive isotopes. Chemical decontamination slashed months from the time necessary for the Accelerated Cleanup. The challenge for workers was to prepare large gloveboxes (the largest was 64 feet long) for shipment to the Waste Isolation Pilot Project in New Mexico.

The fourth example uses metal cutting technology to accomplish size reduction without full decontamination. This is illustrated in Figure 12.4.

Figure 12.4 Plasma-Arc Cutting Technology
In an effort during World War II to improve joining of aircraft materials, a method of welding was employed which used "a barrier of inert gas around an electric arc to protect the weld from oxidation" (Lockhart, 2006). Modern day refinements of plasma-arc cutting was used in the "Accelerated Cleanup" to size-reduce large equipment.

The fifth example is of the use of explosive demolition to implode major production buildings on themselves. This is illustrated in Figure 12. 5.

Figure 12.5 Explosive Demolition
This Figure 12.5 captures the moment of demolition of Building 881. Explosive demolition contractors were used to explosively cut the structural support for the building so it would implode upon itself. Building 881 was mostly underground. It was first decontaminated of uranium and beryllium (it had no plutonium contamination). With regulatory approval, the imploded building was covered with fill dirt.

Major Difference From Hanford In Not Having High Radiation Activity

There were no areas of high-radiation activity requiring remote systems and no high-gamma radiological materials or contamination with the exception of "Infinity Rooms."

There was an extensive list of technologies utilized in different circumstances, and it did not always work out satisfactorily. But technology deployment was central to maintaining the "accelerated pace" of progress.

Documenting Levels of the DOE "Accelerated Cleanup"

Closure Legacy: From Weapons to Wildlife explains what Lockhart (2006) calls the "conditions and influences" that existed for the Rocky Flats Plant manager and staff during the ten-year period of the "Accelerated Cleanup." Contained within its approximately 400 pages plus data Appendices and linkages to key documents Closure Legacy detailed answers to the same general set of questions addressed by the Government Accountability Office (2006) but from the perspective of a goal of presenting the "full history of the project":

With respect to the lessons which have been learned which are applicable to other DOE nuclear weapons production sites, Lockhart (2006) offers a caution with respect to contextual limitations of the "formula":

> There is no formula which can be applied to every site, since each site is different geographically, in terms of cleanup scope and future mission, and with different cultural and political issues.
>
> (Lockhart, 2006, p. 2)

With the caution that no single factor was responsible for the success of the "Accelerated Cleanup" project within its established parameters, Lockhart did identify a constellation of factors deemed important from the perspectives of Rocky Flats managers and staff:

> ...the regulators were cooperative...the contractor was incentivized and motivated...the DOE delivered most of its government furnished services and equipment on time...the budget appropriations were consistent and reliable at $650 million per year...closure was managed as a finite project and using project management principles...stakeholders were involved in project planning...workers were involved in work planning.
>
> (Lockhart, 2006, p. 2)

The "accelerated closure" obviously occurred after the end of the Cold War. The imperative for production at Rocky Flats had ended, and with it the associated twin imperatives of preserving production-related national security secrets, and the adversarial nature of the environmental crime era.

The "Accelerated Cleanup," if it is dated from 1995-2005, followed extensive efforts at site characterization summarized in the ChemRisk

(1991a, 1991b) and ChemRisk (1992). These ChemRisk studies built on the access provided to Rocky Flats site production documents, and earlier studies by Dow Chemical and (to a lesser extent) Rockwell International over the thirty-eight year production era. Rocky Flats had been declared to be a Superfund site under CERCLA in 1989. Remediation of the site under the law was to begin within one year. What began in that time frame was an extensive site characterization drawing on both inter-governmental resources and contractors from the private sector.

With respect to the role of DOE to furnish "government services and equipment on time," this did occur. For example, in terms of facilitating off-site shipment of toxic materials to permanent storage after stabilization, DOE obviously facilitated the safe transportation to locations in at least six different states (Washington, Nevada, New Mexico, Tennessee, Texas, and South Carolina) with broad based prior regulatory approval and a political consensus that the "Accelerated Cleanup" was in the national interest.

The constellation of issues embedded in following "project management principles" and management of a "finite" project may be viewed in the context of the Diamond (2005) framework for why some societies failed. The entire set of issues related to "failure to anticipate" no longer became a reason for disastrous government decisions. The deliberate development and incorporation of innovative technologies reduced the challenges associated with the "Accelerated Cleanup" on the issue in the Diamond framework of "it is beyond the capacity of a society to solve it." Problems could be solved on transportation to a permanent repository (in and of itself stabilization of toxic waste techniques benefitted from a major use of innovative technology).

With respect to stakeholders, starting in 1994 there was a credible effort made to engage the Colorado community across many constituencies in providing consultation on the assumptions and choices underlying the "Accelerated Cleanup." In 1996, DOE's Rocky Flats manager Mark N. Silverman informed members of an engaged civil society that the site would be "cleaned up" to the limit prescribed by law.

The relevant law for establishing the required level of cleanup was the Comprehensive Environmental Response, Compensation, and Liability Act (CERCLA) with its Superfund provisions as the standard for level of cleanup related to projected future uses of the site. That in turn led under the law to the designation of two future site uses: the Superfund site and the Wildlife Refuge.

Finally, the efforts at worker safety in the "Accelerated Cleanup" were addressed in the context of the unknown challenges of dismantling an "ultrahazardous" nuclear weapons production complex. This accomplishment is in stark contrast to the failure by the U.S. government to even pass a unique hazards law until eleven years after Rocky Flats closed for production.

Government Accountability Office On the Accelerated Cleanup

In 1999, responding to Congressional request, the Government Accountability Office issued a report on:

1. DOE's plans for accelerating the site's closure and challenges which could impede closure;
2. The condition of the site at closure and the activities that will remain after closure; and
3. The costs of closing the site and the savings expected from accelerating the closure.

The GAO report in 1999 indicated that both DOE and Kaiser-Hill believed that the mission could be accomplished by 2006. GAO indicated that there was in place an agreement with EPA and the state of Colorado on the condition of the site when the "Accelerated Cleanup" had been completed. The cost was estimated by DOE and Kaiser-Hill to be $7.3 billion.

<u>Nuclear Cleanup of Rocky Flats As Assessed By GAO In 2006</u>

In 2006 the Government Accountability Office (GAO, 2006) completed a study of the "Accelerated Cleanup." The amount of waste that had to be removed from the site was the "equivalent of a 65-story building the length and width of a football field" (p. 1). GAO (2006) credited large financial incentives for Kaiser-Hill to lower its costs, finish its work early and utilize creative solutions. With respect to the cost of the "Accelerated Cleanup," the GAO concluded that "the physical clean up at Rocky Flats is complete at a total cost (including long-term costs) of about $10 billion" (GAO, 2006, p. 26). This estimate extends cost projections through 2080.

<u>Lessons Learned From the "Accelerated Cleanup" to Improve Oversight Cleanup Activities At Other Sites</u>

The title of this section is the subtitle of the 2006 report by the U.S. Government Accountability Office released under the title <u>Nuclear Cleanup of Rocky Flats</u> (GAO, 2006). This report parallels in part the <u>Closure Legacy</u> report (Lockhart, 2006). The parallels are the use of an "Accelerated Cleanup" process and the financial incentives for the contractors to finish the work, as GAO characterized it, "quickly and safely." With respect to financial incentives, the team associated with Kaiser Hill received $560 million for their accomplishments as an incentive bonus (no matter what it is called) of which they in turn disbursed $100 million to employees who performed the often dangerous work (like using a special hardened saw system and glycerin fogs to dismantle an "Infinity Room" in Building 771 (Lockhart, 2006). An "Infinity Room" was so named because the radiation level was off the measurement scale.

The historical context for the GAO (2006) report, and why it is such a bellwether for future approaches to other nuclear weapons production sites, is that in 2001 (GAO, 2001) the Government Accountability Office had concluded that the cleanup was "behind schedule and over budget."

GAO in 2006 had been asked by its Congressional requestors to answer four questions. Given the bellwether nature of this report for understanding the future approach of the U.S. government to other nuclear weapons production complexes, it is useful to examine the questions which framed the report in detail, and the answers which were provided:

Question 1 - What factors contributed to the cleanup's early completion?

There were, for example, a total of 1,475 gloveboxes at Rocky Flats. Originally, it had been thought that they would have to be dismantled individually. Instead, the contractor used an "innovative technique" with cerium nitrate (Cerium is the most abundant of rare earth metals). In oxide form, it has uses in the destruction of chemical warfare agents. In nitrate form it has uses in burn units.) At Rocky Flats it was used to decontaminate gloveboxes so they could be shipped intact (without dismantling) off-site to a permanent repository.

Another innovation cited by GAO was a process approach to involve cleanup staff working "side-by-side" with personnel from DOE, EPA, and the Colorado Department of Public Health and Environment to observe together removal of soil and procedures used in sampling. A DOE official is quoted that this process allowed them to detect data issues "as they arose" (p. 6). The "performance-based contracting strategy" and "innovative application of technology" were cited as approaches that DOE had already applied to other sites. GAO identified two additional lessons: involving "the future site manager [from the U.S. Fish and Wildlife Service] in remedial decisions," and "taking a consultative approach with regulatory agencies on cleanup decisions" (p. 7).

There were numerous difficulties in the "Accelerated Cleanup" which could not be solved easily. GAO citied the "automated plutonium packaging system "as an example which ended up $50 million over budget. The system never worked reliably. The task of packaging plutonium was (mostly) accomplished manually. A conclusion attributed to DOE officials was that "...the pursuit of any alternative to the system would probably have resulted in even greater cost and schedule delays." (p. 18)

A major factor cited by GAO for the early completion, and which is exportable to other DOE sites, is that "DOE provided the contractor with a [major] financial incentive to finish early and under cost" (p. 24). GAO (2006) reported that another key factor which contributed to the early completion of the physical cleanup "was a combination of circumstances--climatic, geologic, chemical, structural and economic--that confined the scope and complexity of the cleanup effort" (p. 22). This affirmed the conclusion of Frazer Lockhart reported earlier.

Question 2 - What work is remaining and at what total cost?

Work remaining follows directly from site specific categories of analysis. For example, in some locations, layers of shale and claystone which were hundreds of feet thick "...prevented contamination from seeping into the deep drinking water aquifer that underlies part of the site" (p. 22). Another example is that two key contaminants, plutonium and americium, are "essentially insoluble in water and, instead, tend to bind to soil and sediment particles..." (p. 23).

A criteria which was used predominantly to answer the question of what work is remaining was "what steps remain before the planned wildlife refuge will open?" (p. 26). There were regulatory steps. Whether there will be additional cleanup needed received a process answer:

> EPA and Colorado will jointly determine whether any further cleanup is needed at the site or whether the cleanup actions already taken are sufficient to render the site safe for its intended purpose (p. 26).

<u>Question 3 - What measures will assess whether the cleanup has achieved a protection of public health and environment consistent with the cleanup agreement?</u>

The GAO (2006) answer contains elements of the civil society challenge to the underlying assumptions of the "Accelerated Cleanup." After the Wildlife Refuge opens, "DOE will maintain groundwater monitoring wells and treatment facilities as long as necessary" (p. 35). The phrase "as long as necessary" will engage formally government regulatory agencies in its determination. The transparency of this process and the evidence used to define what constitutes the definition of "public health and environment" obviously has yet to occur. GAO (2006, p. 35) indicated the need for "...some stewardship for an indeterminate time."

The entire 6,500 Rocky Flats site was originally classified as a Superfund site. The challenge from some in civil society is that the legal process designated for transfer under the Rocky Flats National Wildlife Refuge Act of 2001 (Section 3172 (a)(4)) was not followed. Another issue for some in civil society was that the entire site had not been cleaned up and closed prior to the transfer. A second issue is groundwater monitoring. The Superfund site of 1,300 acres is totally surrounded by the Wildlife Refuge. The monitoring of groundwater immediately off of the current Superfund site is a priority for civil society. Among the reasons for the salience of this issue is the fact that there was Spray Irrigation at five sites, not only the notorious East Spray Fields.

An illustrative major defining issue for civil society on "protection of public health and environment" is the decision of the Department of Energy not to remediate what GAO (2006) called the "Present Landfill" and what is referred to in this volume as the "Second (misnamed) Sanitary Landfill." This second landfill operated from 1968 until the closure of the production era. It was the source of the off-site migration of tritium in 1973 and is the location of the burial of ash from the burning of sheet uranium. According to the GAO (2006, p. 35), DOE negotiated an "environmental covenant" for the landfill to assure protection of "public health and environment" by preventing "intrusions into the landfill or damage to the various engineering structures on adjoining lands and preventing exposure to hazardous wastes." The presence of "hazardous wastes" is acknowledged by (almost) all. The covenant prohibits activities such as "digging, drilling, tilling, grading and excavating, as well as any activities that could damage or impair the proper functioning of the landfill cap and runoff controls" (p. 35).

GAO (2006, p. 35) did note the presence of citizen concern "about what would happen at the site in the future." That statement is accurate as phrased. But it does not convey or reflect the intensity of concern in civil society by those who favored a more precautionary principle approach to "Accelerated Cleanup" at DOE nuclear weapons production sites.

This "intensity of concern" is countered by most influential elements in the U.S. government for whom the Rocky Flats model of the "Accelerated Cleanup" will be the model for the future at other nuclear weapons production complexes dating from the Cold War. The response to that U.S. government position is:

1. The EPA and CDPHE have not responded to the 1970s careful research of very credible U.S. government scientists Krey and Hardy (1970) and the Dow Chemical committee chaired by Seed (1971). The Seed Committee report did not come available to civil society until 2018. It directly contradicts the current position of EPA and CDPHE.
2. The current NEPA challenges in Federal District Court will surely utilize the U.S. government associated very credible science represented by the National Research Council (NRC, 2006) and the Agency for Toxic Substances and Disease Registry (ATSDR, 2010) on special risks to children posed by plutonium and risks from low dose exposure.
3. The Jefferson Toll Road Authority cannot possibly proceed with their role in completing the Denver Regional transportation corridor across the Eastern edge of the Wildlife Refuge without a complete and credible examination of the level of plutonium contamination on at least that portion of the Wildlife Refuge.

Question 4 - What specific lessons may be drawn from the Rocky Flats cleanup for other DOE sites?

There were a number of key lessons. During its production era, Rocky Flats and its contractors never solved the problem of finding a solution to off-site transportation and storage of contaminated waste. Under the "Accelerated Cleanup" it was the responsibility of DOE to identify locations which would receive shipments from Rocky Flats. There were both location and process issues in accomplishing this task. The use of a mix of sites worked.

A key process issue was to engage top management in crafting a solution. Addressing the safety issues, beyond involving top management, included improving the work planning process, involving workers early in analyses of the safety planning process, and the identification of job hazards as well as improving training. A course was developed for employees which provided a simulated work environment in which workers could learn to use tools and ladders in contaminated spaces. None of this process for worker safety was followed during the national security production era, at least based on the archives of Dow Chemical.

Planning within the "Accelerated Cleanup" for off-site transportation for toxics and for safety training contains important lessons for other nuclear weapons production sites.

The Citizen "Precautionary Principle Cleanup" Which Was Not

The "Precautionary Principle Cleanup" which was not is premised on the most profound disagreements with the Department of Energy "Accelerated Cleanup" over the immediate and long-term risk to human and ecosystem health posed by exposure to plutonium. It is possible to quantify the level of plutonium per gram of soil which is acceptable to (some) informed citizens and importantly why that level for them is an essential minimum. A return of the Rocky Flats site close to its original condition did not meet the legal obligation mandated by CERCLA related to future use as a Wildlife Refuge.

Formal Citizen Consultative Initiatives By DOE On Cleanup Standards and Process Choices

The disagreement over plutonium level risk found expression in the 1995 report of the DOE funded Future Site Use Working Group which had been convened in 1994. The purpose of the group was to reflect the goals of the public for the Rocky Flats site in its post-production era. The Future Site Use Working Group wanted the site to become a laboratory for technology which did not yet exist, and which could be exported to other nuclear weapons production sites around the country.

With respect to the level of plutonium contamination in soil, the Future Site Use Working Group called for cleaning the site to "average background level"--from fallout--of 0.04 picocuries per gram of soil. LeRoy Moore, one of the members of the working group, reported (Moore, 2014) that two citizen oversight groups, the Rocky Flats Citizens Advisory Group and the Rocky Flats Local Impacts Initiative (an initiative to assist plant workers when they became unemployed) both supported the recommendations of the Future Site Use Working Group. The Working Group advocated the establishment of a laboratory at Rocky Flats to develop technology to clean up a plutonium-contaminated site. One approach to accomplishing this, according to the Working Group, was to "keep people permanently working to clean to fallout levels. We weren't asking that fallout plutonium be removed" (Moore, 2016).

There was no immediate response to the Working Group Report by the Department of Energy, the Environmental Protection Agency, or the Colorado Department of Public Health and Environment. After an Op-Ed piece written by Dr. Moore was published in January of 1996, the DOE Plant Manager at Rocky Flats, Mark N. Silverman, indicated that the public, as reflected in the recommendation of the Working Group, wanted more than the law required, and that the site would be cleaned up only "to levels prescribed by law."

Earlier in July of 1995, according to Dr. Moore, the Department of Energy and contractor Kaiser-Hill had unveiled a plan which in essence would have covered the site with layers of rock and soil. A consequence of this choice for cleanup would have been to leave basic contamination in place. The negative public response led to a rethinking of this alternative by the government.

The next significant event was the unveiling in 1996 of the first version of the Rocky Flats Cleanup Agreement between the Department of Energy, the Environmental Protection Agency, and the Colorado Department of Public Health and Environment. The standard set in this agreement was 651 picocuries per gram of soil. There was a negative public reaction to the 651 picocuries standard. This negative reaction led Assistant Secretary of Energy Al Alm to instruct the DOE Rocky Flats Manager to fund another citizen study. The Radionuclide Soil Action Level Oversight Panel recommended in February of 1998 that the cleanup standard per gram of soil be set at 35 picocuries. This citizens' recommendation was not accepted by government officials, nor was a standard of between 5 and 10 picocuries recommended by the Rocky Mountain Peace and Justice Center and the non-governmental Institute for Energy and Environmental Research.

There were three additional consultative groups formed to advise on the parameters of the cleanup: the Citizens Advisory Board, the Coalition of Local Governments, and the Cleanup Agreement Focus Group. The Citizens Advisory Board was created formally under the Federal Advisory Committee Act. By the rules under which it was created, the Citizens Advisory Board was to be "broadly representative" of affected and interested parties in the community. The Citizens Advisory Board decided that its recommendations were to be reached by consensus. Their reasoning for this determination was so DOE could not play "divide and conquer" with members of the Board (Moore, 2016). The Coalition of Local Governments was not created under the Federal Advisory Committee Act. The purpose of creating the Coalition of Local Governments group was to give a greater voice to DOE from affected suburban communities. The addition of the Coalition of Local Governments and then the Cleanup Agreement Focus Group was perceived by at least some within the Citizens Advisory Board as a mechanism for reducing their influence.

The Government Accountability Office conducted a survey of members of the Citizens Advisory Board and the Coalition of Local Governments in order to make a judgment about the role and effectiveness of community groups in the Rocky Flats cleanup. One conclusion of GAO (2006) was that "in addition to collaborating on decisions about soil cleanup levels, community groups provided comments and advice on many other aspects of cleanup that influenced DOE in some cases" (p. 16).

GAO (2006) further reported that their survey of the Rocky Flats Citizens Advisory Board and the Rocky Flats Coalition of Local Governments revealed that 21 of the 25 respondents were "very or somewhat satisfied" with how DOE engaged the groups. Also, GAO found that 19 of the 25 respondents said that they were "very or somewhat satisfied" with how the DOE used the input of groups. Nearly all (23 of 25) said they believed "the groups' input had a very or somewhat positive effect on the cleanup process" (GAO, 2006, p. 17).

Irreconcilable Substantive Differences

There were and are irreconcilable substantive differences of opinion on the acceptable level of plutonium which could be left in soil after the cleanup was completed. The limitations on funding available for the cleanup at Rocky Flats remains a matter of major disappointment for some observers and participants, especially community activists. According to LeRoy Moore (2016):

> I myself recall the day we learned that a previous decision on funds and timing had been made. The Rocky Flats Cleanup Agreement Focus Group was created in late 2001. The agenda for these meetings was set by DOE and the regulators, not the public. Some of us were increasingly dissatisfied with what the agencies were putting before us. At one of the meetings several of us said what we needed was cleanup to background, as recommended by the Future Site Use Working Group. We asked those responsible for the cleanup to tell us what it would cost to get the cleanup we wanted. Once we have the numbers, we said, we'll go to Congress and press them to provide the funds.

Moore went on to elaborate on the discussion with DOE spokesman Joe Legare. Legare is quoted as saying:

> You can't get what you're asking for, because an agreement has already been reached that sets a ceiling on what can be spent and puts a limit of the time it will take.

Legare later is quoted as telling a sociologist that the effect of his remarks on the audience was: "…like throwing a dead rat on the table."

Moore indicated that "I learned much later the agreement with Congress to which Legare referred actually occurred in 1995, the year in which Kaiser-Hill began this cleanup contract (Moore, 2016). In fact, in the middle 1990s, a bipartisan consensus in Congress had emerged that the United States had more urgent national priorities that required finite resources than those of environmental restoration at the nuclear weapons complexes after the end of the Cold War.

The appropriations committees of both the House of Representatives and the U.S. Senate were going to enforce fiscal restraint on environmental restoration. The topic of funding for the Rocky Flats cleanup has been a subject of continuing contention given the earlier estimate as part of the Estimating the Cold War Mortgage process. Dr. Peter Johnson, a scientist for the Office of Technology Assessment of the U.S. Congress, estimated that the total restoration costs for the U.S. nuclear weapons production complex could exceed half-a-trillion dollars ($500 billion), and that there was no known technology for solving many of the most compelling cleanup challenges (OTA, 1991).

Final Adoption of a Plutonium Standard "Fixed in Law"

Another influence on government policy, and relevant to a statement of DOE Plant Manager Silverman, was that a standard for cleanup level must be reflected in law; namely, in the Comprehensive Environment Response, Compensation, and Liability Act. While CERCLA does not establish precise cleanup levels, the standard it establishes is that the cleanup must result in a range of lifetime cancer risk which is related to envisioned future uses of the site. CERCLA requires a setting of standards within a range of 1 in 10,000 harmed to 1 in 1 million harmed.

The cleanup standard under CERCLA must protect the "maximally exposed individual." What was ultimately created for the Rocky Flats site was a "refuge worker scenario." The "maximally exposed individual" under the legal requirements of CERCLA would be a wildlife refuge worker exposed for 20 hours per week for up to 20 years (Moore, 2014). It was assumed that no visitor would spend as much time there as a refuge worker. If a refuge worker is protected, everyone else would be protected.

The final result of the extended process for determining cleanup level in soil was the release in 2002 by DOE, EPA, and CDPHE of a revised Rocky Flats Cleanup Agreement. This was finalized in 2003. The 651 picocuries standard was discarded in favor of what was used in the "Accelerated Cleanup:" three standards based on depth in soil. Moore (2016) indicated that because of a limit had been set in an agreement with the U.S. Congress on what could be spent on the cleanup, the result was cleaning up the site to a 3-level standard: the top 3 feet, (50 picocuries); 3 feet to 6 feet (1,000 to 7,000 picocuries); and below 6 feet (no limit).

A Critique of the Results of the "Accelerated Cleanup" From an Informed Citizen Perspective

Moore (2014) identified a series of defects in the cleanup which was actually conducted at Rocky Flats by the U.S. government. Major defects cited by Dr. Moore are as follows:

1. The three level graduated standards adopted by DOE were rejected by 86 percent of the commenting parties when they were put out for public comment as an unwise and unnecessary policy choice.
2. The DOE cleanup standard ignored the fact that soil beneath the 903 Pad was known to contain plutonium which had leaked for over a decade from at least 100 barrels left outdoors (5,000 gallons) from the thousands of barrels stored exposed all-year-round in Colorado weather (Chapter Five).

3. Nearly three miles of underground process waste lines containing plutonium, many below 6 feet, were grouted and left in place.
4. Five large Solar Evacuation Ponds and their contaminated liners were left intact and backfilled with soil.
5. Plutonium processing buildings were imploded onto their slabs without the plutonium being removed, then covered with soil.
6. Plutonium which had been dumped 30 or so feet underground, once mapped, was ignored according to former Rocky Flats worker Jerry San Pietro in an interview for the Maria Rogers Oral History Program hosted at the Carnegie Branch of the Boulder Public Library.
7. The 24,000-year half-life of plutonium provides ample time for some of it to make it into the surface environment.
8. Documents from the FBI raid and the Special Federal Grand Jury were never examined by those responsible for the cleanup.
9. Evidence was not considered which had been provided by former Rocky Flats worker and whistleblower Jacque Brever showing that DOE had provided false information regarding the East Spray Fields to EPA and CDPHE.
10. The research conducted in 1995 by engineer Iggy Litaor was not considered which found real-time migration of plutonium in sub-surface soil.
11. The research in 1996 by ecologist Shawn Smallwood, which was not considered, found 18 species of burrowing animals at the site that dig down as much as 20 feet and are constantly moving soil. Tiny particles of plutonium brought to the surface can be picked up by the strong winds common to the area.
12. Rocky Flats is directly over the western edge of the 7,000 square mile Denver Basin System of four aquifers. It was wrongly asserted that the terrain below Rocky Flats is impermeable, and that the underlying aquifers cannot be contaminated.
13. The exposure standards used at Rocky Flats were based on data from Japanese survivors of the atomic bombing, and thus biased because the stronger survived and the weaker perished. Different exposure requirements were essential.
14. The exposure standards were calculated to protect a "reference man," a 154-pound Caucasian male in his twenties, not fetuses, not young children, not the very old, nor the infirm.
15. The designation of the Rocky Flats site as a Wildlife Refuge allowed DOE to avoid performing the best possible cleanup which was necessary to assure as much as possible human and ecosystem health for the long term.

These fifteen defects are not an exhaustive list of criticisms of the "Accelerated Cleanup" as it was authorized and conducted by the U.S. government. The spirit of these critiques may be characterized as reflecting the "precautionary principle" as practiced in Europe--better safe than sorry. The citizen critics of the "Accelerated Cleanup" will not have their concerns met.

Moore (2019) summarized his in-depth critics of Rocky Flats in a comprehensive statement of the broader rationale for the Rocky Mountain Peace and Justice Center and generalized critiques of nuclear weapons well beyond Rocky Flats.

Understanding the Backstory of the Irreconcilable Versions of the "Cleanup" of Rocky Flats

One of the key areas of inquiry in this study is understanding the response of government to concerns of citizens in areas of sensitive national security imperatives. This chapter recounts a variation of that inquiry: understanding government response to citizen concerns when an "ultrahazardous" Cold War nuclear weapons production site is closed permanently but the facility is surrounded by suburbs of a major metropolitan area.

Frame For Government Response To the Urgency of a Rocky Flats Cleanup

Government response was influenced by a number of factors. No other Cold War era nuclear weapons production site has ever been dismantled. And the U.S. Senate had served notice with Train Wreck Along the River of Money (Blush and Heitman, 1995) that the nation had more pressing fiscal priorities than environmental restoration at former nuclear weapons production sites. There was also no known technology for addressing successfully some of the known toxic risk challenges. This frames the broader context for citizen response.

Frame For Citizen Response: Environmental Justice and Threat Reduction

The frame for citizen concerns was reducing the threats to the Denver metropolitan region and reducing as much as possible the vestiges of environmental injustice. The organizing construct for an influential group of citizens was the "precautionary principle." The goals for civil society at every choice point where they were included were increased levels of cleanup, and to use the site as a laboratory for developing new technology which would apply not only to Rocky Flats but to other site restorations in the future. The GAO (2006) reported that innovative technology utilized as integral to the accomplishments of the "Accelerated Cleanup" is already being deployed at other locations in the U.S. nuclear weapons development complexes.

Chapter Thirteen
"Irreconcilable Conceptions": Continuing Clashes Over the Future of the Wildlife Refuge In the Frame of Obligations Under Law and Assessment of Risks To Children and the Community Posed By Missing and Residual Plutonium

The opening part of the title for Chapter Thirteen, "Irreconcilable Conceptions," was accurate in 2004-2005 when the Fish and Wildlife Service (FWS) submitted its initial Comprehensive Conservation Plan and Environmental Impact Statement and Kathleen Becker of the Sierra Club used provisions of the National Environmental Policy Act to systematically challenge the adherence of the Fish and Wildlife Service to the letter and the spirit of the rule of law.

"Irreconcilable Conceptions" was accurate again in 2015-2016 when the Fish and Wildlife Service intended to use a prescribed burn on 701 acres of the Wildlife Reserve to combat invasive species. This intended action was challenged on a number of grounds including that the Fish and Wildlife Service did not have a legal permit to do so. The proposed prescribed burn did not occur.

"Irreconcilable Conceptions" is again a valid characterization in 2020 on the clash over the meaning of the rule of law for determining the future of Rocky Flats. This time the Fish and Wildlife Service has been joined by the Department of the Interior of which they are a part, the Department of Transportation, and the Federal Highway Administration to open trails across the 5,237-acre Wildlife Refuge. The inclusion of the Department of Transportation and the Federal Highway Administration is because the proposed trail system across Rocky Flats is to be part of a regional network.

The response by elements of civil society was to seek Preliminary Injunctions in Federal District Court for the District of Colorado citing "irreparable harm." In both 2017 and 2018 two different Federal District Court Judges did not grant a Preliminary Injunction. But what is different in 2020 from earlier in the history of the Wildlife Refuge is that the Judicial branch of government, relying heavily on legal precedent, will once again have a deciding role in determining the meaning of the rule of law and the risk to human health at Rocky Flats.

Since the Comprehensive Conservation Plan and Environmental Impact Statements were considered in 2004-2005 there has been a risk assessment alert and two prestigious U.S. government associated groups have issued highly relevant reports. The risk assessment alert is that the Fish and Wildlife Service model of risk was for an adult male refuge worker in 2004-2005. Opening the Wildlife Refuge to a regional network of trails to the public will include families with children.

The U.S. government associated groups are the National Research Council (NRC) of the National Academies of Sciences, Engineering and Medicine and the Agency for Toxic Substances and Disease Registry (ATSDR) of the Department of Health and Human Services. The highly relevant NRC report of 2006 is an analysis of the cancer risk of low dose exposure to ionizing radiation. The ATSDR report of 2010 is on the effects of exposure to plutonium especially for children.

An underlying issue between the FWS and some elements of civil society is whether the law requires a new Environmental Impact Statement. A new Environmental Impact Statement would become a venue for considering the meaning the plutonium present on the Wildlife Refuge, and the meaning for human health in the former lands of the Rocky Flats "buffer zone" of NRC (2006) and ATSDR (2010) as they pertain to the uses of the Wildlife Refuge.

The narrative will now shift to the Congressional legislation which established the Wildlife Refuge in 2001 A.D.

Founding Vision of the Rocky Flats National Wildlife Refuge In 2001

The U.S. Congress established the Rocky Flats National Wildlife Refuge in 2001. Its purposes and significant qualities as enshrined in the Act of 2001 are as follows:

1. Most of the Refuge has remained undisturbed since its acquisition by the government in 1951;
2. The refuge preserves valuable open space and striking vistas at the edge of the Prairie and the Front Range Mountains; and
3. The Refuge provides habitat for many wildlife species, including a number of threatened and endangered species, and is marked by the presence of rare tallgrass prairie communities.

The choice of the phrase "significant qualities" is a rationale for establishing a Wildlife Refuge at the edge of the Prairie and the Front Range of the Rocky Mountains.

The entire Rocky Flats site was evaluated under the provisions of CERCLA and 1,300 acres determined to be a Superfund site under the continuing authority of the Department of Energy. After review under the provisions of CERCLA, the remainder of the site was transferred to the Department of the Interior and put under the management of the U.S. Fish and Wildlife Service.

There have been two initial major flashpoints:

1. In 2004-2005 the Fish and Wildlife Service drafted and finalized the Rocky Flats National Wildlife Refuge Final Comprehensive Conservation Plan and Environmental Impact Statement.
2. Fish and Wildlife Service opened up trails across the Wildlife Refuge. The 2004 EIS contemplated the establishment of multi-use trails accessible to the public.

In between these two major clashes was a skirmish in 2015 over a prescribed burn permit as part of combatting an infestation of non-native plants. The presentation of the structure of the two clashes and the skirmish will serve to underscore the incompatibility of the positions of the Executive branch organizations on the one hand, and elements of organized civil society on the other.

Final Outcome Will Be Determined By the Judicial Branch

As with the Church cluster of lawsuits from 1975, and the Cook class action lawsuit from 1990, the final outcome will be determined by the Judicial branch of government starting in courtrooms of the Federal District Courts for the District of Colorado and such appeals which may arise to the 10th Circuit Court of Appeals and beyond.

Special Concern Over the Low Dose Exposure to Children

In addition to precedents from previous determinations by the Judicial branch, the irreconcilable conflicts include disputes over the toxicity of plutonium in low doses especially to children, and the credibility of previous findings of U.S. government scientists in the 1970s on the levels and risks from known plutonium in soil on the lands which have become the present Wildlife Refuge.

An additional new development since the 2004 EIS was the acquisition of additional land at the southwest corner of the Wildlife Refuge known as the "Section 16 Parcel." This additional land acquisition has received no known assessment for level of plutonium in soil.

"Irreconcilable Conceptions" of Obligations Under Law

"Irreconcilable Conceptions" of obligations under law in the title of this chapter includes to the action of the FWS on March 23, 2018 when it issued an Environmental Action Statement for the purpose of making changes to its Comprehensive Conservation Plan/2004 EIS in order to incorporate the newly planned trail routes with the Federal Highway Administration.

According to Federal District Court Judge Phillip A. Brimmer, the legal rationale for this FWS action is the "categorical exclusions" provisions under 40 C.F.R. §1508.4.

The "categorical exclusions" cited by FWS according to Judge Brimmer allow for:

> minor changes in the amounts or types of public use" [of public lands] "in accordance with existing regulations, management plans, and procedures" as well as categorical exclusions for changes to such plans "when no or minor effects are anticipated.

The "Irreconcilable Conceptions" position of the plaintiffs took the form of a request to the Federal District Court for the District of Colorado to grant a preliminary injunction against trail construction and the opening of the trails to the public. The plaintiffs claimed violations of the Federal

Administrative Procedures Act 5.U.S.C. §701 et seq. and the National Environmental Policy Act §4331 et seq.

At fundamental issue in the courtroom of Judge Brimmer was whether "irreparable harm" will occur if a preliminary injunction is not granted by the Court. The new context is decision-making based on prior legal precedents. His 2018 review of FWS decisions was "generally limited to review of the administrative record" citing a 10th Circuit Court of Appeals decision in *Custer Cty. Action Ass'n v. Garvey*, 256 F.3d 1024, 1027n.1(10th Cir.2001). Citing *American Mining Cong. v. Thomas*, 772 F.2d 617, 626 (10th Cir. 1985), Judge Brimmer indicated that: "The circumstances which warrant consideration of extra-record materials are 'extremely limited'. " The determination by Judge Brimmer was not to issue an "irreparable harm" Temporary Injunction. As a result, trails are open from sunrise to sunset year around except for major national holidays.

The clashes of 2004 and 2018 and the skirmish of 2015-2016 will be presented to illustrate the deep roots of irreconcilable differences which exist. There will be a review of the best U.S. government associated science on the special vulnerability by children to plutonium exposure compiled by the Agency for Toxic Substances and Disease Registry in 2010, and by the National Research Council on the effects of low dose exposure in 2006.

THE CLASH OF 2004

Assessment of the U.S. Fish and Wildlife Service In 2004

The assessment of the U.S. Fish and Wildlife Service concerning the Refuge and their vision for the future in the public interest was conveyed officially in their response to submissions as part of the 2004 public comment period on the then draft Comprehensive Conservation Plan/Environmental Impact Statement:

- The EPA and CDPHE have indicated that subsurface contamination does not exist in the area that will become the Refuge.
- The EPA and CDPHE have indicated that all of the area to become the Refuge, including surface water, will be safe for Refuge visitors and workers.
- The Service does not believe that residual soil contamination has adversely affected wildlife at Rocky Flats.
- The Service is confident that all aspects of the CCP/EIS process have followed NEPA requirements.

Assessment of the Sierra Club In 2004

An alternate assessment of the current condition of the Wildlife Refuge in 2004 was conveyed by Kathleen C. Becker of the Sierra Club in 2004

(Becker, 2004) as part of the same comment period on the then draft Comprehensive Conservation Plan/Environmental Impact Statement:

> ...the FWS has not provided enough information to the public to make an informed decision about public access.

This statement on behalf of the Sierra Club is based on the National Environmental Policy Act (40 CFR 1502.22(a)) which requires that environmental considerations be integrated into federal planning when the action will have a significant effect on the quality of the human environment and "the degree to which the proposed action affects public health and safety." Among the conclusions and the unanswered questions of the Sierra Club were:

> FWS fails to consider whether the human uses proposed in the areas of contamination and cleanup at Rocky Flats under the various alternatives will have 'a significant effect on the quality of the human environments.'
>
> How can FWS manage for native ecosystems without inquiring into the baseline levels of residual contamination that will exist when it receives the property?
>
> Has FWS evaluated the effect of burrowing animals in or near to cleanup sites?

The relevance of rigorous animal population studies is to found in Smallwood, Morrison and Beyea (1998) who investigated the characteristics of burrowing animals which are relevant to hazardous waste management. Of particular relevance to the Wildlife Refuge is the Krey and Hardy (1970) map of plutonium which shows overlap of plutonium concentrations with the Wildlife Refuge site. In 2018, the Seed Committee (Seed, 1971) intensive study of the credibility of the findings of Krey and Hardy (1970) and Martell and Poet (CCEI, 1970a; 1970b) became available. The Seed Committee established that plutonium levels in soil of Rocky Flats on-site and off-site exceeded acceptable levels, and that earlier studies by Krey and Hardy (1970) and Martell and Poet (CCEI, 1970a; 1970b) were scientifically sound.

> Has the FWS or any other agency evaluated the extent to which toxic materials in contaminated soils would be exposed in a controlled burn, or would be dispersed into the environment during a burn?

The answer to this question according to the Sierra Club is "No."

The Sierra Club in 2004 called for a full Environmental Impact Statement:

> ...to summarize and reference information on the potential adverse effects from hazardous substances, and from the effects of human, wildlife and natural actions and interactions on those hazardous substances.

The Sierra Club drew attention to the necessity for the public to have more information:

> The public needs more information regarding residual contamination after remediation and details about what monitoring will occur, who is responsible for it, whether it will occur on the Refuge or only on DOE retained-lands, and how often it will occur.

Juxtaposing the convictions of the U.S. Fish and Wildlife Service about the Wildlife Refuge and the convictions of the Sierra Club highlights why the title of this chapter on the future of the Rocky Flats National Wildlife Refuge begins with "Irreconcilable Conceptions."

Draft Comprehensive Conservation Plan and the Extensive Comments It Received In 2004

The U.S. Fish and Wildlife Service prepared a draft Comprehensive Conservation Plan (CCP) and Environment Impact Statement (EIS) which was released to the public in February of 2004. There were public hearings in March of 2004. By the close of the comment period in April of 2004, they had received 5,307 comments from individuals, organizations and agencies. The final Comprehensive Conservation Plan was completed in 2005. It sets out the management and use plan for the next 15 years.

According to the Comprehensive Conservation Plan, there will be four uses for the Refuge:

1. To restore and preserve native ecosystems;
2. To provide habitat for, and population management of, native plants and migratory and resident wildlife;
3. To conserve threatened and endangered species; and
4. To provide opportunities for compatible scientific research.

The FWS took a number of positions in response to issues raised by individuals:

- Equestrian use would be limited to specific trails.
- Most of the ponds are not suitable for waterfowl hunting.
- The presence of dogs is forbidden. The Service does not believe that dogs would be compatible with the Refuge.
- Habitat restoration would be the focus in the first five years, especially ecological considerations posed by noxious weeds.
- Model glider use would not be compatible.
- Wildlife harassment is against Service policy.
- Trails through sensitive habitat would be subject to seasonal closure.
- There will be no requirement to sign an informed consent statement since a premise of the CCP/EIR is that "the land to

become the Refuge would be safe for the Refuge worker and visitor."

- None of the CCP alternatives include playground facilities.
- Prescribed fire would be one component of a comprehensive vegetation management strategy.
- Grazing would be used as a management tool.
- The contamination levels in the area to become the Refuge are currently low enough (prior to cleanup) to not require any response actions.
- The Service does not propose using prescribed fire on the Eastern portion of the Refuge between Walnut Creek to the north and Woman Creek to the south.

Comment: There was no explanation offered for excluding prescribed fire between Walnut Creek and Woman Creek. This is an area known to have been the location of significant disposal of toxic waste, and to be within an area determined by Krey and Hardy (1970) to be contaminated with plutonium.

Comprehensive Conservation Plan of the U.S. Fish and Wildlife Service

Within the framework of the four general uses of the Wildlife Refuge identified in the original draft Comprehensive Conservation Plan, and three "significant qualities" identified by the U.S. Congress in the establishing legislation, the Final Comprehensive Conservation Plan articulates nine major actions which will be part of the plan for the first fifteen years.

In order to understand the mixed reaction to the plan, the nine major actions will be presented one at a time.

<u>Maintaining Native Grasslands</u>

> Use of variety of tools including prescribed fire, grazing, and 'mowing to stimulate and maintain native grasslands'.

Comment: This proposed action has been highly controversial over the prescribed burn and in fact became the subject of the still on-going major conflict with elements of civil society.

In 2015, representatives of civil society appeared before the Colorado Air Quality Control Commission. The commission subsequently restricted the Fish and Wildlife Service controlled burn permit on February 29, 2015. The Colorado Department of Public Health and Environment reviews smoke permits. For further smoke management applications by the Fish and Wildlife Service, the Commission will require that the Fish and Wildlife Service participate in a public hearing. The Colorado Department of Public Health and Environment will not be allowed to consider any smoke management permits for Rocky Flats until that public hearing has been completed.

Improve Habitat Conditions

> To improve habitat conditions for wildlife species including the Preble's meadow jumping mouse, remove 28 miles of unused road and 13 stream crossings.

Comment: This is a fundamental reason for establishing the Refuge according both to Congress and the U.S. Fish and Wildlife Service. It is the "public use" portion which remains controversial with elements of civil society.

Cooperation With Surrounding Jurisdictions

> Work closely with surrounding jurisdictions to coordinate natural resource management, public use, and the regional protection of wildlife movement corridors.

Comment: This is seen as a key task of the U.S. Fish and Wildlife Service.

Scientific Research

> Provide for compatible scientific research opportunities that focus on wildlife habitat and interactions between wildlife and human use.

Comment: This is one of the four key uses for the Refuge in the Comprehensive Conservation Plan. Independent scientific research has not yet been permitted. Moore (2016) reported that in 2011 permission was asked and denied for Marco Kaltofen to conduct soil sample research on the Refuge.

Prairie Dog Management

> Allow prairie dogs to expand in some areas.

Comment: The U.S. Congress presented this type of activity as a reason to establish the Refuge in the first place, and the U.S. Fish and Wildlife Service presented it as one of the four uses of the Refuge. FWS chose to import prairie dogs from off-site, an action which is being challenged in Federal District Court in 2019.

Provide For Visitor Use

> Provide visitor use facilities including 12.8 miles of multi-use trails with 3.8 miles of hiking-only trails, a visitor contact station, interpretive overlooks, viewing blinds, and parking areas.

Comment: This major action is premature in the critique of the Sierra Club in 2004 (see next section), and is being challenged in Federal District Court in 2019.

Limited Public Hunting

Work with Colorado Division of Wildlife to have a limited public hunt or cull of deer and elk populations. In the EIS, the FWS stated:

> The Service believes that limited public hunting would be compatible with the purposes and management of the Refuge… (34.3)

Comment: This major action is premature in the Sierra Club critique, but not over the question of whether limited public hunting is appropriate in the federal wildlife refuge, nor over the issue of whether or not limited hunting is appropriate when it is a necessary tool to use when herds must be culled. It is a premature action because there are studies of deer killed accidently by automobiles concerning whether the animals had some level of radioactive contamination. It is also a premature action because of the acreage of the Refuge has not been sufficiently characterized under the legal requirements of NEPA and the application of the precautionary principle in the opinion of informed civil society.

Environmental Education and Special-Use Limited Hunting

> Provide environmental programs for high school and college students and a limited hunting program (two weekends a year) for youth and persons with disabilities.

Comment: The comments on the risks associated with hunting in this particular Refuge were covered in the previous section. Offering environmental education programs for stewards of the next generation is a precious opportunity for society. The Sierra Club critique is that it is premature.

Access Restrictions and Safety Concerns

> Address public concerns about the safety of the refuge through public outreach and clear signage about access restrictions and public use opportunities."

Comment: This issue undertaking the base-line assessments called for by the Sierra Club to assure public and worker safety. This Fish and Wildlife Service statement contemplates the need for limited access ("access restrictions") but at the same time is silent on the reasons why access may need to be limited. They indicated without elaborating that there will be no prescribed burns between Woman Creek and Walnut Creek.

Sierra Club Detailed Critique of the U.S. Fish and Wildlife Service Comprehensive Conservation Plan

On April 21, 2004 the Sierra Club indicated that "the FWS has not provided enough information for the public to make an informed decision about public access…" The Sierra Club called for an updated Draft Environmental Impact Statement after the EPA assessment of the site occurred (completed in 2007).

The Sierra Club raised a number of issues which have yet to be addressed. The National Environmental Policy Act requires that environmental considerations be integrated into federal planning when the action will have a significant effect on the quality of the human environment and "the degree to which the proposed action affects public health and safety." (40 CFR 1502.22(a))

Conclusions in 2004 of the Sierra Club were:

1. FWS fails to consider whether the human uses proposed in areas of contamination and cleanup at Rocky Flats under the

various alternatives will have "a significant effect on the quality of the human environment."

2. The CCP/EIS is premature because (1) FWS will not inherit the site for many years (note: it did so in 2007), and conditions may change in the interim, and (2) FWS cannot accurately analyze the impacts of proposed alternatives until the nature and extent of residual contamination on the site is known—something that cannot be known until FWS received jurisdiction to the site.

Comment: Since receiving jurisdiction to the site, FWS has done nothing to advance understanding of the extent of residual contamination.

3. The EIS leaves many specific questions unanswered.

Comment: There were a number of specific questions posed in 2004 which have still not been answered over 16 years later in 2020.

<u>Question Related To the Rocky Flats National Wildlife Refuge Act</u>

For example, the Rocky Flats National Wildlife Refuge Act of 2001 requires that the Refuge be "managed to restore and preserve native ecosystems, provide habitat for native plants and wildlife, and provide opportunities for compatible scientific research." A Sierra Club question of 2004 which remains unanswered in 2020 is:

> How can FWS manage for native ecosystems without inquiring into the baseline levels of residual contamination that will exist when it receives the property?
>
> Has FWS evaluated the effect of burrowing animal in or near to cleanup sites?
> *Note: These questions raise fundamental challenges to the future of the Wildlife Refuge in the opinion of elements of civil society.*

<u>Questions Related To Impacts Due To Fire</u>

On page 103 of the Draft Environmental Impact Statement, it was reported that neither the Atomic Energy Commission nor the Department of Energy had allowed wildfires to burn on the Rocky Flats Site. Further, there had been only one controlled burn conducted on the Rocky Flats site since 1972. It is the stated policy of the Fish and Wildlife Service to have controlled burns (prescribed burns as a management tool). The Sierra Club posed a series of questions in 2004 which are unanswered in 2020:

> Has FWS or any other agency evaluated how controlled burns could affect residual contamination?
>
> Has FWS or any other agency evaluated the extent to which toxic materials in contaminated soils would be exposed in a controlled burn, or would be dispersed into the environment during a burn?

A Sierra Club critique was that the Draft Environmental Impact Statement "is devoid of any information or discussion on the potential adverse effects from hazardous substances."

The Sierra Club went on to state that "it is virtually impossible for the public to adequately assess the impacts of the alternatives." The Sierra Club called for a full Environmental Impact Statement to "summarize and reference information on the potential adverse effects from hazardous substances, and from the effects of human, wildlife and natural actions and interactions on these hazardous substances."

This has not occurred as of 2020.

Necessity For More Pubic Information

It was the position of the Sierra Club that:

> The public needs more information regarding residual contamination after remediation and details about what monitoring will occur, who is responsible for it, whether it will occur in the Refuge or only on DOE retained-lands, and how often it will occur.

Comment: FWS is still silent on this issue of monitoring.

DOE with EPA have made clear that Five-Year Reviews under CERCLA will occur only on DOE-retained lands—the 1,300-acre Superfund Site.

Expectations of an Environmental Impact Statement

The Sierra Club critique of the FWS in 2004 was as follows:

> FWS cannot just assume that the site will be cleaned up to a level that makes any of the activities proposed under the various alternatives "safe" for the environment. If FWS is assuming that cleanup poses no risk to human health and environment, the EIS should say so, rather than pretend this is not a question at all.

Comment: This question has not been addressed by FWS.

Fundamental Requirement of NEPA

A fundamental challenge by Kathleen C. Becker (2004) of the Sierra Club pertained to compliance by the Fish and Wildlife Service with federal law as represented by NEPA:

> Conclusory statements about the impact of these proposed activities do not meet the requirements of NEPA that FWS consider environmental impacts of the proposed federal action.

Comment: There has been no reconciliation of proposed FWS actions and the requirements of NEPA in the view of elements of civil society, and is being challenged to the Judicial branch.

Contested Terrain of Obtaining a New Permit For a "Prescribed Burn" At the Wildlife Refuge Circa 2016

The Colorado Department of Public Health and Environment (CDPHE) is responsible for the permit process for "prescribed burns." In January of 2016 Gordon Pierce of CDPHE spelled out the process for obtaining a permit. Contested terrain regarding the Wildlife Refuge certainly will be over a new permit to conduct a "prescribed burn." It is useful, therefore, to identify both steps in the process and substantive issues at each step.

Public Comment Process

According to Gordon Pierce, the following is the "Public Comment" part of the process:

> F. Public Comment
>
> 1. The Division will evaluate fires proposed in permit applications to determine whether the fire will pose a high smoke risk. The evaluation will include consideration of the size of the fire area, the type and density of fuel, whether or not the fire is a pile burn, expected duration, and the proximity of the fire to smoke sensitive receptors, such as communities and class I areas.
>
> 2. If the Division determines that the fire poses a high smoke risk, the Division will provide appropriate notice to the public, such as by making the information available through the internet and by email, and will issue a draft permit for public comment. The notice will include information about location of the fire, expected burn dates, expected duration, potential emissions, potential air quality and visibility impacts at smoke sensitive receptors and the opportunity for public comment. The procedures in the Commission's Procedural Rules apply to any draft permit subject to the public comment provisions of this paragraph. If the Division receives public comment, it will consider those comments in determining whether to issue a permit and what conditions to impose upon the permittee and within thirty days of the close of the comment period or within thirty days following the close of any public comment hearing, whichever is later, shall either grant or deny the permit.
>
> 3. Draft permits for high smoke risk burns that have been subject to the public notice provisions of paragraph V.F.2 and for which a permit has been issued, but which have not been completed within the period for which the permit is valid are not subject to the public comment provisions of paragraph V.F.2. This exception to the public comment requirements applies only: for the geographical unit for which a permit that has been subject to the public comment provisions was issued; if less than ten percent of the area originally permitted has been burned; and for a maximum of five years from the date of issuance of the original permit. The Division may amend the permit, based on a case-by-case analysis, to accommodate minor changes in the permit area without triggering the public comment provisions of paragraph V.F.2.

The Fish and Wildlife Service did not use the permit they had been granted for a five-year period. It expired at the end of 2015.

Planning Documents

In addition to the extensive requirements identified in point 2 below, Pierce refers to the fact that complying with all of the requirements in planning documents does not "preclude" the user from having to "obtain a permit for each prescribed fire." The expired FWS permit had been for the entire state of Colorado.

B. Planning Documents.

1. Significant users of prescribed fire as a grassland or forest management tool shall submit plans addressing the use and role of prescribed fire and the air quality impacts resulting there from.

2. Planning documents shall:
   a. Specify the area to which they apply, which should be relevant to as large an organizational unit of land as is administratively reasonable;
   b. Specify the time period to which they apply;
   c. Be relevant to vegetative, habitat, and/or fuel management of grassland and/or forest land; and
   d. Explain the decision process and criteria considered or applied to show:
      i. How the significant user of prescribed fire identifies fuel treatment alternatives to achieve the fuel, habitat, and/or vegetative land management goals;
      ii. How the significant user of prescribed fire selects among fuel treatment alternatives;
   e. Include a discussion of the alternatives to prescribed fire considered and a discussion of how prescribed fire, if selected, minimizes the risk of wildfire; and
   f. Demonstrate, for land selected for fuel treatment, how the significant user of prescribed fire considered the state standard set forth in Section VII.C and how the user will achieve compliance with that standard.

Role of the Colorado Air Quality Control Commission

There are quite stringent existing regulations for emissions control regarding particulate matter, smoke, carbon monoxide, and sulfur oxides; for air pollutant emissions notice requirements, control of hazardous air pollutants, and prescribed fire and permitting.

The chair of the Colorado Air Quality Control Commission has already made a commitment to schedule a public hearing on any prescribed burn permit received for the Wildlife Refuge.

The AQCC Regulation on hazardous air pollutants became effective on October 15, 2015 as Regulation 8. It incorporated the Code of Federal Regulations (CFR) promulgated by the U.S. Environmental Protection Agency. It lists a number of unmonitored substances used at Rocky Flats including Beryllium, Benzene, and Vinyl Chloride. Reference is made to both Title III of the Clean Air Act of 1990 to reduce the levels of nationwide air toxics emissions, and to the Colorado Air Quality Control Act.

The AQCC regulation on "prescribed burns" became effective on March 20, 2015. It applies to all open burning activity throughout the state of Colorado. Section VII is quite rigorous. For example, FWS will have the burden of proof that it can conform to the "state standard to minimize emissions..." A public hearing is required.

It is clear that the "prescribed burn" permit process is an important avenue for civil society to use to exert citizen influence over the outcome of a contested action by a division of the U.S. government at Rocky Flats; namely, the U.S. Fish and Wildlife Service.

Invasive Species and Environmental Justice

A standard management tool of FWS is to use prescribed burns to counter excess underbrush and invasive species. A prescribed burn would have been conducted in early Spring of 2015 but was postponed not cancelled in February of 2015. The backstory is that there have been eleven wildfires and one prescribed burn at Rocky Flats since 1993 (DOE, 2013). The only listed threatened species at Rocky Flats is Preble's meadow jumping mouse. FWS determined it would not be at risk because it lives near streams. The selection of 701 acres by the Fish and Wildlife Service for a prescribed burn excluded the land between Woman and Walnut Creeks.

Civil society has been able to mount an exceptional coalition against a prescribed burn on the grounds of risk for environmental injustice from the potential that burning could release plutonium from grasses and soil that could drift into surrounding communities including the nearby subdivisions of Whisper Creek and Candelas. The position of some elements of civil society is that there is simply too much unaccounted for plutonium, and too long a history of lack of adequate monitoring to accept assurances that there is not a serious potential risk to health from a prescribed burn.

THE CLASH OF 2018

The clash of 2018 actually began in 2017.

<u>Motion For a Preliminary Injunction In 2017</u>

There was essentially no common ground between the plaintiffs and the defendants in Case No. 1:17-cv-01210-CMA filed in the United States District Court for the District of Colorado.

<u>The Plaintiffs' Point of View</u>

The plaintiffs such as the Rocky Mountain Peace and Justice Center went to Federal District Court to obtain a Preliminary Injunction to force the Department of the Interior and the U.S. Fish and Wildlife Service to engage in review processes associated with the National Environmental Policy Act of 1969.

The Defendant's Point of View

The defendants claim that the motion of the plaintiffs is premature ("not ripe") and will interfere, if granted, with the planning process. Further, while FWS indicates that while it has not begun the NEPA process it "will conduct any required environmental reviews prior to issuing" any final decisions.

The Ruling in Federal District Court

The Motion for a Preliminary Injunction was not granted.

Motion For a Preliminary Injunction In 2018

In 2018 as in 2017, there was no common ground between the plaintiffs on behalf of some in civil society and entities of the U.S. government. The position of the government was that all legal and regulation obligations including NEPA have been satisfied.

The filing in 2018 is identified as Case No. 1: 18-cv-01017-PAB. While it is beyond the scope of this contribution to characterize fully the scope of the filings in the Federal District Court for the District of Colorado, it is possible to illustrate the gulf between the parties by presenting the major headings of the defendants in their 27-page filing:

- "Plaintiffs will not succeed on their claims that the Service has failed to conduct required environmental analysis prior to opening trails on the Refuge to the public" (p. 12).
- Plaintiff's Claim That the Service Has Violated NEPA Is Without Merit
- The Service Analyzed Environmental Impacts of Trail Use on the Refuge in the CCP (p. 13)
- There Are No New Circumstances Or Information That Would Warrant a Supplemental EIS (p. 14).
- There are No Extraordinary Circumstances That Would Preclude the Use of a Categorical Exclusion (p. 17).
- Plaintiffs Claim That The Service Has Violated the National Wildlife Refuge System Administration Act is Without Merit (p. 18).
- Plaintiffs Are Not Likely To Succeed on the Merits of Their ESA Claims (p. 19).
- Plaintiffs Have Failed to Establish Standing to Pursue Their ESA Claims (p. 19).
- The Service Has In Fact, Conducted Section 7 Consultation, and Plaintiffs' ESA Claims Are Moot (p. 29).
- Plaintiffs Cannot Demonstrate Irreparable Injury (p. 22).
- Plaintiffs Cannot Demonstrate Imminent Harm To Their Health (p. 23).
- Plaintiffs Cannot Demonstrate Any Interest In the Jumping Mouse That Would Be Irreparably Harmed (p. 24).
- The Balance of Harms and Public Interest Favors Denying Plaintiffs' Motion (p. 24).

The Plaintiffs submitted a reply on June 20, 2018 to the Federal Defendants' Opposition ECF No. 14. There were a number of elements related to the following Plaintiff summary of the position of Defendants:

> FWS essentially argues that it can ignore any aspect of the acknowledged plutonium contamination on the Refuge because other agencies made subsequent determination of safety.

The relevant backstory is that the Federal District Court for the District of Colorado had found in *Superior 913 F. Supp.2d* at 1123:

> ...[g]iven the expertise of the DOE, EPA, and the CDPHE in managing the risk of radionuclide exposure in the Refuge...FWS did not act arbitrarily and capriciously by incorporate[ing] the agencies' findings into its analysis.

There are fundamental rejoinders to the outcome of the *Superior* litigation: DOE does not speak with one voice. The "best available" DOE government science which was conducted in 1971 did not become available to civil society until 2018 (Seed, 1971). The CDPHE model of risk assessment downplays the importance of the "best available" science about risk to human health of low dose exposure to a speck of inhaled plutonium established by the prestigious National Research Council (NRC, 2006), and the special vulnerability of children to plutonium established by the Agency for Toxic Substances and Disease Registry (2010). There are irreconcilable differences over the reliance of the EPA on the credible science of Kriging the way it was utilized at Rocky Flats. The difference is over the shallowness of the Kriging sampling. In 1977-78 under the supervision of Federal District Court Judge Matsch, Howard Holme reported that the depth of sampling was eight inches.

In the view of critics of how Kriging was employed, there simply has been too much wind blown sand and soil buildup. The credible science of Kriging (Jourel and Hijbragts, 1978; Bohling, 2005; Chiles and Del Finer, 1999) is not in dispute. The EPA reliance of the contractor is misplaced because, after the passage of time in the context of high winds and soil buildups, the soil sampling by the contractor was too shallow.

The narrative will now shift for presentations of ATSDR (2010) and NRC (2006).

The ATSDR Toxicological Profile On Plutonium and Its Special Relevance To Children At Risk

The mission of the Agency for Toxic Substances and Disease Registry is to: "Serve the public through public health actions to promote healthy and safe environments and prevent harmful exposure." ATSDR was authorized as an independent agency on December 21, 1980 within the U.S. Department of Health and Human Services. The key documents in its establishment and functioning are the Comprehensive Environmental Response, Compensation, and Liability Act of 1980 (CERCLA), the Hazardous and Solid Waste Amendments of 1984 to the Resource

Conservation and Recovery Act (RCRA), and the Superfund Amendments and Reauthorization Act of 1986 (SARA).

In November of 2010, ATSDR released its Toxicological Profile for Plutonium. CERCLA section 104(i)(1) requires the Administrator of ATSDR to "...effectuate and implement the health related authorities" of the statute. CERCLA section 104(i)(3) as amended directs the Administrator of ATSDR to prepare a "toxicological profile" of each substance on the CERCLA National Priorities List of toxic substances in Superfund sites.

Peer review is an essential element for establishing the validity of assertions in the Toxicological Profiles. The Peer Review Panel for plutonium consisted of:

1. Lynn Anspaugh, Ph.D., Research Professor, Department of Radiology, Radiobiology Division, University of Utah, Salt Lake City, Utah;
2. Fletcher Hahn, D.V.M., Ph.D. Scientist Emeritus, Lovelace Respiratory Institute, Albuquerque, New Mexico;
3. Richard Leggett, Ph.D., Research Scientist, Life Sciences Division, Oak Ridge National Laboratory, Oak Ridge, Tennessee;
4. Nicholas Priest, Ph.D., Head Radiobiology Section, Atomic Energy of Canada Limited (AECL) Chalk River Laboratory, Chalk River, Canada; and
5. Ethel Gilbert, Ph.D., M.P.H., Expert (biostatistician), Radiation Epidemiology Branch, Division of Cancer Epidemiology and Genetics, National Cancer Institute, Bethesda, Maryland.

The Toxicological Profile for Plutonium identifies the collective expertise of the Peer Review Panel as covering knowledge of: "plutonium's physical and chemical properties, toxicokinetics (assessment of systemic exposure at toxic dose exposure levels), key health end points, mechanisms of action, human and animal exposure, and quantification of risks to humans" (p. xxii).

There are key sections in the 250-page Toxicological Profile which are particularly germane to understanding the public health issues and risks associated with the determination of whether or not to permit public access, especially by children, to the Wildlife Refuge.

Section 1.6 How Can Plutonium Affect Children?

There are differences between children and adults:

> Studies in young animals have shown that a larger amount of the plutonium deposited in the lung will move to the growing bones. (p. 5)

Effects in unborn children:

> We do not know if plutonium causes birth defects or affects the ability to have children, although some plutonium that reaches the blood can be found in ovaries and testes. (p. 5)

Section 1.7 How Can Families Reduce the Risk of Exposure to Plutonium?

Risk in the air you breath:

> Breathing plutonium-contaminated air is the most dangerous way to be exposed to plutonium. (p. 5)

Section 2.3 Minimal Risk Levels (MRLs)

> The strongest evidence for plutonium exposure-response and radiation dose-response relationships in humans is for cancers of the lung, liver, and bone. (p. 7)

Section 3.2.1 Inhalation Exposure

3.1 Health Effects Introduction

> Internally deposited plutonium, however, possesses the potential to produce significant health effects via transfer of energy from alpha particles to nearby cellular molecules." (p. 20)
>
> The most widely studied plutonium compound 239 Pu O_2, is only moderately soluble, which results in long-term retention in the lung following inhalation exposure. (finding based on dog studies). (p. 23)

Section 6.6 Exposure of Children

> Children are not small adults. A child's exposure may differ from an adult's exposure in many ways. Children drink more fluids, eat more food, breathe more air per kilogram of body weight, and have a larger skin surface in proportion to their body volume. (p. 188)
>
> Children crawl on the floor, put things in their mouths, sometimes eat inappropriate things (such as dirt or paint chips), and spend more time outdoors. Children also are closer to the ground, and they do not use the judgment of adults to avoid hazards (NRC, 1993). (p. 188)

Section 6.8.2 Ongoing Studies

> No ongoing studies pertaining to the environmental fate of plutonium or plutonium compounds were identified in a search of the Federal Research in Progress database (FEDRIP, 2007). (p. 193)

Section 8 Regulations, Advisories, and Guidelines

> The basic philosophy of radiation protection is the concept of ALARA (As Low As Reasonably Achievable). (p. 207)

> As a rule, all exposure should be kept as low as reasonably achievable and the regulations and guidelines are meant to give an upper limit to exposure (p. 207).
>
> ...the EPA Office of Air and Radiation believes that all radionuclides, including the plutonium isotopes, should be considered to be known carcinogens, and has assigned them to Group A. (p. 207)
> (Federal Guidance Report No. 13 Morbidity Risk Coefficients, in Units of Picocuries, (EPA, 2001)
> (*Note:* Group A *means* known *human* carcinogens).

Reflections On Placing Children At Risk of Plutonium Exposure

The expertise of the original authors of the ATSDR Toxicological Profile and the cumulative expertise of the Peer Review Panel is not at issue. The vulnerability of children clearly exceeds that of adults. This raises the paramount issue of what the best available science reveals about the consequences for human health of low dose exposure to plutonium. That is the subject of the next section.

Evidence On Health Effects of Ionizing Relation Which Has Sufficient Energy To Displace Electrons From Molecules In Human Cells

Free electrons damage human cells. The principal risk at Rocky Flats after the "Accelerated Cleanup" is no longer the exposure to high does of radiation from electrons in highly enriched uranium and plutonium from either plutonium igniter fabrication or its toxic aftermath. Rather, the principal risk to human health is from low levels of energy transfer in ionizing radiation. The Biological Effects of Ionizing Radiation (BEIR) reports from the National Academies of Sciences, Engineering and Medicine are the definitive U.S. government associated assessments of basic research on the biological consequence of ionizing radiation.

The National Academies were established by Charter from the U.S. Congress in 1863. The National Research Council in turn was established by the National Academies in 1916. Its purpose is:

> ...to associate the broad community of science and technology with the Academy's purposes of furthering knowledge and advising the federal government.
> (NRC, 2006, p. iii)

The National Research Council assembles expert Peer Review Panels to assure that both process and content of final reports are based on sound assessment of the "best available" science.

The composition of the Peer Review Panel for BEIR VII is:

> Committee to Assess the Health Risks from Exposure to Low Levels of Ionizing Radiation: Richard R. Monson (Chairman), Harvard School of Public Health; James E. Cleaver (Vice

Chairman), University of California, San Francisco; Herbert L. Abrams, Stanford University; Eula Bingham, University of Cincinnati; Patricia A. Buffler, University of California, Berkeley; Elisabeth Cardis, International Agency for Research on Cancer, Lyon, France; Roger Cox, National Radiological Protection Board, UK; Scott Davis, University of Washington and Fred Hutchinson Cancer Research Center, Seattle, WA; William C. Dewey, University of California, San Francisco; Ethel S. Gilbert, National Cancer Institute; Albrecht M. Kellerer, Ludwig-Maximilians-Universität, München, Germany; Daniel Krewski, University of Ottawa, Ontario, Canada; Tomas R. Lindahl, Cancer Research UK London Research Institute; Katherine E. Rowan, George Mason University; K. Sankaranarayanan, Leiden University Medical Centre, Leiden, The Netherlands; Daniel W. Schafer, Oregon State University (from May 2002); Leonard A. Stefanski, North Carolina State University, (through May 2002); Robert L. Ullrich, Colorado State University; Rick Jostes (Study Director), National Research Council

Organization of the Report

For the purposes of understanding the level of threat of plutonium exposure on the lands of the Wildlife Refuge, and from the opening of trails to the public across it, there are four major questions addressed in this summary by quoting exactly the thinking of the experts on behalf of the National Research Council:

- What new evidence requires further investigation?
- Why did the NRC committee not accept the view that low doses are substantially less harmful than estimated by the Linear No-Threshold Model?
- Why did the NRC committee not accept the view that low doses are substantially more harmful than estimated by the Linear No-Threshold Model?
- What are the principal conclusions of the NRC committee about the Linear No-Threshold Model?

What New Evidence Requires Further Investigation?

The context for answering this question is that in the opinion of the NRC committee: "Major advances have occurred during the last decade in several key areas that are relevant to the assessment of risks at low radiation doses" (NRC, 2006, p. 6).

They went on to elaborate as follows:

> These advances have contributed to greater insights into the molecular and cellular responses to ionizing radiation and into the nature of the relationship between radiation exposure and the types of damage that underlie adverse health outcomes.
>
> (NRC, 2006, p. 6)

As an example of advances in understanding molecular and cellular response to ionizing radiation, they cited new research from biology on the "bystander cells effect."

New Evidence From Biology: Cells Do Not Necessarily Have To Be Hit Directly By a Radiation Track For the Cell To Be Affected

> New evidence from biology suggests that cells do not necessarily have to be hit directly by a radiation track for the cell to be affected. Some speculate that hit cells communicate with non hit cells by chemical signals or other means. To some, this suggests doses, where all of the cells in the body are not hit, "bystander" cells may be adversely affected, resulting in a greater health effect at low doses than would be predicted by extrapolating the observed response at high doses. Others believe that increased cell death caused by so-called bystander effects might lower the risk of cancer by eliminating cells at risk for cancer from the irradiated cell population. Although additional research on this subject is needed it is unclear at this time whether the bystander effect would have a net positive or net negative effect on the health of an irradiated person.
>
> (NRC, 2006, p. 6)

Why the NRC Committee Did Not Accept The View That Low Doses Are Substantially More Harmful Than Estimated By the Linear No-Threshold Model?

> ...some materials provided to the committee suggest that the LNT model exaggerates the health effects of low levels of ionizing radiation. They say that the risks are lower than predicted by the LNT, that they are nonexistent, or that low doses of radiation may even be beneficial. The committee also does not accept this hypothesis. Instead, the committee concludes that the preponderance of information indicates that there will be some risk, even at low doses.
>
> (NRC, 2006, p. 9)

It is the role of the NRC committee to make a finding based on the "preponderance of evidence."

Why the NRC Committee Did Not Accept the View That Low Doses Are Substantially Less Harmful Than Estimated By the Linear No-Threshold Model?

> The BEIR VII committee has concluded that radiation health effects research, taken as a whole, does not support this view. In essence, the committee concludes that the higher the dose, the greater is the risk; the lower the dose, the lower is the likelihood of harm to human health.
>
> (NRC, 2006, p. 9)

In rendering this assessment, the NRC committee utilized the "preponderance of evidence" standard but characterized it as basing their judgment on the body of relevant research "taken as a whole."

What Are The Principal Conclusions Of the NRC Committee About the Linear No-Threshold Model?

The basic conclusion is that "...the risk at low doses will be small" NRC, 2006, p. 10). Their standard after extensive study was the "preponderance of evidence."

The basic conclusion of the NRC committee is expressed in three quotes from their report:

<u>The BEIR VII Risk Model Estimating Cancer Risk</u>

> The committee judged that the linear no threshold model (LNT) provided the most reasonable description of the relation between low-dose exposure to ionizing radiation and the incidence of solid cancers that are induced by ionizing radiation.
>
> (NRC, 2006, p. 6)

<u>Simple Risk Calculations</u>

> As the simple risk calculations in the Public Summary show, the risk at low doses will be small. Nevertheless, the committee's principal risk model for solid tumors predicts a linear decrease in cancer incidence with decreasing dose.
>
> (NRC, 2006, p. 10)

<u>Unlikely That a Threshold Exists</u>

> The BEIR VII committee concludes that current scientific evidence is consistent with the hypotheses that there is a linear dose-response relationship between exposure to ionizing radiation and the development of radiation-induced solid cancers in humans. The committee further judges it unlikely that a threshold exists for the induction of cancers but notes that the occurrence of radiation-induced cancers at low doses will be small.
>
> (NRC, 2006, p. 10)

Reflection On the Biological Effects of Ionizing Radiation VII Report In Relationship To the ATSDR Plutonium Profile and the New Role For the Judicial Branch

These two reports represent the "state of the science" which can be mobilized on two vital challenges to human health from exposure to ionizing radiation. Both reports (separately and together) are highly relevant to understanding risks associated with the unrestricted opening of trails to the public across the Wildlife Refuge.

It is not feasible on a practical basis to monitor who gets access to the trails and who does not once they are declared "open to the public." This places children at direct risk if risk is found to exist.

There is clearly an impasse which is not reconcilable between agencies of the Executive branch intending to create trails open to the public through the Wildlife Refuge which are to connect to a broader trail network, and groups of citizens and organizations united against that Executive branch undertaking.

Two attempts by the opponents to obtain Preliminary Injunctions have not been successful in persuading two different Federal District Court Judges that they have met the high burden of proving "irreparable harm." The Judicial branch of constitutional government has assumed (once again) what may become a determining role in an intractable conflict over Rocky Flats and its future. As of 2020 there are carefully crafted civil litigation on behalf of groups from civil society. The final outcome will not be known anytime soon.

Developments At the End of 2019

The end of 2019 is a very active time for continued litigation at Rocky Flats over the opening to visitors on trails of the Rocky Flats National Wildlife Refuge. There are two unresolved NEPA violation lawsuits. In addition, on January 19 2019, seven citizen groups represented by Attorney Patricia A. Mellon filed a legal petition intended to gain "evidence of unreported and underdressed residual plutonium contamination and other ongoing environmental dangers at Rocky Flats National Wildlife Refuge" (Steadman, 2019). The petition seeks access to exhibits presented to the Special Federal Grand Jury.

<u>Civil Action No. 1:18-cv-1746</u>

The town of Superior as Plaintiffs filed Civil Action No. 1:18-cv-1746 against Defendants the U.S. Fish and Wildlife Service and the Wildlife Refuge Manager on July 10, 2018. Tim Gablehouse is the Attorney of Record. Anthony Hahn (2018) quoted Gablehouse that: "What we believe is the Fish and Wildlife Service is trying to carve up the decisions they're making into small pieces in order to avoid doing a complete environmental review."

The Town of Superior lawsuit "is an action for declaratory and injunctive relief under the Administrative Procedure Act, 5 U.S.C. §701 et seq., based on Defendants' failure to comply with the National Environmental Policy Act…"

<u>Civil Action No. 1:18-cv-01017-PAB</u>

The Rocky Mountain Peace and Justice Center and other civil society groups filed Civil Action No. 1:18-cv-01017-PAB on August 9, 2018 against Defendants U.S. Fish and Wildlife Service, the Secretary of the Interior, the Local Manager of Wildlife Refuge, the U.S. Federal Highway Administration, and the Secretary of Transportation.

According to the filing, the Plaintiffs seek "a preliminary injunction against trail construction and the opening of public access to the trails…"

Judge Philip A. Brimmer denied the motion for a preliminary injunction and the Fish and Wildlife Service opened trails within the Wildlife Refuge. Attorney Randall Wiener represents the Plaintiffs.

On September 3, 2019 the Plaintiffs submitted a Supplemental Authority. According to Pritchett (2019), there were several elements to the Weiner filing:

- There must be compliance with NEPA in the form of a reexamination of the 2018 decision of the Fish and Wildlife Service on March 23, 2018 to reconfigure hiking, biking and equestrian trails. The argument is that the reconfiguration of trails through the previously off-limits "wind blown area" should not be construed as a "minor change" to the previous Final Comprehensive Conservation Plan.
- In 2019 the Fish and Wildlife Service decided to import prairie dogs from off-site locations. Weiner argued this is a direct violation of the 2005 Comprehensive Conservation Plan:

  > Deep burrowing animals could bring plutonium and other dangerous materials to the surface, where they could be scattered by wind or heavy rain.
- The third element on the Weiner briefing to the Judge was that a recent soil sample taken in August of 2019 contained a level of plutonium that far exceeds by five times the EPA cleanup standard, and directly contradicts EPA's assertion that the Wildlife Refuge meets safety standards.

Aguilar (2019) reported that the standard of cleanup of the Rocky Flats plant was 50 picocuries per gram at the highest three-feet level of soil. The yet-to-be-built Jefferson Parkway Public Highway Authority had commissioned soil sampling along the Eastern edge of the Wildlife Refuge. The independent contractor had discovered a sample of 264 picocuries of plutonium per gram of soil.

The Toll Road would connect the cities of Bloomfield and Golden. The response of the city of Bloomfield was the following post on its website:

> At this time, based on the soil samples conducted by the Jefferson Parkway Public Highway Authority…Given the recent test results and the Council's feedback, the Parkway is not moving forward at this time.

<u>Civil Action No. 19-cv-00076-RM-KLM</u>

In Civil Action No. 19-cv-00076-RM-KLM seven civil society organizations are represented by attorney Patricia A. Mellon as Petitioners against the United States of America as Defendant in United States District Court for the District of Colorado. The legal form of the Petitioners was a Writ of Mandamus to locate documents of the Rocky Flats Grand Jury.

The statutory authority is 28 U.S.C.A. §1361:

> The district courts shall have original jurisdiction of any action in the nature of mandamus to compel an officer or employee of the United States or any agency thereof to perform a duty owed to the plaintiff.

The Writ of Mandamus was filed on January 10, 2019.

According to the U.S. Department of Justice, Mandamus is "an extraordinary remedy, which should only be used in exceptional circumstances of peculiar emergency or public importance." Mellon cited five immediate threats which prompted the Petition:

- Build out of the trail system at Rocky Flats National Wildlife Refuge,
- Construction of a visitor's center at the refuge,
- Eventual plan to make the trails at Rocky Flats an extension of the Rocky Mountain Greenway Trail,
- Concern that construction of the Jefferson Parkway east of Rocky Flats would disturb contaminated dirt, and
- Fracking and drilling permits.

Steadman (2019) also quoted Jon Lipsky that: "...the documents could assist Rocky Flats nuclear workers with their unique claims and the concerns of residents living downwind of Rocky Flats."

The matter before Judge Raymond P. Moore with *Wilder v. Prokop*, 846F.2d 613, 620 (10th Cir. 1988) as a precedent is:

> The remedy of mandamus is a drastic one, to be involved only in extraordinary situations.

On September 4, 2019 Judge Moore did not compel the U.S. government to expedite its search for missing Grand Jury records.

The purpose of the group represented by Attorney Mellon is to break the stalemate over the determination of EPA and CDPHE that the Wildlife Refuge is "safe" for "unlimited use and unrestricted exposure." The goal of the Mellon group is to obtain Grand Jury records to determine if they assist to "reasonably evaluate the remediation methodology and risk analysis used at Rocky Flats upon which the agencies' determination and imminent threats to the region rely."

The motion of Petitioner's was DENIED by United States Federal District Judge Raymond P. Moore on September 4, 2019. His stated reason was that "Petitioners have not shown that they are entitled to mandamus relief" for three reasons. The Petitioners did not demonstrate:

1. An "extraordinary situation for which the drastic remedy of mandamus would be appropriate...they have not cited any legal authority for the proposition that they are entitled to the business records..."

2. That "they have not shown a plainly defined and peremptory duty on the part of the Defendant to locate and disclose the records at issue here."
3. "Petitioners have not shown that no other adequate remedy is available."

A series of headlines in Denver media from August 16 through December 27, 2019 reflect the status of the Toll Road as 2019 comes to a close and sharpens the chasm separating the differing conceptions of the public interest.

- Elevated Plutonium Detected In Jefferson Parkway Soil Sample Denver.cbslocal.com August 16, 2019
  "The elevated sample came from an area just off Indiana Street in Northern Jefferson County."

- Elevated plutonium levels found near old Rocky Flats plant: How worried should we be? www.9nes August 19, 2019
  "Elevated plutonium levels were found by a group that wants to build a toll road connecting C-470 to Northwest Parkway."
  "We felt it was most efficacious to take one sample about every one-third of one acre for the 65 acres that comprised that portion of the right-of-way…I think knowing the truth is always the better thing." Bill Ray, Jefferson Parkway Authority Executive Director

- Plutonium found in Rocky Flats soil near Denver
  Coloradosun.com, August 20, 2019
  "Elevated levels of plutonium were found in soil…"

- A soil sample hear Rocky Flats had 5 times the cleanup standard for plutonium www.9news.com August 20, 2019
  "A soil sample taken near Rocky Flats has tested for an elevated level of plutonium five times higher than cleanup standard set to protect public health."

- Test show conflicting levels for plutonium at Rocky Flats
  www.cpr.org August 21, 2019 by the Associated Press in CPR News
  "The Colorado Department of Public Health and Environment said it is trying to sort out the conflicting results…"

- Soil sample shows potential plutonium hotspot in Jefferson Parkway Path arvadapress.com Avarda Press, August 27, 2019
  "Soil may have five times more contamination than expected."

- Jefferson Parkway project stalled following test that shows elevated plutonium levels nearby www.9nes.com September 1, 2019
  "The city of Bloomfield said that project is 'not moving forward' while health officials determine if the plutonium found near Rocky flats is dangerous to the public."

- Jefferson Parkway Stalled by Hot Sample Outside Rocky Flats Highway Authority Puts the Brakes on Jefferson Parkway by Rocky Flats www.westword.com September 2, 2019
  "The Jefferson Parkway Public Highway Authority just hit a major roadblock. Last month, a soil sample taken along the proposed parkway's right-of-way along Indiana Street, just east of the former Rocky Flats Nuclear Weapons Plant came back hot."

- Plutonium Test Results Stall Plan for Toll Road near Rocky Flats www.apr.org September 3, 2019 by the Associated Press and Colorado Public Radio in CPR News
  "Broomfield city official are suspending their search for investors..."

- Here's What You should know About Radiation At Rocky Flats Live Radio KUNC October 4, 2019
  "...25 additional samples requested after that unexpected result have come back looking pretty benign."
  "In the case of Rocky Flats—and this is true in many environmentally contaminated sites—you have all the factors that make the perception of risk a lot higher than the reality of risk...There's an element of dread that radiation in particular inspires...It's an uncomfortable thing to think about but it's quire a low risk." L. Masters, Environmental Protection Specialist, Colorado Department of Public Health and Environment

- Numbers don't work for proposed Jefferson Parkway www.denverpost.com December 27, 2019
  "...anticipated (toll) revenue" [is insufficient]
  Comment: A problem is that tax dollars are precluded from the project.

Continuing Litigation In Federal District Court For the District of Colorado

As this study comes to a close, there are three legal actions in Federal District Court as introduced above which are in process—reflecting the continuing concern in civil society that the threat remaining from the Rocky Flats production era is not over.

Immediate Challenges In the Coming Years For a Safe Environment

The "contour lines" established by previous U.S. government researchers using the Health and Safety Laboratory (HASL) 235 data (Krey and Hardy, 1970) and the data from the Colorado Committee for Environmental Information (CCEI) report by Martell and Poet (CCEI, 1970a; 1970b), as well as the data obtained by the Rocky Flats Health Physics Department (representatives were on the Seed Committee), "*show values in excess of recommended limits extending beyond our property boundaries*" [italics added] (Seed, 1971, pp. 9-10).

Subsequently, a member of the Seed Committee, C.T. Illsley, was selected to document the research procedures to identify level of plutonium in soil which were utilized in the 1977-1978 soil sampling conducted under the supervision of Judge Richard Matsch in the Church cluster of lawsuits (Illsley and Hume, 1979). Documenting research procedures is essential to insuring public health.

The definitive answer to the risk of children from plutonium exposure on the lands of the Wildlife Refuge is not yet known. Many of the reasons can be overcome:

1. In assessing the risk of potential harm, the Seed Committee concluded that the data available to them:
   ...show values in excess of the recommended limits extending beyond our property boundaries.
2. The research procedures employed currently in 2109-2020 assessing level of picocuries per gram of soil have not been disclosed at the level of detail associated with the Illsley and Hume (1979) documentation in the soil sampling in the Church cluster of lawsuits.
3. The U.S. government elected to settle the Church cluster of lawsuits without a trial for $9 million with the stipulation that the lawsuit be sealed. The U.S. government clearly had access to the Seed Committee findings and the 1977-78 soil sampling.
4. The U.S. government elected to settle the Cook class action lawsuit for $375 million over the proven fact that plutonium had not been contained on-site where it was supposed to remain contained.
5. The Department of Energy elected to be in Contempt of Court before a Federal District Court rather than reveal that there were more than 677,000 pages of documents which prove that there are 2,600 pounds of missing and unaccounted for plutonium from Rocky Flats.
6. The jury in the Cook class action lawsuit determined "beyond a reasonable doubt" that the actions of both Dow Chemical and Rockwell International had been "willful and wanton" at Rock Flats. Both military-industrial contractors had assured the community around Rocky Flats that plutonium had been contained on-site.
7. The nuclear scientists and engineers from the Atomic Energy Commission, Dow Chemical, and the National Center for Atmospheric Research were experts in radiation in general and plutonium in particular. The Chair of the Seed Committee, for example, earned a Ph.D. in physical chemistry with a specialty in radiation from the University of California, Davis. After graduating from the United States Military Academy at West Point, Edward Martell earned a Ph.D. in radiochemistry from the University of Chicago.
8. The fundamental question of risk to human health from the missing and unaccounted for plutonium can be resolved scientifically by obtaining the science in the public interest resources of the BEIR VII scientists, the plutonium experts utilized by ATSDR, and the National Research Council.

Lessons Learned From Plutonium Soil Sampling For a Safer Nuclear Future For the Former Lands of the Rocky Flats Plant

This concluding section of Chapter Thirteen on irreconcilable futures for the Wildlife Refuge is organized around the following questions:

- Following the Colorado Committee for Environmental Information (CCEI) "Is it the responsibility of citizens to prove health hazards exist from a given pollutant, or is it the responsibility of the polluter to demonstrate a lack of hazard?" (CCEI, 1970b)
- "Are samples of plutonium taken in surface soils (0-3 cm) representative of the total quantity of plutonium in the soil column which exists below the surface?

On January 12, 1070, CCEI (1970a) submitted a report to the Chairman of the Atomic Energy Commission, Glenn T. Seaborg, summarizing its preliminary findings concerning plutonium hazards associated with the operations of the Dow Rocky Flats plutonium plant:

- "Our investigation has...serious implications to the health and safety of Dow Rocky Flats Plant employees and the people of the Denver area."
- "The May 11, 1969 plutonium fire at this plant, involving a very large quantity of plutonium, therefore attracted much local public attention and concern."
- "...in his letter to us dated August 4, 1969, Dr. Joshel [Rocky Flats Dow General Manager] expressed a negative view about a soil measurement program."
- "...the offsite environmental monitoring program conducted by Dow personnel has been inadequate..."
- "...estimates based on independent soil measurements indicate that curies to tens of curies of plutonium from the Rocky Flats plant have been deposited in offsite areas."
- "...the Dow Rocky Flats plutonium plant is a prime example of improper siting of a major nuclear facility."

A recurrent theme across the sixty-nine years of Rocky Flats is that the nuclear age is unforgiving of mistakes. One of the major points in the January 13, 1970 communication to Dr. Glenn T. Seaborg by the CCEI (1970a) was the magnitude of the threat at and near Rocky Flats from uncontained plutonium. But the broader context was the unforgiving mistake of the "improper siting of a major nuclear facility."

On February 24, 1970, the CCEI (1970b) released to the Governor of Colorado and citizens of the state a document "Plutonium Contamination in the Denver Area." Significant observations include:

- "Plutonium oxide dust particles produced in plutonium fires are among the most serious cancer-producing agents known." (p. 1)

- "...since the radioactivity from plutonium oxide dust particles (such as those produced in the May 11, 1969 fire) is millions of times more intense than that from naturally occurring radioactive dust particles of the same size, only minute amounts in the lung are sufficient to cause cancer." (p. 1)
- "The radioactivity of these small plutonium oxide particles is not readily diluted by rain, snow or other natural processes because plutonium oxide is very insoluble." (p. 1)
- "With regard to the question of whether or not a hazard exists, an important question must be raised. Is it the responsibility of citizens to demonstrate that a definite health hazard results from a given pollutant, or is it the responsibility of the polluter to demonstrate that there is no hazard?" (p. 2)
- It seems apparent that quantities of plutonium, hundreds of times the carefully controlled and monitored total Rocky Flats stack effluent, have found their way into the local soils, indicating a serious lack of good containment and waste handling practices." (p. 3)

There is no dispute that imprecisely known quantities of plutonium are in local soils and at depths below the surface level.

Additional critically important findings reported by Poet and Martell (1972) include:

- "Measurement of 239 Pu and 90 Sr in the top 1 cm surface layer of soils show that in this layer the 239 Pu contamination in offsite areas just east of the Rocky Flats Plant ranges up to hundreds of times that from nuclear tests." (p. 1)
- "The depth distribution of 239 Pu in undisturbed soils in the more heavily contaminated areas show that the contaminant is concentrated largely in a thin surface layer." (p. 1)
- "Results for soils of known last date of disturbance make it evident that most of the offsite Pu accumulated between 1966 and 1969. This corresponds to the period in which Pu in an oil spill area on the plant site was deposited and subjected to wind re-entrainment." (p. 1)

It will be recalled that Dr. Martell originally thought that the most likely source of offsite plutonium was the 1969 fire. Informed experts at Rocky Flats told Dr. Martell that more likely sources were the fire of 1957 and leaking drums on the 903 Pad.

The plutonium on the surface in 1970 had been very recently deposited. Since 1970 plutonium has been covered by half-a-century of windblown sand from the Front Range of the Rocky Mountains where winds can exceed 100 miles an hour.

Sources of Plutonium at Rocky Flats

Schmidt (1994) noted that there is no "natural" background for plutonium in the world since "essentially all plutonium in the environment is due to human activities (p. 1)." He analyzed the background concentrations of plutonium in soils around the Rocky Flats Plant distinguishing between what is attributable to global fallout from the atmospheric testing of nuclear weapons, from the global fallout from the atmospheric burnout of a satellite, from Regional fallout from the Nevada Test Site, and from operations at Rocky Flats.

Of direct relevance to the core area around the former Rocky Flats Plant is the question of:

> Is the threat to human and ecosystem health over?

Schmidt provided an analytical process for answering this fundamental question relevant to safe uses of the Rocky Flats land. This analytical process includes depth of soil sampling:

> The depth of soil sampling can be a key factor, and must be appropriate to the goals of the evaluation. Samples taken from surface soils (to a depth of 1 cm or so) are generally not representative of the total quantity of plutonium in the soil. Quantitative comparisons between studies using widely different sample depth will usually be difficult or unwise. (p. 2)

The immediate goal of soil sampling in 2020 is to assure worker safety during the excavation for pilings for the Toll Road, and safety for the surrounding community. The current approach by government entities of shallow sampling is not going to uncover the total quantity of plutonium in the soil column which can put construction workers and the community at risk.

Depth Distribution of Plutonium In Soils

Schmidt (1994) provided an analysis of previous research from the 1970s on the issue of the depth of soil sampling which is essential to identifying risk to human and to ecosystem health:

> ...many of the background soil samples analyzed for Pu content were taken from surface soil of 0-1 cm depth or less, or from relatively thick samples of 0-20 cm or 0-10 cm.

Given the strength of the winds at Rocky Flats, by 2020 soil buildup will have buried deeper than 1 cm plutonium from the 903 Pad, and from the 1957 and 1969 fires.

The overall conclusion of Schmidt (1994) was that:

> ...it is clear from the results that significant quantities of Pu have migrated to depths of at least 10-20 cm. (p. 12)

Approaches to Measuring Plutonium In Soil Near the Rocky Flats Plant Circa 1970-1978

Studies were performed to measure plutonium in soil near the Rocky Flats Plant by the Colorado Department of Health (CDH), by the Health and

Safety Laboratory (HASL) of the Atomic Energy Commission, and by the National Center for Atmospheric Research for the Colorado Committee for Environmental Information (CCEI).

Krey and Hardy (1970) departed from the CCEI approach of the top 1 cm and took samples down to 20 cm. Depending on the location, they found that between 9% and 61% of plutonium was found to be located below 5 cm. The assessment by Schmidt (1994) was that:

> ... the NCAR [National Center for Atmospheric Research] and CDH results cannot be directly compared with HASL results. (p. 13)

CDH and CCEI reported their results as "mass concentrations" based on surface samples. With the passages of half-a-century it is not possible to ask the researchers why they selected such shallow depths for their sample collections. In the case of CCEI and Martell and Poet, it is known that they were investigating for plutonium fallout from the very recent 1969 fire.

Schmidt (1994) identified the negative limitation for inferences which may be drawn from shallow sampling:

> The shallow depths of the NCAR [CCEI] and CDH samples mean that not all of the Pu in the soil column was sampled. (p. 12)

Little and Whicker (1978) investigated the location plutonium from 0 cm down to 21 cm:

0-3 cm	58 percent
3-6 cm	11 percent
6-9 cm	18 percent
9-12 cm	4.6 percent
12-15 cm	5.8 percent
15-18 cm	2.3 percent
18-21 cm	0.82 percent

They provide insights into plutonium in soil at Rocky Flats based on their research:

> Plutonium concentrations in Rocky Flats soil were inversely proportional to distance from the plutonium source, to depth of the sample, and to particle size of sieved samples...the mechanisms of environmental dispersion may have involved the attachment of plutonium oxide to soil particles...
>
> (Little and Whicker, 1978p. 451)

Little and Whicker (1978) identify the relevance of depth of the sample, particle size, and the attachment of plutonium oxide to soil.

Implications for Protection of Public Health On the Lands of the Wildlife Refuge, During Construction of the Jefferson Parkway, and For the Surrounding Community

If the Jefferson Parkway proceeds across the Eastern section of the Wildlife Refuge, this review of literature on plutonium soil sampling at Rocky Flats has revealed that only the Krey and Hardy (1970) methodology is relevant in 2020 to protecting construction workers from plutonium contamination at the depths they will have to penetrate the soil.

By way of review, the Krey and Hardy approach revealed that between 9% and 61% of plutonium was found below 5 cm in depth. Little and Whicker found that 58% of plutonium was found between 0-3 cm, but obviously the other 42 percent was located down to 21 cm. The point for the public health and safety is that excavation for the Toll Road should not proceed based only on shallow soil sampling.

Critical Public Interest Choice For the People of Colorado On Soil Sampling Depth

A critical issue for the people of Colorado in 2020 is to determine the public interest in the depth of soil sampling for plutonium. The official U.S. government standard at Rocky Flats is 50 picocuries-per-gram of soil for the top 3 feet per-gram of soil. The recently elevated sample of plutonium by an independent contractor for the Toll Road Authority was 264 picocuries-per-gram of soil.

Subsequently, the Colorado Department of Public Health and Environment reported a series of "benign" samples without simultaneously disclosing the depth of sampling.

Credible U.S. government soil sampling in the 1970s found some plutonium in the 0-3 cm range but other plutonium down to 20 cm. And the U.S. government standard is 50 picocuries-per-gram of soil in the top 3 feet.

An issue for 2020 for the people of Colorado is to learn how deep construction workers will need to dig the foundations for freeway pilings and the risk to both workers and the public. The public interest also needs to be determined on the question of "should soil sampling be conducted down to the depth of the U.S. government standard of 50 picocuries-per-gram in the top three feet of soil?"

EXAMINATION OF CONTENTIOUS APPROACHES TO CITIZEN ENGAGEMENT AND USE OF THE JUDICIAL BRANCH OF U.S. CONSTITUTIONAL GOVERNMENT FOR REDRESS

Introduction To Part V – Contentious Paths To Citizen Influence Using Community Engagement and Access To the Judicial Branch of Government With Its Power of Subpoena

Key Elements In the Chapter Fourteen Narrative

- There are two very different approaches chronicled in this chapter to exercising citizen influence over the national security imperative of the U.S. government.
- The first approach presented is the Lamm-Wirth Task Force which was a Blue Ribbon group of citizens appointed by the Governor of Colorado and a Congressman who subsequently became a U.S. Senator.
- There were substantive and irreconcilable differences between the U.S. government represented by the Energy Research and Development Administration and the Task Force which lacked the power of subpoena to compel a process of discovery.
- The substantive differences could not have been more basic: for example, there were polar opposite answers to the following question: "Should the known operations carried out at Rocky Flats only be located in an area remote from any population center?"
- The second approach taken to citizen influence in Chapter Fourteen is the Church cluster of lawsuits in Federal District Court for the District of Colorado.
- Attorney Howard Holme used the power of subpoena of the Judicial branch to initiate a process of discovery within the on-site archives of Rocky Flats and Dow Chemical which yielded a 520-page pre-trial statement.
- The U.S. government made an inexplicable decision to settle the Church cluster of lawsuits based on off-site plutonium for $9 million with the reasons for the settlement sealed—then did not require Dow Chemical to make necessary changes to the storage and disposal of plutonium which was to lead over 30 years later to a settlement of $375 million in the Cook class action lawsuit.

Key Elements In the Chapter Fifteen Narrative

- A most notable feature of the Cook class action lawsuit was that "time did not fly"–it was over 26 years between when it was filed in 1990 and when it was settled in 2016 for $375 million just before the Justices of the U.S. Supreme Court were to decide whether or not to grant a Writ of Certiorari filed by Dow Chemical and Rockwell International.

- There were ten Memorandum Opinions and Orders issued from Federal District Court Judges from 1991 to 2004.
- As with the settlement in the Church cluster of lawsuits, it is the taxpayers of the United States who paid the settlement.
- Attorney Merrill Davidoff for the Plaintiffs prevailed in the Cook case by filing an amended appeal to the 10th Circuit Court of Appeals which argued that the Plaintiffs had a valid jury verdict under state of Colorado tort law on trespass and nuisance.
- Another verdict by the Cook class action jury was awarding punitive damages of $200 million using the standard of "beyond a reasonable doubt" that the actions of both Defendants Dow Chemical and Rockwell International had been "willful and wanton."

Chapter Fourteen presents two contentious paths to citizen influence over the national security imperative of the U.S. government for nuclear weapons development. The contentious paths were quite different in both process and outcome. They will be presented one at a time.

The Lamm-Wirth Task Force was a collaboration of the newly elected Governor of Colorado, Richard Lamm, and the newly elected U.S. Congressman and subsequently U.S. Senator from Colorado, Tim Wirth. They planned the Blue Ribbon Task Force before they had been administered their respective Oaths of Office. The Task Force issued their report in 1975. The response of the U.S. government was by the Energy Research and Development Administration. This organization was a transitional entity between the Atomic Energy Commission and the Department of Energy. The contentiousness between the Task Force and ERDA is captured in the opening part of the title of Chapter Fourteen: "Ships Passing In the Night." ERDA elected to stonewall the Task Force.

Seven issues which were especially highly contentious were:

1. Permanent Monitoring Committee
   LWTF - There needs to be a permanent Monitoring Committee to carry forward the work of the Task Force.
   ERDA – No. There is already "unprecedented" agreement with the state of Colorado.
2. Relocation To an Area Remote From a Population Center
   LWTF – Rocky Flats should be relocated to an area remote from a population center.
   ERDA – Disagrees with the basic Task Force conclusion that "the Plant poses a threat justifying such a drastic action."
3. Regular Briefing To the State On Nuclear Weapons Programs
   LWTF – State of Colorado should be briefed on a regular basis on nuclear weapons programs at Rocky Flats.
   ERDA – It would be "improper" to furnish classified information to the state of Colorado.

4. Expansion of the Buffer Zone
   LWTF – This expansion would increase protection from "a possible nuclear disaster."
   ERDA – No. The original buffer zone acquisition did not imply "any existing or potential hazard."
5. EPA Should Regulate Radioactivity At Rocky Flats
   LWTF – EPA should "control and regulate all discharges of radioactivity…"
   ERDA – The Department of Justice and EPA concur that the law does not grant this authority.
6. "Zero Discharge" of Radioactive and Toxic Substances
   LWTF – Rocky Flats should "strive toward zero discharge of any radioactive or highly toxic substance…"
   AEC/ERDA - policy is "as-low-as-is-practicable" and in effect will accomplish "zero discharge" of liquid effluents.
7. Require Contractors To Bear Liability
   LWTF – The Price-Anderson Act should be replaced with a "nuclear industry liability act."
   ERDA – Congress amended PL 94-197 without including this and "We believe no further legislation should be proposed."

ERDA said Rocky Flats is "…inseparably linked to our current defense posture." ERDA also chose to make a ringing endorsement of Rockwell International.

The second approach to citizen engagement chronicled in Chapter Fourteen is the Church cluster of lawsuits. This cluster of lawsuits had two origins. One origin was when Marcus Church himself was sued by someone he had sold land to when contamination with plutonium had been found on the land destroying its value for development. The second origin was when Marcus Church went to see the Managing Partner of Fairfield and Woods law firm in Denver and said, "This time they've gone too far." Howard Holme was assigned to assume responsibility for the process of discovery which resulted in the compilation of a 520-page submission to Federal District Court Judge Richard P. Matsch.

Judge Matsch identified four theories of the Plaintiffs:

1. The corporate defendants are strictly liable for damages from their ultrahazardous operations at Rocky Flats.
2. They are liable in nuisance.
3. The doctrine of res ipsa loquitur is applicable. (Res ispa loquitur is a Latin phrase meaning "the thing speaks for itself." This is a principle that allows plaintiffs to meet their burden of proof by what is in effect circumstantial proof).
4. Both corporate defendants and the United States are liable for trespass without proof of negligence.

The position of Defendants, according to Judge Matsch, was that:

> …the government defendants contend that the contamination of plaintiffs' lands does not present a

significant health risk and without an appreciable health risk, the plaintiffs are unable to show that their lands are damaged or that the government defendants have any legal liability.

Judge Matsch ultimately dismissed the Plaintiffs' case because "the critical question of whether there had been any injury to the plaintiffs could not be litigated because it was an issue which could not be resolved through the adversary process..."

The 10th Circuit Court of Appeals cited Supreme Court Justice William Brennan's opinion in *Baker v. Carr* that "there should be no dismissal for non-justiciability on the ground of a political question's presence." The 10th Circuit Court of Appeals ordered a trial in Federal District Court. The case was settled for $9 million with a stipulation that the settlement be sealed.

An enduring mystery is why the U.S. government paid the $9 million settlement with taxpayer funds, and then did not require that Dow Chemical rectify the reasons why Plant operations led to off-site plutonium contamination.

Chapter Fifteen recounts the twenty-six year odyssey of the Cook class action lawsuit. The first four words in the chapter title, "time did not fly" authentically captures the experience of all the participants. There were ten Memorandum Opinions and Orders from Federal District Court Judges between 1991 and 2004. The Memorandum Opinions and Orders served to narrow the scope of what was at issue legally in the class action lawsuit over the more than fifteen years before the case was submitted to a jury in late 2005.

The jury instructions on state of Colorado tort law by Senior Federal District Court Judge John L. Kane, Jr. were never challenged on appeal, nor was the evidence in support of the jury findings. Judge Kane instructed the jury that the Plaintiffs were making two claims: for trespass and nuisance.

- "Trespass is an invasion of a person's property without his permission." The Plaintiffs claim that Dow and Rockwell caused a trespass "because the Class Area is contaminated with plutonium released from Rocky Flats as a result of various accidents, mishaps, and bad environmental practices..."
- A nuisance is something that "substantially and unreasonably interferes with another person's use and enjoyment of their land." The Plaintiffs claim that Dow and Rockwell "substantially and unreasonably" interfered with their use and enjoyment of their properties "because plutonium contamination in the Class Area has led to increased health risk and because there is a risk of additional future releases of

plutonium and other hazardous substances into the Class Area from the plant site."

The jury was called upon to vote on these issues first for both Defendants separately. The jury was obligated also to assign the loss of value to the Plaintiffs' properties.

There are a number of significant legal issues in this case. The controlling elements in the Memorandum Opinions and Orders were almost always legal precedents cited by the Judges and their relevance to the diametrically opposed positions of the Plaintiffs and the Defendants. A characteristic of current civil litigation as it occurred in this case is the exhaustive nature of discovery. One Judge remarked that no stone should be left "unturned" before a case goes to trial.

A third significant element is that the legal settlement of $375 million was paid by the taxpayers of the United States, not defendants Dow Chemical and Rockwell International. The backstory on this third significant element goes back to 1942-1943. The President of DuPont Corporation in dialogue with General Leslie Groves indicated that DuPont would assume responsibility without corporate profit during World War II for the management of plutonium production at the Hanford Nuclear Reservation. Given the "unique hazards" involved, General Groves accepted the proposal from the President of DuPont that the U.S. government would indemnify liabilities incurred by military-industrial contractors for their national security service on behalf of the United States.

Chapter Fourteen
"Ships Passing In the Night": Contentious Paths to Citizen Influence Over the National Security Imperative of the U.S. Government For Nuclear Weapons Production

> "We believe that operations such as those currently carried out at RFP should be located in areas remote from any population center."
>
> Lamm-Wirth Task Force, Recommendation 4

> "ERDA disagrees with the basic conclusion that the Plant poses a threat justifying such drastic action [relocation to a remote area]."
>
> Comment on 3A and 4, Energy Research and Development Administration

> ""The core question in this litigation [the Church cluster of lawsuits of 1975] is whether private citizens can use the authority of the judicial branch of that government to define a duty of care to be followed by the executive and legislative branches."
>
> Federal District Judge Richard P. Matsch

Two very different contentious paths taken during the Cold War to assert citizen influence and control over the sensitive national security imperative for nuclear weapons production by the U.S. government will be presented in this chapter: the Lamm-Wirth Task Force of 1975 followed by the response on behalf of the U.S. government by the Energy Research and Development Administration (ERDA), and the Church cluster of civil lawsuits which lasted from 1975 until finally settled in 1984. An underlying question about both the Task Force without the power of subpoena and a civil lawsuit with the power of subpoena is whether the case of Rocky Flats during the Cold War provides examples of viable approaches by which citizens can question and influence a sensitive area of national security involving nuclear weapons production.

Lamm-Wirth Task Force of 1975

In the context of pervasive Cold War secrecy, concerns which citizens might have had about Rocky Flats were muted before plutonium and tritium were discovered off-site and downstream from the site between 1970 and 1973. The severity and the risk for catastrophic release of radioactivity from the plutonium fires of 1957 and 1969 were not known to the public. The real reasons behind the decision to acquire the 4,000-acre "buffer zone" in 1974 had not been disclosed to the surrounding community.

Figure 14.1 Richard Lamm

Governor Richard Lamm served three terms as the Governor of Colorado from 1975 until 1987, and has been honored for making "significant and lasting" contributions to Colorado and its history. After earning an undergraduate degree in Accounting, he served as a First Lieutenant in the United States Army. After earning a law degree, he joined the faculty of the University of Denver. His first service as an elected official was as a member of the Colorado General Assembly. He was President of the First National Conference on Population and the Environment. He designated musician and composer John Denver as the first Poet Laureate of the state of Colorado.

Figure 14.2 Tim Wirth
Tim Wirth had a distinguished career of public service to the state of Colorado and to the United States. The Lamm-Wirth Task Force was established when he was a Congressman in the U.S. House of Representatives. He was then elected to the United States Senate from Colorado. Following that service, he was Counselor to the Department of State and Undersecretary of State for Democracy and Global Affairs. From 1998 through 2013 he was the President of the United Nations Foundation.

Citizen concerns acquired a voice with the establishment of the Lamm-Wirth Task Force in 1974. Governor Richard Lamm and Congressman Tim Wirth took great care to establish a Task Force which was chosen for its breath of perspective, its stature in the community, its competence, and its collective credibility with key Colorado constituencies. The membership of the Task Force and their affiliations were:

Name	Affiliation
Robert D. Siek, Chairman	Colorado Department of Health
Wesley Brittin	University of Colorado
William Cohen	Attorney of Law
John Cobb	University of Colorado Medical Center
Patrick Kelly	United Steelworkers of America (a Labor Union at Rocky Flats)
George Lukas	Colorado School of Mines
Margaret Markey	Boulder County Commissioner
John Monaghan	Governor Lamm's Office
Joanne Paterson	Jefferson County Commissioner
Arthur Robinson	National Jewish Hospital
Robert Damrauer	Technical Staff Member, University of Colorado

Charge To the Task Force
The charge to the Task Force was:

> ...developing options, alternatives, and recommendations to enable the Governor and the Congressman to understand and deal with the RF (Rocky Flats) operation. (p. 4)

The Task Force studied: "the operations of the Rocky Flats Plant, the hazards associated with it, its impact on the environment, and the legal and occupational problems associated with the facility" (p. 4).

There were two days of public hearings. The hearings made clear to the Task Force that: "not only that many people had only rudimentary knowledge of the operations at RF, but also that there were grave misgivings about the Plant's safety and the potential for a cataclysmic accident" (p. 4).

The Core Positions of the Lamm-Wirth Task Force Report and the ERDA Comments On Behalf of the U.S. Government
Between the end of the Atomic Energy Commission and the establishment of the Department of Energy, a transitional organization was the Energy Research and Development Administration (ERDA). The release of the Lamm-Wirth Task Force Final Report on October 1, 1975 occurred when it was the responsibility of ERDA to frame a response on behalf of the U.S. government.

In order to reflect the essence of the recommendations of the Lamm-Wirth Task Force and the responses by ERDA, they will be juxtaposed by topic, followed by an analysis. The categories of recommendation and response are not of equal importance. And they vary in their insight and especially their accuracy and credibility into the trajectory of Rocky Flats from 1975 until its closure to production in 1989.

The Reasons For the Gulf Which Existed In 1975

Lamm-Wirth recommendations and the ERDA response provide a prism for understanding the reasons for the gulf which already existed in 1975 between the essence of the views manifested in the Lamm-Wirth recommendations, and the often diametrically opposed positions of the U.S. government.

Another prism provided by the Task Force recommendations and the ERDA response is over the level of the risk Rocky Flats posed to neighboring communities and the broader Denver metropolitan area. As a consequence of years of civil lawsuits, the Congressional inquiry, and numerous Freedom of Information Act filings, there is now a record of facts about Rocky Flats which was unavailable in 1975. This makes possible an assessment of the accuracy of assertions by both the Lamm-Wirth Task Force and the U.S. government as conveyed by the ERDA comments.

Record of Facts Now Available About Risks Posed By Rocky Flats

This record of facts began with the power of subpoena used by plaintiffs in the Church cluster of lawsuits filed in 1975 (Holme, 1978). The next milestones in compiling credible facts about Rocky Flats were the criminal investigation, the FBI raid, the Wolpe subcommittee of the U.S. Congress, the Cook class action lawsuit filed in 1990, the Stone False Claims lawsuit filed in 1989, The Ambushed Grand Jury book of 2004 and the Citizens Investigation of 2005. They have all contributed to the factual record, as have numerous civil society initiatives such as those by the Rocky Mountain Peace and Justice Center. The point is that there is now an extensive factual record available to support or contradict positions taken by both the Task Force and ERDA in 1975. The U.S. government itself starting in 1994 through the Closure Legacy report of 2006 has provided or commissioned extensive characterizations. The subpoena power of the Cook class action lawsuit over its 26-year history obtained exceptional factual information.

Recommendation 1 - Monitoring Committee

The purpose of the Monitoring Committee was "to continue the work of the Task Force," especially in the areas of health, safety and environment.

The U.S. government response to the Monitoring Committee recommendation was two-fold. The first response was that there was already in place an "unprecedented" formal agreement with the state of Colorado providing access to "complete Plant and activities there on." The second response was that funding for a state monitoring activity would not be appropriate.

In terms of facts about health, safety, and environment, it was the Department of Energy itself which issued several negative evaluations. Among those that are publically known at the end of 2019, the characterization by the then secret 1986 Mary L. Walker Briefing Memo of conditions at Rocky Flats as "patently 'illegal'," the "Tiger Teams" dispatched in 1989 to evaluate Rocky Flats by Secretary of Energy James Watkins, the 1989 dialogue between Secretary Watkins and Rockwell Chairman Donald Beall, the 1992 guilty plea by Rockwell to ten felony and misdemeanor violations of the Clean Water Act and RCRA, or the revelation by ChemRisk (1992) that almost all of the 8,000 substances used at Rocky Flats had not been monitored for either on-site or off-site contamination.

The broader point of U.S. government entities, across ERDA responses on other recommendations, is that alarm is misplaced over health, safety, and environment. As the very new military-industrial contractor, Rockwell was characterized as working diligently with ERDA to establish policies and procedures which will advance health, safety and the environment.

ERDA on Rockwell International:

> Rockwell's performance under the present contract has been completely satisfactory to ERDA in all areas of health, safety and the environment.
>
> ERDA Comment on 3C

Finally, the existing relationships between the Federal government and the state of Colorado was already "unprecedented." The bottom line for the U.S. government was that an external Monitoring Committee was not necessary.

Recommendation 2 - Broomfield Water Supply

The Task Force concluded that operations at the Rocky Flats Plant had "resulted in radioactive and toxic material contamination of the Great Western Reservoir." The ERDA reply was that in response to citizen concern, it had commissioned an "...independent study of the relationship between Rocky Flats Plant effluents and the Great Western..." ERDA has adopted all 28 of recommendations of the independent study.

Analysis

ERDA could not have known that in only four years (1979), Rockwell International would initiative the practice of Spray Irrigation to dispose of what was to become 80 million gallons of liquid effluent a year by the time of plant closure in 1989. The point is that disposal of plant liquid effluents continued as an unsolved problem of off-site contamination until Plant closure. ERDA (1975) either did not know of, or chose not to reveal, the Seed Committee (Seed, 1971) findings and evaluation of the significance of on-site and off-site contamination.

Recommendation 3 and Recommendation 4 Regarding Continuation of Nuclear Weapons Components Manufacturing At Rocky Flats

These two distinct recommendations by the Lamm-Wirth Task Force received one integrated response from ERDA and are presented together.

Recommendation 3.A urged the U.S. government to undertake a program of "gradually phasing out its present operation...and possibly transferring those operations to a more suitable site..."

The goal of Recommendation 4 was to shift the purpose of Rocky Flats to "less hazardous uses of energy." Related to Recommendation 3A was the call in Recommendation 4 to relocate current operations to an "area remote from any population center."

ERDA Comments

ERDA focused on "issues of economy and practicality" and to the centrality of the mission of Rocky Flats to national security: the "justification for the facility ...[which is] inseparably linked to our current defense posture..."

ERDA provided a cost estimate of $2 billion for a relocation and a projected time frame of ten years. The bottom line was that "ERDA disagrees with the basic conclusion that the Plant poses a threat justifying such drastic action."

Analysis

Major differences between the Lamm-Wirth Task Force Report and the ERDA comments are crystallized in fundamentally different answers to the question concerning the degree of threat posed by current operations of Rocky Flats. The fundamental issue was subsequently explored with the power of subpoena in a landmark lawsuit (see Chapter Fifteen) which lasted from 1990 until 2016.

In addition, within seventeen years, allegations of environmental crime had been substantiated by a guilty plea in 1992 by Rockwell International, and a spirited clash within and between "separate but equal" branches of U.S. constitutional government between 1992 and 1994.

Recommendation 3.B. - Adequacy of Security

Recommendation 3.B by the Task Force calls for an investigation to determine "whether security at the Rocky Flats Plant is adequate…"

ERDA Comments

The answer from ERDA was simple and straightforward to the adequacy of security: "Yes, security is adequate."

Analysis

This is one of the overwhelming number of Task Force recommendations which went nowhere. The format of two separate position statements without subsequent dialogue between the parties simply did not lead to cooperative action or mutual understanding.

Recommendation 3.C. - Consultation On Short and Long-Range Planning

The Lamm-Wirth Task Force urged the Federal government to advise the state of Colorado "on a regular basis" on the short and long-range planning relevant to nuclear weapons programs affecting both the nature of the mission and the level of production at the Rocky Flats Plant.

ERDA Comments

There were three elements to the ERDA response. The first element was that ERDA is "already committed" to conferring in advance with the state of Colorado on activities relevant to health and safety, and on matters of policy related to "its operations at Rocky Flats."

The second element was that it "would be improper" to furnish additional information on classified nuclear program plans "unless there is a valid national security reason for such action."

The third element was that the level of production "is a sensitive area having national security implications and should not be publically disclosed."

Analysis

Recommendation 3C requesting detailed consultation on forward planning was met with the assertion that in the areas of health and safety, consultation with the state of Colorado was already adequate, and that it was not in the national security interests of the United States to disclose classified nuclear weapons plans or production levels.

These dueling positions obviously went nowhere. ERDA indicated that concerns about health and safety were already being addressed. The national security imperative precluded disclosure of future plans. The implications of the ERDA response echoed prior positions on the level of threat: it was already being satisfactorily addressed.

Recommendations 5.A, 5.B, 5.C, 5.D, 5.E, 5.F - Medical Studies and Occupational Health

The focus of the Lamm-Wirth Recommendations 5.A-5.F and the ERDA comments illustrate the many obstacles which had to be overcome to create, on a national basis, a regime to recognize what Judge Matsch called in 1982 the "ultrahazardous" nature of Rocky Flats as an industrial enterprise, and the unique hazards confronted by nuclear workers in the United States. Such a regime was not enacted into law until 2000, eleven years after the closure of Rocky Flats to production.

Analysis

It was never going to be possible to develop a viable program to compensate injured nuclear workers at only one nuclear weapons production site within just one U.S. state. And in the absence of an on-going process for dialogue between the Lamm-Wirth Task Force and the U.S. government, there was no likelihood that they could have resolved differences over such fundamental issues as the purpose and nature of a registry for nuclear workers.

Recommendation 6.A, 6.B, 6.C - Public Information

The thrust of the three Task Force recommendations on Public Information is for greater transparency by Rocky Flats concerning radiation accidents, radiation goals and progress on reaching those goals, and educating workers at Rocky Flats and the public about the short and long-term effects of plutonium and other radionuclides.

ERDA Comments

ERDA regulations call for the "timely" release of information on all radiation incidents "which may be significant or which might be of public interest or concern." ERDA policy preserves the right to privacy of any involved worker.

New contractor Rockwell International was establishing radiation reduction goals, and developing an "extensive program" to provide Rocky Flats employees with information on the effects of radiation, including the effects of transuranic materials.

Analysis

In the fourteen years after 1975 which remained of the Rockwell era at Rocky Flats, there is no available record of the results from the projects Rockwell was working on to meet the requirements which follow from the ERDA regulation on "timely" release of information, on radiation reduction goals, on the increase in worker understanding of the effects of radiation, or on the results from the employee incentive program.

The broader issue is on-going community concerns over environment, health and safety which clash with the secrecy imperative of the government for preserving national security. There is no evidence in the available record that Recommendations 6.A, 6.B, and 6.C and the ERDA responses led to any lessening of community concern.

Recommendations 7.A, 7.B, 7.C, 7.D, 7.E, 7.F, 7.G, 7.H, 7.I, 7.J - Monitoring of Employees

The Lamm-Wirth Task Force made a significant effort to improve the monitoring of employees. Included in this category of recommendations were:

- Acquiring the best available equipment;
- Maintaining accurate records of individual radiation exposures and providing access to the Colorado Department of Health;
- Top priority should be given to a program for radiation exposure goals;
- Timely provision to employees on exposure levels;
- Improving the radiation protection systems for humans both inside and outside the Plant;
- Re-evaluating present and former employees with new, more sophisticated detection systems;
- Funds should be provided to the state of Colorado to regularly monitor the impact of the Plant on the environment;
- The state of Colorado should be provided ERDA funds to determine "plutonium lung burdens"; and
- The Federal government should provide funds to develop "ultra-sophisticated" bioassay techniques and monitoring equipment.

Across all of these recommendations is the theme that there are numerous opportunities for improving the monitoring of employees, and addressing the suspected negative impacts of the Plant on worker health and the environment.

ERDA Comments

There were a number of areas of ERDA support for individual recommendations such as re-evaluating present and former employees with new, more sophisticated detection systems. A caution offered by ERDA was that re-evaluating employees could only be a voluntary program. ERDA wished more eligible persons would choose to participate in the current program.

Analysis

There was concurrence in having the best available equipment and measures. The compilation of accurate records of each employee's individual radiation exposures is a contractor responsibility according to ERDA.

There were areas of disagreement. While ERDA would welcome expert opinion, it retains "final responsibility for decisions that impact upon the health and safety of persons involved in the operation." According to ERDA, Rocky Flats has spent over $1 million during the past two fiscal years on the development of improved neutron alarm systems and other protection devices. Rockwell is to continue this emphasis. Another area of disagreement was on an environmental surveillance program. This is a direct responsibility of ERDA, and "is carried out by Rockwell in accordance with established requirements…"

Constructive focus on improving monitoring of employees was an area of largely common ground between the Lamm-Wirth Task Force and ERDA. Despite the common ground, this area of recommendations revealed the cost of advancements in monitoring and the inherent difficulties in making progress. ERDA expected its new contractor, Rockwell, to improve old programs and to develop new ones.

Since monitoring of employee health challenges was not unique to Rocky Flats, but was general throughout the U.S. nuclear weapons complex, it is a matter of curiosity why the frame was to be limited only to Rocky Flats. Related to this curiosity is another one: why monitoring of employee exposure did not include a greater focus by ERDA on the welfare of those whose prior monitoring had revealed that they had incurred a significant health risk, from either known strong radiological or chemical prior exposure. The constellation of recommendations 5.A-5.F, Medical Studies and Occupational Health, and the replies by ERDA did not result in progress on improving the welfare of those with known significant health risk from exposures which had been incurred at the Plant.

Recommendation 8.A, 8.B, 8.C, 8.D - Safety

Increasing the safety of Rocky Flats for workers at the Plant and for the surrounding community was obviously a priority for the Lamm-Wirth Task Force. This priority was expressed across a number of domains of activity. In one domain, that of routine comprehensive studies of all aspects of safety, the approach taken by the Task Force was to reach beyond the Operation and Safety Division of ERDA to such potential expert contributors as the Nuclear Regulatory Commission and the Occupational Safety and Health Administration.

Another Task Force concern was the safety of the air space above Rocky Flats and restricting overflights. A further recommendation was for the Colorado Department of Health to monitor the transportation of all nuclear materials. Another related recommendation was to reengage Federal and State regulatory agencies to develop "guidelines for controlling air transportation of plutonium and the transuranic elements."

ERDA Comments

The responses from ERDA were wide-ranging. ERDA indicated that its safety reviews "are far more stringent than those in most areas of industry and government." With respect to restricting airspace, ERDA indicated that it had no objection to the state of Colorado attempting to get the Federal Aviation Administration to prohibit Rocky Flats overflights. ERDA had been unsuccessful in getting the FAA to restrict or prohibit flight in the airspace around certain of its facilities.

More adversarial comments came in response to transportation recommendations. The state of Colorado cannot get access to the information necessary to monitor nuclear material shipments "in the interest of national security..." With respect to the recommendation that State and Federal regulatory agencies should develop guidelines for the air transportation of plutonium and other transuranic elements, ERDA has a long history of cooperation with appropriate Federal organizations on transporting nuclear materials. State organizations are not included in this cooperation.

Analysis

ERDA indicated that it concurred that safety was of vital importance, but was being appropriately handled. And the transportation of nuclear materials was covered by the imperative for secrecy in the pursuit of national security. Such consultations which did occur were with Federal officials not State officials. The area of restricting overflights was the only common ground.

Recommendation 9.A, 9.B, 9.C - Land Use

One of the most significant recommendations of the Lamm-Wirth Task Force was to expand the Plant by an additional buffer zone. The size of the buffer zone expansion was recommended to be one-to-two miles on each side. The purpose was increasing protection "from a possible nuclear disaster at the Plant."

Related recommendations were that appropriate officials in Boulder and Jefferson Counties should declare a moratorium on "further development of land in vicinity of the Plant...," and that Colorado Land Use Statues should be amended to give the State and local government clear authority "to restrict development on contaminated lands and land in the vicinity of hazardous facilities."

ERDA Comments

ERDA opposed the expansion of a buffer zone. ERDA did not comment on possible changes in state of Colorado or local government land use decisions since: "It is not appropriate for ERDA to comment…since it refers to matters internal to the State."

ERDA Explanation For the Original Buffer Zone Acquisition and For Opposing Expansion

ERDA offered a number of distinct reasons for opposing expansion of the buffer zone. One reason was the explanation for acquiring the original buffer zone: "It was acquired solely to provide a logical separation between the Plant's operation and the rapid residential development in the area."

<u>Lack of Existing Hazard Within the Original 2,500 Acres</u>

ERDA was emphatic that the land acquisition of the original buffer zone did not imply "any existing or potential hazard, either to use of the land acquired or to use of land beyond the buffer zone."

<u>Continued Use of the Land By Its Current Owners</u>

It is important to understand that:

> The existing conditions need not preclude the use of this land by its owners under the interim guidelines established by the Colorado Department of Health.

<u>Established Current Plant Boundaries Respect Natural Boundaries</u>

The core argument of ERDA was that the boundaries of Rocky Flats should respect natural boundaries. The current buffer zone boundary takes "advantage of natural topography and the definition provided by adjoining highways." Conversely, a defect of the recommended buffer zone expansion is that it reaches "beyond natural or man-made boundaries."

<u>Preliminary Costs Estimates of the Recommended Buffer Zone Expansion</u>

The Army Corps of Engineers is cited as the source by ERDA for the cost estimate of $50 million for a one-mile expansion and $135 million for a two-mile expansion. What were characterized as "rough estimates" by the Army Corps of Engineers excluded the "rights-of-way for railroads, utilities, public roads and streets, costs of acquiring existing water rights, underground mineral rights, reservoir sites, similar public facilities, and housing developments." This comment concludes with the ERDA overall assessment that "These estimates should be sufficient for evaluation of the recommendation."

<u>New Rocky Flats Construction Meets the Most Rigorous Safety Standards</u>

The safety standards singled out for mention are fire, earthquake, and tornado hazards. An example offered for rigorous safety standards was a new plutonium processing building. Also cited by ERDA are upgrades to process and equipment in older plutonium buildings.

Consequences of New Levels of Safety

A key consequence of the new levels of safety which have been achieved is that: "...the worst credible accident within the Plant, postulated and studied by ERDA to date, will not lead to injuries outside the Plant boundaries." The overall conclusion by ERDA was that the recommended one-to-two-mile buffer zone would not be needed "as a protection against a potential nuclear accident."

Analysis

The year 1975 was in the time period of the beginning of a legal and regulatory transition affecting Rocky Flats. The ultimate impact of that legal and regulatory transition had the following relevant elements:

- Federal District Court Judge Richard Matsch declared Rocky Flats to be "ultrahazardous" in 1982;

- The Church cluster of lawsuits filed under state of Colorado tort law in 1975 led to a $9 million settlement with neighbors. The settlement included the stipulation that the terms of the settlement were to be sealed;

- Fourteen years later in 1989 Secretary of Energy James Watkins declared that the level of accomplishment of the production mission at Rocky Flats had not been matched by a parallel level of accomplishment in protecting health, environment, and safety.

- In 1976, one year after the high profile exchanges reported in this chapter between the Lamm-Wirth Task Force and ERDA, the bipartisan Congress passed the Resource Conservation and Recovery Act which joined the previously enacted Clean Air Act and Clean Water Act, and the recent establishment of the Environmental Protection Agency in protecting the environment. Rockwell International pled guilty to ten felony and misdemeanor violations of RCRA and the Clean Water Act in 1992.

The principal point is that the Lamm-Wirth Task Force recommendations contained a premise that there were valid health, safety and environmental concerns about a small (6,500 acres) nuclear weapons production complex sited in such close proximity to a major metropolitan area. Expanding the buffer zone would provide a level of protection in case of a possible nuclear disaster at the Plant.

The ERDA commentary made the principal point that the contractor had in place, or was developing, a comprehensive safety program. The possibility of a nuclear disaster at the Plant was so remote that another buffer zone was unnecessary. According to ERDA, the acquisition of the first buffer zone in 1974 was not related to any existing or potential hazard, but to establishing a "logical separation" between the Plant and the rapid residential development in the area.

A number of reasons were offered by ERDA in support of not acquiring additional buffer zone. The current owners of the land could continue the beneficial use of their land. The new proposed boundaries would not respect either the natural topography or the physical definition provided by adjoining highways. On the issue of cost, ERDA's position was that the expenditure of $50 million for a one-mile expansion or $135 million for a two-mile expansion was unnecessary based on lack of hazard. Also, these "rough estimates" of cost did not include many of the known expected final costs.

The arguments made by ERDA did not include positive presentations of what use might have been made of the existing and proposed new buffer zones for Spray Irrigation, for example, which could be utilized to contain liquid effluent within the larger Plant site, or to plan more orderly and less compacted space for the ultimately 178 known waste disposal sites.

On the issue of the safety of Rocky Flats, the ERDA comments represent a different model of risk assessment which was subject to challenge based on subsequent developments at the Plant. For example, the new plutonium production building offered by ERDA as a model of safety never opened for production. Building 371 was a multi-floor building. But too much plutonium became struck in the pipes where it flowed between production floors. An argument for Building 371 before the U.S. Congress was that the original Building 771 was no longer safe for operations. When the new Building 371 did not function properly, the original compromised Building 771 was required to remain in service.

The "new levels of safety" argument on behalf of not expanding the buffer zone is problematic. That a "worst credible accident" could be contained within the Plant was an assertion of ERDA. But by 1975, there had already been two catastrophic fires in 1957 and 1969 which had not been contained within the Plant, and which could have been even worse. The spreading of 903 Pad contamination was well known within Rocky Flats. And Major General Edward Giller, the head of weaponizing the atom within the AEC, had testified under oath before the U.S. Congress that if the 1969 fire had breached the roof of Building 776/777, which it had come close to doing, it would have contaminated portions of metropolitan Denver.

The principal point is that there had been dangerous accidents during the Dow Chemical era. The Rockwell International era had just begun. There is no evidence in the available record of the positive impact of Rockwell's safety initiatives which were just beginning. There is ample evidence, however, that safe management did not become a major feature of Rocky Flats in the Rockwell era. This is confirmed by the "Tiger Team" reports commissioned by Secretary of Energy James Watkins in 1989, and Rockwell receiving only 40 percent of its performance bonus for 1988 as determined under the authority of Secretary of Energy James Watkins.

The acquisition of an enhanced buffer zone for "protecting against a potential nuclear accident" was for ERDA an unwise choice based on the existing high level of safety, and was evaluated to be an unnecessary expenditure of public funds. A constituency did not develop to support an expanded buffer zone in advocacy to the relevant Appropriation Committees of the U.S. Congress.

Recommendations 10.A, 10.B, 10.C, 10.D - Emergency Response Plans

The thrust of the Emergency Response Plans recommendations by the Lamm-Wirth Task Force was to assure that there were credible Emergency Response Plans for a "wide variety of potential accidents" with participation beyond Rockwell/ERDA.

ERDA Comments

The thrust of the ERDA comments are captured in the opening sentence of their commentaries: "A coordinated emergency response plan for Rocky Flats is in existence and is currently under review and revision by Rockwell in coordination with appropriate Federal, State, and local agencies."

Analysis

There was not a major emergency in the remaining fourteen years so most fortunately adequacy of the plans was not tested.

Recommendations 11.A, 11.B, 11.C, 11.D, 11.E, 11.F - Environment

One of the Lamm-Wirth recommendations in this section posed a fundamental challenge to the legal and regulatory regime which had governed Rocky Flats since its inception. This challenge was that authority should be granted to the Environmental Protection Agency "to control and regulate all discharges of radioactivity by contractors of Federally operated nuclear facilities." An indication of the depth of this recommendation is that "EPA should establish standards for plutonium and americium in soil..."

A second key recommendation was "to strive toward zero discharge of any radioactive or highly toxic substance at the Rocky Flats Plant." Other recommendations were to assess all plutonium buildings for the purpose of ascertaining whether or not they could "...withstand all natural and human caused credible disasters...," to support the release prevention and detection systems, and to study the system of ambient air sampling.

ERDA Comments

ERDA indicated that "...it is the Federal Government's position, prepared by the Justice Department and concurred in by EPA, that the law does not give EPA authority to regulate radioactive discharges."

AEC/ERDA policy on release of toxic pollutants is that releases should be "as-low-as-practicable." This policy has been in effect at all AEC/ERDA facilities. Resources have been committed at Rocky Flats to provide "complete recycle capability" for liquid effluents which "in effect will accomplish 'zero discharge' of liquid effluents." It is highly ironic that this assurance of "zero discharge of liquid effluents" was to be followed only four years later by the beginning by Rockwell International of Spray Irrigation year around in all Colorado weather.

On the issue of the safety of plutonium buildings, the Albuquerque Office of ERDA has a program to assess the adequacy of existing facilities "with respect to operational accidents and those resulting from natural phenomena." There are already "alarmed, alpha-selective air monitors in major effluent ducts which may contain plutonium." Ambient air sampling utilizes the "best available technology."

Analysis

It was ERDA's position that existing Federal law did not provide authority for EPA to regulate radioactive discharges. This was written in 1975 before the passage of the Resource Conservation and Recovery Act (RCRA) in 1976 and the beginning of a spirited fourteen-year dispute over the applicability of RCRA to Rocky Flats which did not end until 1990.

Subsequently, in connection with the FBI raid of 1989, the Grand Jury investigation, and criminal charges under RCRA and the Clean Water Act, Rockwell International had retained the notable criminal defense firm Williams and Connolly. Williams and Connolly argued that the 1980s had been a period of "regulatory confusion." A related fact confirmed by this ERDA response is that the Department of Justice often supported the position of the Department of Energy; namely, that the EPA lacked authority to regulate radioactive discharges at Rocky Flats during the 1970s and 1980s.

On the subject of "zero discharge," the topography of the site with what became legally under the Clean Water Act two "navigable waters of the United States" made it difficult if not impossible to contain liquid effluents within the Rocky Flats site. When Rockwell responded to the massive volume of effluent from the industrial mission by introducing Spray Irrigation in 1979, it became impossible to achieve "zero discharge" given the topography of the site.

With the respect to the installation of alpha-selective air monitors, ChemRisk (1992) concluded that the overall monitoring system at Rocky Flats was grossly deficient as had DOE (1986a) earlier but in secret in the Mary L. Walker Briefing Memo. There are no records available to substantiate that ambient air monitoring by Rockwell had been effective over the next fourteen years.

Recommendation 12.A, 12.B - Formal Environmental Impact Statements

The Task Force recommended that the preparation of an Environmental Impact Statement be required at the time that the Rockwell contract was up for renewal, or when bids were to be open for other potential contractors.

The Task Force recommended a related EIS for the "removal of all nuclear and non-nuclear toxic materials buried or otherwise contaminating the soil at or near the Plant site."

ERDA Comment

An Environmental Impact Statement was being prepared on the entire operation at Rocky Flats.

Analysis

Rockwell ceased being the contractor on December 31, 1989. EG&G became the successor contractor at the beginning of January, 1990 with no Environmental Impact Statement on the entire operation having been prepared prior to the contract award.

Recommendations 13.A, 13.B, 13.C - Legislation

There were three major recommendations with respect to new legislation. The first recommendation was that Congress should pass legislation reimbursing the state of Colorado and county and local governments for "impact costs incurred by any state government units because of the existence of Rocky Flats..."

The second recommendation was to replace the Price-Anderson Act with a "nuclear industry liability act" which would require contractors and licensees to bear the risk of doing business in the industry.

The third recommendation was internal to the state of Colorado: to pass the Nuclear Liabilities Act which had passed the Colorado House of Representatives in 1975 but which was subsequently defeated in the Colorado State Senate.

ERDA Comments

ERDA disagreed fundamentally with the first two recommendations and indicated that it was inappropriate to comment on an internal state of Colorado determination with respect to the state legislation recommendation.

The essence of the ERDA objection to subsidizing impact costs by the Federal government was that: "we do not agree that monitoring the Plant mission and evaluating Plant management are appropriate matters for a public monitoring committee."

With respect to replacing the Price-Anderson Act, the position of ERDA was that Congress just recently had considered "all the points discussed in this recommendation [and they] received careful Congressional consideration." PL 94-197 was amended dated December 31, 1975. The bottom line for ERDA was that: "we believe no further legislation should be proposed in this area until meaningful experience has been gained under the recently amended statute..."

Analysis

It is clear that a foundation belief embedded in the Lamm-Wirth Task Force recommendations was that the governance and regulatory regime at Rocky Flats required fundamental change.

The purpose of the state of Colorado advocacy to Congress requesting authorization of reimbursement of impact costs resulting from the existence of Rocky Flats was, of course, to gain federal subsidy. But the issues of Lamm-Wirth concern started with health, safety and environmental monitoring and went beyond to establish and fund a Monitoring Committee. The point is that the Lamm-Wirth Task Force lacked confidence in the model of ERDA and the contractor governing and regulating themselves.

A key purpose of replacing the Price-Anderson Act was to require contractors "to bear the risk of doing business in the industry..." This change would reverse a decision of General Leslie Groves during World War II to indemnify contractors from what the President of DuPont Corporation called the "unpredictable hazards involved" in the production of nuclear weapons.

A consistent theme in the Task Force Report was that, given the risks inherent in the very essence and location of Rocky Flats, far greater attention should be given to environment, health and safety. The contractor should be liable legally for its Plant operations.

An important issue not introduced in this exchange between the Task Force and ERDA was the original status of the Price-Anderson Act as a section of the Atomic Energy Act. The Atomic Energy Acts of 1946 and 1954 provided legal authority for the preeminence of the Atomic Energy Commission over all aspects of weaponization of the atom.

The establishment of the Environmental Protection Agency and the suite of new bipartisan environmental legislation was already seen by the Task Force as a positive alternative to what was essentially self-governance by the AEC and ERDA over Rocky Flats (and the rest of the nuclear weapons development complex). This specific tension was not to be resolved, at least partially, until 1990 when Judge Lewis Babcock ruled in the *Sierra Club* lawsuit that RCRA applied to Rocky Flats. But that partial resolution only occurred after Rockwell had ceased being the contractor and Rocky Flats had closed effectively for production.

Recommendations 14.A, 14.B, 14.C, 14.D - Medical Facilities Related To Rocky Flats

The thrust of the four recommendations on Medical Facilities was to enhance the preparedness and training of the medical and nursing staffs, to improve the medical facility, and to conduct regular drills "for treating various health emergencies..."

Analysis

The area of improving the quality of facilities and providers was one of the least controversial.

Recommendation 15 - Contract Provisions

This recommendation was deceptively harsh. It singled out the Occupational Safety Division of ERDA for implied criticism. The core recommendation was to amend the Rockwell contract to allow for "unannounced inspections without notice to the contractor..."

The inspections were to be independent of the Occupational Safety Division of ERDA. The unannounced inspections were to be conducted by such Federal groups as the Nuclear Regulatory Commission and the Occupational Safety and Health Administration.

ERDA Comment

The most important comment by ERDA was a strong endorsement of Rockwell: "Rockwell's performance under the present contract has been completely satisfactory to ERDA in all areas of health, safety and the environment."

Analysis

There is no evidence that anything came of this recommendation. And as to Rockwell's performance, DOE under the authority of Secretary James Watkins in 1989 assigned only 40 percent of the possible performance bonus for 1988.

Recommendation 16 - Information On Plant Operations

This recommendation was "to ask Congress to authorize the Government Accountability Office to conduct an investigation into the facts surrounding this past policy…"

The past AEC policy referred to was presented by the Task Force as "withholding and distorting information from the public with respect to plant operations." A Task Force comment was that the AEC and now ERDA has "the responsibility and the mandate" to monitor and regulate the contractor at Rocky Flats in the public interest.

ERDA Comment

ERDA chose to limit its response to only two sentences: "ERDA has been advised that the GAO will conduct an investigation concerning the Rocky Flats Plant. ERDA will cooperate with that investigation."

Analysis

It is a limitation of the format of engagement that one side can state a position (the Lamm-Wirth Task Force) and the other side (ERDA) can avoid a substantive response, and nothing in the exchange advances the dialogue or leads to constructive change.

The Task Force position was that the AEC during its years of stewardship for weaponizing the atom had a policy of withholding and distorting information. This was a serious charge. It was not addressed by ERDA in any way which led to constructive change.

Conclusions

Core positions were revealed by the Task Force in its recommendations and by ERDA in its comments. They provide a prism for understanding the depth of the gulf which already existed by 1975 between citizen concern for health, safety, and the environment, and the imperative of the

U.S. government to preserve national security through manufacturing nuclear weapons components at the Rocky Flats site.

A second prism is understanding what were the facts about environment, health, and safety at Rocky Flats in the years after 1975. The Task Force had been constrained by the format of engagement and its lack of subpoena power, as well as by the fact that ERDA opposed the establishment of a continuing external Monitoring Committee.

The drafters of the ERDA comments are not known. But ERDA assertions about the status of environment, health, and safety usually were not supported by facts in the years which followed. The drafters of the ERDA comments obviously labored in the constraints of a short time frame in which to prepare and issue their comments. It is not known what access they had to Dow Chemical archives, particularly the Putzier (1970), Seed (1971), and Owen and Steward (1974) reports.

It is possible to illustrate the depth of the fundamental disagreements between the Lamm-Wirth Task Force and ERDA on behalf of the U.S. government by examining seven core and consistent positions of each. Seven core and consistent positions of the Lamm-Wirth Task Force (LWTF) are followed by the linked core and consistent positions of ERDA on behalf of the U.S. government:

Permanent Monitoring Committee

1. LWTF-There needs to be a permanent Monitoring Committee to carry forward the work of the Task Force, especially in the areas of health, safety, and the environment.
   ERDA's position was that there is already an "unprecedented" agreement with the state of Colorado providing access to "complete activities." The Task Force concern over health, safety and the environment is misplaced. A Monitoring Committee is not needed.

Relocation To a Remote Area

2. LWTF-Rocky Flats should be relocated to an area remote from a population center.
   ERDA's position was that it disagrees with the basic Task Force conclusion that "the Plant poses a threat justifying such drastic action."

Regular Briefing To the State On Nuclear Weapons Programs

3. LWTF-The state of Colorado should be briefed on a regular basis on short and long-range nuclear weapons programs at Rocky Flats.
   ERDA's comment was that unless there was a valid national security reason for such action, it would be "improper" to furnish information on classified nuclear program plans.

Expansion of the Buffer Zone

4. LWTF-The original buffer zone acquired in 1974 should be augmented by another one. An argument in favor of expansion was increased protection "from a possible nuclear disaster" at the Plant.
   ERDA opposed the expansion recommendation for the buffer zone starting with the explanation that the only reason for acquiring the original buffer zone was to provide "a logical separation" between the Plant's operation and the rapid residential development in the area. There was no mention of the off-site plutonium which was discovered in 1970 or the 1973 off-site tritium discovery. ERDA was emphatic that the original buffer zone acquisition did not imply "any existing or potential hazard." It is not known based on existing available records if the ERDA drafters knew about the amount of missing and/or unaccounted for plutonium and uranium (see Chapters Seven and Fifteen).

EPA Should Regulate Radioactivity At Rocky Flats

5. LWTF-Authority should be granted to the Environment Protection Agency "to control and regulate all discharges of radioactivity" by contractors of Federally operated nuclear facilities. EPA should also establish standards for plutonium and americium in soil.
   ERDA - "...it is the Federal government's position, prepared by the Justice Department and concurred in by EPA, that the law does not give EPA authority to regulate radioactive discharges."

"Zero. Discharge" of Radioactive and Toxic Substances

6. LWTF-Rocky Flats should "strive toward zero discharge of any radioactive or highly toxic substance..."
   ERDA – AEC/ERDA policy is "as-low-as-is-practicable." Resources have been committed at Rocky Flats to provide "complete recycle capability" for liquid effluents which "in effect will accomplish 'zero discharge' of liquid effluents."

Require Contractors To Bear Liability

7. LWTF-The Price-Anderson Act should be replaced with a "nuclear industry liability act" which would require contractors and licensees to "bear the risk of doing business in the industry..."
   ERDA - Congress had considered all points discussed in this recommendation and amended PL 94-197: "We believe no further legislation should be proposed..."

By way of identifying a further prism on fundamental disagreements, Rocky Flats was for the U.S. government a vital and central contributor to national security. [Rocky Flats is] "...inseparably linked to our current defense posture." With less than a year of experience, ERDA chose to make a ringing endorsement of Rockwell International as the contractor: "Rockwell's performance under the present contract has been completely satisfactory to ERDA in all areas of health, safety and the environment."

If the Lamm-Wirth Task Force had prepared a rejoinder to these ERDA comments with access to the findings of the Seed Committee, it would of course have strengthened their conclusion that Rocky Flats should never have been sited where it was. Until relocated to a remote area away from population centers, the very nature of its mission makes it too risky for a nuclear accident when sited upwind and uphill from a major metropolitan area.

The Task Force would have found the description by Judge Matsch seven years later of Rocky Flats as "ultrahazardous" compatible with their own thinking and overall risk assessment. And whatever they thought of the management of the production mission by Rockwell at Rocky Flats after just months of responsibility, it would have seemed premature to the Task Force to conclude that in all areas of environment, health and safety stewardship by Rockwell merited the recognition accorded to it by ERDA.

Afterword By U.S. Senator Tim Wirth of Colorado (Ret.) In 2018

Lamm and I decided to form the Task Force while we were traveling to and from the Kansas City mid-term convention in 1974 - we had both been elected, but not yet sworn in. We recruited Task Force Members, and I was able to obtain funding from the Office of Emergency Preparedness in Washington. We hired a staff director, and I managed to recruit help from Professor Harold Smith, a senior scientist at Berkeley and an old friend.

Lamm and I had to follow a careful balancing act throughout. There were some strong protests and sit-ins from advocates wishing to close down the plant - an anti-nuclear response. They were occasionally joined by the chief health officer of Jefferson County, who was increasingly alarmed by the potential health impacts on the workers.

The Steelworkers organized the plant when it came to Colorado, and had a strong organization headed by Pat and Jim Kelly, the leaders of the union. They were particularly active politically (Pat), while Jim Kelly played a careful role within the union. We had constant contact with the two brothers, and they were very helpful even as they were trying to protect their workers.

Another major and helpful player was Bob Williams, a veteran Rockwell executive. I spent many many hours with Williams, who also had a balancing act between Rockwell and the overall community. He was cheerful, helpful, and became a good friend - I was deeply saddened when he retired for health reasons, and soon after died of a brain tumor. Was that caused by his exposure at Rocky Flats?

As the years rolled along and we tried to increase pressure on Rockwell and the Department of Energy, the central issue of the health of the workers became more and more apparent. This too, was a balancing act issue, as most of the workers refused to recognize the health threat, but reports of sickness began to circulate. On a number of occasions, I asked for copies of health records (with agreement from the workers) and after years of silence, the Department finally told me that they could not release the records because they were very technical and we would not understand them. It did no good for me to argue that we had close relationships with the medical school at the neighboring University of Colorado. This was for me the last straw in our work, with the help we had from Professor Smith, and recurring issues like the dangerous and extremely careless transport of nuclear materials (often on small airplanes chartered from private companies, to carry materials from Rocky Flats to Los Alamos). Also the volume of "MUF" (Material Unaccounted For) became clearer - was the plutonium really lost in the plant, or was there another explanation?

All of these issues made the central building at Rocky Flats more of a focus. Rockwell and the government made improvements and began to develop plans for building a major new facility at Rocky Flats. By this time, I had been elected to the Senate and was a new member of the Senate Armed Services Committee. I had explained my concerns and the increasing issues at Rocky Flats to the able Chairman, Senator Sam Nunn (Georgia), and my staff began to work with the senior and very talented professionals working for Sen. Nunn.

The greatest opportunity for taking final action on Rocky Flats occurred when Admiral James Watkins came to testify on the Department's budget. Included was the first phase of a major multi-billion-dollar project for building a new facility, for more thoroughly hardening Rocky Flats, and to begin a major cleanup of the site. By this time, international arms control agreements suggested that the further development of new weapons was a low priority, and that the chain of nuclear material development facilities from Flower Mound to Hanford to Rocky Flats could be closed and cleaned up. When Watkins came to testify, and the discussion moved to the large Rocky Flats appropriation request, Senator Nunn recognized me to tell Secretary Watkins that a new facility would not be funded, and that the Committee looked forward to working with the Department in cleaning up the site.

The Secretary - a large, affable and highly decorated Naval officer - was clearly surprised and stunned by the finality of the Committee's decision. Rocky Flats was dead, and a whole new phase began. After a few multi-billion dollar, but unsuccessful, efforts were launched to handle the cleanup, the contract was finally given to a Denver-based contractor, CH2M Kaiser-Hill, who over many years completed the jobs ahead of schedule and under budget. They were very impressive in a culture of huge over sums and slow results in most of the government's other cleanup efforts.

Over the last 30 years, clean up of much of the site has been completed. A large area to the North is a newly opened wildlife refuge; the central location of [the proposed] Building 360 is completely closed off and will be permanently off-limits to any visitors [as a Superfund site]. And on the Southern edge of the old reservation, land has been sold to private developers, and new houses are beginning to appear. Critics call this development "Radiation Acres" and argue that health issues remain. It sounds very familiar.

For more than forty years, a catalogue of good citizens developed and implemented the strategy of shutting down Rocky Flats:

- The Task Force members who met frequently and offered careful advice on how to move forward. They provided great cover.
- The Jefferson County Health Department was careful, very well informed, and persistent.
- Rockwell was remarkably helpful - maybe they knew all along that the jig was up.
- Harold Smith's consultancy proved valuable, as was the consistent help from Lamm and Wirth staffs, supported in part by the early grant from the Carter White House.

I suppose this is a great case study about citizen action, anchored by our citizen task force, a strong union, and good politics. Getting Rocky Flats to close was a significant accomplishment - lengthy, frustrating, but successful.

T. Wirth
Hon. Timothy E. Wirth
U.S. Senate, Colorado (Ret.)
Vice Chairman and President Emeritus
The United Nations Foundation
(Personal communication, December 2018)

The Task Force As a Viable Cold War Approach

Without the power of subpoena in a Cold War context when the Atomic Energy Commission and successors the Energy Research and Development Administration and the Department of Energy and Contractors Dow Chemical and then Rockwell International were determined to be non-disclosive, short-term change did not occur. The imperatives of national security required an uninterrupted flow of plutonium igniters. Secrecy was imperative. By 1975, the investment in infrastructure and human capital was so large at Rocky Flats that it was considered prohibitively expensive for the U.S. government to choose to relocate.

In the long-term, the issues identified by the Task Force have dominated the escalating conflict over the future of Rocky Flats site beyond its closure to production. The choice made by the U.S. government to be unresponsive to inquiry about two workers irradiated in

July of 1975 set the tone for suspicion that the government was not according enough attention to worker safety—a bitter theme now with the continuing conflict over the 2000 unique hazards law.

The Task Force in its recommendations was largely ignored at the time. But the credibility of their work contributed to helping galvanize long-term citizen action. Stonewalling and denial, however, were effective tactics with the Lamm-Wirth Task Force in 1975 during the Cold War.

The Church Cluster of Civil Lawsuits With Subpoena Power

> The core question in this litigation [the Church cluster of lawsuits of 1975] is whether private citizens can use the authority of the judicial branch of that government to define a duty of care to be followed by the executive and legislative branches.
>
> Federal District Judge Richard P. Matsch
> Church Cluster of Lawsuits, May 27, 1982

In 1975 a cluster of civil lawsuits empowered by the subpoena power of the Judicial branch of government set in motion an unprecedented process of discovery at Rocky Flats. The civil litigation filed in 1975 is most associated with the name of Marcus F. Church, a descendant of the family who originally owned the 2,500 acres of ranchland on which Rocky Flats was built. Howard Holme prepared the 520-page submission which had to have been influential in the decision by the U.S. government to elect to settle the Church cluster of lawsuits for over $9 million and then insist that the terms of the settlement be sealed (Holme, 1978).

The first of three civil lawsuits was actually filed against Marcus F. Church. The lawsuit (75-M-1111) against Marcus F. Church and Dow Chemical filed on October 22, 1975 related to the 1973 purchase from Church of property in Jefferson County. The thrust of this lawsuit was that since the land was contaminated substantially by plutonium, uranium, and americium, it was not fit for residential or commercial development.

On October 30, 1975 Plaintiffs Marcus F. Church and Marcus F. Church, Trustee filed a lawsuit (75-M-1162) against the United States of America, Dow Chemical, and Rockwell International regarding 1,400 acres he and his family owned adjacent to the Rocky Flats Plant. The thrust of this lawsuit was that since 1953 defendants had deposited radioactive material on their land which made the property a health hazard and unusable for development.

On December 4, 1975, Great Western Venture filed a lawsuit (75-M-1296) against the defendants alleging that their land purchase of 243 acres on November 9, 1970 was destroyed in value because of the contamination by radioactive substances.

Since the three lawsuits had common questions on both fact and law, they were consolidated by an Order dated June 8, 1976. Judge Richard P. Matsch decided that the complexity of the litigation "could be reduced by placing the primary focus on the factual questions of what was on the land, in what amounts, and with what effects on the use of that land."

Between January 8, 1977 and September 13, 1977 there were some 22 days of evidentiary hearings on the results of a soil sampling and testing program conducted by the U.S. government. The traditional burden of proof on the plaintiffs was altered by agreement between the parties and the Court because the necessary testing was beyond the resources of a private party. The sampling and testing program continued in 1978 and 1979.

Additional Plaintiffs William C. Ackard, Samuel Butler, Jr. and Butler Investments, Ltd. ("Ackard-Butler") were permitted by the Court to join the lawsuit on March 9, 1978. They had acquired property in 1955 which they had subsequently sold to Good Fund in 1973. "Ackard-Butler" had reacquired the property from Good Fund in lieu of foreclosure.

Judge Matsch's Summary of the Facts Asserted By Church and Ackard-Butler

The facts asserted by Church and Ackard-Butler were summarized by Judge Matsch. Judge Matsch cited the submission by the Fairfield and Woods law firm (Holme, 1978). There were four theories submitted by lawyers for the plaintiffs:

1. The corporate defendants are strictly liable for damages from their ultra hazardous operations at Rocky Flats;
2. They are liable in nuisance;
3. The doctrine of res ipsa loquitur is applicable,[3] and
4. Both corporate defendants and the United States are liable for trespass without proof of negligence.

The facts summarized by Judge Matsch of the Church/Ackard-Butler plaintiff case are important to document as the defendants chose to settle rather than dispute the facts and defend the plaintiffs' case in court.

The following are the plaintiffs' points in Judge Matsch's presentation of the factual basis:

1. In the manufacture of nuclear weapons, the Rocky Flats Plant handles large quantities of hazardous materials, including the radioactive materials plutonium, americium, uranium, thorium, tritium, curium, neptunium, and polonium, and non-radioactive materials beryllium, lithium, cadmium etc.

---

[3] "res ipsa loquitur" is a Latin phrase meaning "the thing speaks for itself." According to a webpage of the Cornell Law School, in tort law, "res ipsa loquitur" is a principle that allows plaintiffs to meet their burden of proof with what is, in effect, circumstantial evidence. The plaintiff can create a rebuttable presumption of negligence by the defendant by proving that harm would not ordinarily have occurred without negligence, that the object that caused the harm was under the defendant's control, and that there are no other plausible explanations." Cornell Law School website explanation.

2. The site which was chosen for the plant by Dow, the United States, and the Austin Company, in 1951, is not suitable for a nuclear plant.
3. Its disadvantages include its proximity to Denver, a large metropolitan area, and geographical features peculiar to it such as strong prevailing winds blowing toward Denver from the West.
4. The plant generates substantial quantities of radioactive and non-radioactive waste materials, the handling and disposal of which has permitted contamination of the adjacent property.
5. Waste oil, laden with radioactive material was stored in drums, which, in time, corroded, leaking the contents into the soil.
6. Some of that soil was resuspended into the air and some of the contamination was redistributed through the water table.
7. Additional waste contamination has resulted from waste burning and sanitary landfill activities.
8. At least two major fires have occurred at the Rocky Flats Plant.
9. At 10:10 p.m. on September 11, 1957, a fire was discovered in a Plexiglas glovebox in which plutonium was stored in Building 771. Apparently, the plutonium spontaneously ignited. The heat burned through the Plexiglas wall and into the filtering system. The fire in the filtering system was not extinguished until 11:28 a.m. the next day.
10. A second major fire occurred on the afternoon of Sunday, May 11, 1969, in Building 776/777 in a glovebox containing plutonium.
11. There have been other fires over the years.

Judge Matsch also identified additional submissions pertaining to health effects and public fear.

12. This analysis suggests a significant correlation between cancer incidence in the Denver area and exposure to radioactivity from the Rocky Flats Plant.
13. Public fear of the plant impacting the market value of the parties' lands.

Judge Matsch summarized the position of the defendants as follows:

> ...the government defendants contend that the contamination of plaintiffs' lands does not present a significant health risk and without an appreciable health risk, the plaintiffs are unable to show that their lands are damaged or that the government defendants have any legal liability.

Judge Matsch's Analysis of the Church Cluster of Lawsuits Case, and the Ultra-Hazardous Ruling

Judge Matsch ruled that "The Rocky Flats Plant can be characterized as an "ultrahazardous enterprise." He indicated the following in his statement of decision:

> These cases present unique questions. The Rocky Flats Plant is a singular facility within the peculiarly governmental function of providing for the national defense. Only the government has the authority to possess and use special nuclear materials in the context of a weapons system. The core question in this litigation is whether private citizens can use the authority of the judicial branch of that government to define a duty of care to be followed by the executive and legislative branches.
>
> In Colorado, as elsewhere, negligence is the failure to observe reasonable standards of care required by the circumstances. The degree of care necessary is commensurate with the foreseeable risks. Plaintiffs assume that the risks associated with handling special nuclear material at Rocky Flats are great and therefore that the standard of care should be high.
>
> ...I said that I had been persuaded that the critical question of whether there had been any injury to the plaintiffs could not be litigated because it was an issue which could not be resolved through the adversary process and was, therefore, beyond the court's competence."
>
> It is my considered view that the determination of the existence of such hazards [health hazards] and the acceptability of them are also political decisions for the Congress and the President. They have placed responsibility for the collection and evaluation of the relevant data in the designated agencies and in the absence of showing that there has been a violation of the agency standards, the court has no power to intervene. Accordingly, all the plaintiffs' claims against the United States., Dow and Rockwell must be dismissed for lack of jurisdiction.
>
> ORDERED that all of the plaintiffs' claims in these cases against the United States, Dow Chemical Co. and Rockwell International Corporation are dismissed, and it is...Matsch, May 27, 1982

As part of his order for summary judgment, Judge Matsch indicated that:

> ...for more than six years I have directed the parties to proceed in particular phases, with differing emphases on the facts and the law...I have come to the conclusion that most of these claims must be dismissed for lack of jurisdiction...it may well be that ultimately the plaintiff's present dilemma will be resolved through other branches and levels of government.

The response of the Plaintiffs was to appeal Judge Matsch's summary judgment dismissal to the U.S. Court of Appeals for the 10th Circuit.

The U.S. Court of Appeals For the 10th Circuit—703 F. 2nd 464 (1983) March 23, 1983

The 10th Circuit Appeals Court cited Supreme Court Justice William Brennan's opinion in *Baker v. Carr* that "there should be no dismissal for non-justiciability on the ground of a political question's presence" 369 US at 217 82 S. Ct at 710.

The 10th Circuit Appeals Court ruled that after six years of pretrial activity and after the lengthy hearings that the matter should not be summarily dismissed:

> Plaintiffs are entitled to a trial. Neither the preemption doctrine nor the political question theory justify the dismissal of the case...
>
> There are viable remedies present and the facts that we have been able to read in the record argue in favor of presenting the evidence to a court.

The 10th Circuit Court of Appeals overturned the summary judgment by Judge Matsch and ordered that the Plaintiff's case be heard in Judge Matsch's court.

The Settlement One Month Before the Trial and Three Reflections of Plaintiff Attorneys

After the reversal of Judge Matsch's summary dismissal, and one month before a trial was to commence in his courtroom, three months of settlement negotiations overseen by Federal District Judge Sherman Finesilver resulted in a pre-trial settlement which was notable for the fact that the settlement was ordered sealed.

In Secrets of a Bomb Factory by Frontline, it was revealed who the recipients were of payments made by the U.S. government in return for secrecy:

$2,531,103.86 to Perry and Charles McKay
$2,531,103.86 to Ruth McKay (McKays were descendants of Marcus F. Church who had passed away)
$1,000,000.00 to Land Title Guarantee Co.
$1,000,000.00 to Kotin, Regan & Mouchley, Inc.
$2,092,506.03 to Wm. Ackard, Samuel Butler, Jr., Butler Investment Ltd., and Great Western Venture
$9,154,713.60

Lawyers For Marcus Church Speak

Eileen Welsome of Westword was able to obtain quotes from some of the lawyers for Marcus Church, and published them on July 27, 2000.

Waste Disposed Was Like By a Minor In the Old West

According to attorneys for Marcus F. Church, the Plant behaved like a miner in the Old West, simply shoveling the waste down the mountain and forgetting about it:

> Waste Management was a low priority at Rocky Flats. It was plagued by bad planning, stop-gap solutions, and attempts to save a few dollars.

Hard To Imagine a Better Way of Contaminating the Countryside

Referring to open pit burning in 1957, 1961, and 1965 which consumed 1,082 barrels of oil contaminated with traces of uranium:

> a better means of contaminating the countryside with large quantities of insoluble particles of uranium oxide of respirable size can hardly be imagined.

The lack of records, lack of monitoring and lack of separation of depleted uranium, enriched uranium and plutonium-bearing wastes at Rocky Flats give little assurance no plutonium oils were burned.

Monitoring Was Notable For What Was Not Monitored

"Rocky Flats' monitoring practices have, in several accident situations as well as during routine operations, been notable not for what was measured but for what was not measured. Rather than make attempts to evaluate the extent of such releases through material balance calculations, environmental sampling or other means, Dow and the AEC rather classified whatever information does exist, assumed releases were 'negligible'; made 'estimates' with little basis in fact; and/or made carefully worded, frequently misleading statements with a minimum of information in press releases and reports to outside agencies such as Colorado Department of Health, the governor of Colorado, and members and committees of Congress."

Reflections On Contentious Paths of Citizen Engagement At Rocky Flats During the Cold War

At least during the context of the Cold War, citizen initiatives to assert some level of democratic control over the most sensitive national security priorities of the U.S. government for nuclear weapons production met with great resistance.

After off-site contamination was found in 1970 at Rocky Flats, the sitting Governor of Colorado and a member of the Colorado delegation to the U.S. House of Representatives jointly established in 1974 a very credible broad-based Task Force which offered advice in 1975 on Rocky Flats. A generous assessment is that they were stonewalled by the U.S. government and its contractors from the military-industrial sector.

In the same time frame of 1975, neighbors filed a civil lawsuit using Colorado state tort law. Nine years later they received a settlement of over $9 million with a stipulation that the settlement be sealed. A problem which haunted Rocky Flats until it closed for production in 1989 was that government authorities had done nothing substantial to rectify the reasons for citizen concern which had led to the creation of the Task Force, or which caused the government to settle the Church cluster of civil lawsuit without a trial.

The response by the Energy Research and Development Administration (which was the transition agency between the Atomic Energy Commission and the Department of Energy) was to choose not to engage in the opportunity for meaningful and constructive dialogue with either the Task Force or with citizens in general. Rocky Flats continued to fulfill its national security mission through the effective close of production in 1989.

The Church cluster of lawsuits with the power of subpoena took over eight years to reach a settlement. But the effectiveness of the power of subpoena was augmented by the soil sampling which occurred. It had been determined by Federal District Judge Richard P. Matsch that it was beyond the resources of a private party to conduct soil sampling. The point is that plaintiffs learned much from this Court sanctioned undertaking.

At some point the U.S. government decided it was in the national security interest of the country not to go before a jury and have to disclose in open court the extent of off-site contamination by radionuclides. In vigorously opposing an expansion of the buffer zone, ERDA advanced the narrative that Rocky Flats was not a threat to the community. Having the court record sealed prevented the surrounding community from becoming fully informed and a constituency developing for relocation to a remote area as recommended by the Lamm-Wirth Task Force.

The broader context was that a new suite of environmental laws in the 1970s had ultimately criminalized many of the practices which had led to Task Force concerns and to the sealed settlement. In terms of the question of viable approaches within American democracy by which citizens can maintain some level of democratic influence over such a sensitive area of national security as nuclear weapons production, the answer is that on their own they cannot be effective, at least in the context of a Cold War. The secrecy imperative of nuclear weapons production is why there is such a tension within U.S. constitutional democracy when citizen concerns are expressed and left unresolved.

In May of 1982, Federal District Judge Richard P. Matsch, the presiding judge in the Church cluster of lawsuits, posed the question of whether citizens can use the judicial branch "to define a duty of care by the executive and legislative branches." The history after 1982 definitively answers that question positively. Further, while there have been continuing disputes within and between the "separate but equal" branches of government over Rocky Flats, both the Legislative and Judicial branches have empowered pivotal moments of responsiveness to concerns of citizens.

Chapter Fifteen
"Time Did Not Fly": The Odyssey of the Twenty-Six Year Cook Class Action Lawsuit

The Cook class action lawsuit was filed on January 30, 1990. This chapter reveals why "time did not fly." A contentious fifteen years passed from the filing of the lawsuit in 1990 until the case finally went before a jury on October 23, 2005. From the time the jury reached a verdict on February 13, 2006 it was more than a decade before there would be a resolution in the form of a settlement, finally in May, 2016.

The Cook case reached the U.S. Supreme Court twice in the form of requests that the Court grant a Writ of Certiorari. The first request for a Writ was by the plaintiffs which resulted in remand to the Federal District Court of original jurisdiction for further proceedings. Judge Kane on remand ultimately dismissed the plaintiffs' case in 2014.

The plaintiffs successfully appealed the dismissal to the 10th Circuit Court of Appeals. The 10th Circuit opinion, written by Judge Neil Gorsuch in 2015, reinstated the jury verdict which was based on state of Colorado tort law. This time the request to the U.S. Supreme Court for a Writ of Certiorari was by the defendants. The case was settled only days before the Supreme Court was to decide whether or not to grant the Writ.

The Cook Class Action Lawsuit As Described In 2015 To the Jury In the Instructions It Received From Senior Federal District Court Judge John L. Kane, Jr.

The ten Memorandum Opinions and Orders by Federal District Court Judges between 1991 and 2004 narrowed what the jury was asked to decide in their deliberations. The 10th Circuit Court of Appeals observed in its decisive ruling in 2015 that neither the jury instructions by Judge Kane on state of Colorado tort law, nor the evidence in support of the jury findings, had been challenged on appeal.

Therefore, it is informative to consider the instructions to the jury after the initial fifteen-year narrowing period:

> The Plaintiffs make two claims against Dow and Rockwell on behalf of themselves and the rest of the Class. The first claim is for trespass. Trespass is an invasion of a person's property without his permission. Plaintiffs claim Dow and Rockwell caused a trespass because the Class Area is contaminated with plutonium released from Rocky Flats as a result of various accidents, mishaps, and bad environmental practices that occurred during each defendant's period of operation.

> Plaintiffs' second claim is for nuisance. A nuisance is something that substantially and unreasonably interferes with another person's use and enjoyment of their land. Plaintiffs claim Dow and Rockwell substantially and unreasonably interfered with Class members' use and enjoyment of their properties because plutonium contamination in the Class Area has led to increased health risk and because there is a risk of additional future releases of plutonium and other hazardous substances into the Class Area from the plant site.
>
> Plaintiffs claim that Dow and Rockwell's trespass and nuisance have depressed the value of properties in the Class Area. The Plaintiffs seek damages to compensate the Plaintiff Class for this lost property value. Plaintiffs also seek punitive damages from both Dow and Rockwell.
>
> Defendants deny Plaintiffs' claims and contend that they each operated Rocky Flats in a safe and responsible manner. Defendants admit that plutonium from Rocky Flats is present in the Class Area, but dispute that it is located throughout this area or that they are liable for trespass. They also deny that plutonium is present in the Class Area at levels that pose any significant health risk or that there is a threat of future releases of plutonium or other hazardous substances from the Rocky Flats site to the Class Area. Defendants further deny that any interference with Class members' use and enjoyment of property is substantial and unreasonable enough to constitute a nuisance or that they are otherwise liable for nuisance. Finally, Defendants deny that the alleged trespass or nuisance depressed property values in the Class Area or that plaintiffs and the other Class members suffered any loss of property value as a result of Defendants' operations at Rocky Flats.

The scope of the lawsuit was narrowed between the original filing in 1990 under three causes of action, and the final settlement of the case in 2016. The following sections report on the major narrowing which occurred before the start of the trial in the Memorandum Opinions and Orders from Cook I in 1991 through Cook X in 2004.

The Over Fifteen Years After the Cook Class Action Suit Was Filed In 1990 Before It Went To Trial In 2005

The principal reason it took over fifteen years before the start of the trial was the series of ten lawsuits and associated legal proceedings leading to ten key Orders and Memorandum Opinions by Federal District Court Judges from February 13, 1991 through December 17, 2004. Each took the plaintiffs and defendants immense time to prepare. The Court compounded the nightmare with scheduling. Appeals to the 10th Circuit Court of Appeals and two petitions to the U.S. Supreme Court caused further delay.

In its 2015 Opinion, the 10th Circuit Court of Appeals acknowledged that it is time-consuming for a case in civil litigation to reach a jury since the necessity is to meet the requirement to:

> ...ensure before any trial is held every stone will be overturned in the discovery—even if it means forcing everyone to endure the staggering delay and (no doubt) equally staggering expense the parties endured here.

The 10th Circuit Court of Appeals in 2015 also referred to "the lethal gauntlet of pretrial proceedings." In answering the question of why the "staggering delay" of fifteen years before trial which occurred in this case, it is instructive to review the contentious issues which were the subject of each of the ten key Orders and Memorandum Opinions.

This review identifies what the plaintiffs and defendants were seeking with their briefs and motions which in the aggregate led to what the 10th Circuit referred to as the "titanic fifteen years for the case to reach a jury." The 10th Circuit is again helpful in understanding that these lawsuits and the accompanying briefs and motions were a part of what is described as "contemporary civil litigation practices."

Cook I (*755 F. Supp. 1468*) Decided By Federal District Judge Lewis T. Babcock On February 13, 1991.

<u>Three Causes of Action</u>

Judge Babcock summarized three types of causes of action with which began the case:

1. Under the Comprehensive Environmental Response, Compensation, and Liability Act *42 U.S.C. § 9607*, the plaintiffs cited section 107 to seek to recover "response costs;"
2. Under the Price-Anderson Act, *42 U.S.C.A. § 2211-2284*, plaintiffs seek damages which also utilize laws of Colorado because Colorado is the location where the alleged "nuclear incident" occurred; and
3. Plaintiffs alleged "under Colorado common law: negligence; strict liability; private nuisance; trespass; misrepresentation and concealment; outrageous conduct; and punitive damages." Plaintiffs alleged that: "they have incurred injury and damages caused by releases of hazardous substances from Rocky Flats."

The substantive law of the state where the alleged "nuclear incident" occurred is utilized by the Price-Anderson Act. Therefore, common law claims under Price-Anderson and state of Colorado tort law are identical.

Standard For Deciding Motions To Dismiss and For Summary Judgment

Articulated in this opinion were the criteria for deciding what Judge Babcock characterized as "numerous motions" to dismiss or for summary judgment by Dow Chemical and Rockwell International and "various motions" to amend by plaintiffs. Judge Babcock stated, "For the purposes of a motion to dismiss, I accept all factual allegations as true and resolve all reasonable inferences in favor of the plaintiff." Citing the *Tri-Crown,*

Inc. v. American Federal Savings and Loan Association 908 F.2d 578, 582 (10th Circuit, 1990), he elaborated the standard for his courtroom with a quote from Tri-Crown:

> A case should not be dismissed for failure to state a claim unless the court determines beyond doubt that the plaintiff can prove no set of facts which entitle them to relief.

Citing *Merrick v. Northern Natural Gas Co., 911 F.2d 426, 429* (10th Circuit, 1990) there was a different standard for deciding the summary judgment motions:

> In deciding a motion for summary judgment the evidence and any possible inferences are viewed in the light most favorable to the party opposing summary judgment.

He granted plaintiffs the right to amend a filing to better state "a claim upon which relief may be granted."

After a discussion of provisions of claims under CERCLA and medical monitoring under Price-Anderson and Colorado common law (and assessing all of the claims of plaintiffs under Colorado tort law), Judge Babcock granted a number of the defendants' motions to dismiss. The defendants' motion for summary judgment was granted for damages arising from a "nuclear incident." A broader observation about Cook I is that both sides were contesting vigorously their positions, and were committed to allocating the time and research necessary to prevail. There was simply almost no common ground or points of agreement on law or fact between the parties.

Cook I was the first of ten opinions which were issued by Federal District Court Judges prior to the jury trial in 2005. It is instructive for understanding how the case ultimately ended to identify the initial causes of action of the plaintiffs prior to motions in opposition by the defendants and rulings by Federal Court Judges at the levels of the Federal District Court and of the 10th Circuit Court of Appeals.

<u>CERCLA Claims of Plaintiffs</u>

Judge Babcock was very critical of the CERCLA claims by plaintiffs and denied some, but they were "granted leave to amend." Judge Babcock's discussion of the CERCLA claims occupied over three pages. Congress, in order to facilitate the cleanup of environmental contamination, "created a private cause of action where certain 'response costs' could be recovered against those who contributed to dumping hazardous waste at a site, *42 U.S.C. § 9607(a)*."

Judge Babcock cited *Brewer v. Ravan, 680 F. Supp. 1176,* where the court had denied a claim for medical treatment:

> CERCLA's legislative history clearly indicates that medical expenses incurred *in* the *treatment* of *personal injuries* or *disease* caused by an unlawful or discharge of hazardous substances are not recoverable under *section 9607(a).*

There had been different decisions made in courtrooms by other jurisdictions on the substance of *Brewer v. Ravan*. The point is that this was not an area of settled law in 1991. The action by Judge Babcock was to deny dismissal of that portion of the CERCLA claim of the plaintiffs for medical costs which are necessary "to monitor the environmental effects of defendants' alleged releases." All of the other CERCLA based claims for medical testing were dismissed.

Price-Anderson and Colorado Common Law Claims

Judge Babcock was critical of the Price-Anderson and Colorado common law claims. He found that: "genuine questions of material fact" existed on some claims. Other claims, such as that for "misrepresentations and concealment," were "fatally defective because they can prove no set of facts that would entitle them to relief and any effort to amend would be futile."

There were claims for medical monitoring by some plaintiffs under Price-Anderson and Colorado common law. Particularly contested were those causes of action for medical monitoring. Judge Babcock reported that the theory behind a claim for monitoring is simple:

> When a plaintiff is exposed to a hazardous substance, it is often sound medical practice to seek periodic medical monitoring.

He reported that: "Courts have generally accepted tort claims for medical monitoring."

Non-Traditional Tort Claims

On the common law claims under state of Colorado tort legislation, Judge Babcock indicated that courts have recognized "non-traditional" tort claims. This category is one "which afford relief even though a plaintiff has not yet manifested present physical injury."

Distinction Between Medical Monitoring and Enhanced Risk of Future Harm

There is an important legal distinction between medical monitoring and enhanced risk of future harm which was important to the case of the plaintiffs. The following are the differences between them according to Judge Babcock:

- Medical monitoring: "An action for medical monitoring seeks to recover only the quantifiable cost of periodic medical examinations necessary to detect the onset of physical harm."
- Enhanced risk claim: "seeks compensation for the anticipated harm itself, proportionately reduced to reflect that it will not occur."

The plaintiffs had chosen not to seek damages for anticipated harm with an "enhanced risk claim."

Theory of Medical Monitoring

The reason a defendant may be required to pay for medical monitoring is because of "defendant's tortuous acts or omissions." What Judge Babcock called the "simple" theory behind medical monitoring is to ascertain "whether the plaintiff has contracted a disease...when a plaintiff [has been] exposed to a hazardous substance."

With respect to medical monitoring in the Cook case, Judge Babcock wrote that the Colorado Supreme Court had yet to rule on the topic. But he said "it would probably recognize, in an appropriate case, a tort claim for medical monitoring."

Court Critique of Individual Tort Claims

1. Individual Medical Monitoring

   Plaintiffs' complaint is deficient in pleading a claim...because it fails to allege that Plaintiffs have been exposed to a toxic substance.

   Plaintiffs have 20 days to file an adequate amended complaint which alleges [specific plaintiffs] have each been significantly exposed to a proven hazardous substance.

2. Fund For Scientific Studies

   I do not believe the Colorado Supreme Court would recognize as cognizable plaintiffs' claims for generalized scientific studies.

   Plaintiffs have cited no authority for their common law claims to recover the costs of generalized scientific studies.

   Accordingly, Rockwell's motion to dismiss the portions of the complaint that seek such studies will be granted.

3. Outrageous Conduct

   In *Rugg v. McCarthy, 173 Colo. 170, 476 P.2d 753, 1970,* the Rugg court "adopted verbatim the elements of the tort from the *Restatement (Second) of Torts § 46*: "One who by extreme and outrageous conduct intentionally or recklessly causes severe emotional stress to another is subject to liability..."

   Defendants' motions to dismiss "...will be denied provided plaintiffs file an adequate amended complaint within 20 days.

D. Misrepresentation and Concealment

   Defendants contend that: "plaintiffs failed to plead these claims with particularity as required by *Fed. R. Civ. P. 9(b).* Because the complaint is fatally defective in failing to plead detrimental reliance, defendants' motion to dismiss this claim will be granted."

E. Punitive Damages

The Senate Reports make clear that this section [of the Price-Anderson Act] was meant to prohibit punitive damages against the DOE contractor indemnified under Price-Anderson.

Sovereign Immunity Issues

The Court cited the U.S. Supreme Court decision in *Brady v. Roosevelt Steamship Co., 317 U.S. 575, 63 S. Ct. 425, at 429-430* that: "if [plaintiff] had a cause of action against [the private party defendant], it is difficult to see how she could be deprived of it by reason of a contract between [defendant] and the [United States]...At least in the absence of a clear Congressional policy to that end..."

The ruling by Judge Babcock was that: "Even assuming that DOE must indemnify defendants for an eventual award of punitive damages, summary judgment is inappropriate."

F. Statute of Limitations

Dow moves for summary judgment on all claims against it based on the statute of limitations. Because Dow has not met its burden, this motion is denied.

G. Injunction To Prevent Further Releases

Citing the authority of *Thournir v. Buchanan, 710 F.2d 1461, 1463* (10th Circuit 1983) and *United States v. Parish of St. Bernard, 756 F.2d 1116, 1123 (5th Circuit 1985)* ("it is black letter law that an injunction will not issue when it would be ineffectual" *474 U.S. 1070, 106, S. Ct. 830* (1986)), the plaintiffs sought an injunction "preventing further releases of plutonium and radioactive and non-radioactive substances from Rocky Flats." The motion of Rockwell to dismiss these claims was granted.

H. Mootness and Lack of Ground For Injunctive Relief

The Court dismissed the Plaintiffs' claims to fund scientific studies. The defendants' motions "are made moot by my dismissal of those claims."

<u>Key Orders By Judge Babcock On February 13, 1991</u>

The ten Orders by the Court from 1991 through 2004 shaped the conduct of the trial. Cumulatively, the Orders narrowed the claims under the Price-Anderson Act and state of Colorado common law which could be contested at the trial. For that reason, the key Orders from each of the ten lawsuits will be summarized by cause of action as part of reviewing the highlights of Cook 1-10. The following are the highlights of Cook I:

CERCLA
The plaintiffs' claims for post-judgment costs are DISMISSED in response to defendants' motion, and for all other CERCLA claims for medical testing costs except for the medical testing cost to establish the environmental effects of the alleged Rocky Flats releases.

Price-Anderson Act
The Price-Anderson Act was not specifically mentioned in the Orders, but the 1988 amendments resulted in summary judgment dismissing claims for punitive damages for "nuclear incidents" which occurred on or after August 20, 1988. The Court DENIED summary judgment for the punitive damages claims before August 20, 1988.

Colorado Common Law
The defendants' motions to dismiss for outrageous conduct was DENIED but granted for misrepresentation and concealment.

Rockwell had filed separate motions to dismiss:

- on scientific studies (GRANTED),
- the entire claim for medical monitoring (DENIED), and on
- plaintiffs' claim for an injunction to "prevent further releases" (GRANTED).
- Dow filed a separate motion to dismiss on the statute of limitations (DENIED).
- The defendants had filed a motion for summary judgment based on "mootness and lack of grounds for equitable relief" was DENIED because the claims themselves had been dismissed by the Judge.

Cook II Was Decided By Federal District Judge Lewis T. Babcock (*778 F. Supp. 512*) On November 26, 1991.
The Memorandum Opinion and Order by Judge Babcock was only three and a half pages in length. Its comparative brevity was not an indication of a lack of intense struggle between plaintiffs and defendants. After reintroducing the three causes of action of the plaintiffs, Judge Babcock presented three motions Rockwell made to dismiss or strike:

- the retrospective and prospective population based health surveys claim by the plaintiffs;
- the request for "injunctive relief" for medical monitoring of the plaintiffs;
- the claims of two banks (Banks) of violations of the Price-Anderson Act and the common law of Colorado.

Population Based Health Surveys

Rockwell had moved to strike or dismiss:

- Plaintiffs claims for "retrospective and prospective population based health surveys"

Judge Babcock had stated in Cook I that:

> Although Colorado has yet to do so, I conclude that the Colorado Supreme Court would probably recognize...a tort claim for medical monitoring.

He went on to state that: "I do not believe that the Colorado Supreme Court would recognize as cognizable plaintiffs' claim for generalize scientific studies." The plaintiffs "have cited no authority for their common law claims..." Judge Babcock granted Rockwell's motion to dismiss the claim for population based health studies.

Medical Monitoring

The plaintiffs sought "medical monitoring and surveillance services..." Rockwell moved to dismiss this claim but Judge Babcock considered that: "dismissal of this claim is not appropriate."

In his Memorandum Opinion, Judge Babcock stated:

> A medical monitoring claim compensates a plaintiff for diagnostic treatment, a tangible and quantifiable item of damage caused by a defendant's tortuous conduct. Such relief is akin to future medical expense.

He indicated that a reasonable complement to normal diagnostic testing is the pooling of examination results and "further the objective behind the tort—to assure the early diagnosis of latent disease."

Injunctive Relief For Medical Monitoring

The Court had been asked to dismiss or strike this claim. In their Second Amended Complaint Prayer for Relief para b, plaintiffs' state:

> If and to the extent that complete medical surveillance relief is unavailable at law...[the plaintiffs seek] medical monitoring and surveillance services in the form of injunctive relief.

According to Judge Babcock, Rockwell moved to dismiss this claim for two reasons:

- This first Rockwell claim was not in the first Amended Complaint. Judge Babcock disagreed.
- The second Rockwell claim was that a medical monitoring claim is not "cognizable." The conclusion which Judge Babcock made was that: "dismissal was not appropriate."

Creation of a Court Administered Fund As Part of "Injunctive Relief"

Citing Ayers v. Township of Jackson, 106 N.J., 557 A.2d 287, 313-14 (N.J. 1987)

Judge Babcock indicated there is authority for a court administered fund. In his opinion, "plaintiffs state a valid claim for relief." The argument against a court administered fund (presumable like that in *Ayres*) by Rockwell, according to Judge Babcock, was that such a fund was not one to be established by "injunctive relief." The authority for this position under class certification was *Federal Rule of Civil Procedure 23(b)(2)*.

A spirited difference of opinion existed between plaintiffs and defendants over the medical monitoring claim in the form of "injunctive relief." Judge Babcock indicated that: "This questions is not ripe for decision."

The Tort Claims of Two Banks

Two banks made claims under the Price-Anderson Act and Colorado common law for negligence, strict or absolute liability, private nuisance, trespass and exemplary damages. Judge Babcock decided that the entity making trespass claims must have either title or possession of a property to be eligible to file a trespass claim. After citing numerous cases in Colorado substantiating this position, he granted the motion of defendants.

Judge Babcock did not dismiss the nuisance claims:

> ...a nuisance claim does not require special ownership or possession of land to state a valid claim.

The Economic Loss Rule Issue

Citing *Jardel Enters., Inc. v. Triconsultants, Inc., 770 P.2d 1301, 1303 (Colo. App. 1988)*, Judge Babcock quoted from the opinion of the Court that:

> ...the economic loss rule prevents recovery for negligence when the duty breached is a contractual duty and the harm incurred is the result of failure of the purpose of the contract.

The motion from the defendants was to extend this doctrine [economic loss rule] to bar all of the tort claims of the banks:

> Because such an extension is inconsistent with Vaughn [(*Vaughn v. Grigsby, 8 Colo. App. 373, 46 P.624, 625, 1896*] and its progeny, "which allow a mortgagee to recover for injury to its security," the response by Judge Babcock was: "I decline to extend the economic loss rule to bar the Bank's claims."

Rockwell Vulnerable To Double Recovery Issue

Rockwell made an argument that it was vulnerability to "double recovery" in the context of a motion to dismiss. Judge Babcock ruled that: "there is no risk of double recovery."

Key Orders By Judge Babcock On November 26, 1991

Judge Babcock outlined the motions which had been made by Rockwell. Rockwell wanted the Judge to dismiss or strike claims which had been made by the plaintiffs:

- Plaintiffs' claim for "retrospective and prospective population based health surveys," (DISMISSAL GRANTED)
- Plaintiffs' request for medical monitoring "injunctive relief," (NOT RIPE FOR DECISION)
- The Banks' [Bank Western and Fields Savings Corporation] claims for violations of the Price-Anderson Act (TRESPASS CLAIMS DISMISSED, NUISANCE CLAIMS NOT DISMISSED).

In addition, Judge Babcock did not extend the "economic loss rule" to bar the claims of Banks, and ruled that the "vulnerable to double recovery" motion of Rockwell was without merit.

Cook III Decided By Senior Federal District Judge John L. Kane, Jr. On March 26, 1993 (147 F.R.D. 237).

Cook III unfolded on three levels. The first level occurred when the case went to Magistrate Judge Abram. A difficulty for the plaintiffs was, as Judge Kane put it in his Memorandum Opinion, "The discovery history of this case is not pretty." The second level was the contentious struggle over discovery itself. The third level followed from the characterization of the lawsuit itself by Judge Kane as "technically complex" with the factual information "to support Plaintiffs' allegations has been largely within the control of the Defendants and third parties."

Discovery Was Effectively Stopped

Plaintiffs first served discovery on April 19, 1990, shortly after Dow and Rockwell moved to dismiss the case or for summary judgment. Dow Chemical and Rockwell International also moved for a protective order on May 14, 1990. This motion effectively stopped discovery.

As part of resolving those motions (the motion for a protective order was never adjudicated), there were proposals from both sides for how to do discovery. When plaintiffs moved for what was called a "case management" order, defendants responded by claiming that since there was no factual basis for the plaintiffs' claims, they moved (again) for a stay of discovery. The court did not make a ruling on these motions. Judge Kane observed that as a consequence, "discovery did not commence until last summer [of 1992]."

On April 16, 1992, Magistrate Judge Abram was tasked to establish a discovery schedule. Plaintiffs filed a petition for discovery of certain Grand Jury records on May 15, 1992. Chief Judge Finesilver denied the petition on May 20, 1992, finding that the plaintiffs had not established "good and sufficient" reasons for granting access to the documents.

On June 29, 1992 Judge Abram conducted a scheduling conference. On June 30, 1992, he ordered the defendants to produce a series of databases and indexes relevant to the plaintiffs' discovery interests. He also ordered the plaintiffs to provide a statement of facts which supported each of the individual allegations in all three causes of action. (This Order by Judge Abram, without plaintiffs being accorded the opportunity for discovery to obtain those very facts, was perceived by the plaintiffs as especially disadvantaging to the development of their case.)

Objects For Discovery Controlled By U.S. Government

This period in the history of the Cook class action case was especially contentious. The relevant targets of discovery were all in the hands of the U.S. government or the contractors. Dow and Rockwell objected to Plaintiffs' Statement of Factual Basis. There were numerous motions for reconsideration during the rest of 1992.

Judge Abram issued two orders on December 30, 1992 which denied plaintiffs' discovery motions. On the timetable for expert discovery, he ruled that:

> ...the plaintiffs have had nearly two years to obtain their own soil, air and water tests as well as the determination of any lost market value of property.

On the "Document Discovery Order" he granted defendants' motions to quash the subpoenas of the plaintiffs. He also ruled that plaintiffs on several occasions had not complied with his orders to define the factual basis of their claims.

Most of 1992 was lost to skirmishes as summarized briefly above. Plaintiffs filed objections to, and an appeal from, Judge Abram's Expert Discovery Order and the Document Discovery Order. Judge Kane found that: "Reports available in the public sector document the routine mishandling of hazardous substances at Rocky Flats." These reports also convey that: "a central problem at Rocky Flats has been the lack of adequate monitoring of the handling and disposal of hazardous substances."

Judge Kane determined that some rulings by the Magistrate Judge are "based on some misstatements of substantive law...Reviewing the entire record, I am left with the definite and firm conviction that error occurred."

Judge Kane ordered that:

1. Plaintiffs are permitted "additional discovery of environmental records generated at Rocky Flats";
2. The "blanket limitations on all future document discovery are lifted";
3. The Magistrate Judge's denial of access to Grand Jury materials is upheld;
4. The closure of Rocky Flats after the FBI and EPA raid for safety reasons over the handling of hazardous substances,

and the Grand Jury and known reports "are sufficient to provide a foundation for Plaintiffs' claims."

Judge Kane went on to observe that:

> ...much of the information necessary to calculate accurately Plaintiffs' exposure to hazardous substances was classified and had been in Defendants' control or the control of third parties. Plaintiffs were denied any meaningful discovery...

Cook III revealed a major reality of citizen initiatives at Rocky Flats to influence health, safety, and environment: the obstacles which had to be overcome, with great difficulty, even with access to the Judiciary and its powerful role in U.S. "separate but equal" constitutional government.

By way of summary, Judge Kane identified the need to "strike a balance" between plaintiffs and defendants:

> In sum, ... there is need to strike a balance between giving Plaintiffs a fair chance at discovery and protecting Defendants from a lawsuit which may have no factual basis. If Plaintiffs' case is as meritless as Defendants' suggest, Defendants can easily prevail on summary judgment.

He remanded the case to Magistrate Judge Borchers with "directions to convene the parties and enter a new case management order...appeals of discovery orders, if any, will be handled by me on an expedited basis."

Factual Contentions Before Trial and Discovery Rules

Citing *Hickman v Taylor, 329 U.S. 495, 500 (1947), 67 S. Ct. 385,* Judge Kane observed that the purpose of discovery rules was so that:

> civil trials in the federal courts no longer need be carried on in the dark.

With respect to rejecting the contention of defendants that plaintiffs had no factual bases for their claims under Rule 11 of the Federal Rules of Civil Procedure, Judge Kane gave five reasons for his ruling.

The first reason was that an advisory committee note to the Federal Rules of Civil Procedure stated that:

> The court is expected to avoid using the wisdom of hindsight and should test the signer's conduct by inquiring what was reasonable to believe at the time the pleading was submitted.

The second reason was a precedent from *Mary Ann Pensiero, Inc. v Lingle, 847 F.2d 90, 95 (3d Cir. 1988).* The precedent was proving a claim in the context of the complexity of factual issues in the circumstance that proof is in the control of defendants. The parallel referred to by Judge Kane in Cook was that the plaintiffs could not initially identify "the specific substances to which they were exposed."

The third reason was that since:

> Plaintiffs have demonstrated a Rule 11 basis for their case, they should be permitted a fair contest with the basic issues and facts disclosed to the fullest practicable extent.

Judge Kane in denying the defendants' motion to dismiss cited as precedent *United States v. Procter & Gamble Co., 356 U.S. 677, 682-83, 78 S. Ct. 983 (1958).*

The precedent for the fourth reason was *Franz v. United States Powerlifting Foundation 836 F.2d 1063, 1067-68 (7th Circuit 1987).* Judge Kane determined that: "Defendants import a requirement into *Rule 11* that does not exist:"

> Citing Rule 8(a)(2), the *Franz* Court ruled that: "The complaint should contain a 'short and plain' statement of the claim showing that the pleader is entitled to relief…Bloated, argumentative pleadings are a bane of modern practice."

A fifth reason cited by Judge Kane drew on a precedent from *Pensiero*:

> A requirement that counsel, before filing a complaint, secure the type of proof necessary to withstand a motion for summary judgment would set a pre-filing standard beyond that contemplated by *Rule 11*.

<u>Key Orders By Judge Kane On March 26, 1993</u>

> Plaintiffs' motion for the scheduling of further proceedings is GRANTED…
>
> Plaintiffs and Defendants' earlier motions for the entry of a pretrial or case management order are DENIED as moot…
>
> Plaintiffs' objections to and appeal from the magistrate judge's December 30, 1992 orders are GRANTED IN PART and DENIED IN PART…
>
> Defendants' motions to dismiss under *Rules 11* and *37* are DENIED.

Cook IV Decided On October 8, 1993 By Senior Federal District Judge John L. Kane, Jr. (151 F.R.D.378).

Plaintiffs made a motion to certify a medical monitoring class and a property class. Citing Fed.R.Civ.P.23(a), Judge Kane spelled out the four requirements which had to be satisfied for certification:

1. the class is so numerous that joinder of all members is impractical;
2. there are questions of law or fact common to the class;
3. the claims or defenses of the represented parties are typical of the class; and

4. the representative parties will fairly and adequately protect the interests of the class.

Judge Kane indicated that:

> Plaintiffs must establish that the questions of law and fact common to the class predominate over any questions affecting only individual members (predominance), and that a class action is superior to other available methods for the fair and efficient adjudication of controversy (superiority).

Dow and Rockwell criticized the medical monitoring and property class definitions as "overbroad." They had numerous other complaints, for example, including the meaning which should be assigned to plutonium and VOC doses. Judge Kane did find the parties agree that:

> ...from time to time, hazardous radioactive and non-radioactive materials have escaped beyond the borders of Rocky Flats. Defendants downplay the significance of the releases...

Judge Kane indicated that: "...for purposes of this class action motion, there is preliminary evidence that radioactive and non-radioactive materials have left Rocky Flats." According to Judge Kane, defendants' argument:

> ...is not focused on the fact of dispersion of hazardous radioactive and non-radioactive substances, but on the amount and effect of dispersion of such substances... dispersion of the materials to class members attributable to the plant are minimal and not significant enough to cause either a reasonable risk of health effects or any loss of property value.

This core "argument by defendants has been consistent throughout the lengthy proceedings."

After subjecting the four requirements noted above to the presentation and analysis of the positions of both plaintiffs and defendants and the relevant body of law and findings in other cases, Judge Kane found that the prerequisites for certifications of the property class and the medical monitoring class had been met.

<u>Key Order By Judge Kane On October 8, 1993</u>

> IT IS ORDERED THAT Plaintiffs' motion for class certification is GRANTED.

Cook V Decided On April 12, 1995 By Senior Federal District Judge John L. Kane, Jr. (161 F.R.D. 103)

Plaintiffs were asking for reconsideration of a minute order entered on February 22, 1995. The reconsideration request had been filed by plaintiffs in response to defendants prevailing on their motion to compel plaintiffs to respond to discovery. The position of the plaintiffs was:

1. Rockwell's "contention-style discovery" was "premature";

2. and, as a consequence, "unduly burdensome";
3. most of the requests will be "the subject of expert testimony"; and
4. production of the requested documents would be "invasive of Plaintiffs' counsel's work product" or "violation of attorney-client privilege."

As part of deciding to compel plaintiff discovery, Judge Kane quoted a passage from Charles A. Wright et al.'s (1994) Federal Practice and Procedure defining the purpose of discovery:

> The purpose of the discovery rules is not only to elicit unknown facts, but also to narrow and define the issues and for this purpose it is often necessary to use discovery about known facts (p. 2014).

Plaintiffs argued that they can identify hazardous substances used at Rocky Flats "only through their experts." Judge Kane found that plaintiffs do not cite "any authority in support of their position." Citing *Bohannon v. Honda Motor Co., 127 F.R.D. 536 (D. Kan. 1989)* and *King v. E.F. Hutton & Co., 117 F.R.D.2 (D.D.C. 1987)*, the Court indicated in ruling against plaintiffs that these cases, according to Judge Kane:

> ...hold that even if complete answers to discovery requests may require the answering party to consult with experts such considerations do not transform permissible factual discovery into 'expert discovery'.

Key Order by Judge Kane On April 12, 1995

> IT IS ORDERED THAT Plaintiffs' motion for reconsideration of the minute order granting Rockwell's motion to compel production of documents is DENIED.

Cook VI Decided By Senior Federal District Judge John L. Kane, Jr. On November 13, 1995 (907 F.Supp.1460)

Before getting into the substance of the case, Cook VI is notable for the sheer numbers of lawyers involved as counsel. Rockwell International had 21 lawyers, Dow Chemical had 14, and the Department of Energy had 2. For the plaintiffs, Bank Western had 11. Merilyn Cook had 22. The Department of Law of the Attorney General of Colorado had 1. Sally Barnett had 8. Some lawyers were listed for several plaintiffs.

The backstory of Cook VI is that in pretrial discovery in 1993 plaintiffs served subpoenas duces tecum[4] on the Department of Energy for documents related to Rocky Flats. According to Judge Kane, DOE did not

---

[4] a "subpoena duces tecum" is a court summons expressed as a Latin phrase meaning "you will bring with you under penalty of punishment." It is also referred to as a summons for the production of evidence.

file objections nor submit motions to oppose the subpoenas. Plaintiffs began in January of 1994 to compel compliance. Magistrate Judge Borchers ruled in March of 1994 that since DOE is not a party to the litigation, the only remedy available to plaintiffs was to use Federal Rules of Civil Procedures 45 to initiate contempt of court proceedings.

From the history assembled by Judge Kane, on July 8, 1994 DOE agreed to a proposal which the Magistrate Judge made a court order on September 13, 1994. On May 30, 1995, the Magistrate Judge certified the contempt motions to the Federal District Court (Magistrate Judges themselves do not have the authority to hold anyone in contempt. They may certify the presence of contempt to a Federal District Court Judge, in this case Judge Kane). Judge Kane's response was:

> I have heard the evidence and find DOE has violated the Order and is in contempt of this court.

Plutonium and Uranium Missing From Rocky Flats

The substance of the DOE Contempt of Court was documents pertaining to plutonium and uranium missing from Rocky Flats. The origin of this knowledge was DOE itself in June of 1994. In December of 1994 plaintiffs requested any classified documents to be declassified under the authority of the Magistrate Judge's Order of September 13, 1994.

The plaintiffs made repeated follow-ups between November 23, 1994 and March 3, 1995. Plaintiffs offered to send document reviewers to inspect the documents. According to Judge Kane, "DOE does not claim it has begun a classified review...I find DOE has failed substantially to comply with the Order in this regard." He went on to conclude that:

> DOE seems to be contending if it cannot comply with the specific deadlines in the Order it may disregard the Order altogether. This argument is hardly persuasive.

Documents Requested From the Environmental Measurements Laboratory In New York

There are unclassified documents in the Environmental Measurements Laboratory in New York under the control of DOE. After plaintiff letters of February 2, and March 1, 1995, DOE produced documents four months later. Judge Kane's judgment was that: "Plaintiffs have not established DOE has failed substantially to comply with the Order."

Failure To Identify Computer Tapes

EG&G disclosed to plaintiffs in March of 1995 that there were 257 computer tapes located at the Plant. They contained millions of pages of documents. These tapes may be relevant to the subpoenas. Judge Kane concluded:

> I find DOE, in failing to identify the computer tapes, has not complied substantially with the Order."

Documents In the Custody Of Los Alamos National Laboratory
According to Judge Kane, plaintiffs contend DOE violated the Order with respect to research on the effects of plutonium on human health. At issue was access to research conducted on the human health effects of plutonium by Los Alamos National Laboratory. LANL had begun the "Plutonium Workers Study" in 1976 which included 5,400 workers at Rocky Flats.

DOE made two assertions. The first was they had understood that plaintiffs only sought to inspect the documents not have them produced. The second assertion was the documents had been available for inspection since August of 1994. Judge Kane's assessment was that:

> I find this assertion incredible.

Rocky Flats Database
There was a database on Rocky Flats in the custody of Los Alamos National Laboratory. On July 22, 1994 defendants filed objections based on "relevance, privilege and time limitations." Magistrate Judge Borchers overruled all of the defendants' objections. The assessment of Judge Kane was:

> DOE asserts the Order applies only to privileges asserted by DOE and it cannot waive any privilege of the Defendants. This assertion is entirely fatuous. Plaintiffs contend the Order applies to all documents under DOE's control. I agree with Plaintiffs and find DOE has failed substantially to comply with the Order.

Judge Kane further wrote that: "Paragraph 19 of the Order expressly prohibits delay for privilege review and provides neither DOE nor defendants waive any right to assert privilege by producing documents pursuant to the Order."

By way of conclusion, Judge Kane indicated that:

> DOE's assertion that it is unable to comply with the Order is untenable...I find DOE has failed substantially to comply with the Order in the several respects delineated above.

<u>Key Orders By Judge Kane On November 13, 1995</u>

1. DOE shall pay Plaintiffs their reasonable attorney fees, costs, and expenses for all their efforts to secure production of documents in DOE's control...
2. I have found DOE in contempt.
3. DOE shall comply with the Order in all respects and in particular with regard to Plaintiffs' requests for documents relating to missing materials [plutonium and uranium].

Cook VII Decided By Senior Federal District Court Judge John L. Kane, Jr. On August 8, 1996 (935 F. Supp. 1452)

Judge Kane indicated that at a status conference on November 14, 1995, he had ordered the Office of the United States Attorney to contact the Chief of Staff at the Department of Energy (or failing that) the Secretary of Energy "for the purpose of designating an official of DOE to assure compliance with the Stipulated Order and the Contempt Order."

The outcome was that DOE Chief of Staff Richard Rosenzeig appointed DOE Chief Counsel at Rocky Flats Dana C. Lindsay to be directly responsible for compliance. Lindsay appointed twelve additional declassifiers and four administrative support staff, and subsequently increased that number when more relevant documents were discovered. Dana Lindsay of DOE submitted an affidavit attached to DOE's Corrected Status Report of February 14, 1996 that ten to twelve types of MUF had been identified with a final tally of 677,211 pages (not 11,000 pages). (Subsequently there was reference to fifteen types of MUF documents, with no further elaboration).

Judge Kane wanted legal arguments on whether the Court had the authority "to require the Department of Energy to take over the defense of this case under the Price-Anderson Act *42 U.S.C.A. §2210(h)*," and "the authority of this Court" to order all of these documents impounded in the custody of the U.S. Marshall.

The position of the plaintiffs was that Judge Kane should order DOE "to purge itself of contempt by exercising its lawful authority..." Plaintiffs wanted DOE to take over the defense of the case: DOE has "wrongly and substantially interfered with the judicial process," and has acted "willfully and in bad faith..."

On the question of whether the Court had the authority to require DOE to take over defense of the case, DOE "asserts the Price-Anderson Act provides the Secretary of DOE with absolute discretion to determine whether to refer the defense of the case to the Attorney General." DOE agreed that the Court has the authority to order the U.S. Marshall to impound discovery-related documents but that "such impoundment is not warranted in this case." The outcome was that Judge Kane neither ordered DOE to take over the defense of the case, nor did he order the U.S. Marshall to impound the subpoenaed documents.

Judge Kane ordered the parties within 45 days to file a status report "regarding compliance with this Order." He indicated that he would defer "any further order on issues of contempt or sanctions until I have reviewed such status report." DOE was ordered to pay plaintiffs "their reasonable attorney fees, costs, and expenses for all their efforts to secure production of documents in DOE's control..."

Judge Kane specifically ordered DOE to comply "with the Order in all respects and in particular with regard to Plaintiffs' requests for documents related to missing materials." This Order on missing plutonium and uranium materials makes the Status Conference of

November 14, 1995 of special importance to the ultimate resolution of the case in May of 2016.

This Memorandum Opinion and Order focused on numerous follow-ups to the Contempt of Court issues. Plaintiffs wrote on February 5, 1996 that:

> ...it appears that for a better part of a year, DOE has permitted its counsel to serve, unwittingly, as the conduit for testimony and representations to this Court that DOE knew, or should have known, to be false.

DOE's Economic Stake In the Defense Of the Lawsuit

There was a fundamental disagreement identified between DOE and plaintiffs over the issue of the consequences, if any, of DOE's economic stake in the outcome of the lawsuit. As Judge Kane presented it:

> DOE argues, although Plaintiffs may allege DOE has the real economic stake in the defense of this lawsuit, they have not and cannot claim DOE was the principal representative of Dow and Rockwell since DOE neither principally represents nor owns and controls Dow or Rockwell. Accordingly, DOE submits *Mammoth Oil* does not provide authority for me to impose preclusive sanctions on Defendants for the actions of DOE.

Judge Kane indicated that Rockwell had argued differently in a motion filed on March 4, 1996. Rockwell also opposed the motions by plaintiffs for supplemental sanctions against DOE. The Rockwell position was that: "only a party to an action may 'submit' to evidentiary sanctions, the striking of defenses or a default judgment and DOE is not a party to the lawsuit." Rockwell made a related argument that Rockwell and Dow themselves would not pay a judgment because of the indemnity relationship with the U.S. government is not a reason for ignoring this critical point, and as Judge Kane phrased it, "treating Rockwell and DOE as if they were one entity for the plaintiffs' motion."

Dow also argued in a motion in opposition to plaintiffs' Motion for Sanctions that the contractual relationships between Dow, Rockwell and DOE means that the Court of Claims under the Tucker Act is the appropriate forum for adjudication. A final Dow argument was that plaintiffs lack standing, as Judge Kane phrased it, "to seek relief under DOE contracts with Dow and Rockwell as they are not parties to those contracts."

The Issue of Classification of Missing and Unaccounted For Plutonium (MUF) Documents

The DOE position on access to MUF documents was based on the Atomic Energy Act. MUF documents are classified as "restricted data." According to Judge Kane, the DOE position was that a significant amount of "restricted data" in MUF documents cannot be declassified. Further, disclosure may harm national security:

> The Atomic Energy Act defines "restricted data" as "all data concerning (1) design, manufacture, or utilization of atomic weapons; (2) the production of special nuclear material; or (3) the use of special nuclear material in the production of energy... *42 U.S.CA. §2014.*

The DOE argument was that plaintiffs, according to Judge Kane:

> ...may not be able to use a significant amount of information in the MUF documents for any purpose in this case. Therefore, DOE does not believe any interest ultimately will be served by a classification review of the voluminous MUF documents.

DOE filed a protective order on MUF documents and plaintiffs opposed it on May 21, 1996. Plaintiffs indicated that the DOE arguments had already been used in opposing the motion for contempt and pointed out that DOE had entered into the Stipulated Order.

Key Orders of Judge Kane On August 8, 1996

An indication of the chasm which existed between DOE and plaintiffs over the MUF documents ("materials unaccounted for") is reflected in some of Judge Kane's Orders:

1. Plaintiffs' Supplemental Motion for Sanctions against DOE is DENIED;
2. Dow's Motion for a Protective Order is DENIED;
3. DOE's Motion for a Protective Order is GRANTED "insofar as it seeks relief from the March 15, 1996 deadline for review of "materials unaccounted for" ("MUF-related") documents;
6. IT IS FURTHER ORDERED THAT DOE's Motion for a Protective Order is DENIED in all other respects;
7. IT IS FURTHER ORDERED THAT Plaintiffs' Motion that the DOE be held in Contempt RE: Document Requests dated June 23 and August 24, 1995 is DENIED.

Both DOE and plaintiffs were ordered to return with a status report on production of documents "relevant responsive to Plaintiffs' requests, including that for MUF-related documents, since, according to DOE, the search for and production of responsive documents has been ongoing." Judge Kane also said that: "I do not consider further sanctions are warranted pursuant to the Contempt Order."

Cook VIII Decided By Senior Federal District Judge John L. Kane, Jr. On July 28, 1998 (181 F.R.D. 473)

On October 8, 1993 Judge Kane issued an Order certifying two classes in the litigation which partially overlapped: a Property Class, and a Medical Monitoring Class. The declassification of both classes was sought by Dow and Rockwell. The defendants claim that the case of Rocky Flats, based on recent district and circuit court decisions, fails to meet the prerequisites of *Fed.R.Civ.R.23(a)* or *b(3).* Judge Kane described the position of the defendants as claiming that the case fails to satisfy any prerequisites for class action status:

> ...it presents highly individualized claims and no common course of conduct, destroying commonality, predominance, adequacy of representation and superiority of class adjudication.

There had been other significant claims by the defendants: There had been no history of previous trials containing the "novel" and "immature" claims of the plaintiffs.

There was another line of attack on the Medical Monitoring Class according to Judge Kane:

> ...the prevailing laws holds that the Medical Monitoring Class should not be certified under Rule 23(b)(2) as such a claim is essentially a suit for damages. Finally, they submit, the class definitions are inadequate.

In commenting on the claims of the defendants, Judge Kane said that once a class is certified:

> ...the parties can be expected to rely on it...on the assumption that in the normal course of events it will not be altered except for good cause.

The original certification had used the Rule 23 requirements. Judge Kane had determined that the plaintiffs had arisen "from the same set of circumstances, the release of hazardous and non-nuclear substances from the [Rocky Flats] plant during its operation by Dow and Rockwell."

Bar of Class Certification In Mass Tort Actions

A core argument of the defendants was that recent Supreme Court and Circuit Court decisions on mass tort cases bar class certification. They cited numerous cases in support of their arguments. An example is the citation of *Amchem Products, Inc. v. George Windsor, 117 S.Ct. 2231 (1997).* The conclusion of Judge Kane was that:

> the reliance [on the Amchem Products case]...is misplaced in that it concerned the legitimacy under Rule 23 of a class-certification sought to achieve global settlement of current and future asbestos-related claims." With respect to Rule 23 and its application to the Cook case, "this case concerns neither the role of settlement under Rule 23 nor settlement classes." A conclusion of Judge Kane after reviewing a series of defendant-cited cases is that: "There is no absolute bar to class certification in mass tort cases.

Decertification of the Medical Monitoring Class Under Rule 23(b)(2)

The core of the argument of the defendants was that the goal (purpose) of the medical monitoring claim was obtaining money damages not equitable relief. Citing *Building Construction Dept. v. Rockwell Int'l Corp. 7 F. 3d 1487,*

1492 (10th Cir. 1993), the argument of the defendants was that this decision by the 10th Circuit on medical monitoring at Rocky Flats was "essentially a suit for damages." Judge Kane responded by electing to decertify the medical monitoring class.

Revisiting the Property Class Certifications

Judge Kane then made a detailed analysis of whether or not the Property Class conformed to the new rulings from the 1993 decision of the 10th Circuit:

> Because I decertify the medical monitoring class, I consider whether the requirements of Rules 23(a) and 23(b) are still met with regard only to the Property Class.

In applying Rule 23(a)(2) and 23 (b)(3), Judge Kane analyzed four requirements:

1. *Commonality and Predominance*

   Rule 23(a)(2) requires that: "there are questions of law or fact common to the class..." The position of defendants, according to Judge Kane, was that: "the case presents too many individualized issues and no single common course of conduct."

Judge Kane rejected this claim for two reasons:

a) ...the claims of class members need not be factually identical.
b) Although some of the facts underlying each Plaintiff's claims may vary, the claims arise from the same set of broad circumstances, the release of hazardous substances from Rocky Flats during its operation by Defendants.

2. *Typicality and Adequacy of Representation*

   For the defendants, individualized evidence destroys the Rule 23(a)(3) requirements that: "the claims or defenses of the representative parties are typical of the claims or defenses of the class."

   For the defendants, the rule 23(a)(4) requirement is nullified by the tension caused by the different interests of class members. The requirement which in not met is to "fairly and adequately protect the interests of the class."

Judge Kane advanced three level of argument:

- ...the threshold for typicality is low" and the claims asserted by the class representative "need only be typical of, not identical to, those of other class members (Cook IV 151 F.R.D.378 at 385).
- With respect to property class representatives, "the claims of the Property Class representatives are based on the same legal theories as those brought on behalf of the classes. They will advance the interests of the Property

Class because all representatives own property allegedly exposed to hazardous substances due to Defendants' conduct."

- ...although the alleged harm suffered by the representatives may differ in degree from that suffered by other class members and require individual proof, it is essentially of the same type and does not destroy typicality.

Judge Kane's overall conclusion on typicality was that: "I remain of the view that the Property Class representatives' claims satisfy the typicality requirement and that the class is adequately represented."

3. *Superiority*

   Under Rule 23(b)(3) according to Judge Kane, "a class action [must] be superior to other available methods of the fair and efficient adjudication of the controversy..."

Judge Kane's opinion of the submission by defendants was that some courts have avoided class actions out of concern for "the differences between individual plaintiffs on issues of liability, as well as damages, would result in 'mini-trials'..." This would "overshadow the common disposition of the class."

The opinion of Judge Kane was that: "the common issues will materially advance the resolution of the case itself." He also noted that: "...the cost of retaining experts alone demonstrates that none of the property claims could be brought individually." This supports a finding of superiority.

4. *Adequacy of Class Definition*

   Defendants position in Cook IV, 151 F.R.D.378 at 382-84 was different. After observing in Cook IV that the defendants claimed the class definitions were over-inclusive, they "apparently argue [now] they are under inclusive." Now they argue that: "the class definitions would be fatally inadequate if any person outside of the class area were exposed to hazardous substances..."

The reasons Judge Kane rejected this position of the defendants were two:

- "...the plutonium contour as defined by the plaintiffs" adequately defines a class.
- This plutonium contour provides "a differentiation from more distant members of the public."

At trial, an issue will be "whether the evidence shows that the properties within the contours, not without, were so exposed...I deny decertification of the Property Class."

Defendants' Motions For Summary Judgment

The position of the defendant Dow Chemical was to seek summary judgment. Citing *Maughan v. SW Servicing, Inc., 758 F.2d 1381, 1387 (10th Circuit, 1985)*, "Summary judgment cannot be granted unless the evidence is so clear that there is no genuine factual issue." Further, again citing *Maughan*, Judge Kane indicated that:

> ...while cases involving statute of limitations defenses frequently lend themselves to summary disposition, a court should not grant summary judgment for the defendant if there is a viable issue of fact as to when the limitations period began. If the statute of limitations depends on disputed facts, summary judgment is inappropriate.

Dow Chemical and Rockwell International together sought summary judgment because the claims of plaintiffs, according to Judge Kane, "are based merely on the publicity and the controversy generated by the 1989 FBI raid of Rocky Flats." Judge Kane identified major differences between plaintiffs and defendants:

> Defendants argue Plaintiffs cannot satisfy the requirement of a physical invasion of and entry upon property, because their trespass claims is based on stigma and public perception, nor can Plaintiffs satisfy the element of causation. Plaintiffs assert a prima facie case of trespass exists against both Dow and Rockwell in that each engaged in conduct which resulted in the entry and deposition, without permission, of plutonium and other hazardous substance upon the surface of Plaintiffs' properties.

Citing *Burt v. Beautiful Savior Lutheran Church, 809 P.2d 1064, 1067 (Colo. Ct. App. 1990)* Judge Kane provided the accepted definition of trespass:

> ...the physical intrusion upon property of another without the permission of the person lawfully entitled to the possession of the real estate."

Citing *Allison v. Smith, 695 P.2d 791, 793-94 (Colo. App.1984)*, Judge Kane provided a definition of private nuisance:

> Private nuisance is a non-trespassory invasion of another's interest in the private use and enjoyment of his land.
>
> and
>
> [A] plaintiff must show that the defendant unreasonably and substantially interfered with the use and enjoyment of his property.

According to Judge Kane, defendants argued that the plaintiffs cannot prove that: "Defendants unreasonably and substantially interfered with the use and enjoyment of their property." The position of the plaintiffs as conveyed by Judge Kane was that: "by creating an independent threat to their health, safety and property, through physical invasion of their land, and the resultant property stigmatization and loss in value" defendants "unreasonably and substantially" interfered with their enjoyment and use of their property.

The decision by Judge Kane after analyzing the above was:

> I find genuine issues of material fact exist in the record to date as to the actual physical invasion of Plaintiffs' land through contamination emanating from Rocky Flats as well as to the severity and impact of the threat from contamination emanating from the plant.

Therefore, he ruled that: "these complex issues of causation preclude summary judgment both on the trespass and nuisance claims."

Judge Kane also denied:

- Plaintiffs' Motion for Partial Summary Judgment,
- Defendants' Motions for Summary Judgment on Statute of Limitations, Trespass and Nuisance Claims, and Expert Discovery and to Strike Plaintiffs' Experts (stricken), and
- Defendants' Motion to Preclude Testimony of Expert Witness John Radke.

Cook VIII was exceptionally influential in shaping the scope of the jury trial which was not to occur for over seven years. As part of delivering an opinion on the defendants' position on the nature of the Property Class, Judge Kane indicated that at trial an issue will be "whether the evidence shows that the properties within the [plutonium] contours, not without, were so exposed." At one-point Judge Kane referred to "complex issues of causation" and at another point to "genuine issues of material fact exist in the record to date as to the actual physical invasion of plaintiffs' land."

Issues of material fact dominated the actual jury trial when it occurred. While not intended (since plaintiffs did not wish to lose the Medical Monitoring Class), a positive outcome of Cook VIII for the plaintiffs was being able to focus expert witnesses on addressing the "material facts" within the plutonium contour of the Property Class. That is just one legacy of Cook VIII. Another legacy of Cook VIII was the denial by Judge Kane of the challenges by defendants to the expert witness of the plaintiffs. As the 10th Circuit Court of Appeals stated in 2015, there had been no challenges to the sufficiency of the evidence underpinning the jury verdict.

<u>Key Orders By Judge Kane On July 28, 1998</u>

> Defendants' Motion for Class Decertification is GRANTED with regard to the Medical Monitoring Class, which is hereby decertified, and is DENIED with regard to the Property Class.
>
> IT IS FURTHER ORDERED THAT Dow's motion for summary judgment on statute of limitations grounds is DENIED.

> IT IS FURTHER ORDERED THAT Defendants' combined Motion for Summary Judgment on Plaintiffs' trespass and nuisance claims is DENIED

Cook IX Decided By Senior Federal District Judge John L. Kane On July 24, 2003 (273 F. Supp. 2d 1175)

This Memorandum Opinion and Order, according to Judge Kane, "decides the issues raised by parties and thus clarifies the scope of the trial in the Property Class phase of this action." Judge Kane had bifurcated the trial with the Property Class phase on trespass and nuisance preceding the loss of property values which could be attributable to future invasions of plutonium.

The proceeding in Cook IX was again notable for the number of representatives for both plaintiffs and defendants. Three groups of plaintiffs were represented by 31 lawyers (several represented more than one plaintiff). Rockwell International was represented by 13 lawyers. Dow Chemical was represented by 9 lawyers.

<u>Insight Into Why "Time Did Not Fly"</u>

The arguments presented in Cook IX are the culmination of pretrial evidentiary motions and the Memorandum Opinions and Orders from Cook I through Cook VIII. As Judge Kane observed, there had been "failed efforts to prepare a stipulated pretrial order, [and] the parties have argued and extensively briefed their disagreements regarding various elements of the Property Class claims and the issues to be tried in the first phase of this action."

As elements of understanding "why time did not fly," the time involved in preparation for and scheduling of Cook I - Cook VIII plus what Judge Kane characterized as the plaintiffs and defendants having "argued and extensively briefed their disagreements" contributed to the delays.

But there were substantive other proceedings beyond Cooks I-VIII:

> The parties' arguments regarding the issues addressed in this memorandum opinion and order were set forth in connection with all or some of the following proceedings: briefing and oral argument regarding Defendants' Motion to Strike Plaintiffs' Experts, filed August 29, 1997; the June 25, 1999 hearing before Judge Matsch and pre- and post-hearing submissions filed in May and July, 1999; the February 28, 2001 status conference; the parties' proposed pretrial plans and Class Plaintiffs' Response to Defendants' Report Concerning Opinion Testimony Pertinent to Trial Issues, filed March 21, 2001. In some instances, the parties and the court also addressed these issues in whole or in part in connection with dispositive motions filed earlier in this action.

The proceedings noted immediately above, and what were characterized as "dispositive motions," all took time beyond the preparation of the formal records provided by Cooks I-VIII.

Violation of Federal Nuclear Safety Standards

Judge Kane reported that:

> None dispute this is a 'public liability action' arising under the *Price-Anderson Act, 42 U.S.C.A. § 2210* because it is an action in which Plaintiffs seek to impose liability arising out of or resulting from a 'nuclear incident'...

The importance of the "public liability action,' further, as Judge Kane noted, is that defendants admit:

> ...this statutory provision [that substantive rules for decision in such action shall be derived from the law of the State in which the 'nuclear incident' involved occurs] allows the Plaintiffs to assert claims in Colorado tort law...

According to Judge Kane, the defendants argued that claims under Colorado law "are inconsistent with federal nuclear safety standards for the release of plutonium in air and water." The defendants also argued that Congress intended for federal regulatory standards to preempt state law-based tort claims. Judge Kane did not agree. He devoted over fourteen pages to an analysis of the Atomic Energy Act, the Price-Anderson Act and its amendments, and the legislative history of the Price-Anderson Act itself in order to determine the intent of Congress with respect to preemption.

In support of his position on the legislative intent of Congress, Judge Kane cited three sources: the plain language of the Congress in the 1988 Amendments Act and related legislative history, *Good v. Fluor Daniel Corp., 222 F. Supp. 2d 1236, 1247 (E.D. Wash. 2002)* and Dan Guttman, "Price-Anderson Act Reauthorization: Due Diligence Is In Order", *32 E.L.R. 10594, 10600 (May 2002).*

The Validity of a Trespass Claim Under Colorado Tort Law

The position taken before the trial by the plaintiffs and the defendants could not have been more different on the standard which must be met to prove a claim of trespass. The Colorado Supreme Court had ruled in *Public Service Co. of Colorado v. Van Wyk, 27 P.3d 377 (Colo. 2001)*, according to Judge Kane:

> ...that an intangible intrusion onto another's property may constitute a trespass, but only upon proof that the intangible intrusion caused physical damage to the property.

The trespass in the *Van Wyk* case was noise, electromagnetic fields and radiation waves emitted by power lines. These were determined to be "intangible rather than physical invasions." Colorado law had been that the tort of trespass was "a physical intrusion upon property of another without the proper permission from the person legally entitled to possession of that property." Judge Kane reminded of the ruling in Cook VIII when denying defendants' motion for summary judgment:

> I denied summary judgment on Plaintiffs' trespass claim on the ground that genuine issues of fact exist as to the actual physical invasion of Plaintiffs' land through contamination emanating from Rocky Flats. 181 F.R.D. 473 at 485-86.

By way of explanation, Judge Kane indicated that the "tort of trespass is predicated upon and intended to protect the plaintiff's right to exclusive possession of its property." Later in this section, Judge Kane indicated that: "under the traditional rule, proof that the trespassory invasion caused actual damages is not required to establish liability" citing *Adams v. Cleveland-Cliffs Iron Co., 237 Mich. App. 51, 602 N.W.2d 215, 219 (Mich. App. 1999)* and others.

Judge Kane indicated that: "litigants have pushed the envelope of what constitutes a trespassory invasion of property" by asserting trespass claims based on intangible phenomena such as light, noise, electromagnetic fields and airborne gases. The law is not settled nationwide on whether "intangible intrusions are actionable as trespass."

This unsettled law nationally makes the *Van Wyk* and *Hoery* rulings decisive for tort claims in Colorado. Judge Kane noted that: "the Colorado Supreme Court recently clarified in *Hoery 64 P.3d 214 at 221-222* that: "the migration and presence of contaminants upon another's property constitutes a physical invasion that is actionable under the traditional rule of trespass." The trespass which was alleged to have occurred in *Hoery* was TCE in soil and groundwater beneath the plaintiff's property.

A conclusion of Judge Kane based on the Colorado Supreme Court ruling was that:

> Plaintiffs need not demonstrate that plutonium and other Plant-derived contaminants are present on their properties at levels of toxicological concern or are otherwise causing damage to their properties in order to prevail on their trespass claim and the *Hoery* decision.

The defendants offered a diametrically opposed conception based upon their reading of the Price-Anderson Act.

<u>Elements of Private Nuisance Claims</u>

There were two elements of private nuisance claims singled out for extensive analysis: health risk and other alleged limitations on nuisance claims. Of the two elements, health risk was covered most extensively.

Judge Kane was "not persuaded that verifiable health risk is required to prove nuisance in cases involving a physical intrusion or detrimental change to the property resulting from the presence of contamination there." The Colorado Supreme Court decision in *Hoery* omitted the concept of health risk. In *Van Wyk*, Judge Kane quoted the Colorado

Supreme Court on the definition of "the significant invasion of another's interest":

> ...if normal persons living in the community would regard the invasion in question as definitively offensive, seriously annoying, or intolerable.

As an authority, Judge Kane cited *Exxon Corp. v. Yarema, 69 Md. App.124, 516 A.2d 990, 1005 (Md. Ct. Spec. App. 1986)* to the effect that:

> Actual contamination that does not pose a demonstrable health risk to property occupants might also substantially interfere with the use and enjoyment of property in other ways...

With respect to another category of other alleged limitations on nuisance claims, a concern of defendants was that plaintiffs were impermissibly seeking recovery based on proximity to Rocky Flats. Judge Kane cited *Staley v. Sagel, 841 P.2d 379, 382 (Colo. Ct. App. 1992). Staley* is relevant for several reasons. First, the loss in value of a property which might be attributed to proximity [in the case of Staley to a hog factory] is insufficient to establish liability for nuisance. Second, Judge Kane made clear that:

> proof of actions at the facility that result in substantial interference with a plaintiff's interests are sufficient to establish nuisance.

Judge Kane cited *Allison v. Smith, 695 P.2d, 791, 794 (Colo. App. 1984)* with the related precedent that: "legitimate activity may become nuisance if it is unreasonably operated." The conclusion reached by Judge Kane based on the *Staley* and *Allison* precedents was that:

> Plaintiffs' allegations that Defendants' actions at the Plant resulted in actual contamination of their properties and a risk of future contamination that substantially interfere with their use and enjoyment of property meet this test.

Judge Kane also stated that he was not persuaded that: "verifiable health risk is required to prove nuisance..."

<u>Proof of Decrease Property Values Gives Rise To Nuisance Liability</u>

Citing three relevant cases, *Adkins v Thomas Solvent Co., 487, N.W. 2d at 725, Oliver v. AT&T Wireless Servs., 76 Cal. App. 4th 521, 90 Cal. Rptr.2d 491, 500 (Cal. Ct. App. 1999)* and *Edgcomb v. Lower Valley Power & Light, 922 P.2d 850, 860 (Wyo. 1996)*, evidence "that affected property's value has depreciated may be evidence, however, that actual contamination or other claimed factors of interference are substantial and unreasonable enough to give rise to nuisance liability."

Determination and Measure of Damages

This section of the Memorandum Opinion addressed separately compensatory damages and exemplary damages. Judge Matsch had addressed some of the compensatory damages issues at a Status Conference Hearing on June 25, 1999. According to Judge Kane, Judge Matsch had rejected the attempt by the defendants to "limit potential damages to any decrease in the value of class properties immediately before and after the FBI raid."

Citing *Board of County Commissioners v. Slovek, 723 P. 2d 1309 (Colo. 1986)*, Judge Kane quoted from the opinion: "...the measure of damages for injury to real property is not invariable." The goal according to Judge Kane based on the Colorado Supreme Court's decision in *Board of County Commissioner v. Slovek* is "in all cases is to compensate the property owner for the actual loss suffered."

Exemplary Damages

On exemplary damages, it is the position of defendants that section 2210(s) of the Price-Anderson Act precludes seeking exemplary or punitive damages. Judge Kane quoted from section 2210:

> No court may award punitive damages in any action with respect to a nuclear incident...against a person on behalf of whom the United States is obligated to make payments under an agreement of indemnification covering such incident.

This provision applies only to "nuclear incidents" occurring on or after August 20, 1988 based on the 1988 Price-Anderson amendments. Judge Kane ruled that: ...it remains the law of the case that Plaintiffs may seek punitive damages, to the extent available under Colorado law, with respect to occurrences, including releases of plutonium, they prove took place before August 20, 1988. (See Cook I 755 F. Supp. at 1481).

Conclusion

Using the heading "conclusion" rather than ORDER, Judge Kane made rulings very relevant to the upcoming trial:

- Under Price-Anderson, plaintiffs do not have to prove defendants violated federal regulatory standards.
- Colorado law does not require plaintiffs to prove that contamination poses a health risk...
- Under Colorado law proximity to neighboring properties does not constitute a nuisance in and of itself. The depreciation in property values can support a tort claim based on "substantial and unreasonable interference with the use and enjoyment of neighboring properties."
- A jury will determine whether "any legitimate factors of interference are substantial and unreasonable enough to constitute a nuisance and

to its determination of damages if it finds liability for nuisance and/or trespass."

Cook X Decided By Senior Federal District Judge John L. Kane On December 17, 2004 (358 F. Supp. 2d 1003)

Cook X was the last Memorandum Opinion and Order before the jury trial. Perhaps it was the looming of the trial itself combined with the focus on jury instructions which assembled so many lawyers. There were 31 lawyers representing 14 plaintiffs (with some lawyers representing more than one client). Rockwell International was represented by 13 lawyers. Dow Chemical was represented by 8 lawyers.

The primary focus of the Cook X was the resolution of the argument between plaintiffs and defendants over the content of instructions to the jury on the question of whether the plutonium deposited from Rocky Flats on plaintiffs' property allegedly by the defendants constituted a permanent tort (the position of defendants) or a continuing tort (the position of plaintiffs).

The legal context for the argument between plaintiffs and defendants was that state of Colorado tort law had been unsettled on some critical details of what defined a continuing tort versus a permanent tort. In Cook IX, Judge Kane had cited two very relevant precedents. Citing *Gill v. LDI. 19 F. Supp. 2d 1188, 1197-98 (W.D. Wash. 1998):*

> Where there is an actual entry upon the land, either by the defendant or some tangible thing intruding on the land because of the defendant's actions, interference with the exclusive right of possession is assumed.

Citing *Adams v. Cleveland Cliffs Iron Co., 237 Mich. App.51, 602 N.W. 2d 215, 219 (Mich. App. 1999)* and others, Judge Kane indicated that:

> ...under the traditional rule proof that the trespassory invasion caused actual damages is not required to establish liability, and the plaintiff is always entitled to recover at least nominal damage.

A conclusion of major significance by Judge Kane in Cook IX was that: "overtime, however, litigants have pushed the envelope of what constitutes a trespassory invasion of property..." The point of this observation for the Cook case was that *Hoery v. United States, 65 P.3d 214, 217 (Colo. 2003)* had become a controlling precedent in Colorado. *Hoery* had sued the United States under the Federal Tort Claims Act (28 U.S.C. §1346 (b)(1), 2671-80). In the *Hoery* case, the chemical TCE had been found in the groundwater and soil beneath the property of the plaintiff as a result of migration from the defendant's property. According to Judge Kane, the original trial court had held that that this "constituted a trespass because the toxic pollution released by the [defendant] physically intruded upon the [plaintiff's] property without his permission."

As the presiding judge of the original *Hoery* trial court, Judge Kane had ruled that the trespass claim constituted a permanent tort under

Colorado law and had dismissed Hoery's trespass and nuisance claims based on:

> ...statute of limitations grounds based on its findings that: (1) these claims were for permanent rather than continuing torts under both federal and Colorado law; (2) these claims occurred no later than 1995, when Hoery knew or should have known that his property might be contaminated as a result of the defendant's activities; and (3) as a result of the foregoing, Hoery's claims were time-barred because Hoery asserted them after the limitations period, measured from the 1995 accrual date, had passed.

The trial court had decided that the trespass was permanent and time-limited. Plaintiff Hoery appealed the dismissal to the 10th Circuit Court of Appeals in 2001.

<u>The 10th Circuit Court of Appeals Certification To the Colorado Supreme Court On the Hoery Case</u>

The 10th Circuit Court of Appeals certified to the Colorado Supreme Court unresolved questions of law that, according to Judge Kane, controlled a 2001 appeal from his court on *Hoery, 64 P.3d 214 at 215 (Colo. 2003):*

> The Hoery decision resolved or clarified a number of difficult issues of Colorado trespass and nuisance law that had previously perplexed litigants and this court and contributed to the delay and uncertainty in proceeding with diversity environmental contaminations actions here.

Judge Kane indicated that the relevant question certified to the Colorado [Supreme] Court was precisely the question presented in the Cook case:

> 'Does the ongoing presence of ...toxic chemicals on plaintiff's property [allegedly caused by the defendant] constitute continuing trespass and/or nuisance under Colorado law.'

According to Judge Kane, the Colorado Supreme Court distinguished between "continuing" and "permanent" torts under Colorado law. A key determination was that: "so long as the offending object remains and continues to cause the plaintiff harm," a tortuous property invasion is a continuing tort. The statue of limitations does not apply until the tortuous intrusion has ceased. A trespass or nuisance such as a railway line or irrigation ditch constitutes a permanent tort. The conclusion of the Colorado Supreme Court was that:

> Colorado law recognizes the concepts of continuing trespass and nuisance for those property invasions where a defendant fails to stop or remove continuing, harmful physical conditions that are wrongly placed on the land.

Where the property invasion should continue indefinitely because the invasion was both socially beneficial and lawful was a factual situation constituting a permanent tort.

The underlying opinion on the Colorado Supreme Court ruling in *Hoery* was devastating to the position of the defendants:

> After further review of the *Hoery* decision, I find Defendants' contentions cannot be reconciled with the Colorado Supreme Court's analysis and holdings in that case.

He indicated that the *Hoery* decision of the Colorado Supreme Court "stands for the rule that a tort based on a property invasion that continues in fact is a continuing tort..."

The only exception is based on two conditions:

> ...(1) the invasion will continue indefinitely; and (2) the invasion should continue indefinitely because it is integral to an enterprise vital to the development of the state. If these two conditions are met, then the property invasion is deemed a permanent tort in spite of its on-going nature.

The definitive Colorado Supreme Court ruling, Judge Kane remarked, had been repeated three times within its opinion. The permanent tort exception only applies to invasions that: "will *and* should continue indefinitely."

At trial, according to Judge Kane, the plaintiffs must prove "not only liability in trespass or nuisance," but also that the tortuous invasion "will probably continue indefinitely..." The defendants may defeat this showing, and thus preclude Plaintiffs from recovering prospective damages, by demonstrating that they or others have abated or remedied the trespass or nuisance or are about to do so.

Judge Kane observed that the plaintiffs have elected to seek prospective damages for the continuing invasion. This circumstance is permitted by *Restatement section 930*:

> the decrease in the value of the land caused by the prospect of the continuance of the invasion measured at the time when the injurious situation became complete and comparatively enduring.

General Instructions To the Cook Jury In Reaching a Verdict In the Civil Trial

There were three categories of challenging general instructions for the jury relating to reaching a verdict: understanding the burden of proof in a civil case; the weight which should be assigned to an affirmative defense by the defendants in relationship to the claims of the plaintiffs; and what consideration should be given by the jury to Rockwell's guilty plea in making a determination about whether Rockwell had committed any particular course of conduct.

Preponderance of Evidence

This jury instruction on preponderance of evidence described the difference between an evidence-based verdict in a civil trial versus in a criminal trial. Davidoff (2016) reported that the parties had stipulated "to accept a less-than-unanimous verdict (i.e. one supported by at least eight of the ten jurors who ultimately remained by the close of deliberations)."

The essence of the instruction was:

> ...if you were to put all of the evidence in favor of the Plaintiffs and all of the evidence in favor of Dow or Rockwell on opposite sides of the scales, Plaintiffs would have to make the scale tip to their side in order for your verdict to be against Dow or Rockwell. If Plaintiffs fail to meet this burden, your verdict must be for Dow and Rockwell.

Affirmative Defense

In defense against the claims made by plaintiffs, both defendants made "affirmative defenses":

> In defense to Plaintiffs' claims, Dow and Rockwell have each asserted an affirmative defense... An affirmative defense is more than a denial of the claim. You should treat Dow and Rockwell's affirmative defenses in the same way you treat the Plaintiffs' claims. That is, Dow and Rockwell, as the parties asserting the affirmative defense, have the burden of proving that defense by the same standard, that is, of proving that the affirmative defense is more likely true than not true.

The judgment for the jury to make is whether "the affirmative defense is more likely true than not true."

Evaluating Rockwell's Plea Agreement

No further evidence is required to establish that Rockwell committed the acts to which it pled guilty. The role of the of the jury, therefore, is clearly defined:

> You should not consider the plea agreement or its content, however, or the fact that Rockwell did not plead guilty to additional crimes and that the government did not seek to prosecute Rockwell for any other crimes, in determining whether Rockwell committed any particular conduct at Rocky Flats in addition to the misconduct to which it pled guilty. You should make your determination regarding Rockwell's other alleged conduct based solely on the other evidence presented at this trial.

The essence of this instruction bases the verdict on evidence presented during the trial.

Specific Instructions To the Jury On Trespass and Nuisance

On May 3, 2004 at a status conference, Judge Kane accepted the suggestion of the plaintiffs' counsel (Davidoff, 2016) that: "the remaining unresolved legal issues be addressed in a carefully structured pretrial process devoted to formulating the jury instructions."

Judge Kane directed the parties "to submit proposed instructions in three principal phases: (1) trespass instructions; (2) nuisance instructions; and (3) damages instructions. Judge Kane had bifurcated the case thereby splitting the trespass and nuisance claims from the loss of property values which could be attributable to future invasions of plutonium.

The Trespass Instructions

At the beginning of jury instructions, Judge Kane indicated that one claim against Dow and Rockwell was for trespass: "Trespass is an invasion of a person's property without his permission." The essence of the trespass claim is that the area of the Property Class is "plutonium released from Rocky Flats as a result of various accidents, mishaps, and bad environmental practices that occurred during each defendant's period of operation."

Later in the jury instructions Judge Kane elaborated further on elements of trespass as it applied to the Cook case:

> Elements of the Claim of Trespass:
>
> A landowner has the "right to exclusive possession and control of his property, which includes the right to keep the property free from contamination deposited there by others without the landowner's consent."

Judge Kane indicated that in the Cook case the parties had stipulated that plaintiffs had not consented to plutonium being on their properties.

Plaintiffs had to prove at trial the following three elements which combine to constitute a preponderance of evidence:

1. Plutonium from Rocky Flats is present on the Class Properties. Judge Kane said that:

   > To prove this element, Plaintiffs are not required to show that plutonium is present on the Class Properties at any particular level or concentration, that they suffered any bodily harm because of the plutonium or that the presence of plutonium on the Class Properties damaged these properties in some other way.

2. Dow or Rockwell or both of them intentionally undertook an activity or activities that in the course of events caused plutonium from Rocky Flats to be present on the properties of the class.

3. It appears plutonium will continue to be present indefinitely:

   > ...it is not necessary for you to find that the plutonium will be there forever. Instead, in deciding this element, you should consider whether there is any reason to expect that the plutonium present on Class Properties will be removed at any definite time in the future. If you

> find there is no reason to expect it will be removed by a definite time, then you must find it appears that plutonium will continue to be present on the Class Properties indefinitely.

The Jury Verdict On Trespass Claims
The following are key votes taken by the jury during their seventeen days of deliberation on trespass claims against Dow and Rockwell individually and the exact wording of the ballot.

<u>Trespass Claim Against Defendant Dow Chemical</u>

> Do you find that plutonium from Rocky Flats is present on the Class Properties?
> ANSWER: Yes
>
> Do you find that Dow intentionally undertook an activity or activities that in the usual course of events caused plutonium from Rocky Flats to be present on the Class Properties?
> ANSWER: Yes
>
> Do you find it appears that this plutonium will continue to be present on the Class Properties indefinitely?
> ANSWER: Yes

<u>Trespass Claim Against Defendant Rockwell International</u>

> Do you find that plutonium from Rocky Flats is present on the Class Properties?
> ANSWER: Yes
>
> Do you find that Rockwell intentionally undertook an activity or activities that in the usual course of events caused plutonium from Rocky Flats to be present on the Class Properties?
> ANSWER: Yes
>
> Do you find it appears that this plutonium will continue to be present on the Class Properties indefinitely?
> ANSWER: Yes

<u>Nuisance Instructions</u>
At the beginning of the jury instructions Judge Kane had indicated that a second claim against defendants was for nuisance:

> A nuisance is something that substantially and unreasonably interferes with another person's use and enjoyment of their land. Plaintiffs claim Dow and Rockwell substantially and unreasonably interfered with Class members' use and enjoyment of their properties because plutonium contamination in the Class Area has led to increased health risk and because there is a risk of additional future releases of plutonium and other hazardous substances into the Class Area from the plant site.

Elements of Proof of Nuisance By a Preponderance of Evidence

There were four elements of nuisance which must be proven by a preponderance of evidence. Element One was to prove that the class members use and enjoyment of their properties had been interfered with in one or both of:

- Causing exposure to plutonium and placing them "at some increased risk of health problems as a result"; and/or
- Causing objective conditions that pose "a demonstrable risk of future harm to the Class Areas."

On the second objective condition of a "demonstrable risk" of future harm, jurors need not find that future harm *will* occur but only that: "conditions exist that present the *potential* for such class-wide harm to occur."

Element Two was to prove that interference with the use and enjoyment of the properties was both "unreasonable and substantial." Judge Kane defined "substantial interference" as follows:

> ...the interference is significant enough that a normal person in the community would find it offensive, annoying or inconvenient. In this case, that means you must determine whether a reasonable landowner of normal sensibilities would find the proven interference caused by Dow or Rockwell to be offensive, annoying or inconvenient. "Normal sensibilities" for these purposes means a person who is neither unusually sensitive nor unusually insensitive to the interference you are considering.

The jury was to consider "only the magnitude or level of interference that is common to the Class as a whole..."

Judge Kane described the balancing test required of the jury with respect to whether interference with "a person's right to use and enjoy their land is 'unreasonable'." The challenge for the jury is to determine "if the gravity of the harm outweighs the utility of the conduct which caused it."

The challenge more specifically for the jury is to determine:

> ...whether a proven interference is unreasonable in this case, you must consider and balance the gravity of the harm to Class members against the utility of the Dow and Rockwell's conduct at Rocky Flats and determine whether the gravity of this harm outweighs the utility of this conduct.

The gravity of harm is to be assessed by five factors:

1. The extent of the harm involved.
2. The character of the harm involved.
3. The social value of the type of use or enjoyment of property that has been harmed.
4. The suitability of the particular use or enjoyment harmed to the character of the locality.
5. The burden on the Class members of avoiding the harm.

Point 5 was explained further to mean "when it is possible for the landowner to take some action to avoid the harm."

The utility of the conduct by the defendants which caused the harm was another area for the jury to assess. The factors focusing on the conduct causing the harm were four:

1. The social value of the primary purpose of this conduct.
2. The suitability of the conduct to the character of the locality.
3. The impracticability of preventing or avoiding the interference.
4. The financial burden to compensate others for any interference caused by Dow and/or Rockwell's activities.

Judge Kane observed that:

> A nuisance action for damages seeks to place the financial burden for any interference with the use and enjoyment of property on the actor that caused this harm.

Element Three was to determinate if the conduct of defendants was "intentional" or "negligent" conduct. The challenge for the plaintiffs according to Judge Kane was that they had to prove to substantiate their nuisance claim "that the activity or activities causing the 'unreasonable' and 'substantial' interference were either 'intentional' or 'negligent'." By way of elaboration, Judge Kane stated:

> You do not need to find that all of the conduct that caused any substantial and unreasonable interference was intentional or that all of it was negligent in order to find that Plaintiffs proved this element of their nuisance claim. Proof that a substantial and unreasonable interference resulted from a combination of intentional conduct and negligent conduct is sufficient to prove this element.

There were three tests, any one of which would be proof, for what is considered to be "intentional":

(1) The defendant knew that its conduct would interfere with others' use and enjoyment of their property; or
(2) The defendant knew it was substantially certain that its conduct would interfere with others' use and enjoyment of their property; or
(3) The defendant learned that its conduct was interfering with or was substantially certain to interfere with others' use and enjoyment of their property and yet continued this conduct.

The definition of "negligent" was integrally related to the danger at Rocky Flats. Related to the nuisance claim and negligence, Judge Kane articulated the meaning of "inherently dangerous activities" in the context that operating a nuclear plant is an inherently dangerous activity:

> As a result, Dow and Rockwell were required to exercise the highest possible degree of skill, care, diligence, and foresight in

> conducting these activities, according to the best technical, mechanical and scientific knowledge and methods that were practical and available at the time. If either Dow or Rockwell or both of them did not fulfill this duty when they performed any activities that caused or contributed to a substantial and unreasonable interference (as defined in these instructions), then their conduct was negligent.

The "facts and circumstance" around an accident, not the fact of the occurrence of an accident, contribute to the establishment of negligent conduct.

Element Four of nuisance which must be proven by the plaintiffs was that: "negligent" or "intentional" conduct will continue indefinitely:

> ...it is not necessary for you to find that the interference meeting these requirements will last forever. Instead, you should consider whether there is any reason to expect that this interference will end at any definite time in the future. If you find there is no reason to expect this interference to end by a definite time, then you must find it appears the interference will continue indefinitely.

<u>Nuisance Claim Against Defendant Dow Chemical</u>

The following are the key votes taken on the nuisance claims against Dow individually and the exact wording of the ballot:

1. Do you find Dow interfered with Class members' use and enjoyment of their properties in the Class Area in one or both of these ways: (A) by causing Class members to be exposed to plutonium and placing them at some increased risk of health problems as a result of this exposure *(see Instruction Nos.* 3.7,3.18); and / or (B) by causing objective conditions that pose a demonstrable risk of future harm to the Class Area *(see Instruction Nos.* 3.7, 3.18)?
   ANSWER: Yes

2. Do you find this interference with Class members' use and enjoyment of their properties was both "unreasonable" and "substantial" *(see Instruction Nos.* 3.8-3.12)?
   ANSWER: Yes

3. Do you find the activity or activities causing the unreasonable and substantial interference by Dow were either "intentional" or "negligent" *(see Instruction Nos.* 3.13 - 3.16)?
   ANSWER: Yes

4. Do you find it appears the unreasonable and substantial interference with the use and enjoyment of property caused by Dow's intentional or negligent conduct will continue indefinitely *(see Instruction No.* 3.17)?
   ANSWER: Yes

Nuisance Claim Against Defendant Rockwell Corporation

1. Do you find Rockwell interfered with Class members' use and enjoyment of their properties in the Class Area in one or both of these ways: (A) by causing Class members to be exposed to plutonium and placing them at some increased risk of health problems as a result of this exposure *(see Instruction Nos.* 3.7,3.18); and/or (B) by causing objective conditions that pose a demonstrable risk of future harm to the Class Area *(see Instruction Nos.* 3.7, 3.18)?
   ANSWER: Yes

2. Do you find this interference with Class members' use and enjoyment of their properties was both "unreasonable" and "substantial" *(see Instruction Nos.* 3.8 -3.12)?
   ANSWER: Yes

3. Do you find the activity or activities causing the unreasonable and substantial interference by Rockwell were either "intentional" or "negligent" *(see Instruction Nos.* 3.13 - 3.16)?
   ANSWER: Yes

4. Do you find it appears the unreasonable and substantial interference with the use and enjoyment of property caused by Rockwell's intentional or negligent conduct will continue indefinitely *(see Instruction No.* 3.17)?
   ANSWER: Yes

Trespass Verdicts Against Both Dow and Rockwell and Aggregate Damages

The following are the key votes taken against Dow and Rockwell collectively on trespass, the aggregate damages, the degree of responsibility assigned, and the exact ballot wording:

11. Do you find the injurious situation resulting from the trespass by Dow and Rockwell became "complete" and "comparatively enduring" some time between January 1, 1988 and December 31, 1995 *(see Instruction No.* 3.22)?
    ANSWER: Yes

12. As of the time you find the injurious situation became "complete" and "comparatively enduring," do you find the actual value of the Class Properties was less than the value these Properties would have had but for the trespass committed by Dow and Rockwell *(see Instruction No.* 3.22)?
    ANSWER: Yes

13. As of the time you find the injurious situation became "complete" and "comparatively enduring," what is the amount of the difference between the actual value of Class Properties and what their value would have been but for the trespass by Dow and Rockwell? For each of the three types of

property below, please state your answer as follows *(see Instruction No. 3.23):*

(a) in the first column, state the average percentage by which Class Properties were diminished or depressed in value, relative to what their value would have been, without the trespass; and

(b) in the second column, the corresponding total dollar amount by which Class Properties, as a whole, were diminished or depressed in value, relative to what their value would have been, without the trespass. For purposes of this answer, you should not consider Defendants' affirmative defense of setoff or any "prior market discount" at which Class Members may have purchased their properties.

	Percentage Undervaluation		Aggregate Damages
RESIDENTIAL	7 %		$144,199,088.00
VACANT LAND	30%		$ 27,000,000.00
COMMERCIAL	53.03%		$ 5,651,252.00
		TOTAL:	$176,850,340.00

14. Taking as 100 percent the combined trespass that caused the damages you have found, what percentage, if any, was caused by the trespass by Dow and the trespass by Rockwell *(see Instruction No. 3.19A):*

ANSWER:	Percentage, if any, charged to Dow:	90%
	Percentage, if any, charged to Rockwell	10%
	MUST TOTAL:	100%

Nuisance Verdicts Against Both Dow and Rockwell and Aggregate Damages

The following are the key votes taken against Dow and Rockwell collectively on nuisance, the aggregate damages, the degree of responsibility assigned, and the exact wording of the ballot:

11. Do you find the injurious situation resulting from the nuisance by Dow and Rockwell became "complete" and "comparatively enduring" some time between January 1, 1988 and December 31,1995 *(see Instruction No.* 3.22)?
    ANSWER: Yes

12. As of the time you find the injurious situation became "complete" and "comparatively enduring," do you find the actual value of the Class Properties was less than the value these Properties would have had but for the nuisance committed by Dow and Rockwell *(see Instruction No.* 3.22)?
    ANSWER: Yes

13. As of the time you find the injurious situation became "complete" and "comparatively enduring," what is the amount of the difference between the actual value of Class

Properties and what their value would have been but for the nuisance by Dow and Rockwell? For each of the three types of property below, please state your answer as follows *(see Instruction No. 3.23):*

(a) in the first column, state the average percentage by which Class Properties were diminished or depressed in value, relative to what their value would have been, without the nuisance; and

(b) in the second column, the corresponding total dollar amount by which Class Properties, as a whole, were diminished or depressed in value, relative to what their value would have been, without the nuisance.

For purposes of this answer, you should not consider Defendants' affirmative defense of setoff or any "prior market discount" at which Class Members may have purchased their properties.

	Percentage Undervaluation	Aggregate Damages (Entire Class)
RESIDENTIAL	7 %	$144,199,088.00
VACANT LAND	30 %	$ 27,000,000.00
COMMERCIAL	53.03%	$ 5,651,252.00
	TOTAL:	$176,850,340.00

14. Taking as 100 percent the combined nuisance that caused the damages you have found, what percentage, if any, was caused by the nuisance by Dow and the nuisance by Rockwell *(see Instruction No. 3.19A):*

ANSWER:

Percentage, if any, charged to Dow:	30%
Percentage, if any, charged to Rockwell	70%
MUST TOTAL:	100%

15. Do you find that Dow and Rockwell…nuisance caused a diminution in the value of Class Properties in one or more specific time periods before June 7, 1989?

ANSWER: Yes

Punitive Damages For "Willful and Wanton" Conduct Proven "Beyond a Reasonable Doubt"

In order to receive punitive damages, plaintiffs must prove "beyond a reasonable doubt" that the behavior of defendants was "willful and wanton." Defendant behavior may only be considered up to August 20, 1988, "including conduct occurring before this date that resulted in harm on or after that date."

The definition of "willful and wanton" conduct in the jury instructions was:

> ...an act or omission purposefully committed by the Defendant in question, who must have realized that the conduct was dangerous, and which conduct was done heedlessly and recklessly, either without regard to the consequences, or without regard to the rights and safety of others, particularly the Plaintiff class.

The standard of proof was not the usual "preponderance of evidence" of a civil case but rather "beyond a reasonable doubt." The definition of "reasonable doubt" is:

> ...an uncertainty of mind in which your judgment is not at risk and your can explain this uncertainty based on a fair and thoughtful consideration of all of the evidence, or lack of evidence, in this case.

The purpose of punitive damages is not to provide compensation to the plaintiffs but "to punish the Defendant, and to deter that Defendant and others from committing such acts in the future."
<u>Punitive Damages Against Dow</u>

1. Do you find beyond a reasonable doubt that Dow's conduct in committing the trespass and/or nuisance was "willful and wanton" as defined in Instruction No. 3.27? In deciding this question, you may only consider Dow's conduct up to August 20, 1988, including conduct that resulted in harm on or after this date.
   ANSWER: Yes
2. What amount of punitive damages do you find should be awarded against Dow?...
   ANSWER: $ 110,800,000.00

<u>Punitive Damages Against Rockwell</u>

1. Do you find beyond a reasonable doubt that Rockwell's conduct in committing the trespass and/or nuisance was "willful and wanton" as defined in Instruction No. 3.27? In deciding this question, you may only consider Rockwell's conduct up to August 20, 1988, including conduct that resulted in harm on or after this date.
   ANSWER: Yes
2. What amount of punitive damages do you find should be awarded against Rockwell? ...
   ANSWER: $ 89,400,000.00

Situation Created By Defendants Is Capable of Causing Fear, Anxiety, Or Mental Discomfort

Judge Kane made it clear that the situation created by Dow and Rockwell need not be proven to pose "an actual or verifiable health risk." The reason was that the "fear, anxiety or mental discomfort can also be based on concerns that may be without scientific foundation or other support in fact."

> 5. Do you find that any intentional or negligent conduct by Dow or Rockwell or both of them at Rocky Flats, and/or actual or threatened harms caused by such conduct, created a situation that is capable of causing fear, anxiety, or mental discomfort in individual Class Members *(see Instruction No.* 3.28)?

		DOW	ROCKWELL
ANSWER:	YES	X	X

The 2006 Jury Verdict and the "Slow Motion" Roller Coaster Aftermath of a Decade

It had taken over fifteen years to get to trial followed by four months of the trial itself. During the trial there had been 330 docket entries between opening statements and the closing arguments which concluded on January 20, 2006. After seventeen days of deliberation with approximately thirty questions voted on, the jury advised the Court on February 13, 2006 that it had reached a verdict which was read the next day. The "slow motion" roller coaster aftermath of a decade began with unpredictable twists and turns.

In 2008, Senior Federal District Judge Court John L. Kane, Jr. awarded Plaintiffs $926 million with penalties, fines, and interest. Kirkland and Ellis LLP, on behalf of the defendants, successfully appealed the award to the 10th Circuit Court of Appeals which vacated the award on the basis of the Price-Anderson Act. The plaintiffs appealed this determination to the U.S. Supreme Court which remanded the case back to the Federal District Court of original jurisdiction for further proceedings.

Following Remand By the U.S. Supreme Court, Senior Federal District Court Judge John L. Kane, Jr. Dismissed Plaintiffs' Case in 2014

After Judge Kane had held a status conference, he directed the parties to brief the following two issues:

1. May plaintiffs who bring a Price-Anderson Act claim simultaneously pursue a freestanding state-law claim based on the same facts?
2. Does the existing jury verdict in this case support entry of this Court's judgment of Plaintiff's nuisance claim under Colorado law and diversity jurisdiction?

After receiving the briefs, Judge Kane decided that the answer to the first question is "no" and consequently "I do not reach the second question." While indicating that he thought that the plaintiffs had a valid claim under Colorado common law, Judge Kane dismissed with prejudice the plaintiff's refiled case in February of 2014.

Judge Kane then went on to make the following assessment:

> I arrive at my conclusion with reluctance and sympathy for Plaintiffs. I remain convinced that the Colorado jurors in this case correctly found Plaintiffs to have suffered a nuisance under Colorado state-law based on the nuclear contamination for which Defendants are responsible, but I have come to terms with the fact that the law, which I am bound to uphold does not permit independent Colorado state-law claims based on the same facts.

Judge Kane ruled that: "the Price-Anderson Act ("PAA" or "Act") preempts state-law tort claims arising from atomic energy production and establishes an exclusive federal remediation scheme for such claims."

Cook Case Ruling Against Plaintiffs Appealed To 10th Circuit In 2014
In March of 2014 a notice of appeal was filed by the plaintiffs with the 10th Circuit Court of Appeals.

The 10th Circuit Court of Appeals Decision In 2015
The opinion of the 10th Circuit Court of Appeals in June of 2015 would have been the definitive outcome of the Cook civil lawsuit if the defendants had not filed a Writ of Certiorari to the U.S. Supreme Court. It is therefore timely to examine the reasoning of the 10th Circuit Court of Appeals.

The Refiled Plaintiffs' Case On Remand From the U.S. Supreme Court
Instead of refiling their original Price-Anderson claim, the core of the refiled case by the plaintiffs was that the jury verdict was valid from the 2005-06 trial under state of Colorado tort law.

The Arguments of the Defendants and Their Acceptance By the Original Federal District Trial Judge On Remand From the U.S Supreme Court
According to the opinion of the 10th Circuit, defendants offered two arguments to counter the position of the plaintiffs concerning why it would be impermissible to enter a judgment on the existing state law nuisance verdict. The first argument was that: "the Price-Anderson Act prevents it." This argument had been accepted by the Federal District Court in Judge Kane's dismissal of the plaintiffs' case. The second argument of the defendants was that the 10th Circuit mandate in the first appeal "independently barred the plaintiffs from securing relief on their state law nuisance verdict." The plaintiffs had filed an appeal of dismissal.

Analysis of the Arguments of the Defendants By the 10th Circuit Court of Appeals

Congress had not explicitly expressed "its intent to supplant state law." Further, citing a number of cases:

> ...the Supreme Court has instructed us to start with the assumption that the historic police powers of the states were not to

> be superseded, unless that was the clear and manifest purpose of Congress.

The defendants had never on appeal suggested that: “a freestanding state tort judgment would be preempted and precluded." The bottom line for the 10th Circuit was that: “...it's important to raise a preemption defense or risk losing it."

The Appeals Court has long explained that:

> jurisdictional arguments a party forfeits on appeal 'may not be asserted on remand...' Furthermore, the defendants had no business attempting a field preemption affirmative defense following the first appeal, and the district court had no business adopting it. The Appeals Court also said that they do not see "field preemption in the {Price-Anderson} Act of the sort the defendants describe.

The second argument of the defendants was that: “...the federal government has occupied the entire field of nuclear safety concerns." The response of the Appeals Court was that:

> ...this court identified no error associated with (the plaintiffs} state law nuisance verdict.

The defendants had been successful in arguing that the Rocky Flats case did not involve a "nuclear incident" under the Price-Anderson Act. They now had to argue a version of "go-big-or-go-home": the Price-Anderson Act in this version "preempts and precludes" any state law recovery "where (as here) a nuclear incident is asserted but ultimately unproven."

<u>10th Circuit Opinion On the Validity of the State Law Nuisance Verdict</u>

The opinion of the 10th Circuit is that:

> ...nothing prevented the district court from entering a new judgment on their state law nuisance verdict standing alone.

The position of the plaintiffs in district court was that the "background state tort laws" operate normally and "everything needed for a judgment on a state law nuisance claim already existed."

The opinion of the 10th Circuit was that:

> At trial, the district court instructed the jury on Colorado nuisance law and the jury returned a state law nuisance verdict according to those instructions." Further, "And no one has ever challenged the sufficiency of the evidence in the record.

Unforeseen Consequences of Legal Tactics By the Defendants

The 10th Circuit made a very interesting commentary on the (possibly) unforeseen consequences of some tactical choices of strategy by the defendants:

> Perhaps the defendants' push in appeal for a narrow definition of what qualifies as a nuclear incident won them the battle, but it lost them the war--failing to eliminate the plaintiffs' state law claim...

The Downside of Filing a Civil Lawsuit In Federal Court On the Substance of Environmental Justice Related To National Security Imperatives

The 10th Circuit opinion of June of 2015 referred specifically to the problem of delay:

> It took a titanic fifteen years for the case to reach a jury. No doubt a testament to contemporary civil litigation practices that ensure before any trial is held every stone will be overturned in discovery--even if it means forcing everyone to endure the sort of staggering delay and (no doubt) equally staggering expense the parties endured here. Somehow, though, this case managed to survive the usually lethal gauntlet of pretrial proceedings and stagger its way to trial.

This process did allow the plaintiffs to use the power of the subpoena to successfully obtain in the first six years after the lawsuit was filed 2 million pages of previously secret DOE documents on the Rocky Flats case.

The Issues Defined By the Defendants In the Appeal of the Tenth Circuit Ruling To the U.S. Supreme Court: 2015-2016

Dow Chemical Company and Rockwell International Corporation filed a petition with the U.S. Supreme Court for a Writ of Certiorari[5] in December of 2015. The question presented to the Supreme Court was:

> Whether the Price-Anderson Act...preempts state-law claims by plaintiffs arising out of alleged radioactive release and, if so, whether the decision below can stand.

The petition provided the statutory and factual background, described the Trial and First Appeal, the Proceedings on Remand, the Second Appeal, and then offers their core argument:

> The PAA preempts state-law claims by plaintiffs who assert a claim for injuries arising out of alleged radioactive release.

The Tenth Circuit decision is "manifestly incorrect," created an acknowledged Circuit conflict, and contains a wrong ruling about

---

[5] This is a legal form to request that the U.S. Supreme Court order a lower court (in this case the 10th Circuit Court of Appeals) to deliver the documents in a case so that the Court may review them.

"alternative forfeiture." The "alternative forfeiture" incorrect ruling by the Tenth Circuit was that Dow Chemical Company and Rockwell International Corporation "...forfeited their preemption appeal in this case."

Dow Chemical Company and Rockwell International Corporation were supported in their petition by "Friend of the Court" briefs (Amicus curiae)[6] filed on their behalf by American Nuclear Insurers, the Nuclear Energy Institute, the U.S. Chamber of Commerce, and Louisiana Energy Services LLC. The plaintiffs in the Cook Case filed a rejoinder on March 4, 2016.

American Nuclear Insurers

American Nuclear Insurers (ANI) is a non-profit unincorporated joint underwriting association of insurance companies. ANI provides insurance for public liability as defined by the Price-Anderson Act. While ANI insures all 100 nuclear power plants, the Price-Anderson liability and preemption provisions in the Cook case apply to every nuclear power plant they insure. Their core argument is that the 10th Circuit ruling "misreads both the plain language and the history and purposes of the PAA." The stakes are high because of "the approximately $1.3 billion at stake."

The conclusion of this brief was that the "The court should act now, reverse the 10th Circuit's decision, and re-affirm the common-sense, plain-language understanding of PAA preemption that ANI, the nuclear industry, plaintiffs and the courts have heretofore shared."

Nuclear Energy Institute

The Nuclear Energy Institute, Inc. (NEI) establishes policies affecting the nuclear energy industry. It also represents the nuclear energy industry in litigation, and on regulatory, technical, and legal issues.

At issue for NEI in the Cook case is the "comprehensive federal statutory and regulatory framework that Congress has established during the past sixty years..." The core argument is that: "...the Tenth Circuit stands alone..." in ruling that: "...plaintiffs alleging injury from releases of ionizing radiation may avoid the preemptive effect of the Price-Anderson Amendments Act of 1988...despite their failure to prove the predicate facts for a compensable injury under this Act."

U.S. Chamber of Commerce

The Chamber of Commerce of the United States is "the world's largest business federation with 300,000 members it represents directly." It also represents 3 million businesses indirectly. The core arguments are that the 10th Circuit ruling in the Cook case:

---

6 Petition to a court by a party with a strong interest or views on a case but who is not a party to the lawsuit and who is asking to submit a brief for the purpose of influencing the court's decision.

1) ...Upsets the balance between providing adequate compensation to injured individuals and limiting unpredictable liability; and

2) ...Upsets the balance between the federal government and the states.

In the viewpoint of the Chamber of Commerce, the issues in the Cook case are "...a common question facing American businesses–whether a federal regulatory or remedial regime supplants, or merely supplements, state law."

Louisiana Energy Service, LLC

Louisiana Energy Service, LLC owns and operates a commercial uranium enrichment facility and a nuclear power plant within the jurisdiction of the Tenth Circuit.

The core arguments were that:

1. ...Congress thoughtfully balanced competing societal interests and created a comprehensive scheme for regulation of nuclear liability.
2. The Tenth Circuit improperly substituted its views for the judgment of Congress. and
3. The Tenth Circuit's novel interpretation of the PAA is not supported by the text of the Act or its legislative history.

The effects of the Tenth Circuit decision is to "...undermine the federal nuclear liability regime and has implications for the federally mandated system of nuclear liability coverage..."

Rejoinder By the Plaintiffs In the Cook Case

On March 4, 2016, the plaintiffs in the Cook case responded by framing different questions for the U.S. Supreme Court. On the Price-Anderson Act, they quoted a section which stated that the Price-Anderson Act provides for federal jurisdiction arising out of, or resulting from "a nuclear incident" which is defined to include radioactive discharges causing "bodily injury, sickness, disease, or death, or loss of or damage to property, or loss of use of property."

They then went on to make a key claim derived from quoting another section of the Price-Anderson Act:

> ...the substantive rules for decision in [a public liability] action shall be derived from the law of the State in which the nuclear incident involved occurs.

The Cook plaintiffs defined different questions from those submitted by the amicus curiae briefs:

1. Whether the court of appeals correctly concluded that Dow and Rockwell forfeited the preemption claim on which they

now seek this Court's review by failing to timely raise it in the proceedings below.

2. Whether Price-Anderson preempts state-law claims for radioactive contamination that constitutes a nuisance under the state law but does not amount to a "nuclear incident" as the court of appeals defined the term.

The questions defined in the petitions to the U.S. Supreme Court were starkly different. The (perhaps) final ruling by the Judicial branch was being awaited. The eight justices of the U.S. Supreme Court had scheduled a meeting on May 26, 2016 for the purpose of starting a discussion of whether or not to grant a Writ of Certiorari. The settlement occurred less than two weeks before the scheduled meeting on whether to grant the Writ.

Meanings Derived From the Multiple Viewpoints On the Cook Civil Litigation In 1996

Mark Obmascik (1996) provide multiple perspectives on the Cook cases reported by in "Rocky Flats Sued, tax payers pay; ex-operators' legal fees hit $20 million."

<u>View of a Department of Energy Attorney On the Cook Case in 1996</u>

In order to respond to the subpoenas in 1996, 80 Rocky Flats workers were copying documents at a cost of $11 million. Lisa Schiavo Blatt, Assistant General Counsel for the Department of Energy said:

> We believe these cases have no merit and should not have been filed in the first place...But what can you do when there is a lawsuit asking for a zillion dollars without merit? If the plaintiffs were to prevail, taxpayers would be forced to pay the bill.

<u>View of a Plaintiff's Attorney In the Cook Case</u>

Bruce Deboskey is an attorney in Denver who has represented injured Rocky Flats workers for years. He sought out the Berger and Montague law firm in Philadelphia for the Cook case because they had obtained a Three Mile Island Nuclear Plant class-action settlement for $25 million.

Deboskey made the following comment about the Cook case:

> Why should I feel guilty?...Dow and Rockwell came into this community and poisoned the environment and damaged the real estate values. People were misled about what was going on at the weapons plant. People whose lives were adversely affected should be compensated for that.

<u>View of a Lead Attorney For Defendant Dow Chemical</u>

Dow Chemical's lead attorney was David Bernick of the Kirkland & Ellis law firm in Chicago. As of 1996 his firm had collected $5.9 million from the Department of Energy for their defense work on the Rocky Flats case. A Freedom of Information filing revealed that the ultimate cost to taxpayers exceeded $100 million.

> The best evidence of whether there is a problem at Rocky Flats is that in a whole series of scientific investigations conducted over the past 25 years…not a single one of the outside experts, or state or federal agencies, has ever found that Rocky Flats poses a health risk…
>
> My role is to conduct an effective and cost-efficient defense…I think the taxpayers of the United States deserve to have their interests represented effectively. The only parties in this matter who aren't serving the taxpayers are people insisting on bringing 25-year old claims in court.

View of the DOE Plant Manager At Rocky Flats

The position of Mark Silverman, the DOE Plant Manager in 1996, was that the U.S. government should require contractors to assume more legal responsibility. He went on to state:

> I believe that we have an obligation to share information with the public and those who live downwind of our site. But we've spent more than $2 million just to produce records for this lawsuit, which is an awful lot of money that I'd rather have spent on cleanup of the site…I think this is turning into a horrendous waste of taxpayer dollars. The only ones who are benefiting are the lawyers...It's hard to see how this helps the taxpayer.

The Final Outcome of the Cook Case In 2016

The Cook class action lawsuit was settled in May of 2016 for $375 million with as many as 15,000 homeowners dating from 1989 as potential beneficiaries. The affected homeowners lived in the vicinity of Standley Lake east of Rocky Flats.

This was one of the areas most affected by toxics which flowed off-site via the Woman Creek drainage through the Rocky Flats site. The area was especially impacted by the massive spray irrigation across the East Trenches dumpsite.

The defendants in the case, Dow Chemical and Rockwell International, each made statements about the settlement, as did the Department of Energy on behalf of the U.S. government.

Dow Chemical Statement

Rachelle Schikorra on behalf of Dow Chemical said Dow's share was $131,250,000. She stated:

> The U.S. Department of Energy authorized the settlement, and Dow fully expects to be indemnified for the full cost of the settlement…This settlement resolves 26 years of litigation, and Dow believes this settlement is the right decision for the company and its shareholders.

Rockwell International Statement:

In a press release reported in The Denver Post (Aguilar, 2016), Rockwell indicated that its share was $243,750,000. The press release stated:

> Under Rockwell International's contract and federal law, it is entitled to indemnification by the DOE for the settlement amount.

Department of Energy Statement:

The Department of Energy released the following statement to The Denver Post (Aguilar, 2016):

> While the Department of Energy was not a party to the case, the department is pleased that the parties have reached a settlement resolving long-standing litigation between property owners in the vicinity of the former Rocky Flats plant and Dow Chemical Company and Rockwell International Corporation, DOE's former contractors at the plant.

It is, indeed, the long-standing policy of the U.S. government to indemnify contractors from the military-industrial complex for liability arising out of their work in nuclear weapons production.

The rationale for this policy was the "unknown hazards" involved for contractors and the central importance of nuclear weapons to the national security of the United States. The fact of the matter is, however, that the $375 million settlement has been paid by the taxpayers of the country.

ANALYSIS OF DECISION-MAKING AND WHETHER THE "SYSTEM" OF CONSTITUTION GOVERNMENT WORKED

Introduction to Part VI: Evaluating Decision-Making and the Functioning of the American "System" of Constitutional Government at Rocky Flats

Key Elements of the Chapter Sixteen Narrative

- The focus is on what stewards of Rocky Flats were thinking when they made decisions which had such fateful and disastrous consequences.
- "Disastrous" is used in the usual sense of an action resulting in a ruinous outcome or causing great damage or harm.
- "Fateful" captures decisions of stewards which constrained positive choices available to future stewards.
- In some of the "fateful" decisions, the stewards did not believe that they had any other choice in order to maintain the production mission at the level required for the War Reserve.
- The choice of location for Rocky Flats was a "failure to anticipate" the nature and physical requirements of the Rocky Flats industrial enterprise.
- The storage, treatment and disposal of toxic chemical and radioactive waste was "beyond the capacity" to solve for Rocky Flats as it was sited and governed during the Cold War.
- The rational decision to burn in open pits over 1,000 barrels of cooling oil contaminated with traces of uranium was an example of "rational bad behavior"—"good for Dow Chemical" and "bad for everyone in the unsuspecting surrounding community."
- Many such decisions in the Diamond framework of analysis reflected a lack of assessment of the long-term consequences of decisions.

Key Elements of the Chapter Seventeen Narrative

- The focus is the functioning of the U.S. "system" of constitutional democracy across the sixty-nine year history of Rocky Flats.
- The intellectual structure of this approach to evaluation is based on the contribution by Leslie Gelb and Richard Betts in their study of the workings of the "system" of constitutional democracy in the Vietnam War.
- The first part of their analytical approach is to make an independent assessment of policy choices (which failed in the case of the Vietnam War).
- The second part of their analytical approach is to make an independent assessment of decision-making in the constitutionally based "system" of U.S. government decision-making.

- Prior to April 10, 1970, it was not a constitutionally based "system" of decision-making in the sense that the Atomic Energy Commission and military-industrial contractor Dow Chemical governed and regulated themselves.
- The eleven years between when plutonium was discovered in 1940 to when the Rocky Flats site was selected in 1951 had been shrouded in secrecy.

- The discovery on April 10, 1970 of unauthorized burial of barrels with traces of radioactivity at the Mound Area mobilized the Congressional Joint Committee on Atomic Energy to take the first meaningful decisive action at Rocky Flats by the Legislative branch of the U.S. government.
- The Chapter considers:
  - instances where the "system" did not work,
  - instances where the "system" worked, and
  - instances where the "system" may yet work.
- The Judicial branch of government has become increasingly the deciding force at Rocky Flats to enable the "system" to work when intractable conflicts developed between the Executive branch and an empowered civil society.

Chapter Sixteen investigates decision-making at Rocky Flats. There were three categories of decisions. The first category related to the design of the process for creating new plutonium parts and for reprocessing returned plutonium igniters from actual deployment in the defense of freedom. As was diagramed in Chapter Three Figures 3.1, 3.2, 3.3 and 3.4, first stage plutonium igniters were never faulted for their quality and reliability.

The second category of analysis was whether decisions were disastrous or fateful. Disastrous decisions were only ruinous. Fateful decisions foreclosed what could have better choices for stewards in the future.

The third category of decision analysis was created by Jared Diamond to gain insight through history of why some powerful societies collapsed at the height of their power and glory. The Diamond categories of analysis were introduced in Chapter One and will not be re-reviewed.

The early history of Rocky Flats was characterized by "failure to anticipate" initial problems and "failure to perceive" subsequent problems which had actually arrived. As production accelerated, and the toxic buildup of the site increased, there was an appearance of "too little, too late," "some solutions backfired and made the problem worse," and "some problems in their context did not have solutions."

"Rational bad behavior" often appeared when the military-industrial contractor encountered an imperative to keep igniter production on track to meet War Reserve targets.

In Chapter Seventeen, the basic approach of Leslie Gelb and Richard Betts in The Irony of Vietnam: The System Worked is to separate policy choices from the constitutionally based process of decision-making. With respect to the Vietnam War, their judgment was that the foreign policy choices by the United States had failed, but the constitutionally based processes of decision-making worked.

Application of the Gelb-Betts system to policy choices at Rocky Flats resulted in a rich framework for analysis, as does decision-making in the constitutionally based system. At Rocky Flats there are nuances to consider. The first nuance is that the policy choices worked to fabricate plutonium parts, and to reconstruct nuclear parts which had been previously deployed in defense of America. The technical sequences were never questioned for the quality and reliability of the products. The technical sequences had been developed initially at Los Alamos and the Hanford Nuclear Reservation. An unanswered question is whether those responsible for the initial technical design transferred knowledge about the requirements of the industrial enterprise for toxic waste disposal.

A second nuance was uncovered by Len Ackland and reported in Making a Real Killing: there was a concerted effort in Washington, D.C. to portray the nature of the missions of Rocky Flats as safe and lacking risk in causing "radiation sickness." The point is that by the time in 1982 when Federal District Court Judge Richard Matsch declared Rocky Flats to be "ultrahazardous," it was far too late to redo the drafting of site selection criteria and the flawed application of those criteria to the actual selection of a site for Rocky Flats.

The third nuance is that Rocky Flats was not governed by a constitutionally based system until April 10, 1970. For the first nineteen years of operations at Rocky Flats, the Atomic Energy Commission of the Executive branch under the authority of the Atomic Energy Acts of 1946 and 1954 along with military-industrial contractor Dow Chemical governed and regulated themselves in extreme secrecy.

Based on the background provided by these three nuances, the presentation of whether or not the "system" of constitutional democracy worked at Rocky Flats is organized to present: first the instances when it did not work; second, where based on all the relevant information being in, especially from subpoenaed sources, the "system" of constitutional democracy worked; and third, where it may (hopefully) yet work.

Chapter Sixteen
"Fateful and Disastrous Decisions": Evaluating Decision-Making At Rocky Flats and What Stewards Were Thinking When They Made Them

This investigation of decision-making by stewards of Rocky Flats utilizes the framework created for Collapse by Pulitzer Prize Winner Jared Diamond (2005). This framework was designed to illuminate why such past societies as Easter Island, the Maya, and Angkor failed at the height of their power and glory. The focus in this research approach is on what stewards of Rocky Flats were thinking when *they* made decisions which had such disastrous consequences. "Disastrous" is used in the usual sense of an action resulting in a ruinous outcome or causing great damage or harm. This investigation evaluates decisions in site selection; on the storage, treatment, and disposal of toxic waste; and on the actions before and after the catastrophic plutonium fires of 1957 and 1969.

At Rocky Flats some of the decisions which are disastrous in retrospect were also *fateful*. The definition of "fateful" captures decisions of stewards which constrained positive choices available to future stewards. In some of the fateful decisions, the stewards did not believe that they had any other choice. This was because either earlier decisions had themselves narrowed their options--and if the choices they were making were not ideal, they at least did serve to maintain the production mission at the level required for the War Reserve.

For each of the decisions which were so disastrous or fateful in retrospect, if the reasoning behind a decision is not apparent from the available archives, or if the "best evidence" is still sealed, a sense of historical context will be provided, or collaborative evidence offered to provide an informed observation.

Adapted Diamond Framework For Evaluating Decision-Making

In the introductory chapter to Collapse, Diamond observes that the collapsed societies he studied in detail all left behind monumental ruins which were "spectacular" and "haunting." In parallel, the ruins at Rocky Flats are also "spectacular" and "haunting." Diamond made a significant observation about collapsed societies relevant to understanding the case of Rocky Flats:

> The scales of the ruins testify to the former wealth and power of the builders...yet the builders vanished, abandoning the great structures they had created.

Rocky Flats made a vital contribution to the national security of the United States during the Cold War. That accomplishment in defending America's way-of-life is a further parallel to Diamond's construct of "former wealth and power."

Diamond's statement that "the builders vanished…abandoning the great structures" could be used to describe Rocky Flats ending in a haze of environmental crime and pollution from an industrial enterprise which had used the most deadly substance created by humankind, plutonium, along with other radionuclides and 8,000 substances, almost all hazardous to human and ecosystem health.

Diamond posed two thoughtful questions:

> What were the fates of its individual citizens? – Did they move away and (if so) why, or did they die there in some unpleasant way?
>
> Lurking behind this romantic mystery is the nagging thought might such a fate eventually befall our own wealthy society?

Taking the "romantic mystery" first, it was (and is) actually an unromantic nightmare. Rocky Flats was in too close proximity to a major metropolitan area. It had two catastrophic fires which could have contaminated one of America's great cities and its suburbs. DOE itself determined in 1994 that the two most dangerous buildings in the entire U.S. weapons development complex were at Rocky Flats.

With respect to Diamond's asking what happened to the citizens of these once powerful societies, he asked if they might have moved away. Actually, the former employees of Rocky Flats are like the residents of the immortal song by the Eagles "Hotel California":

> …you can check out but you can never leave.

Workers from Rocky Flats labored in ultrahazardous conditions starting in 1951, but there was not a law passed to compensate them for the development of adverse health effects until 49 years later.

Origins of the "What Stewards Were Thinking" Question

The inspiration for this question is what Diamond is trying to understand:

> What were Easter Islanders thinking when they chose to cut down the last tree of a previously forested Pacific island?

The answer to this question concerning Rocky Flats will be presented by using the explanation of stewards for their choices about decision challenges over the time frame of the production era in terms of categories of the adapted Diamond evaluation framework (presented in Chapter One).

The safe handling by Rocky Flats of completed plutonium igniters and their quality and reliability were never questioned critically by the Special Federal Grand Jury, the U.S. Attorney's Office handling the prosecution of environment crime, or the investigation conducted by the Wolpe subcommittee of the U.S. Congress (Wolpe, 1993).

Rocky Flats also accomplished successfully special classified missions on assignments from the national weapons laboratories at

Lawrence-Livermore, Sandia, and Los Alamos. The exact nature of these special assignments has not been disclosed in the interest of national security. The point here is that Rocky Flats successfully accomplished highly technical new industrial challenges in further weaponizing the atom.

Failure To Anticipate the Magnitude of the Radioactive Nature of Rocky Flats In the Choice of Location

Failure to anticipate is a category in the Diamond framework of analysis which begins at Rocky Flats with the choice of criteria for site selection. Staff members of the Congressional Joint Committee on Atomic Energy considered operation of the industrial enterprise to be "perfectly safe..." None of the site selection criteria contained requirements to consider potential impacts on human health, a cornerstone of the National Environmental Policy Act of 1969 which was to be passed eighteen years later.

Len Ackland (Ackland, 1999, p. 101) quoted a press release from the Atomic Energy Commission which stated that "radioactive sickness" was not a factor at Rocky Flats. This was of course dead wrong. To the extent that it was believed within the AEC, it may be one of the reasons that the monitoring regime was so deficient at the inception for both worker safety and for off-site contamination. In the Diamond framework of analysis, this is an example both of a failure to perceive a problem before it arrived, and failure to perceive a problem which had actually arrived.

Failure To Anticipate the Toxic Waste Burden From the Industrial Enterprise

The records are very clear concerning what stewards were thinking when they made choices in site selection criteria. These choices foreclosed future options for operating the industrial enterprise safely in proximity to a metropolitan area. They were fateful because the small size of the site selected, and its topography with two creeks running through it, and a drop in elevation of 528 feet in the 2 miles from the Western boundary to the Eastern boundary made it impossible to contain the toxic byproducts of the industrial enterprise within the site.

Planners failed to anticipate the sheer volume, dangerous volatility and complexity inherent in the operation of the industrial enterprise. This lack of anticipation was fateful. But once the problems associated with the storage, treatment, and disposal of toxic waste were known, stewards made disastrous decisions about what to do about acute problems as they occurred and continued to occur. For example, a Dow Chemical report (Dow Chemical, 1969) stated that waste management tends to be a "low priority" project because it was "desirable to do as little as possible to get rid of any waste streams."

A further indication of the failure to anticipate problems which actually had arrived was provided by Biles (1970) when he reported to the Colorado Health Planning Council that when Rocky Flats began, "the amount of contaminated waste that would be generated was not predictable with precision." Windahl (1952) indicated that one approach under consideration for "hot waste" initially had been to encase it in concrete for potential burial at sea. Calkins (1970) provided further insight into what stewards were thinking concerning what to do with organic fluids contaminated slightly with radioactive materials: "Little consideration was given to this in the initial design of the plant..."

The pattern of risky and unwise "solutions" to unanticipated problems continued with bad choices made to dispose of toxic waste in steel barrels which were buried either in the Mound Area (1954-1958), or stored above ground on the 903 Pad (1958-1969) or in the Triangle Area (1966-1974). Insights into what stewards were thinking about the toxic waste burden were revealed in a meeting on April 3, 1970 of Dow Chemical management and members of the Rocky Flats Steelworkers Labor Union. In answer to a question of why drums containing toxic waste were buried in the Mound Area, Bill Lee of Rocky Flats management said, "And we didn't know what else to do with them..." Bill Lee also said that while it was not against regulations, "Whether or not it's a good idea is another subject."

In the same meeting, Jim Willging of Dow Chemical management cautioned that: "You see if you start using a place for a burial ground, you can't ever use that place again for anything else." This predated the Superfund legislation. But the magnitude of the problem they were creating for the future clearly was known and understood. The scope of their approach to waste management was put in context by Bill Lee: "I think we have the world's barrel capitol in Rocky Flats."

Calkins (1970) recounted initial efforts by Dow Chemical to avoid either burial or large scale outdoor storage. Recognizing emerging "problems of permanent disposal and increasing quantities," development studies undertaken in 1956 to create a new approach were unsuccessful. Promising initiatives in 1963 were hampered by "design flaws" and "a low financial priority." What began as a failure to anticipate in facility selection and infrastructure design became a "problem beyond their capacity to solve" during the production era.

Failure To Design Effective Executive Branch Solutions

Since the Atomic Energy Commission controlled the ARCO site in Idaho and the Nevada Test Site, there could have been a greater effort to solve the problem of off-site storage of toxic contaminants. It may well be that the contractors from the military-industrial complex did not have either the technology or the on-site space to sufficiently stabilize the toxic materials prior to shipment. Some of the obstacles to off-site shipment were from Department of Transportation regulations.

Years later during the "Accelerated Cleanup," contractor Kaiser-Hill found creative solutions to transportation challenges. The relevant point is that the available archives are silent on attempts at problem solving within the Executive Branch over transportation of Rocky Flats generated contaminants to appropriate off-site storage during the production era. The archives of Dow Chemical documents reveal many instances where immense effort was devoted to trying to find an economical approach to getting small amounts of plutonium out of large volumes of contaminated oil. The choice to start open pit burning of uranium contaminated cooling oil was a recognition that there was not a technology which could accomplish the separation. It may well be that during the complexities and newness of the nuclear age in Cold War context, this problem was not adequately solvable.

Letting Known Problems Get Worse Rather Than Prioritizing Decisive Actions To Solve Them

The "Pondcrete"[7] and "Saltcrete"[8] fiascos were another much criticized and well publicized area of failure to solve a problem. The idea behind Pondcrete and Saltcrete was to mix sludge and concrete into hardened concrete bricks for shipment to the Nevada Test Site for permanent storage. The idea reportedly originated within the Department of Energy, not with the contractor. Even the Second Special Federal Grand Jury Report did not call for the indictment of any of the individuals involved in these fiascos.

But the solution to mix sludge and concrete in a revised formula (with less cement) after initial success (using more concrete in the formula) resulted in a play-doh mush which did not harden into concrete. And it reportedly took two years for the DOE Albuquerque Office to request funding to cover the slumping cardboard containers of Pondcrete and Saltcrete from the effects of open exposure to the elements.

This raises another possible reason why the problem of storage, treatment and disposal of toxic waste proved to be beyond the capacity of Rocky Flats as it was governed to solve: the urgency of the problem could not be solved without Congressional appropriations of adequate financial resources. The contractor could not solve the disposal problem of the failed approach to Pondcrete and Saltcrete without necessary supervisory vigilance. This did not occur, nor did timely corrective action. And after two years, the DOE Albuquerque regional office request for funding was only for an appropriation to fund a cover to shield the play-doh mush from the elements.

---

[7] Pondcrete is a mixture of cement and sanitary sewage sludge. As a result of design errors, some process lines from the production area contaminated the sewage sludge with radioactivity. Therefore, Pondcrete was classed as a mixed-waste. The Nevada Test Site did not have a permit for storing mixed-waste.

[8] Saltcrete is a mixture of low level radioactive waste, cement, nitrate salts and brine. It is a RCRA hazardous waste.

Decision-Making In Toxic Waste Management

The Diamond framework applied to societies which failed separates failure to anticipate a problem before it arrived from failing to perceive a problem that has actually arrived. The approach taken to waste management by Dow Chemical in the first years of Rocky Flats was to under-anticipate the problem but also very importantly the sheer volume, its dangerous volatility, and the complexity of the waste from 8,000 substances and long-lived radionuclides.

Management in the early years found it "...difficult to predict what changes will be made..." In retrospect, it is unfortunate that the meaning of this phrase was not more fully explained. But the rest of the sentence is very clear about the result: "...this tends to develop a crisis-type situation." A consequence of going from crisis to crisis was that:

> ...effort is expended to solve a specific problem, followed by lack of activity until the next crisis.

The adapted Diamond framework will now be applied to what is known about what the stewards were thinking when they made decisions about major waste disposal sites.

The Mound Burial Site

Within the Diamond decision-making evaluation framework, a solution was required that did not exist for the failure to anticipate the magnitude of waste disposal, and the disastrous choice to assign it low priority and then to do as little as possible when the problem was finally perceived. So in the immediacy of the fact that there was even greater quantities of toxic waste coming behind, with no option to halt production until there was a solution, they simply chose to bury it.

Another insight into the failure to develop a solution to waste disposal in the Dow Chemical archives was that originally only three staff people made up the entire waste management group. E.S. Ryan (Ryan, 1953) stated in a report that:

> This is a new group with little background knowledge of problems presented by the material to be handled.

The bottom line is understandable: the problems were new to a small group with "little background knowledge..."

Finally, the Martin C. Biles (1970) communication stated that by the time that it was determined in 1954 that shipment of radioactive liquid waste to Idaho was not a solution, 900 drums of wastes had accumulated "making storage a problem." And the reason storage was a problem was that full production was underway making the generation of more drums of toxic waste a certainty.

Another bottom line of the Biles communication (1970) was that:

> ...drums of waste material were originally sent to the Mound because it was uncertain at the time what the disposition of Rocky Flats waste would be.

This was 1954. The theme will be recurrent in analysis of decision-making on other waste disposal practices: there was no known solution to remove radioactive trace amounts from contaminated oils and solvents. This raises another of the elements in the Diamond evaluative framework for decision-making:

> ...the problem may be beyond the capacity of a society to solve it.

<u>The Trenches</u>

It became routine to dispose of plutonium contaminated oils and solvents in 55 or 30-gallon drums and bury them in the Trenches.[9] D.M. Bassler twice authorized E.S. Ryan to bury on the Rocky Flats site different drum groups on April 12, 1961 (Bassler, 1961a, 1961b):

> Also due to the safety hazard of transporting these chips, it is recommended that they be buried on site [fourteen 55 gallon drums containing uranium chips].

and

> The drums [twenty-seven 55 gallon drums and sixteen 30 gallon drums] should be buried on site since they do not meet shipping requirements.

An additional factor in the burial decision was in the contents of the drums: "...because of probable hydrogen pickup, it is considered a hazard to bring it back in for roasting [in an ore roster] and reduction to oxide."

In terms of the adapted Diamond evaluation framework for decision-making, this was another example of:

> ...the problem may be beyond the capacity of a society to solve it.

The toxic groundwater plumes under the Mound Area and the Trenches are still reported to exist by the Agency for Toxic Substances and Disease Registry in 2005. In terms of the Diamond framework, "some attempted solutions backfire and make the problem worse."

There have been fourteen different burial Trenches identified. But those known as the East Trenches have become the most notorious because of them becoming a prime example of "rational bad behavior"—"good for me and bad for everyone else." In this case, the decision by Rockwell International to spray irrigate 80 millions of gallons of contaminated effluent a year over 17 acres of the East Trenches in all weather conditions solved the problem of one type of waste disposal but exacerbated further on-site contamination by moving the contaminants in the Trenches closer to, if not down to, ground water, and spreading

---

[9] There were ultimately at least 14 confirmed unlined Trenches. Some were as large as 300 feet by 50 feet.

contaminates off-site since the ground above the Trenches could not possibly absorb the massive amounts of contaminated water.

ATSDR (2005) reports that contaminated groundwater remains under the East Trenches site despite the removal of contaminated soil: an example in the Diamond framework of "efforts may be too little, too late."

Radioactive Sewage Sludge

The complexity of the radioactive sewage sludge problem is the reason that it was introduced in a partial transcript of a meeting between Dow Chemical management and Rocky Flats Labor Unions. The transcript revealed that there was radioactivity in North Woman Creek and in a holding pond at a level of 30,000 to 40,000 disintegrations per minute per kilogram of sediment. The source of the radioactivity was drains in the production area which were not supposed to connect to the sanitary drain system but in fact did.

It was revealed also in the transcript of the meeting that radioactive sewage sludge had been "thrown in the trenches and mounds..." By the time that sewage sludge was reclassified and placed in 55 gallon drums and shipped to Idaho in 1972, it was a classic example in the Diamond framework of "too little, too late."

The 903 Waste Storage Pad

Twelve years after the opening of the 903 Pad, K.W. Calkins sent a letter dated August 19, 1970 to Rocky Flats Plant Manager L. M. Joshel detailing the witches brew of plutonium, uranium and deadly chemicals which earlier in the history had been deposited on the exposed 903 Pad with the dry comment that:

> ...leakage of the oil was recognized early...

The failure to act to solve this known problem early combined aspects of "rational bad behavior," and the circumstance that there was no known technology to remove radioactive traces from oils. The "good for me, bad for everyone else" rational choice to burn over 1,000 barrels of contaminated oil in open pits is a related example of "good for Dow Chemical" and "bad for everyone in the unsuspecting surrounding community."

It was short term thinking with long term negative effects on the small site—unmonitored buildup of a toxic waste burden which was in turn "beyond the capacity to solve" for Rocky Flats and still operate the plant in close proximity to a metropolitan area. But the action taken by Dow Chemical to add ethanolamine to the oil "to reduce the corrosion rate of the steel drums" was, in the adapted Diamond evaluative framework, "effort may be too little, too late." By January of 1964, archives of Dow Chemical reveal that:

> Large scale deterioration of contaminated drums was reported with soil contamination increasing.

The 903 Pad became, in the Diamond framework, "beyond the capacity of a society to solve it." A Site Survey Monthly Report (January 12, 1965) indicated, "Some spread of contamination from the waste solvent storage areas [the 903 Pad] from high wind velocities occurring in December."

The amount of contaminated oil which was leaked into the soil of the 903 Pad was estimated in the Freiberg Report (Freiberg, undated) to be 100 barrels or 5,000 gallons. It was not until October of 1969 that monitoring wells were drilled. When Ed Martell first detected off-site plutonium contamination, he thought it had come from the 1969 fire. Instead, Dow Chemical experts told him that it more likely came from the 903 Pad and the 1957 fire.

The Triangle Area

The Triangle Area takes its name from it being literally a triangle of land located between the Northeast Perimeter Road and Spruce Avenue. The Triangle Area opened as a repository for "a large number of drums" when it became necessary to move them from a field "north of Building 883 due to construction of a new decontamination facility in that area" (Barker, 1974). Then in 1969 "accumulated fire waste and residues" were placed in drums and moved to the Triangle Area. All of the drums were stored in the open air until 1971 when they were placed in rail/truck type cargo containers.

The 1968 Waste Management Report of November by R.M. Vogel (1968a) stated that:

> The rapidly increasing number of drums which contain plutonium bearing wastes and residues, and which have been placed in outside storage is a matter of concern to Health Physics. There are 6,000 such drums.

And this was a year before the debris and residues from the 1969 fire were sent to the Triangle Area. Again writing in 1968, Vogel (1968a) stated that: "...no environmental contamination has occurred to date but it is certainly inevitable if the storage becomes long-term." (Storage in the Triangle Area went on for another 6 years). And only one month later than the Vogel report of November of 1968, the Health Physics Report of December of 1968 indicated that: "Some...contamination has resulted from leaking drums."

In the Diamond framework, given all that had gone before, the safe disposal of solvents and oils contaminated with traces of plutonium and uranium were "beyond the capacity...to solve it." The December 1968 Health Physics Report indicated that "high winds have blown over as many as 150 drums at one time." A warning in the 1968 Health Physics Report went unheeded: "The potential of contamination to the environment is of special concern..."

Disastrous Decisions In Building Design and Management Operations Which Contributed To the Ferocity of the 1957 and 1969 Fires

The AEC Fire Investigation Committee Report on the causes of the 1957 fire observed that:

> ...[by hindsight] it appears irrational that facilities were ever designed in such a manner as to permit hazards from a relatively insignificant amount of material to so strongly influence the chain of events which led to a major loss...

The Diamond framework of failure to perceive a problem which has already arrived is applicable. But it is difficult to understand how such disastrous decisions in plant operations could have been made.

The Los Alamos facility, operating originally under the authority of the Manhattan Project, had developed a model of safe facility design for handling radioactive substances, and the operational procedures which employees were mandated to follow on pain of dismissal. Those design and operational procedures were not transferred to Rocky Flats. The Dow Chemical archives are silent on why the "best practices" from Los Alamos were not transferred to Rocky Flats.

The 1957 Fire Investigation Committee Report identified well understood safety procedures in the explosives industry which were clearly applicable to safety at Rocky Flats. These included: limiting the amount of material handled at one location, and limiting the number of people permitted at one location. Certainly the amount of plutonium at one worksite, and how it was stored after the close of production at the end of a workshift, were central causes of the initial ferocity of the fire. The obvious risk had not been identified by stewards.

"What stupid engineer did that?" The question posed by Senator Gordon Allott is one of the most iconic in the entire history of Rocky Flats. It was a response to being told that an engineer, as part of routine maintenance in 1969, had failed to put back a heat detection device where it belonged. Therefore, the 1969 fire which originated from spontaneous combustion was able to accelerate undetected. The engineer made a one-time mistake that turned out to have disastrous consequences.

In the years leading up to 1969, the stewards as a collective made decisions about fire prevention which are disastrous in retrospect. The most inexplicable decision was not to incorporate the known lessons from the 1957 fire into subsequent production operations. The mandate for sprinkler installation in Building 776/777 was not implemented. The combined Building 776/777 was made less fire safe by dismantling the wall separating the two buildings which created one large area that did not impede structurally the spread of fire. The stewards rejected a proposal to create internal structures to impede the spread of fire.

A common denominator of these fateful decisions was not to take seriously in the design of production facilities known lessons from the past, especially those identified in the AEC First Investigation Reports from 1957. It is still a mystery why the first responders to the 1969 fire had not been informed, as a matter of routine policy and procedure, that water had been used successfully in 1957 to quell the ferocious spread of the beginning fire.

The stewards of Rocky Flats between 1957 and 1969 chose to act as though the risk of another fire did not warrant a different response: a classic example of failure to perceive a problem which had actually arrived. The risk was well known regarding spontaneous combustion of plutonium.

In the case of Rocky Flats, the incentive structure disproportionally rewarded production and the allocation of resources toward that end. Five years before the catastrophic 1969 fire, the Manufacturing Manager at Rocky Flats had recommended a reconfiguration of internal space in Building 776/777 to make it more fire safe. Edward Putzier (1982), who is the recounter of the story, indicated that the recommendations by the Manufacturing Manager would have made internal space in Building 776/777 resemble the space configuration at Los Alamos at the end of World War II. The recommendations by the Manufacturing Manager were also consistent with lessons from the 1957 fire. These recommendations were not acted upon. Rejecting the proposal to make the building more fire-safe put workers at greater risk: they could be trapped inside one of the new fire prevention structures and not be able to escape danger.

Five years after the Manager's recommendations were dismissed, the danger in Building 776/777 was indeed deadly. In quantitative terms, the 1969 fire cost $22 million in lost plutonium and $48 million in damages to Building 776/777. This expense brings up Diamond's explanatory category of short-term versus long-term motives. A major reason for not making Building 776/777 more fire safe five years before the 1969 fire was the cost of the retrofit. The cost of the proposed retrofit was not estimated in available documents, but it would have been miniscule in the short-term compared with the damage and long-term loss in 1969 conservatively estimated at $70 million dollars plus the time lost from productivity for the War Reserve. One Rocky Flats manager was concerned about the retrofit impacting operations, and was disastrously wrong in underestimating the risk of fire.

Identifying a Lesson Not Learned

The lesson which Thompson drew and which was not learned in waste management and in fire prevention was that:

> ...the potential for fire to start in any one box may be small, but many of our recent incidents have been events with small probability of occurring.
>
> (Thompson, 1969)

Decision-Making About Fires and Explosions

Fires and explosions were a routine occurrence at Rocky Flats. Between 1954 and mid-1971, 615 fires were officially reported of which 384 were plutonium fires. The Rocky Flats Executive Safety Committee memorialized the officially reported fires in their minutes which were chronicled by Putzier (1982). But the astounding fact is that Dow Chemical made structural choices after the 1957 fire that created the conditions for the 1969 fire to escalate from being a serious fire into being a catastrophic fire. And the brave first responders to the 1969 fire had not been given access to the lessons from the 1957 fire. Further, in the years leading up to the 1969 fire, Dow Chemical management had disregarded many thoughtful recommendations in the Atomic Energy Commission Fire Investigation Report on the 1957 fire.

The 1957 Fire

The 1957 fire began by spontaneous combustion in a flammable Plexiglas plutonium storage glove box which was not equipped with a fire detection device. The fire detection device in the nearby filter plenum had been disconnected to facilitate "operations." There were known design defects in Building 771 including few firebreaks. It was overcrowded. The original gloveboxes at Rocky Flats were modeled on those at Los Alamos with fire resistant steel frames and laminated safety glass. In terms of the Diamond framework, the fabrication decision to install (more) flammable Plexiglas to replace fire resistant steel frames and laminated safety glass was intended to reduce neutron exposure to workers, and not based on some "disastrous value" determination. A consequence of that construction choice of more flammable Plexiglas, however, was to increase overall glovebox flammability.

Two disastrous choices were made to expedite production: first by storage of plutonium in close proximity to the work areas to facilitate production; and second to leave nickel carbonyl cylinders in close proximity to the work station for ease of applying a film coating to the finished plutonium pit. Both of these value choices benefitted ease of production at the expense of fire safety.

Another disastrous choice was to disable heat detection equipment in the filter plenum in order to prevent what AEC (1957) called "operating difficulties" which were being experienced. A consequence was that when the beginning of the fire was discovered by plant protection staff, no alarm had sounded in the Superintendent's Office, nor was the fan system automatically shut down (Dow Chemical, 1957d).

The concern over the use of water on a plutonium fire which could lead to a criticality incident is a prime reason Building 771 had not been designed for water use. Fans were ordered to be turned on high shortly after the fire had been discovered, presumably to clear out smoke; but this led to an acceleration of the fire and an explosion which severed the connection to electrical power. This was a disastrous decision made worse by the earlier decision to utilize flammable Plexiglas in glovebox construction.

The 1969 Fire

Recommendations in the 1957 Fire Report regarding design simplicity were not followed limiting the amount of plutonium to be handled in any one location, and limiting the number of people in one work space. Recommendations for safety in plutonium storage areas were also not followed. No effort was made to reduce the flammability of glovebox construction when the decision was made to not to have the steel frames made the standard which had been used successfully in Los Alamos.

A disastrous decision was made to replace Plexiglas with not the least flammable model which was available. There is in the archives of Dow Chemical a letter from the Dow Chemical Plant Manager, dated December 20, 1957. In it he says that: "...where undue hazard was deemed to exist" substitution was made of glass or metal panels. But he indicated that:

> Plexiglas in the majority of the areas in Building 776/777 does not present an undue fire hazard.

This assessment of risk was to prove disastrously wrong, and was important in turning the serious start of the 1969 fire into one of the causal factors creating a catastrophe.

Three factors about the 1969 fire stand out immediately. First responders were at greater risk than they already were because they had not been given access to the lessons from fighting the 1957 fire. Irrespective of decision category for the failure, this lapse is inexcusable.

A second factor is that sprinklers had been mandated by the AEC for use in all new construction. This requirement applied to Building 776/777 but was ignored. Dow Chemical had made a commitment to install sprinklers, then did not honor that commitment. There is nothing in the available archives to explain this lapse in code enforcement, or why the construction of the roof was so shoddy that it was held in place by concrete blocks. This is a clear instance of "failure to perceive a problem that had already arrived."

The third factor was the almost universal failure to incorporate the extensive lessons from the 1957 Fire Report including the inexplicable failure to incorporate drains to reduce the likelihood of a criticality event.

Utility of the Decision-Making Prism For Disastrous Decisions

The Diamond framework spotlighted categories of decision-making at Rocky Flats which were in retrospect catastrophically wrong. The initial decision-making regarding site selection revealed a lack of understanding of the nature of Rocky Flats as an industrial enterprise, and the appreciation of known industrial processes from Los Alamos and the Hanford Nuclear Reservation.

Many decisions on production practices lacked a long-term perspective regarding consequences. But the cumulative build up of a burden of unmonitored contamination had as a consequence that the site could not be operated legally by the 1980s under the suite of new environmental legal regimes enacted (mostly) in the 1970s.

In terms of Diamond's framework for evaluating decision-making, there was certainly a failure to anticipate the problem of disposition of the sheer volume, complexity, and volatility of contaminated waste from the industrial processes associated with nuclear weapons production. Attempts to solve the problem were not successful once the extent of it was perceived.

Summary of the Applicability of the Diamond Framework On Decision-Making

A Solution Backfiring and Making the Problem Worse

The Mound Area as a site for burying drums of toxic mixed-waste with the steel drums corroding and leaking their contents into the ground was clearly an instance of a solution backfiring and making the problem worse.

Some Efforts Too Little, Too Late

The storage, treatment, and disposal of toxic waste, in Diamond's decision-making framework, is an instance both of some solutions backfiring and making the problem worse, and some efforts being "too little, too late." Plutonium let loose in the environment is toxic for 240,000 years.

Belief In the Power of Science, Engineering and Technology

Based on a belief in the power of science, engineering and technology to address successfully unsolved problems encountered in production of nuclear weapons, Rocky Flats expected to be able to overcome previously unknown problems of nuclear weapons production. An illustrative case where that did not happen involved uranium scrap. In order to be able to ship uranium scrap to ARCO in Idaho, it first had to be converted to uranium oxide. The problem which could not be overcome technologically was that the uranium was so contaminated with numerous inert materials that it was "difficult to convert to oxide."

Problem Beyond the Capacity of Rocky Flats To Solve

In the Diamond decision-making framework, the problem with uranium scrap and uranium traces in cooling oil was a problem beyond the capacity of Rocky Flats to solve. With respect to the uranium scrap which had continued to be generated, the solutions to the uranium waste problem which were adopted only served to make matters worse. Burying the uranium scrap contaminated the ground. Burning the traces of uranium in cooling oil in secret in open pits contaminated the air without notification to those who could use the warning to protect human and ecosystem health (as if no other solution had been possible but to burn uranium in oil in open pits).

Failure To Anticipate a Problem Before It Arrived Is Reflected In the Low Priority Assigned To Early Waste Management

The low priority assigned to early waste management was a quite unexpected finding. An early meeting of the Waste Management Group was held with operating units for the "...purpose of obtaining a working knowledge of waste disposal operation over which the unit is to have control." Further, the first meeting focused only on "tentative procedure" regarding "...how the waste disposal problem should be handled." The date of this meeting was January 8, 1953, after the industrial production operations were well underway.

In the sequence of time, the depth and scope of waste management was clearly an instance of a failure to anticipate a problem before it arrived. The time to anticipate waste management was during the formulation of site selection criteria. This was a massive failure of process. Selecting a site suitable for waste management was not among the site suitability criteria.

Another example of the failure to anticipate a problem before it arrived was in original plant design. This went beyond the site selection errors. K.W. Calkins wrote to Dow Chemical Plant Manager L.M. Joshel on August 19, 1970 (Calkins, 1970) documenting early approaches to waste management. He highlighted one of the subsequent worst problems, "...organic liquids contaminated slightly with radioactive materials," stating that "little consideration was given to this in the initial design of the plant." The problem of the storage, treatment, and disposal of organic liquids contaminated slightly with radioactive materials was considered neither in site selection nor in facility design and constitutes a classic "failure to anticipate."

Disastrous Choices Do Not Necessarily Reflect Disastrous Values

The Diamond framework addresses disastrous values. What happened in waste management in the early years at Rocky Flats is more nuanced than the label "disastrous values": disastrous choices did not necessarily reflect disastrous values. A case in point was recounted in a letter from L.R. Drake to R.R. Harrison on January 5, 1955 (Drake, 1955). At issue was the disposal of radioactive waste (D-38 oxide) with an isotope level too high to send to ARCO in Idaho. The disposition was to place this waste in metal drums for storage at Rocky Flats, and then bury it.

Consequences of Not Recognizing the Danger of High Winds In Site Selection

The Owen and Steward Report (Owen and Steward, 1974) on leaking drums of mixed-waste recounted that high winds had blown the lids off waste drums, thereby releasing uranium onto a loading dock, a driveway, and the surrounding ground. This revealed that there had been a failure to recognize high winds (and their direction) in site selection. This is a classic illustration of failure to anticipate in the Diamond framework.

New "Species of Trouble" For Humankind
This chapter is framed by the global context and consequences of the imperative for providing for the security of the country and to weaponize the atom. The thinking of Dow Chemical was examined in the beginning years of Rocky Flats that led to disastrous decisions, and superimposed upon that record the approach to analysis by Jared Diamond of why previous memorable societies in history made bad and disastrous decisions and subsequently collapsed.

Of course the United States is not going to collapse as did earlier societies of the ancient past such as Easter Island, the Maya, and Angkor. But this approach to analysis reveals that there are common threads of failure to anticipate problems before they arose, failure to address them satisfactorily once they were recognized, and that for some problems, given their historical context and decisions that had been made before, there simply may have been no feasible alternative solutions.

Technological Disasters Have Grown In Their Capacity for Catastrophe
In Kai Erikson's concept of a "new species of trouble" for humankind, technological disasters have grown in number and potential for catastrophe as humans have tested the limits of their competence. Technological disasters which have occurred over the years at Rocky Flats relate in important ways to extending the limits of knowledge and competence under the Cold War pressure to meet the imperative to produce ever increasing numbers of plutonium igniters.

Technological disasters such as those which have occurred over the years at Rocky Flats when systems failed, human beings made errors, essential apparatus misfired, and designs proved to be faulty in actual operations, is to name just a few of the causes of disasters at Rocky Flats as humans overreached their competences in striving to achieve a level of mandated production required for national security.

Distinction Between Natural and Technological Disasters
As introduced in Chapter Two, Kai Erikson points out a key distinction between natural disasters and technological disasters. While natural disasters can be thought of as caprices of nature, technological disasters are of human manufacture, and in concept, are preventable. A consequence, as Erikson points out, is that there is a story to tell about them, a moral lesson to be drawn from each of them, and a share of blame to be assigned. At Rocky Flats, once the veil of secrecy was (only partially) lifted, there have been many moral lessons drawn, and there has been even more blame assigned.

"Radiation's Lingering Dread" and the Diamond Framework
The Diamond notion of "rational bad behavior," coupled with Erikson's notion of "radiation's lingering dread," is relevant to two elements of decision-making at Rocky Flats. The first is the limited perception by nuclear experts of the risks associated, for example, with

open-pit burning of uranium contaminated oils. It was an acceptable choice only from the point-of-view that open pit burning of toxic chemicals plus radionuclides in close proximity to suburbs of a major metropolitan area does indeed reduce the volume of contaminated oil.

The decision to embrace open-pit burning was done in secret. Those down-wind did not even know to protect their own health, and to the extent possible the health of their ecosystem, and water supply for cities in the Denver metropolitan area.

Chapter Seventeen
The American "System" of Constitutional Democracy: Where It Worked, Where It Did Not Work, and Where It May Yet Work

The U.S. "system" of constitutional democracy as it has functioned across the sixty-nine year history of Rocky Flats is the subject of Chapter Seventeen. As stated in Chapter One, the idea to utilize this evaluative framework of whether or not the "system" worked was inspired by the approach of Leslie Gelb and Richard Betts in their 1979 book: The Irony of Vietnam: The System Worked.

Evaluating Policy Choices Separately From Constitutional Decision-Making

The basic Gelb and Betts (1979) approach is to separate policy choices on foreign policy (in the Vietnam War which failed) from the constitutionally based processes of decision-making (which worked). This approach applied to policy choices at Rocky Flats over the sixty-nine years of its history (and counting) results in a rich framework of analysis. For example, under the legal frameworks provided by the Atomic Energy Acts of 1946 and 1954 in a context with mandatory Cold War secrecy, there was an imperative to establish the U.S. nuclear weapons multi-state production complex.

At Rocky Flats the "system" failed to consider in site selection criteria requirements of the industrial enterprise to produce plutonium igniters. This failure was based on decision-making only within a subset of the Executive branch. But James R. Seed (1971) reminded his audience in the time frame of 1971 that plutonium had only been discovered in 1940. Rocky Flats had been sited only eleven years later. Those intervening years between 1940 and 1951 were shrouded in extreme secrecy.

Contrasting Governance Structures for Nuclear Power and Nuclear Weapons

In the Atomic Energy Act of 1954, the more public processes of the Nuclear Regulatory Commission to supervise the development of nuclear power were deemed by both the Legislative and Executive branches of U.S. "separate but equal" government to be different from the proper supervision model for nuclear weapons development. Seed (1971), in a most thoughtful reflection, indicated that he and his colleagues were making decisions in a field of specialization which was only five years older in terms of technology and understanding than the context where they were forced by circumstances to take action. If Rocky Flats stewards learned more from experience they might have made different policy decisions than they had in the early years.

Policy Makers Confronted a Reality That the Nuclear Age Is Unforgiving of Mistakes

Rocky Flats policy makers were faced with a dilemma that the "new species of trouble" posed by the nuclear age was unforgiving of mistakes. Mistaken policy choices in the early years, with the perspective of hindsight, foreclosed the range of policy choices available to subsequent stewards on protecting environment, health and safety.

Further, after 1970 when plutonium was discovered not to have been contained where it was supposed to be contained, the capital investment at Rocky Flats in the industrial enterprise by then was deemed by the U.S. government to be too large to relocate the facility. The requirements of the War Reserve necessitated an uninterrupted flow of plutonium igniters.

The Date of April 10, 1970 Was a Watershed For the First Beginnings of "Separate But Equal" Constitutional Engagement

The reader will recall the secret meeting of April 3, 1970 between representatives of Dow Chemical management and Labor Unions at Rocky Flats. One theme in the meeting was that early policy decisions would have been made differently based on subsequent knowledge. But, of course, with long-lived radionuclides there was not an option to redo an earlier bad policy choice.

The imperative of keeping plutonium igniter production uninterrupted surfaced in another context. Attorney Howard Holme, from the Church cluster of lawsuits, recalled that he had taken a statement from a long time Dow Chemical manager whose employment with the company predated Rocky Flats. When Mr. Holme showed the long term Dow manager memos he had written ordering the burial of barrels of contaminated mixed-waste on-site in the Mound Area, the manager was appalled at what he had done. He had a background for understanding the negative consequences. But that incident is a glimpse into the power and pressure of the imperative to keep production uninterrupted. Secrecy in the years leading up to the discovery of off-site plutonium in 1970 meant that accountability to the Legislative branch had been limited to general presentations as part of the appropriations process.

There was an abrupt shift in accountability over actions at Rocky Flats on April 10, 1970. That was the day that Jim Kelly of the Steelworkers Union told the Congressional Joint Committee on Atomic Energy in Washington, D.C. that the Mound Area had been the location for the burial of barrels containing traces of radionuclides along with cooling oil. The full story is recounted in Chapter Five. Within several weeks, the Mound Area had been excavated, and ceased to exist. This was meaningful Congressional oversight of the Executive branch supervised military-industrial complex contractors at Rocky Flats.

In the same general time frame of 1970, a shift in societal consciousness was reflected in the bipartisan consensus to establish a new legal framework for environmental protection with both civil and criminal penalties. Applicability of these new laws remained contested terrain into this current era, especially the meaning of provisions of the National Environmental Protection Act of 1969 in determining future uses of the former 6,500 acres of Rocky Flats.

Both the Legislative and Judicial branches have been meaningfully engaged since the 1970s in determining the applicability of special environmental laws in particular situations. It is possible to determine after 1970 where the "system" of U.S. constitutional government worked, where it did not work, and where it may yet work in different time periods with April 10, 1970 as a watershed.

Identifying Locations Within This Study Where the Background Is Developed Fully Regarding the "System"

Most of the instances when the "system" failed at the policy level and at the decision-making level occurred when in secret the AEC and Dow Chemical governed and regulated themselves. These instances have been described in earlier chapters.

The Legislative branch became more actively involved at Rocky Flats on April 10, 1970. When the Executive branch was not cooperating with their constitutionally authorized oversight investigation, the Wolpe subcommittee (Wolpe, 1993) made very effective use of its power of subpoena to compel disclosure of previously secret information in 1992. The Judicial branch became involved with its power of subpoena in the Church cluster of lawsuits in 1975. Attorney Howard Holme (1978) prepared a document of over 520 pages using a process of discovery within the archives at Rocky Flats

After the FBI raid of June 6, 1989, the Department of Justice of the Executive branch asked a Federal District Court Judge to empanel a Special Federal Grand Jury. This entire environmental crime Epoch is described in detail in Chapters Eight through Eleven. As reported in Chapter Fifteen, it took 26 years to resolve the Cook class action lawsuit from 1990 until 2016 with ten Memorandum Opinions and Orders of Federal District Court Judges with appeals to the 10th Circuit Court of Appeals and to the U.S. Supreme Court. The U.S. "system" of constitution government has been active in varying degrees after April 10, 1970.

An instance after April 10, 1970 when the "system" actually failed was the corrupt relationship related to the bonus culture involving the Department of Energy at Rocky Flats and Rockwell International. Another instance where the "system" failed was the choice made by a Federal District Court Judge in concert with the Department of Justice to actively misrepresent findings of the Special Federal Grand Jury.

INSTANCES WHERE THE "SYSTEM" FAILED

The "System" Failed Between the AEC and Dow Governing Themselves At Both Policy and Decision-Making Levels

The "system" of policy choices and decision-making failed at Rocky Flats in the establishment of criteria for site selection and in the application of flawed criteria used in the site selection process.

The Manhattan Engineering District and the Atomic Energy Commission already had extensive knowledge about the siting of nuclear weapons production facilities. The most relevant was the design of the criteria and their applications to the Hanford Nuclear Reservation. In part reflecting the concerns of the President of the DuPont Corporation in 1942, the "unknown hazards" involved influenced key decisions about size of the site and layout of the industrial enterprise. The site selected for Hanford was 670,000 square miles, not four square miles. The hazardous production area was 200 square miles, not 334 acres. Key production areas were separated at Hanford so that a disruption in one key plutonium creation or processing site would not impact other sites.

While the missions of Hanford and Rocky Flats are quite different, Hanford fabricated plutonium igniters until 1965 at which time Rocky Flats became the sole mass production supplier. The point is that the AEC had direct experience with igniter fabrication and necessary knowledge of the nature of the industrial enterprise, and toxic byproducts which were not reflected in the criteria used to select Rocky Flats and the design of its industrial enterprise. Once the flawed site selection criteria were established without incorporating past relevant AEC experience, fatal flaws emerged in how they were applied. Rocky Flats was in fact upwind and uphill from Denver and its suburbs, not downwind and downhill as the site selectors believed.

In violation of the criteria mandated by General Leslie H. Groves, Rocky Flats was selected to be sited in a location closer than 25 miles from a metropolitan area. Denver in 1951 was only 16 miles from the Rocky Flats site and rapidly expanding.

Site selection criteria recognized the need for a reliable source of 1 million gallons a day of water as essential for the functioning of the industrial enterprise. But there was a disconnect in considering the topography of the site which dropped 528 feet from West to East. The presence of two creeks which traversed the site essentially guaranteed that off-site contamination was inevitable. The genesis of environmental crime had its origins in the disposal of contaminated water.

The AEC's and Dow's Governing Themselves Failed To Work At Policy and Decision-Making Levels from 1951-1974 Over Storage, Treatment and Disposal of Toxic Waste

The "system" of Executive branch government failed on both policy and decision-making at Rocky Flats over the lengths which Dow Chemical had to go to in the storage, treatment and disposal of toxic waste. Operation of the Mound Area, the 903 Pad and the Triangle Area are but examples of the broader challenges which had to be met to keep the production mission on track. The "system" of the Executive branch governing itself in decision-making did not address toxic waste with appropriate concern:

- Atomic Energy Acts of 1946 and 1954 provided a framework for weaponizing the atom. Neither establishing Act provided environmental protection
- Research of Len Ackland as reported in Making a Real Killing revealed that staff of the Congressional Joint Committee on Atomic Energy considered "Project Apple" (the code name for what became Rocky Flats) so safe that it could be located on 16th Street in Washington, D.C.
- Site selection criteria noted the 1 million-gallon daily water requirement with no consideration of disposal requirements.
- The choice of sites chosen as finalists were reviewed by the same group that made the original choices.
- The Austin Company from Ohio in secret made wind measurements at Stapleton Airport where the wind conditions were very different from those immediately next to the Front Range of the Rocky Mountains where the Rocky Flats site was located.

It was a different era in a dangerous Cold War global context when early disastrous decisions were made. There is no reference to consultation with those responsible for siting decisions and facility layout and design at Los Alamos or Hanford in the available Dow Chemical archives or in the discovery conducted by Howard Holme (Holme, 1978) for the Church cluster of lawsuits.

The "System" At Both the Policy and Decision-Making Levels Failed To Make Building 776/777 More "Fire Safe" After the 1957 Fire (1957-1969)

The "system" failed at many levels. One high-level manager for Dow Chemical was concerned that a retrofit to more fire-safe Plexiglas would interrupt production. He also made a judgment which was to prove disastrously wrong: he underestimated the risk of fire. He and both AEC and Dow Chemical management failed to heed a significant lesson across the production era identified by Dr. M.A. Thompson in 1969: "...the potential for fire to start in any one box [glovebox] may be small, but many of our recent incidents have been events with small probability of occurring" (Thompson, 1969).

The Manufacturing Manager five years before the 1969 fire had recommended a more fire-safe reconfiguration of internal space in Building 776/777. Putzier (1982) indicated that this plan was comparable to some internal space arrangements at Los Alamos. Rejection of this plan was in part over concern that workers could be trapped. The opposite was done. To facilitate operations, the wall between Building 776 and 777 was removed which created one large space in which fire could spread uninterrupted.

There was no constituency or known corporate incentive for requesting more money for safety related expenses. The 1957 Fire Investigation Committee Report said the use of flammable Plexiglas was one of the major factors contributing to the spread of the fire and its severity:

> The use of Plexiglas in glovebox construction without metal separators between Plexiglas panels contributed to the spread of the fire. Those units most completely destroyed were those composed of Plexiglas construction without metal frames.
>
> (AEC, 1957)

The reason for this fateful choice of flammable Plexiglas was explained:

> The original dry box equipment was stainless steel-laminated safety glass; however, in some areas, Plexiglas was used due to the difficulties of fabricating enclosures…"
>
> (AEC Fire Investigation Committee Report, 1957)

Very relevant lessons were there to be learned from explosives manufacturers which could have been followed, but were not:

> Explosive manufacturers commonly limit the maximum possible risk at any single location by limiting the amount of material handled and the maximum number of people permitted at one location.
>
> (AEC Fire Investigation Committee Report, 1957)

Dow Chemical after the 1957 fire made a number of commitments it did not carry out:

1. Markedly decrease the quantity of residue created per unit weight of parts produced,
2. Eliminate or minimize the use of flammable materials in construction,
3. Due to the problems of roof construction…Building 76 [776] will be sprinkled.

This latter commitment of roof sprinkler installation did not occur. Available records do not substantiate that the other commitments were met.

The "System" Between the AEC, Dow Chemical, Department of Energy and Rockwell International Failed to Prevent Three Preventable Catastrophes From 1951-1993

The "system" of Executive branch governance at both the policy and decision-making levels failed to prevent three preventable situations from becoming unthinkable occurrences. These unthinkable occurrences were not "acts of God." Quite the contrary. All three were preventable with their own trajectories of causation. It is possible to assign blame and a level of moral culpability.

The preventable situations which became transformed into unthinkable occurrences were the catastrophic plutonium fires of 1957 and 1969, and the discovery that over 2,600 pounds of plutonium was either missing or unaccounted for—a discovery made possible only after a Federal District Court Judge had found that the Department of Energy was in Contempt of Court.

Catastrophic Fire of 1957

The fire of September 11-12, 1957 in Building 771 was the ninth most expensive loss in the history of the Atomic Energy Commission. Building 771 was the main plutonium building. Dow Chemical had replaced the original gloveboxes made of fire resistant steel and safety glass with a form of flammable Plexiglas. The fire began by spontaneous combustion in a flammable Plexiglas plutonium storage glovebox which was not equipped with a fire detection device. The fire detection device in the nearby filter plenum had been disconnected for "operational" reasons; namely, to facilitate production.

The Fire of 1969

An astounding fact is that significant lessons from the 1957 fire were not known to the first responders to the 1969 fire including ones relevant to their own safety. The use of water to quell successfully the initial ferocity of the 1957 fire was not known to the first responders who were already risking their lives to respond to the 1969 fire.

The fire of 1969 was not an event with a "small probability of occurring." There are two sources of documentation for the fact that two different levels of the Atomic Energy Commission in March of 1969, two months before the May 1969 fire, had independently established the presence of an unacceptable level of fire risk.

Between March 11 and March 15, 1969 an official of the AEC Industrial Safety and Fire Protection Branch reported that there was "practically no built in fire protection or suppression equipment." Also in March 1969, safety personnel from the AEC Regional Office in Albuquerque had discussions with Dow officials on fire risk who acknowledged that "Building 776-777 represented a serious fire hazard and inquired when AL [the AEC Regional Office] was going to require that something be done about it." A cryptic note in the AEC 1969 Fire Investigation Committee Report states that: "no report or recommendations were made to AL [Albuquerque AEC] management."

The key point is that the "system" clearly failed with respect to whether the "system" worked to produce action within the Executive Branch and between the Executive Branch and its contractor Dow Chemical to reduce the risk of a serious fire becoming catastrophic.

"What Stupid Engineer Did That?"

This iconic question posed by a U.S. Senator from Colorado in the aftermath of the 1969 fire reflected U.S. "system" of constitutional government constructively engaging the Executive branch of government over Rocky Flats. The U.S. Senate Subcommittee on Deficiencies and Supplementals of the Committee on Appropriations held a hearing on November 21, 1970. Senator Gordon Allott asked if there was a backup fire detection system in the plutonium storage areas. He was told that:

> ...the heat detecting devices that are normally inside had been placed elsewhere, and were insulated so that they did not give the warning in time.

Senator Allott asked AEC top management: "What stupid engineer did that?"

Urgency of Upgrading the Facilities At Rocky Flats

On October 1, 1970, General Edward Giller on behalf of the AEC appeared before a subcommittee of the House of Representatives Committee on Appropriations seeking funding for an urgent upgrade for Rocky Flats after the 1969 fire. In his testimony, General Giller reported on a "gradually deteriorating situation."

Are We Subject To the Charge of Negligence?

A thoughtful answer to his own question by Congressman Glenn Davis spoke to the immense challenge of making the "system" work.

> Congressman Glenn Davis: Are we fairly subject to the charge of negligence in failing to have some of these basic things [fixed] at these installations? Nobody would permit operations to go on without firewalls and automatic sprinklers and things of this kind unless they were under the jurisdiction of the Federal Government where they can do nothing about it, I suppose.

Missing and Unaccounted For Plutonium

It was only the authority of the Judicial branch of government which compelled the Department of Energy to reveal the extent of missing and unaccounted for plutonium at Rocky Flats between 1952 and 1993. This revelation about the total amount was made only after the Department of Energy had been found to be in Contempt of Court by a Federal District Court Judge as part of the Cook class action lawsuit. It took about 3 years of litigation before the total was revealed.

The ultimate revelation that 2,600 pounds of plutonium were missing or unaccounted for is especially disquieting in the context that for years there had been public assurances by military-industrial contractors that plutonium had indeed remained on-site where it was intended to remain.

The citation for Contempt of Court originated because the Department of Energy itself had revealed in June 1994 that there was plutonium and uranium missing and unaccounted for from Rocky Flats. The Department of Energy had first indicated that there were only 11,000 pages of relevant documents and they were "restricted data" under the Atomic Energy Acts that could not be used in judicial proceedings. Nearly 3 years later it was disclosed that there were over 677,000 pages of documents detailing 15 different types of missing and unaccounted for plutonium. The crisis over the future uses of the lands of the Wildlife Refuge at the end of 2019 is exactly about where the uncontained plutonium now resides, and at what level of picocuries per gram of soil.

The "System" Did Not Work: Incentives Within the Bonus Culture

The Second Citizen Special Federal Grand Jury Report was very critical of the "bonus culture" relationship between DOE Plant Management and Rockwell International Plant Management. The backstory is that under the governing contract, the amount of the annual bonus for Rockwell International was to be determined by the DOE Albuquerque Office (available records do not indicate how the bonus system, if any, worked for Dow Chemical). According to the second report of the Grand Jurors, Rockwell International would basically "grade itself." They would forward their performance evaluation document to the DOE Rocky Flats Manager. He would make such adjustments as he saw fit, then forward his recommendations (in the opinion of the Grand Jurors mostly Rockwell International's) on his DOE letterhead to the DOE Albuquerque Office.

That "system" was overturned by Secretary of Energy James Watkins in the Fall of 1989. The bonus for the previous year was determined by the Washington, D.C. Office of DOE. When Rockwell International received only 40 percent of the possible bonus, they filed an unsuccessful lawsuit on the grounds that it was a term of their DOE contract that the amount of the bonus was to be determined by the DOE Albuquerque office.

Because the Second Citizen Special Federal Grand Jury Report did not become known until the publication of this study, and the era of Rockwell International at Rocky Flats ended at the end of 1989 along with the production era, it is not possible to reconstruct with any precision what the various factors were in the role of the bonus system in shaping contractor behavior. The system of performance evaluation at Rocky Flats did fail with respect to environment, health, and safety, but did succeed in the successful production of plutonium igniters.

The "System" Did Not Work: An Opportunity Was Lost to Inform the Public About Conditions At Rocky Flats When the Grand Jury Exhibits (1992) and Results of the FBI Raid (1989) Were Sealed

It has never been disclosed if there were national security concerns behind the decision by the government to seal the more than 2,000 exhibits presented to the Grand Jury, or to seal the results of the FBI raid. The point is that quite workable procedures exist to inform the public while protecting national security, and they were not enacted.

A mystery remains as to the motivation behind the decision by the U.S. government first to choose to seal the contents of the FBI search warrant, then abruptly to unseal it while the FBI raid was still in progress, therefore informing Rockwell International and DOE officials at Rocky Flats about what the FBI and EPA were seeking to find in support of their "probable cause" affidavit. But the release of the Search Warrant while the raid was in progress served to notify officials at Rocky Flats about what was being investigated.

Top officials of both DOE and Rockwell International had signed affidavits under penalty of perjury attesting to their first hand knowledge that what they were affirming were true facts known personally to them that Rocky Flats was in compliance with the environmental laws and regulations of the country. The "system" under the rule of law did not hold these individuals to account.

The "System" of U.S. Constitutional Government Failed to Work At the Policy and Decision-Making Level When a Federal District Court Judge From the Judicial Branch and the Department of Justice From the Executive Branch Deliberately Misrepresented the Findings of Citizens Empaneled Under Oath On a Special Federal Grand Jury

It is absolutely a failure of the "system" of U.S. constitutional government for a Federal Judge from the Judicial branch, in apparent concert with the Department of Justice of the Executive branch, to release a highly misleading version of the report of the Special Federal Grand Jurors. The Federal Judge claimed the Grand Jury report failed to meet "necessary standards." While the person or persons unknown who leaked what is called in this study the First Special Federal Grand Jury Report exposed the misrepresentations, it was the Legislative branch which made the "system" work as well as it did (see the next section).

INSTANCES WHERE THE "SYSTEM" WORKED

The "System" Worked: Contributions of the Federal Civil Court System and Its Subpoena Power (1975-Continuing)

The civil court system, as an integral part of settling disputes within American democracy, eventually functioned to produce a settlement of disputes in the 1975 Church lawsuits. Plaintiffs in the 1990 Cook case were able to obtain millions of pages of documents from the previously closed vaults at Rocky Flats.

The "System" Worked: On the Vital Role of Citizen Non-Governmental Organizations (1970-Continuing)

Organized civil society preserved archives of Dow Chemical documents from 1951 through 1974. These archives and Holme (1978) reveal that the "Achilles heel" of Dow Chemical stewardship was over the storage, treatment, and disposal of hazardous waste, the failure to inform first responders to the 1969 fire of the lessons from fighting the 1957 fire, and the inability to take effective measures to prevent events with a low probability of occurring from turning, for example, small spontaneous combustion fires in 1957 and 1969 into ferocious catastrophic fires. Civil society had not been informed enough to mobilize until after the 1970 revelations by Martell and Poet (1970).

The "System" Finally Worked: In the Struggle Over the Rule of Law at Rocky Flats (1975-Continuing)

The stakes have been extremely high in the struggle over the rule of law at Rocky Flats. The entire American nuclear enterprise covering both the development of nuclear weapons and of nuclear power had been established under the rule of law as embodied under the Atomic Energy Acts of 1946 and 1954. The provisions of the Atomic Energy Act of 1946 gave great latitude to the Atomic Energy Commission and the successor Department of Energy when substances involved were radioactive (even trace amounts). In the climate of Cold War secrecy, the de facto circumstance was that citizens and independent scientists knew very little about what went on inside of nuclear weapons production facilities. Provisions of the Atomic Energy Act were essentially unchallenged as matters of law and practice for most of the production era.

The consciousness of American society began to change with the 1962 publication of Silent Spring by (then) little known author Rachel Carson. Ms. Carson's two previous books, The Edge of the Sea (Carson, 1955) and The Sea Around Us (Carson, 1951) had been well received by a more specialized readership, but had not touched the national consciousness with the message in Silent Spring.

Carson warned in Silent Spring of the dangers to all living things posed by new pesticides and insecticides developed by the chemical industry when their use was indiscriminate. Efforts to discredit her by the chemical industry stating that she lacked scientific substance and intellectual integrity had the unintended consequence of bringing even more attention to her message. Although Rachel Carson died of cancer in 1964 less than two years after Silent Spring was published, her work became a central wellspring of the transformation of consciousness in America symbolized by the first Earth Day in 1970. Establishment by the Nixon administration of the Environmental Protection Agency and a suite of new bipartisan environmental laws starting with the National Environmental Policy Act of 1969 further advanced awareness related to protecting the environment. President Ronald Reagan bestowed the Presidential Medal of Freedom posthumously on Rachel Carson in 1981, an indication of the significance and lasting contribution of her work.

The struggle over the application of the rule of law began slowly over the application of the Resource Conservation and Recovery Act and the Clean Water Act to the Nation's nuclear weapons production complexes and to Rocky Flats in particular. This struggle was not present during the Dow Chemical era. The Dow Chemical archives do not contain references to the applicability of the new legal regime under their stewardship of Rocky Flats.

There are at least three sources documenting the legal struggle during the Rockwell International era. The first source of documentation is in the draft of the Second Citizen Special Federal Grand Jury Report (Chapter Eleven). The Grand Jurors wanted to indict Rockwell lawyers at Rocky Flats for developing the rationale that a part of the ruling in a case known as *American Mining Congress* exempted Rocky Flats from RCRA because plutonium residues were part of a "continuous manufacturing process." The counter-argument in the struggle over the application of RCRA was that plutonium residues had been in storage for decades and could not possibly be considered part of a "continuous manufacturing process." This controversy had not been settled at the time Rockwell International ceased being the contactor on December 31, 1989.

The second source of documentation begins with the historical record that neither Dow Chemical nor Rockwell International had been able to develop an economical method to extract traces of plutonium or uranium from contaminated oil. It is an environmental crime under RCRA to store unpermitted waste beyond a short time period. The short time period had long since expired. The bind the contractor encountered was that liquid waste could not be shipped to Idaho in an unstabilized form. Without the ability to stabilize it, they could not have obtained a legal permit to store it on the Rocky Flats site even if they had tried.

The third source of documentation on the struggle over the applicability of the Nation's new environmental laws was provided by the preeminent criminal defense firm, Williams and Connolly of Washington, D.C. Williams and Connolly had been retained by Rockwell International. In 1991 they wrote to leaders of the Department of Justice with copies to lead prosecutors in the U.S. Attorney's Office in Denver, Colorado.

Williams and Connolly made two essential points in defense of Rockwell International. The first point was indisputable: the decade of the 1980s had been a period of "regulatory confusion" about the applicability of provisions of RCRA to Rocky Flats. The second point was equally indisputable: Rockwell International had ceased being the contractor at the end of 1989, and the definitive ruling on the applicability of RCRA to Rocky Flats did not come until Judge Lewis Babcock made a final ruling on April 12, 1990 in the Sierra Club case.

The U.S. Attorney's Office in Denver already had issued "target" letters to individuals under scrutiny by the Special Federal Grand Jury. A decision had been made by officials in the Department of Justice in Washington, D.C. to accept a plea bargain from Rockwell International as a corporation. Through its lawyers in 1991 and 1992, Rockwell had signaled that it would mount a spirited national security defense that it was doing exactly what the Department of Energy, on behalf of the U.S. government, had wanted it to do at Rocky Flats, and that it had the documentation to substantiate that fact. Rockwell subsequently proved this fact as part of the discovery process in the Stone whistleblower civil case in 1999. The Department of Energy listed seven instances where Rockwell International had not informed it of important matters. Rockwell International's reply was to identify one hundred and sixty-five times when the Department of Energy had been fully informed of Rockwell's actions at Rocky Flats.

The point of Rockwell International's lawyers was that the production era ended in 1989 under Rockwell's stewardship without a legal resolution of the applicability of RCRA to Rocky Flats. The settlement of charges of criminal violation of environmental laws ended with a plea bargain which excluded individuals and which allowed Rockwell International to continue receiving military-industrial contracts.

The "System" Worked: Transformation of the Case of Rocky Flats By Constitutional "Checks and Balances" Between the Executive, the Legislative, and the Judicial Branches of the U.S. Government (1970-Continuing)

The Legislative branch made a determined foray into the thicket of Rocky Flats when the Wolpe subcommittee began a series of extensive public hearings in 1992. With respect to insights into functioning of American constitutional democracy, the Wolpe subcommittee exercised its power of subpoena to force the disclosure of documents being withheld by agencies of the Executive branch. The Wolpe subcommittee published invaluable volumes of analysis, documents and testimony based in large part on its power of subpoena. This book on The Saga of Rocky Flats has been informed importantly by documents subpoenaed by the Wolpe subcommittee.

The next "separate but equal" foray into Rocky Flats was by the Department of Justice. The Department of Justice obviously objected to the Congressional characterization that it had denied the right of the American people to "know the truth" about Rocky Flats. Another key accusation by the U.S. Congress was that the Department of Justice had acted in such a manner that the momentum for educating the public to the need for fundamental legal and governance reform had been disrupted, and therefore diminished public support for needed new legislation and regulation.

The response by the Justice Department was to take nearly two years to document its basic contention that such a high profile case as the criminal investigation of Rocky Flats had raised unrealistic expectations in the minds of the American public. Attorney General Janet Reno used her authority to declare in 1994 that after an investigation of the underlying facts of the case of Rocky Flats by the Department of Justice inquiry, the appropriate conclusion was that prosecutors, in exercising their sworn duties to uphold the law, had used appropriate discretion in their pursuit of justice.

Under the U.S. Constitution, it has fallen to the third "separate but equal" branch of the U.S. government, the Judicial branch, to exercise a frequently deciding role in the continuing clash over Rocky Flats. Many of the continuing disputes are civil in nature. A recent example of contention is over what is known as the Jefferson Toll Road. A public good intended for the Denver metropolitan area is to complete an express roadway to expedite traffic around the region. Several municipalities objected that an extension of the existing but incomplete Toll Road would negatively affect their quality of life and opposed the toll road extension through their jurisdictions. Still others opposed granting a 300-foot right-of-way across the Eastern edge of the former Rocky Flats reservation.

The legal battle initially ended in favor of toll road supporters. The Denver Regional Council of Governments added the Jefferson Parkway to its master plan for the Denver area on the condition that no federal or states tax revenue may be used to construct the road. The biggest obstacle as reported in Chapter Thirteen is the discovery of a sample of plutonium in the 300-foot right-of-way which is five times higher than the U.S. government standard for plutonium in the soil for the first 3 feet of the Rocky Flats site.

Some of the most vexing challenges for the Judicial branch have been on-going since 1990 when the Cook lawsuit was filed containing some of the general elements of the Church constellation of lawsuits of 1975—allegations relating to the negative effect of off-site plutonium contamination. The Cook case and its resolution in a large way illustrates the increasingly important role of the Judicial branch in the functioning of the institutions of American constitutional democracy at Rocky Flats.

The "System" Worked: Identifying Viable Approaches Within American Democracy By Which the Public Can Maintain Its Democratic Influence On Such a Sensitive Area of National Security As Nuclear Weapons Production When There Is a Curtain of Secrecy Covering the Government's Actions During the Cold War

Limited to the case of Rocky Flats, there was no way during the early years of the Cold War by which the public had a means of even learning what was going on inside Rocky Flats until Dow Chemical reportedly declined to study possible off-site contamination from the 1969 fire. Independent scientists Martell and Poet confirmed off-site plutonium contamination in 1970. Then the unexpected happened. Rocky Flats experts indicated to Dr. Martell that the most likely source of the off-site contamination was plutonium from the 1957 fire and plutonium residues leaking with contaminated oil from the 903 Pad open air storage area. The independent findings by Martell and Poet (CCEI, 1970a;1970b) were confirmed by U.S. government scientists Krey and Hardy (1970).

The next major glimpse into the extent of off-site contamination was the discovery of tritium off-site in 1973. A government explanation for the source of the tritium was that it had been an unknown contaminant in a shipment from Lawrence Livermore National Laboratory in California. [That shipment was buried in an unlined landfill above the drainage into Walnut Creek which emptied into Great Western Reservoir, a source of drinking water for Denver suburbs.]

The Lamm-Wirth Task Force of 1975

The work of the Lamm-Wirth Task Force in 1975 provided both the U.S. government and a Blue Ribbon Commission of carefully chosen citizens and knowledgeable scientists an opportunity to engage in meaningful dialogue. One vignette reveals the lost opportunity to build a beginning of public understanding. There had been rumors that two workers had been injured in an industrial accident by exposure to radioactivity. Officials representing the U.S. government decided to be non-disclosive. At least during the period of the Cold War, a high profile public task force was not a viable means in the short term for asserting meaningful public and citizen engagement at Rocky Flats, let alone exercise some modicum of control. There are two enduring legacies of the Task Force. First, it was overwhelmingly correct on the concerns it raised. Second, it outlined many of the issues subsequently placed before juries of citizens to decide. Both the verdicts of the Special Federal Grand Jury and the Cook class action jury reflected the broad consensus of the Lamm-Wirth Task Force that with its missions Rocky Flats should have been located elsewhere.

The Church Cluster of Lawsuits of 1975

This constellation of civil lawsuits beginning in the same time frame as the Lamm-Wirth Task Force has been commented upon before with respect to the futility for the U.S. government of making a $9 million settlement with surrounding landowners and then not compelling the contractor to rectify the deficiencies which caused the government to be forced to settle the case in the first place. But the lawsuit was obviously an instance of citizen influence over a sensitive area of national defense at Rocky Flats. And it came about because of the discovery phase of a civil lawsuit.

The Influential GAO Reports of the 1980s

Many of the invaluable fact-finding reports by the Government Accountability Office (GAO) were commissioned by individual members of the U.S. Congress. They were part of an effort by the Legislative branch to become more informed themselves about Rocky Flats, and to inform beginning fact-finding in support of the legislative process.

The reports were not classified. They did serve importantly in the process of creating a more informed citizenry. Beyond that, the 1988 GAO report on Rocky Flats was cited as part of the "probable cause" case in the 1989 FBI search warrant application.

The "System" Worked: There Was an Advancement of Democratic Input Over a Sensitive National Security Asset of the U.S. Government By the Clashes Between "Separate But Equal" Branches of Government

Members of the Legislative branch made a decision in 1992 to challenge the handling of Rocky Flats by the Department of Justice and the Department of Energy of the Executive Branch. This challenge by the U.S.

Congress marked a watershed in democratic control. Testimony before the Congress under oath by numerous officials was augmented by dedicating the extensive time involved to publish testimony and previously unavailable documents. This greatly increased public knowledge. When the Congress determined that they were not receiving necessary and vitally important cooperation from the Executive Branch, they exercised their independent power of subpoena. Previously unavailable primary source documents from the U.S. Congress are a most valuable source for this study.

The "System" Worked: The Discovery Phase of the Stone False Claims Civil Lawsuit

Whistleblower Jim Stone died one month after he lost his civil lawsuit under the False Claims Act in the U.S. Supreme Court. This made his case moot legally. But utilizing the subpoena power of a civil lawsuit under the authority of the Judicial branch, this case served to increase greatly public understanding of actions and their meaning by Rockwell International and the Department of Energy at Rocky Flats.

The "System" Worked: The Powerful Educational and Advocacy Role of an Informed and Organized Civil Society (1970-Continuing)

Once Rocky Flats emerged from its era of almost total secrecy with discovery of off-site plutonium contamination by Martell and Poet in 1970, an informed and organized civil society became a powerful force for democratic influence. Organized civil society has been especially effective in educating the public about risks to public health posed by the conversion of most of the previous "buffer zone" into the Rocky Flats National Wildlife Refuge administered by the U.S. Fish and Wildlife Service without meaningful assessment of the level of picocuries of plutonium per gram of soil.

The "System" Worked: The DOE "Culture" Was Challenged Effectively From the Top of the Department of Energy (1989-1992)

The corporate "culture" in effect at the DOE nuclear weapons complex was characterized eloquently by Admiral James Watkins in his confirmation hearings to become the U.S. Secretary of Energy. He was insistent about the necessity for change at Rocky Flats in his exchanges with Rockwell International Chairman Donald Beall in August and September of 1989, just months before Rockwell International ceased being the contractor and the production era ended.

The "System" Worked: In Presenting the Limitations and Strengths of Science, Engineering and Technology (1951-1989)

The archives of Dow Chemical document the existence of an underlying belief at Rocky Flats that science, engineering and technology can solve difficult and complex problems. Testimony to the accuracy of this belief is the reliability of the 70,000 plutonium igniters which were the principal product of the 38-year production era.

Cost was often a factor, however, in making less of an effort to solve difficult and complex problems of environment, health, and safety. Unwise decisions about infrastructure in the earliest years made it extremely difficult, if not impossible, to rectify mistakes in later years. Two examples suffice to illustrate this assessment.

All of the Solar Evaporation Ponds were built with bottoms that were semi-permeable to a greater or lesser degree. It was not possible the way they were constructed to stop contaminated effluent in them from percolating down to ground water. The monitoring wells associated with the ponds were not placed either to serve adequately their intended monitoring purpose at the time they were constructed, or to meet the assessment monitoring standards of RCRA which was passed years later.

The shoddy condition of the roof on Building 776/777 is striking, as well as the failure to install mandated sprinklers. At the time of the 1969 fire, shoddy construction amplified the risk of plutonium escaping through the roof. It was a disastrous decision to permit such a risk to occur.

The extraordinary scientific accomplishment, the technological expertise, and the standards for quality and workmanship at Rocky Flats which combined in the design and construction of plutonium igniters were not directed evenly across the interrelated missions of waste disposal, fire prevention, environmental protection, health, and safety.

The "System" Worked: At the Time of Its Closure Rocky Flats Could Not Operate Legally (1989)

A valid critique of the Rocky Flats enterprise at the time of the closure of the production era in 1989 is that it was not possible to operate it legally within the framework of the suite of environmental laws and associated regulations of the 1970s. Judge Lewis Babcock ruled in 1990 in the Sierra Club lawsuit that RCRA definitively applied to Rocky Flats. A 1987 document by Rockwell International on spray irrigation and the Solar Evaporation Ponds (Rockwell International, 1987) made it clear that the corporation knew that operations at Rocky Flats were not in compliance with the new suite of environmental laws.

The Mary L. Walker Briefing Memo (DOE, 1986a) was clear in 1986 that DOE knew that portions of the Rocky Flats operation were not in compliance with the law. Two Citizen Special Federal Grand Jury Reports coalesced around the juror's contention that Rocky Flats was an "on-going criminal enterprise," that EG&G had become complicit, and that in their opinion the commission of environmental crime could only be stopped by the closure of Rocky Flats. Rockwell International plea bargained to ten felony and misdemeanor violations of the Clean Water Act and RCRA. Critics of the settlement find fault with the prosecutors by disputing that prosecutors had charged all of the environmental crimes they had found, or that they could prove. among other major deficiencies of the outcome of the criminal investigation and prosecution.

INSTANCES WHERE THE "SYSTEM" MAY YET WORK

The "System" Has Yet to Work Fully For Justice Due Nuclear Workers At Rocky Flats

It is beyond the scope of this study to review workings of the unique hazards law and its failures for former nuclear workers at Rocky Flats. The immediate justice challenge is to workers who began at Rocky Flats in the closing years. Irrespective of the circumstance that the U.S. Congress did not establish the unique hazards law until eleven years after the production era closed at Rocky Flats, the way life unfolds, cohorts of workers are either close to retirement age or already retired. For those who are ill from Rocky Flats related radiation exposures, there is an urgency to getting them reviewed fairly for eligibility under the law.

The "System" Has Not Yet Worked: On Determining What the Extent Is of the Long Term Danger of Migrating Nuclear and Chemical Pollution to Public Health and Ecosystem Health Near the Rocky Flats Site (Continuing)

The answer to whether there is risk of long term danger cannot currently be known. Lessons can be learned, however, from the patterns of toxic contamination migration toward the Columbia River which is at the edge of the Hanford Nuclear Reservation. There may also be lessons to be learned from the patterns of toxic migration at other nuclear weapons complex sites as well. Differences of opinion exist in assessing whether there is risk of contaminated ground water migration from Rocky Flats to Colorado aquifers.

The approach used by the Department of Energy in establishing parameters for the "Accelerated Cleanup" has been well understood. Several constraints upon the Department of Energy are less understood. One constraint is the sobering estimates of costs across the U.S. nuclear weapons production complexes and the years it would take for restoration and stabilization contained in the DOE study Estimating the Cold War Mortgage (DOE, 1995).

Another constraint is the declaration of intent behind a U.S. Senate study with the inspired title of Train Wreck Along the River of Money (Blush and Heitman, 1995). This staff study from the U.S. Senate became a bipartisan declaration that the Nation had more pressing priorities for the expenditure of finite financial resources than restoring the environments in and around former nuclear weapons production sites.

Further constraints are the unknown risks to worker health and safety and the relative lack of proven technology for successfully removing plutonium, carbon tetrachloride, and other dangerous chemicals and solvents from the soil more than six feet under the surface of Rocky Flats, and the stabilization of known plumes of groundwater contamination deep under the soil.

The "System" Has Not Yet Worked: On Determining the Level of Plutonium Contamination and Risk On the Lands of the Wildlife Refuge

The Contempt of Court finding by a Federal District Court Judge against the Department of Energy took three years to resolve as part of the Cook class action lawsuit. The broad context is that for decades there have been assurances by officials of the U.S. and Colorado state government to the Colorado community that plutonium has remained contained where it had been expected to remain contained.

The two reports in early 1970 by the Colorado Committee for Environmental Information (CCEI, 1970a; 1970b) revealed that plutonium had been found off-site. Atomic Energy Commission scientists Krey and Hardy (1970) confirmed the basic findings of CCEI.

Additional plutonium in soil research from 1970-1978 established that while some plutonium had remained by 1978 in the very top layer of soil between 0-3 cm, other plutonium had been found by independent and credible researchers down as far as 10 cm and 20 cm.

Since 1978 there have been two major developments related to the location and depth of plutonium in soil. The first development occurred as part of the resolution of the Contempt of Court citation against the Department of Energy in the Cook class action lawsuit.

The Department of Energy revealed that there were 15 different types of missing and unaccounted for plutonium in the 677,000 pages of relevant documents about 2,600 pounds of plutonium. The point is that it is a reality established by the U.S. government itself that 2,600 pounds of plutonium is missing and unaccounted for around Rocky Flats.

The second major development is that high winds from the Front Range of the Rocky Mountains have continued to deposit sand and soil on top of the lands of Rocky Flats for over 40 more years since 1978.

The result is that plutonium in soil will be buried further than it was between 1970 and 1978. Therefore, the soil column must be investigated below the current shallow government sampling down at least to the depth of the piling excavations. Otherwise, construction workers and the community will be placed at risk. The "system" of constitutional government has yet to work in the public interest.

Reflections On Policy Choices and Decision-Making In U.S. Constitutional Context At Rocky Flats (1951-1992)

Cloaked in Cold War secrecy up until April 10, 1970, and authorized by the Atomic Energy Acts of 1946 and 1954, Rocky Flats was governed and regulated almost totally by the Executive branch and its military-industrial contractor.

The Church cluster of lawsuits from 1975 successfully used the power of subpoena of the Judicial branch for discovery in order to compile extensive information of off-site contamination which led the U.S. government to choose to settle several cases with a stipulation that the terms of each settlement be sealed.

The more than 26-year legal history of the Cook class action lawsuit from 1990 to 2016 was defined by the role of legal precedent in both Federal District Court and in the 10th Circuit Court of Appeals. An ultimate settlement was based on the most thorough processes of discovery. The Judicial branch of government has become increasingly the deciding force to enable the "system" to work when intractable conflicts develop between the Executive branch and an empowered civil society.

Chapters Eight through Eleven on the environmental crime era and its aftermath reveal that the "system" of constitutional government can work when all three branches of government exercise their "separate but equal" authorities. Accomplishments of constitutional government, however, had to overcome the "regulatory confusion" of evolving legal regimes in the 1980s when the authorities between them had not been fully adjudicated.

LESSONS LEARNED DEFENDING THE "PRINCIPLES AND PRECEPTS" OF THE AMERICAN WAY OF LIFE AND DEMOCRACY

Introduction To Part VII – Lessons Learned Defending the American Way of Life and Preserving Freedom Without Compromising Our Fundamental Principles and Precepts

Key Elements In the Chapter Eighteen Narrative

- Chapter Eighteen addresses five organizing questions:
    - why preventable situations became catastrophes,
    - why the "Accelerated Cleanup Concept worked,
    - were there "breaches of the public trust?",
    - what the relationship was of decision-making to governance structure, and
    - did Rocky Flats defend freedom without violating American principles and precepts?
- Unthinkable tragedies were not prevented because of:
    - draining of toxics off-site,
    - irrational facility design,
    - failures of a system which policed itself,
    - lack of built-in fire protection or suppression equipment, and
    - failure to contain plutonium where it was supposed to be contained.
- There was a corrupt relationship between Department of Energy incentives and incentives for military-industrial contractor Rockwell International.
- Rocky Flats was transformed by the clash between production-era imperatives and the criminal justice system.
- Incentives of the U.S. Department of Energy and Rockwell International did not include priorities of environmental protection and safety.
- Grand Jurors concluded that government employees and corporate employees breached the public trust through a clever public relations campaign of distraction, deception and dishonest denial.
- Grand Jurors determined that DOE and its operators "acted with haughty indifference to environmental laws..." with the reason being "the firm conviction that they were pursuing a goal that was more important than compliance with the law..."

Key Elements In the Epilogue Narrative

- Negative legacies of the production era remain in the especially compelling form of unresolved (and under-addressed) issues of environmental health and justice.

- There are two NEPA based lawsuits which have yet to reach trial in Federal District Court for the District of Colorado.
- A "unique hazards law" was not passed until eleven years after the closure of Rocky Flats for production—severely compromising the health and welfare of workers exposed to ultrahazardous conditions.
- Five questions are addressed in the Epilogue based on Jared Diamond's framework for understanding what stewards of past societies which collapsed were thinking:
    - did they fail to see the dangers?,
    - was the collapse their own fault?,
    - were the societal problems unsolvable?,
    - was environmental damage unintentional and imperceptible?, and
    - was environmental damage caused by people acting perversely when they knew the consequences?
- Stewards at Rocky Flats did not believe it was in the national interest to compromise meeting production goals.
- Many brave and heroic individuals at Rocky Flats responded to danger by acting extraordinarily to preserve the mission of keeping America safe.
- Stewards of Rocky Flats were not as responsive to danger as they could or should have been.
- Some of the most disastrous decisions from an environmental point of view were made in an era before Environmental Impact Statements were required by the National Environmental Policy Act of 1969.
- There were ultimately unsolvable environmental problems such as what to do with 80 million gallons of contaminated water a year on a small site uphill from municipal water supplies.
- Environmental damage created during one era was likely unknown to stewards in subsequent eras.
- The five greatest contributors to environmental damage at Rocky Flats were:
    - the choice of criteria for site selection,
    - design choices for Buildings 771 and 776/777 (locations for the catastrophic fires of 1957 and 1969),
    - flawed approaches to fire prevention,
    - flawed approaches to containing plutonium where it was supposed to remain contained, and
    - persistent failure to address satisfactorily the unmet challenges posed by the storage, treatment and disposal of contaminated waste from the operation of the industrial enterprise.

Chapter Eighteen juxtaposes the distinctive contributions which Rocky Flats made to preserving the American way of life during a very dangerous Cold War with President Eisenhower's caution to not destroy from within what we are trying to protect from without: the very "principles and precepts" of the American way of life.

There are a number of key elements in Chapter Eighteen that collectively provide major conclusions based on exceptional access to previously unavailable primary source documents. Answers are provided to the following questions:

Why did three preventable situations end up becoming unthinkable catastrophes?

On the plutonium fires of 1957 and 1969, and the missing and unaccounted for plutonium, the overall lack of monitoring of toxics both off-site and on-site raised the risk of technological disaster and reduced the margin of safety for plant operations.

The locations of the fires in both 1957 and 1969 had been constructed in such a manner that hazards from a relatively insignificant amount of plutonium could so strongly influence the chain of events which caused the fires to become catastrophic.

One influential commentator said that the 1969 fire was caused by the routine incompetence and negligence found "in any system which polices itself." An example from two months before the 1969 fire was that experts from the AEC had discovered that there were "practically no built-in fire protection or suppression equipment." Further, Dow Chemical had failed to implement mandated AEC policy requiring that sprinklers be installed in new construction. After the 1957 fire, Dow Chemical made a commitment to install sprinklers, then did not.

With respect to the missing and unaccounted for plutonium, while in Contempt of Court to a Federal District Court Judge, DOE ultimately finally admitted that there were 15 categories of missing and unaccounted for plutonium, not recorded in 11,000 pages as they had originally claimed, but in over 677,000 pages. The broader point is that lack of monitoring is a defining fact of the production era.

Have there been "breaches of the public trust" at Rocky Flats? The <u>Los Angeles Times</u> *in 1993 quoted a Grand Juror that a "breach of trust" was a dominant element at Rocky Flats.*

The Armed Services Court of Appeals found that Rockwell, in a time period overlapping their time of stewardship at Rocky Flats had been implicated in "criminal, fraudulent, or improper acts." The Grand Jurors had found that Rocky Flats was an "on-going criminal enterprise."

Why did the "Accelerated Cleanup Concept" work to ultimately remove the staggering amount of toxics which by volume would have filled a 65-story building on a site the size of a football field?

The answer is that the "Accelerated Cleanup Concept" employed a new incentive-based contractor system while involving all levels of governance and regulation at every stage of the cleanup. This is detailed in Chapter Twelve.

Is there a relationship between quality of decision-making and governance structure?

The detailed answer to this question is provided in Chapters Seventeen and Eighteen. Prior to April 10, 1970 the Legislative branch was not involved at Rocky Flats except for routine Congressional appropriations. Based on available archives, the Legislative branch including the Congressional Joint Committee on Atomic Energy were mostly unengaged at Rocky Flats. The record is extensive of disastrous and fateful decisions while the AEC and Dow governed and regulated themselves.

On April 10, 1970 the Congressional Joint Committee on Atomic Energy became actively involved at Rocky Flats. The Judicial branch became involved with its power of subpoena in 1975 in the Church cluster of lawsuits. The authority of the Judicial branch empowered Attorney Howard Holme in a successful discovery yielding a 520-page document. This document finally became available in 2018.

Rockwell International pled guilty to ten felony and misdemeanor violations of the Clean Water Act and the Resource Conservation and Recovery Act. The Mary L. Walker Briefing Memo (DOE, 1968a) and Rockwell International (1987) reported on the extent of contamination and environmental crime. The broader point is there were disastrous decisions after 1970 and criminal behavior, but the U.S. "system" of constitutional democracy became deeply involved at Rocky Flats. There were no more catastrophic fires after 1969.

Did the United States pay the "full price" of defending freedom without violating the "principles and precepts" of the American way of life?

This is answered for the two elements of the question. In Dwight D. Eisenhower's phrase, the "care of freedom" was entrusted to Rocky Flats. The 70,000 first stage fission plutonium igniters for thermonuclear bombs have never been questioned for their contribution to defending the country in the Cold War.

On the "principles and precepts" challenge of President Eisenhower, the "system" of U.S. constitutional democracy may yet work. Based on the case of Rocky Flats so far, however, there has been an all too heavy price paid for defending the American way of life.

The Epilogue answers a series of five fundamental questions posed by Jared Diamond (Diamond, 2005) in order to identify why societies which had exceptional accomplishments subsequently collapsed. The decision-making evaluation for this study using the Diamond framework is introduced in Chapter One and the analysis presented in Chapter Sixteen.

How could a society fail to have seen the dangers that seem so clear to us in retrospect?

The short answer is that many brave and heroic individuals responded to danger by acting extraordinarily to preserve the mission of keeping America safe. These were individuals who acted decisively to reduce danger.

There are also four general reasons for stewards being unresponsive to obvious danger:

1. There was not an incentive structure for requesting additional funding for the mitigation of not yet fully understood dangers with new atomic era industrial processes.
2. Contractors were indemnified from civil liability for any consequences of their national security work.
3. Incentives were only for reaching targets for completing plutonium igniters.
4. Each disastrous decision about toxic waste management foreclosed positive choices for the future and increased danger.

Can we say that the end of Rocky Flats was the inhabitants' own fault?

The answer to this question as phrased is "No!" The inhabitants defined as the patriots of the Cold War who worked at Rock Flats in ultrahazardous conditions were in fact victims who lacked fault.

Did the stewards of Rocky Flats at times make especially bad decisions?

"Absolutely!"

Were they instead tragic victims of unsolvable problems?

Employees at the Plant were far too often tragic victims of ultrahazardous conditions.

How much past environmental damage was unintentional and imperceptible?

Spray Irrigation off of the Rocky Plants site in all weather conditions was intentional and created perceptible environmental damage. The same is true for filing false information about environmental conditions in pursuit of corrupt bonuses. Damage may have been created in one era but stewards in subsequent eras had no knowledge of past flawed disposal practices.

How much past environmental damage was perversely wrought by full awareness of the consequences?

The steward of Spray Irrigation asked Rockwell management to discontinue Spray Irrigation in some weather conditions. Rockwell management refused.

Five great contributors to environmental damage at Rocky Flats were:

1. The choice of criteria for site selection which included no criteria for human and ecosystem health or for managing the physical operational requirements of the industrial enterprise,
2. The flawed design choices for Building 771 and Building 776/777,
3. Flawed approaches to fire prevention,
4. Flawed approaches to containing plutonium where it was supposed to stay confined, and
5. Unmet challenges posed by the storage, treatment and disposal of contaminated waste from the operation of the industrial enterprise.

There was a lack of initial understanding of the complexity and sheer volume of the byproducts of new industrial processes required by the further weaponization of the atom.

Chapter Eighteen
Lessons Learned Defending the Principles and Precepts of the American Way of Life Across Evaluation Frameworks and Insights For U.S. Constitutional Democracy and Governance

President Eisenhower made a number of observations related to maintaining American freedoms nearly three-quarters-of-a-century ago which provide a timeless context for under-standing the importance of lessons learned from The Saga of Rocky Flats (see Chapter Two).

Excerpts From President Eisenhower's First and Second Inaugural Addresses

"For history does not long entrust the care of freedom to the weak or timid."
Dwight D. Eisenhower, First Inaugural Address, January 20, 1953

"The building of such a peace is a bold and solemn purpose. To proclaim it is easy. And to attain it, we must be aware of its full meaning –and ready to pay its full price."
Dwight D. Eisenhower, Second Inaugural Address, January 21, 1957

The Saga of Rocky Flats is appearing during an especially dangerous time in human history for preserving peace, freedom, and democracy in the world. It will be necessary to continue to be "ready to pay its full price" for securing freedom now and in the future. This chapter is based on available evidence from the case of Rocky Flats and addresses the question of, "Did the United States pay the full price of defending freedom without violating the principles and precepts of the American way of life?

Five Organizing Questions

This chapter addresses five organizing questions:

- Why did the "system" of policy choices and decision-making lead three preventable situations at Rocky Flats becoming unthinkable catastrophes?
- Have there been "breaches of the public trust" at Rocky Flats?
- Why did the "system" of the "Accelerated Cleanup Concept" work in the U.S. constitutional system given both the fiscal constraints and the toxic burden in such close proximity to a major metropolitan area?
- Is there a relationship between quality of decision-making and governance structure within the context of U.S. constitutional democracy?

- Based on evidence from the case of Rocky Flats, did the United States pay the "full price" of defending freedom without violating the principles and precepts of the American way of life?

Why Did The "System" of Policy Choices and Decision-Making By Stewards Lead To Three Preventable Situations At Rocky Flats Becoming Unthinkable Catastrophes?

There are many sources of insight into why preventable situations were not prevented.

The missions of Rocky Flats were well known to stewards from the Atomic Energy Commission and the Department of Energy, and with that knowledge the nature of the ultrahazardous industrial enterprise was understood. But there was an overall lack of monitoring of dangerous on-site radioactive and chemical substances which brought harm to workers and created risk to surrounding communities. This situation reduced the margin of safety and greatly raised the risk of catastrophic disaster.

<u>Draining of Toxics Off-Site</u>

In contrast to Hanford's 670,000 square miles, site selection for Rocky Flats was incredibly flawed, covering only four square miles and located upwind and uphill from Denver. Additionally, disturbing, pronounced altitude differences from West to East of 528 feet caused the draining of toxic substances off-site toward encroaching residential communities.

The fifteen toxic waste disposal sites characterized in Chapters Five and Six were clustered largely in the high security production area. This restricted policy choice for subsequent stewards because of the original fiasco of site selection and the early failure to plan for disposal of the toxic byproducts from the industrial enterprise.

<u>Irrational Facility Design</u>

The AEC Fire Investigation Report in 1957 concluded that it was irrational that:

> ...facilities were ever designed in such a manner as to permit hazards from a relatively insignificant amount of material to so strongly influence the chain of events which led to a major loss...

<u>Failures of a System Which Polices Itself</u>

With respect to the causation of the 1969 fire, H. Peter Metzger (1972) concluded:

> ...it was caused by the everyday incompetence and negligence inevitably found in any system which polices itself.

In the very limited time period just before the catastrophic fire of 1969, there are especially insightful exchanges which demonstrated that the AEC knew of the acute risk based on the lack of "built-in fire protection or suppression equipment":

> On March 14 we visited a Manufacturing Building No. 776-777. This is a very large undivided complex with glovebox lines. This facility is one of the most important to the program, but had practically no built-in fire protection or suppression equipment...
>
> Building 776-777 complex was not designed or constructed to meet AEC fire safety standards – nor does it appear that any concerted efforts over the years have been made to achieve these standards.
>
> 2. Since original construction, decisions were made to add new equipment and structures on ad hoc bases. This "Topsy-like" growth led to overcrowded conditions that made compliance with fire safety regulations even more difficult if not impossible.
>
> ...Building 776-777, a vital weapons production facility, did not meet even the minimum AEC fire safety standards.
>
> (AEC Report after a visit by experts on March 14, 1969 two months before the fire)

Major General Edward Giller was responsible for weaponizing the atom in his role as Assistant General Manager of the Atomic Energy Commission for Military Applications. This excerpt is from General Giller's testimony before the Subcommittee on Public Works of the House Committee on Appropriations on October 1, 1970.

> GENERAL GILLER. They have sufficient material in them during a day's operations and heat generating flammables that are part of the process so that if a major fire were to break out and break through the building, that is breach the roof, then hundreds of square miles could be involved in radiation exposure and involve cleanup at an astronomical cost as well as creating a very intense reaction by the general public exposed to this. Our problem is first to prevent the fire from starting, and then if it starts to keep it within the building.
>
> In the fire we had last year we kept it in the building. If the fire had been a little bigger it is questionable whether it could have been contained.

General Giller testified on November 21, 1970 before the U.S. Senate Subcommittee on Deficiencies and Supplementals of the Committee on Appropriations.

> General Giller. It is certainly true in reviewing the operating practices at our Rocky Flats plant, which was done by our committee after the fire, we identified what we considered to be serious deficiencies in organizational management practices. They have changed the safety operation by giving it an independent status and charging it with a much more visible responsibility.

The broader point is that risk of a catastrophic fire was well known to experts within the AEC and Dow Chemical. The fragmented governance system and the incentives structure precluded necessary decisive action to reduce risk.

Corrupt Relationship Between Incentives of the Department of Energy and Incentives for Military-Industrial Contractor Rockwell International

The case of Rocky Flats was transformed by the criminal justice system and the clash between the "separate but equal" institutions of U.S. constitutional government between 1986 and 1994. In the process, it became clear that there was an unexpected corrupt relationship revealed between the incentives that the Department of Energy and its military-industrial contractor Rockwell International were found to be operating under from 1975 through 1989. These incentives did not include the priority of environmental protection and safety.

A series of different issues at Rocky Flats concerned members of the Grand Jury. These included:

- "...criminally false statements regarding environmental conditions and compliance at the Plant";
- Defrauding the U.S. government and taxpayers through "improper performance bonus awards...";
- The U.S. Attorney "granted immunity to virtually every person who testified before the Grand Jury without consulting in advance with the Grand Jury."
- DOE and its operators "acted with haughty indifference to environmental laws..." The reason was "the firm conviction that they were pursuing a goal that was more important than compliance with the law..."
- The assertion that Rockwell and EG&G--"with the full knowledge and blessing of DOE--have excused their illegal conduct by asserting that they cannot operate the Plant without violating one or more environmental laws."

The Grand Jurors determined that Rockwell International had graded itself falsely on issues of environment, health, and safety for purposes of generating bonus fees with the collusion of the Rocky Flats Office of the Department of Energy.

A broader finding of the Grand Jury was that preventing potential preventable catastrophes was not even on the allocation priorities or incentive structures for stewards of Rocky Flats. The opposite was the case.

The Missing and Unaccounted For Plutonium and the Contempt of Court Citation By a Federal District Court Judge

Under the Atomic Energy Acts of 1946 and 1954, inventory of plutonium was "restricted data." While DOE itself mentioned missing and unaccounted for plutonium in 1994, it is not clear that they had a grasp on the magnitude of the problem, or the fact that there were 15 types embedded in over 677,000 pages of documents ultimately determined to be relevant.

Were There "Breaches of the Public Trust" At Rocky Flats?

The short answer to this question is, of course, "yes." A sad example is the determination of Administrative Law Judge David W. James, Jr of the Armed Services Court of Appeals in 2005 about the conduct of Rockwell International. Rockwell had been charged with or implicated in "criminal, fraudulent, or improper acts." Judge James ruled that:

> Government should not pay for wrongdoing, the defense of wrongdoing, or the results or consequences of wrongdoing by contractors...

Viewpoint of the Grand Jurors

The viewpoint of the Grand Jurors is that they had determined that Rocky Flats was an "on-going criminal enterprise." The U.S. Attorney had sent target letters to individuals who were being investigated for criminal conduct. The Grand Jurors had identified individuals who in their judgment should have been indicted, but it was the Grand Jurors themselves who were investigated criminally to determine if one or more of them had leaked the first version of the Special Federal Grand Jury Report.

Rockwell Did What DOE Wanted

A problem for the Department of Justice was that a top Rockwell official had challenged the Deputy Secretary of Energy by asserting that in essence Rockwell as a corporation and Rockwell employees were doing what DOE on behalf of the U.S. government were asking/directing them to do.

The problem for the Department of Justice in a national security defense by Rockwell would be to convince a jury of citizens "beyond a reasonable doubt" that individual Rockwell employees were guilty of criminal behavior as charged, and Rockwell also for that matter. There is only an insignificant chance that a jury of citizens would have convicted Rockwell employees of individual criminal conduct "beyond a reasonable doubt" after a national security defense.

Three Individuals Could Have Acted Differently

Returning to the "breach of public trust" issue, three individuals could have chosen to act differently. It is not known what was in their minds as reasons motivating the choices they made.

- A Federal District Court Judge lent the credibility of his office to declaring that the First Report of the Special Federal Grand Jury lacked "necessary standards," then lent his credibility again to the release of a heavily redacted version which distorted the conclusions of the Grand Jurors.
- The top officials at Rocky Flats of both DOE and Rockwell signed false statements in 1985 under penalty of law that Rocky Flats was in compliance with relevant environmental law when it was in violation of RCRA.

The point is that there were actions, some unexpected, which clearly rise to the level of a "breach of the public trust."

Insights From the Second Citizen Special Federal Grand Jury Report

Insights into the real incentives in evidence at Rocky Flats in the Second Citizen Special Federal Grand Jury Report are noteworthy for the magnitude of evidence cited regarding priorities motivating recurrent choices by the Department of Energy and Rockwell International. The Cook class action jurors in 2006 determined "beyond a reasonable doubt" that the behavior of both Dow and Rockwell had been "willful and wanton." The jurors were not asked if the military-industrial contractors had breached the public trust, but Judge James clearly did.

The Grand Jurors also offered harsh assessments of the performance of Federal, Colorado, and local regulators:

- Federal, Colorado, and local regulators and elected officials have demonstrated an unblemished record of ineptitude, failure, and bureaucratic incompetence in their "feeble efforts over the last 40 years to make the DOE and the corporate operators of the Plant obey the law."
- "...the Plant has been and continues to be operated by government and corporate employees, who have placed themselves above the law and who have hidden their illegal conduct behind the cloak of 'national security.' "
- "...government employees and corporate employees, who have operated the Plant, have committed the most serious crime that can ever be committed in a democracy...breached the public trust

through a clever public relations campaign of distraction, deception, and dishonest denial."

Why Did the "System" of Policy Choices and Decision-Making Related To the "Accelerated Cleanup Concept" Work In the U.S. Constitutional System Given Both the Fiscal Constraints and the Toxic Burden In Such Close Proximity To a Major Metropolitan Area?

The "Accelerated Closure Concept" is presented by Lockhart (2006, p. I-5) as a "new vision" with the following elements:

> Establishing and implementing the accelerated closure concept was only possible through innovative and groundbreaking strategies for political support (among the DOE site leadership, contractor leadership, DOE political leadership and the key Congressional committees), regulatory applications and relationships, project management and control, and contract development and support.
>
> (Lockhart, 2006, I-5)

As Lockhart (2006, p. I-5) noted: "These four areas does not mean that these areas or activities occurred independently of each other."

In 1994, the DOE site managers at Rocky Flats came out with a new strategic plan (Silverman, 1994). DOE (1995) estimated that cleanup at Rocky Flats would cost $37 billion and take 65 years. A bipartisan U.S. Senate consensus was that the Nation had far more important fiscal priorities than cleanup at Cold War nuclear arms race weapons production sites.

The Clinton-Gore initiatives known as Making Contracting Work Better and Cost Less resulted in a new incentive-based contracting structure. And it was not politically tenable on multiple grounds for the U.S. government not to do something substantive at Rocky Flats given the known risks to environment, health, and safety and the closeness of the site to down-wind and down-hill communities.

The "Accelerated Cleanup Concept" and the cleanup itself have been covered in Chapter Twelve with a focus on the disposition of special nuclear materials to repositories in (mostly) six states spread from Washington to the state of South Carolina, and the use of innovative technologies. It is unlikely anyone could have forecast that the 334-acre high security area would yield toxic waste in the (staggering) amount sufficient to fill a 65-story building on a site the size of a football field.

Is There a Relationship Between Quality of Decision-Making and Governance Assessed In the Context of U.S. Constitutional Democracy?

This section of Chapter Eighteen addresses the levels of evaluation of decision-making in relation to the "system" of constitutional democracy which was operating when disastrous decisions were made.

Siting of Nuclear Weapons Production Sites

Siting decisions predating the Atomic Energy Acts of 1946 and 1954 were made by the Manhattan District under the leadership of General Leslie Groves. The President of the DuPont Corporation had warned of the "unique hazards" involved in 1942. The requirements of General Groves for siting Oak Ridge and Hanford were ignored at Rocky Flats. There was no one obviously involved in the site selection process for Rocky Flats with the perspective of Groves, who had retired in 1948.

The governance structure of site selection was headed by the Atomic Energy Commission, Dow Chemical and the Austin Company from Ohio in early 1951. They were wrong in their calculation of the wind direction. The same group of decision-makers did not engage outside experts, presumably for national security concerns. The decision-makers did not address the operational scope of the industrial enterprise regarding safe waste disposal. Basically, they governed themselves in formulating site selection criteria and in selection of the site, which was so wrong on many levels.

Impact of New Bipartisan Environmental Legislation

Under the new bipartisan environmental legislation which came twenty year later, environmental crime was unavoidable, given what had gone before. There was a negative and direct interaction between governing themselves, and inevitable disastrous decisions and consequences.

Under the governance structure whereby the AEC and Dow regulated themselves, the fifteen toxic waste disposal sites characterized in Chapters Five and Six were created, the fires of 1957 and 1969 occurred, and plutonium was not contained where it was supposed to be contained. Yet ironically, this flawed governance structure administered the creation of plutonium igniters with technical efficiency, thereby contributing to weaponizing the atom as presented in Chapter Three.

After April 10, 1970 the Legislative and Judicial branches played a more active role in constitutional governance. But as presented in Chapters Twelve and Seventeen, a much more robust governance structure among and between the "separate but equal" branches of constitutional government resulted in fewer disastrous decisions.

Where the "System" (Hopefully) May Yet Work

There are (at least) four entries in this category. The "system" has not yet worked to determine the extent of long term danger related to nuclear and chemical pollution migrating from Rocky Flats to the detriment of public and ecosystem health. There is a legally mandated process for monitoring the Superfund site every five years, but only the Superfund site.

When the Fish and Wildlife Service was trying to do a prescribed burn, they chose to exempt the area between Walnut and Woman Creeks. The Seed Committee (Seed, 1971) reanalyzed data from Krey and Hardy (1970) which only strengthened the validity of the basic finding of the presence of off-site plutonium. Holme (1978) established the credibility of the soil sampling under the direction of Federal District Court Judge Richard Matsch. There is no legally mandated requirement for soil sampling on the lands of the Wildlife Refuge, and currently the Fish and Wildlife Service does not permit research. Thus, the extent of the problem is currently unknown.

With respect to environment and health justice for Rocky Flats workers, work on the site began in 1951. The "unique hazards law" was not passed until 2000. With so much passage of time, it was and is too late for most Rocky Flats workers and their families to seek redress or compensation.

Exceptional peer-reviewed science exists related to risk of exposure to low doses of plutonium and on the effects of plutonium on children. This knowledge is yet to be applied to risk assessment related to future uses of the 5,237 acres of the Rocky Flats National Wildlife Refuge.

Based On the Evidence Available From the Case of Rocky Flats, Did the United States Pay the "Full Price" For Defending Freedom Without Violating the "Principles and Precepts" of The American Way of Life? This is really two interrelated questions.

Was There a High Price Paid By Rocky Flats In Defending Freedom?

Insights provided by Linking Legacies in 1997 as presented in Figures 3.1, 3.2, 3.3, and 3.4 of Chapter Three documented the accomplishments at Rocky Flats in helping to preserve freedom across its domains of science, engineering, and technology. These accomplishments are directly in the spirit of both President Eisenhower's first and second Inaugural Addresses: the "care of freedom" was entrusted to Rocky Flats, which was an indispensable contributor to paying freedom's "full price." The 70,000 plutonium igniters for first stage fission bombs were never questioned for their quality and reliability.

Were There Violations of the "Principles and Precepts" of the American Way of Life?

Representative answers to this question are found in the areas where the "system" of U.S. constitutional democracy has not yet worked.

Workers at Rocky Flats were not covered by a "unique hazards law" until 11 years after the close of production. Then the initial rollout of the new law from 2000 was imperfectly handled within the Executive branch. The quest for environmental and health justice for workers serving in the Cold War at Rocky Flats has not been achieved.

This "system" has yet to work for determining the extent of long term danger of nuclear and chemical pollution migrating from Rocky Flats to the detriment of public and ecosystem health. Entities within the Executive branch of government have conducted controversial shallow soil sampling on the lands of the 5,237 acres of the Wildlife Refuge and used that imperfect approach to tell Judges in Federal District Court that the lands of the Wildlife Refuge are safe for public access on trails. At the same time, they have blocked independent scientific investigation—one of the core purposes of establishing the Wildlife Refuge.

As this report concludes, it is the Judicial branch of government which will determine the ultimate outcome of the future uses of lands of the Wildlife Refuge. Two NEPA lawsuits are on-going in Federal District Court for the District of Colorado. The findings of the National Research Council (2006) on low-dose exposure to plutonium and risks to children from plutonium by ATSDR (2010) have not been addressed in Federal District Court. Nor have the scientific findings by U.S. government scientists from the 1970s on plutonium in soil which sharply contradict the current positions of EPA and CDPHE and the U.S. Fish and Wildlife Service.

The "system" of U.S. constitutional government may yet work. That outcome will provide the final answer to the "Principles and Precepts" challenge of President Eisenhower. Based on the case of Rocky Flats so far, there has been an all too heavy and unnecessary price paid for defending the "principles and precepts" of the American way of life.

Epilogue: Compelling Questions Across the Epochs of Rocky Flats: "The Past is Never Dead, It's Not Even Past."

"The past is never dead. It's not even past."
William Faulkner, Requiem for a Nun, 1951

The first part of the quote, "The past is never dead," refers to all the unresolved issues which are legacies from across the Epochs of Rocky Flats. Especially compelling are the environmental health and justice issues which remain. The issue of residual contamination in the Wildlife Refuge has no legally mandated process to address it. Residual contamination is at the heart of two NEPA lawsuits which were filed as the second decade of the 21st century came to a close.

The second part of the quote, "It's not even past," is directly relevant to the fact that the "unique hazards law," within the Energy Employees Occupational Illness Compensation Program Act of 2000 was not even passed until eleven years after production at Rocky Flats ceased. The top Department of Energy official at Rocky Flats in 1996 authenticated the destruction of records. This circumstance did not negatively influence the successful outcome of the Cook class action lawsuit which was settled in May of 2016. It is not yet known if the destruction of records is relevant to the continuing dispute over the future of the Wildlife Refuge, or if it impacts the continuing quest for inclusion in the "unique hazards law" by former workers during the concluding years of production at Rocky Flats.

Insights Into the Juxtaposition of the Legacy of Sustained National Security Accomplishments and the Closure of Rocky Flats Mired In a Quagmire of Toxicity and Environmental Crime

Jared Diamond posed a series of five questions to consider in order to identify why societies which had exceptional accomplishments subsequently collapsed. The answer to these fundamental questions, when applied to Rocky Flats, helps understand more fully the irony of why a facility which attained such a high level of accomplishment in weaponizing the atom for the Nation's national security ended in a quagmire of toxicity and in the ignominy of environmental crime.

The five questions are:

- How could a society fail to have seen the dangers that seem so clear to us in retrospect?
- Can we say that the end of Rocky Flats was the inhabitants' own fault?
- Were they instead tragic victims of unsolvable problems?

- How much past environmental damage was unintentional and imperceptible?
- How much past environmental damage was perversely wrought by people acting in full awareness of the consequences?

Stewards of Rocky Flats did not believe that it was in the national interest to compromise meeting production goals and time-lines for plutonium igniters, even when it was no longer possible to operate the facility as other than, in the words of the Special Federal Grand Jurors, an "on-going criminal enterprise."

How Could a Society Fail To Have Seen the Dangers That Seem So Clear To Us In Retrospect?

The short answer is that many brave and heroic individuals at Rocky Flats responded to danger by acting extraordinarily to preserve the mission of keeping America safe during a harrowing Cold War against nuclear-armed adversaries.

Bravery in Response to Extreme Danger

"At 10:30 a.m. I left for home." This nonchalant statement was the last comment from someone who fought the 1957 fire all night at great personal risk. This comment is noteworthy in two contexts. The first context is that it was the concluding statement in a matter of fact recounting of one of the two most harrowing nights in the entire history of Rocky Flats. The author of the statement was one of the saviors of the plant from even worse damage during the ferocious fire. The second context is that the fire was just one instance of the ever-present danger for those working with radioactivity and carcinogenic chemical substances.

Explosions presented another ever-present danger, often caused by radioactive substances mixed with toxic chemicals such as carbon tetrachloride. Every production area worker on a moment's notice could be called upon to be a first responder. In one explosion caused by a chemical reaction gone wrong, two plant protection workers rushed to aid injured production workers only to become unknowingly contaminated themselves. The point is that while dangers are clear in retrospect, they were also obvious on a daily basis to those whose work put them at risk for known unique hazards.

The archives of Rocky Flats contain countless stories of bravery in response to danger. Ordinary workers and first responders acted to save their companions in danger; yet they were not at a level of operational authority where they could have created policies and protocols to prevent the danger. The stewards across the production era were much more accomplished in meeting production requirements than they were in environmental protection, monitoring of chemicals or radioactivity threats which posed acute dangers, or thoughtfulness regarding the health and welfare of workers.

Individuals Acting Decisively To Reduce Danger

Many individuals acted decisively to reduce danger when they learned about it, and were in a position to take decisive action. Two examples suffice to make the point. One involved the burial of steel drums in what was called the Mound Area between 1954 and 1958. This preceded the above ground storage of steel drums filled with toxic mixed waste (chemical and radioactive) in the 903 Pad and the Triangle area.

It was not until a routine consultation meeting on April 3, 1970 that representatives of Rocky Flats Labor Unions learned the nature of toxic substances in the barrels buried in the Mound Area. Furthermore, no one could remember who ordered the burial to start. A week later Jim Kelly of the Rocky Flats Steelworkers Union was in attendance in Washington, D.C. at a routine meeting on April 10, 1970 of the Congressional Joint Committee on Atomic Energy chaired by Congressman Chet Holifield of California. Within weeks of Congressman Holifield and the CJCAE learning of the drum burial practice, action was taken and the Mound Area ceased to exist. All of the barrels had been dug up, though most of them by then had corroded and their contents lost into the soil. Toxic groundwater plumes still exist from contents of the Mound Area reported in 2005 by the Agency for Toxic Substances and Disease Registry (ATSDR, 2005). No record remains of the strategies employed in removal or the degree of risk to workers who performed the removal.

Another example of decisive action being taken at Rocky Flats was by Major General Edward Giller, Assistant General Manager of the Atomic Energy Commission for Military Programs. After the 1969 fire, he championed obtaining additional appropriations for Rocky Flats from the U.S. Congress, established Fire Investigation Committees, and took decisive action which contributed to preventing catastrophic fires during the last 20 years of the production era.

As with Congressman Holifield, General Giller was in a particularly powerful role but with numerous other responsibilities beyond Rocky Flats. The small size of the AEC staff at Rocky Flats, and their distance from the Congressional Appropriations Committee processes, precluded the AEC from being a direct conduit for routine safety upgrades. The point is that real authority was at too great a distance for regular Rocky Flats safety and environmental protection operations to have meaningful advocacy.

Reasons For Stewards Being Unresponsive To Obvious Danger

Numerous reasons explain why stewards of Rocky Flats were not as responsive to danger as they could and should have been:

1. Contractors were compensated in an incentive structure based on successful production of plutonium igniters. No incentive structure was formulated for bonus compensation related to requesting additional funding for mitigation of not yet fully understood dangers associated with new atomic era industrial processes.

2. Contractors were indemnified from civil liability for any consequences from their national security work on behalf of the people of the United States. They were compensated with bonuses for what they did in the direct national interest—providing vital components for the nuclear weapons deterrent in the Cold War.
3. Prior to exchanges between Secretary of Energy James Watkins and Rockwell International Chairman Donald Beall in the Fall of 1989, there had been no documents in the available archives where incentives were for other than reaching AEC or DOE targets for completed plutonium igniters. That is strong testimony to the power of the incentive structure in shaping contractor behavior.
4. Starting with the flawed siting decision which itself foreclosed positive choices for protecting human and ecosystem health, each disastrous decision about waste management made it more difficult to have less polluting choices in the future. As an informed commentator remarked, Rocky Flats had become the "barrel capitol of the world."
5. Site topography foreclosed the ability to prevent off-site contamination from routine plant operations. To criticize choices made in the 1950s and 1960s on environment, health, and safety fails to recognize that choices by stewards were made under the legal authority of the Atomic Energy Acts of 1946 and 1954, and in a dangerous phase of the Cold War with new and untried industrial processes. Nonetheless, these choices polluted the Rocky Flats site so deeply that there is no known technology to support environmental restoration of the contaminants in the groundwater.

Following the Diamond Framework, Can We Say That the End of Rocky Flats Was the Inhabitants' Own Fault?

The answer to the question as phrased is "No!"

Did those who worked to produce plutonium igniters contribute to the closure of Rocky Flats in the epoch of environmental crime? "No!" In fact, the workers themselves and their work product were key to helping to preserve national security.

Did stewards of Rocky Flats at any one time in the thirty-eight-year production era make especially bad decisions? "Absolutely!"

Did stewards of each succeeding period of years inherit increasingly flawed choices within the production imperative? "Yes!"

Did it reach a point in the 1980s when the plant as a complex could not be operated without violating criminal provisions of both the Resource Conservation and Recovery Act and the Clean Water Act?" "Yes!"

It is not clear in the available archives how long it was deemed necessary in 1951 for Rocky Flats to be in operation when it was initially sited and opened for actual operations. Clearly, little long-term planning was conducted. Even at the Hanford Nuclear Reservation, the single shelled tanks for high level nuclear waste had not been constructed to last for the length of time that the Cold War nuclear arms race ended up lasting. And many of the single shelled tanks became leakers, just as barrels did at Rocky Flats.

Some of the most disastrous decisions from an environmental point-of-view, like building the Solar Evaporation Ponds with porous liners, were made in an era before Environmental Impact Statements were required by the National Environmental Policy Act of 1969.

A cumulative output of 70,000 plutonium igniters is an extraordinary accomplishment—but the toxic waste burden and cost associated with each igniter is large and has not been calculated. Simply blaming the "inhabitants" is wrong when each generation of workers and managers inherited narrowed choices for waste disposal, increased volumes of toxic water, and waste streams which could not be disposed of off-site to either ARCO/INEL in Idaho or the Nevada Test Site (which were under the same legal authority structure of the Atomic Energy Commission/Department of Energy).

Were The Inhabitants Instead Tragic Victims of Unsolvable Problems?
The answer to the question as phrased is that employees at the plant were far too often tragic victims of contamination from explosions and fires and the inability of plant managers to reduce neutron exposure to glovebox workers, and to many other dangers at Rocky Flats.

Managers were faced with a spectrum of unsolvable problems such as what to do with 80 million gallons of contaminated water a year on a small site uphill from municipal water supplies. It proved to be unsolvable to craft safe disposal practices for cooling oil and solvents mixed with trace amounts of plutonium or uranium. One environmental crime under RCRA was storage without a permit of cooling oil with radioactive elements in it. A permit for such on-site storage would not have been issued without a Presidential exemption.

Burning over 1,000 barrels of contaminated oil in open pits was at best "rational bad behavior." It disposed of the oil on-site but it put unsuspecting off-site communities at risk (as well as workers on-site). The practice resulted from one narrow perspective: "good for plant operations of waste disposal and bad for everyone else."

Contaminated oil could not be shipped off-site to ARCO/INEL in Idaho without complying with understandable requirements established by the Department of Transportation that there be a method for absorbing any toxic liquid which spilled during shipment. No available archives exist which document efforts at problem solving possible transportation solutions between the Executive branch entities of the Departments of Transportation and AEC/DOE. In the "Accelerated Cleanup" there are references to the creation of innovative shipping methods associated with the off-site transportation of the debris from the dismantling of Rocky Flats. It is possible that technology at the time it was generated did not exist to ship contaminated cooling oil off-site during the period of the Cold War. The point is that there were numerous unsolved problems, many with negative impact on the "inhabitants."

How Much Past Environmental Damage Was Unintentional and Imperceptible?

Applied to the creation of environmental damage at Rocky Flats, "unintentional" and "imperceptible" are separable questions. Except for Spray Irrigation in all weather in excess of the capacity of the soil to absorb the moisture, almost all environmental damage at Rocky Flats was imperceptible, or invisible, or both. Further, damage may have been created during one era, and those working in another era had no knowledge of past flawed disposal practices. The creation of the Mound burial area between 1954 and 1958 is an example. Subsequent workers did not even know about it until 1970, except perhaps in the files of top management, and even then it could not be determined who ordered it or why. Frequent turnover among the stewards of Rocky Flats who were there on behalf of either the military-industrial contractor or the U.S. government contributed to the lack of shared knowledge over time.

When the Waste Management Group was instructed in the early years to bury drums on-site because the contents represented a fire hazard which precluded shipment off-site, the intent was to solve a problem interfering with production, and to reduce danger. The choice of the methods did intentional environmental damage which was at the same time (mostly) imperceptible.

But environmental damage also was done by deliberate disposal practices, most unmonitored, as documented by the Mary L. Walker Briefing Memo of 1986 (DOE 1986a), Rockwell International (1987), and ChemRisk (1992). The cumulative site burden at Rocky Flats was estimated by the Estimating the Cold War Mortgage (DOE, 1995) to cost $37 billion to clean up, with numerous exclusions in the cost estimates. No known technology exists to remediate some of the pollution; for example, the existing groundwater plumes.

A final note. The ChemRisk (1992) study identified 25 "Materials of Concern" for off-site contamination. Almost all of them are known carcinogens. Environmental damage in the case of almost all of them was imperceptible. For 19 of the 25 "Materials of Concern," they were not monitored at all despite their known toxicity.

How Much Past Environmental Damage Was Perversely Wrought By People Acting In Full Awareness of the Consequences?

The answer to the question as phrased is "None!" Five great contributors to environmental damage at Rocky Flats were:

1. the choice of criteria for site selection which included no criteria for human and ecosystem health;
2. design choices for Building 771 and Building 776/777;
3. flawed approaches to fire prevention;
4. flawed approaches to containing plutonium where it was supposed to stay confined; and

5. the persistent failure to address satisfactorily the unmet challenges posed by storage, treatment, and disposal of contaminated waste from the operation of the industrial enterprise.

But the context for the choices which led to massive environmental damage was a lack of initial understanding of the complexity and sheer volume of byproducts of these new industrial processes required for weaponizing of the atom.

Did Judge Finesilver cause environmental damage directly by sanctioning the misrepresentation of the findings of twenty-three citizens on a Special Federal Grand Jury? Certainly not. But this relates back to the critique by the U.S. Congress that the American people were denied their right to "know the truth" about what happened at Rocky Flats. The American people were also denied the opportunity to have momentum continued for the process of legal and regulatory reform which the U.S. Congress believed was essential.

The Cold War was over by the time that Rocky Flats was declared to be "officially" closed in 1992 by President George H.W. Bush. The U.S. government was able in the national interest to leave the production era at Rocky Flats closed forever, with off-site risk to human and ecosystem health needing continuing vigilance. It still is not known the location of the 2,600 pounds of missing and unaccounted for plutonium. The discovery of a sample of plutonium in soil on the Wildlife Refuge at over five times the U.S. government's own standard reveals the urgency for more independent soil sampling.

Individuals who acted knowingly against the law were top managers of the Department of Energy and Rockwell International at Rocky Flats who falsely attested that the facility was in compliance with environmental laws based on their personal knowledge when in fact they knew it was not. Did that breach of the public trust "perversely" bring consequential environmental damage? "No." It did, however, serve to keep the plutonium igniter fabrication process going in the context of continuing to violate the laws of the United States. As the anonymous Grand Juror stated in 1993: "It's a breach of public trust. That is the biggest crime of all."

Among great questions remaining about the future of Rocky Flats are:

- Is the threat over to the communities surrounding Rocky Flats?

- What is the extent of the long term danger of nuclear and chemical pollution migrating from the Rocky Flats site to the detriment of public and ecosystem health near Rocky Flats?

- Is the unique hazards law passed by the U.S. Congress in 2000 fulfilling its intended purpose for Rocky Flats workers?

These important questions are not likely to receive definitive answers any time soon.

Legacy of Mark N. Silverman For the Rocky Flats Stakeholders of This Era

It is possible to improve cooperation among stakeholders in working toward solutions on the future of the lands of Rocky Flats in the public interest. Mark N. Silverman, the DOE Manager for Rocky Flats in the mid-1990s, spoke about the need for a new approach at Rocky Flats to address its problems. He did so in an address to the Class of 1961 at the United States Military Academy at West Point from which he himself had graduated before serving in the U.S. Army paratroopers.

A powerful legacy of Mark N. Silverman for the stakeholders at Rocky Flats today is the conviction of his that the future of Rocky Flats will be more productively addressed in a climate of trust and open and honest communication in order to gain the public's support. This insight is both a powerful legacy and a direction to the stewards of Rocky Flats in the coming years of the 21st Century. Silverman's words remain as guideposts for the new era ahead:

> Little did I know when I graduated in 1961 that the skills I would need in the post-Cold War era involved negotiations and building political coalitions. The most daunting tasks I inherited in 1993 was to overcome the distrust, anger and hostility between Rocky Flats and its stakeholders including the regulators, citizens, news media and elected officials at all levels. In order for Rocky Flats to get on with the task of reducing the risk and cleaning up the site, I would have to forge an alliance of environmentalists, peace activists, elected officials, and regulators.
>
> Perhaps this is the biggest lessons we need to learn. As we continue to ensure our nation's security, as we live up to our motto of "Duty, Honor, Country" we need to develop and employ different skills than just those needed on the battlefields for which West Point so ably trains us. Now they are many different battlefields. We must have the public's trust and confidence, we must communicate with them frequently, openly, and honestly. We must never forget that our mission is to serve our nation, which we cannot do unless we have the public's support. We will only get and keep their support if we have earned their trust and confidence.

Mr. Silverman was laid to rest with full honors in 2005 in the Military Academy Cemetery at West Point.

Appreciations

A favorite rendition of mine of the American classic "The Saints Go Marching In" is by the Weavers and Pete Seeger at Carnegie Hall in 1955. Their lead-in begins with "We are traveling in the footsteps of those who have gone before, and we'll all be reunited on a new and sunlit shore."

There is an exceptional community of dedicated individuals whose prior work makes this study possible. They come from many different backgrounds, but all of them have made distinctive contributions to this Saga of Rocky Flats. I am blessed to have this opportunity to acknowledge my indebtedness to be a traveler "in the footsteps of those who have gone before..."

The Appreciations strive to capture the persons behind the contributions. The magnitude of what they have contributed is so apparent throughout the book. This chronicle of the history of Rocky Flats covers sixty-nine years (and counting). It was not possible to know personally some important contributors. More complete records of some of their work became available only in 2018, as did the Holme (1978) pretrial statement in the Church cluster of class-action lawsuits, and the extraordinarily important Seed Committee report from 1971.

Carl J. Johnson, M.D., MPH: A Tireless Medical Scientist For Documenting Health Risk: "Rocky Flats: Death Inc."

"...the actual number of people who have been injured or died because of the operations of Rocky Flats can never be fully known."

After service in the United States Army from 1946 until 1948, and then becoming an Army Reservist, Carl J. Johnson attended Michigan State University as an undergraduate and Ohio State University for Medical School. He subsequently earned a Master's degree in Public Health from the University of California, Berkeley. As Director of the Jefferson County Department of Health, he utilized his research skills in public health and medicine to determine that Rocky Flats posed a demonstrable risk to the health of the surrounding area. After publishing his findings in peer-reviewed scientific journals, and speaking out in public forums, he was forced out of office by Jefferson County officials in 1981. He received subsequently a legal settlement from Jefferson County.

One of his most important peer-reviewed studies was entitled: "Cancer Incidence in an Area of Radioactive Fallout Downwind from the Nevada Test Site." It was published in the prestigious Journal of the American Medical Association, 251(2), January 13, 1984. Dr. Johnson's final contribution was an Op. Ed. in The New York Times for December 18, 1988 entitled: "Rocky Flats: Death Inc." He passed away prematurely 11 days later from complications arising from coronary bypass surgery. Colonel Carl J. Johnson was laid to rest in Fort Logan National Cemetery.

Edward A. Martell, Ph.D. Pioneering Discoverer of Off-site Plutonium At Rocky Flats:
"Implications of plutonium releases to the public health and safety."

Dr. Martell's first professional positions were as a group leader at the Fermi Institute for Nuclear Studies at the University of Chicago, followed by service at the Air Force Cambridge Research Laboratory in Bedford, Massachusetts. In this latter position he managed radiation-effects tests at the Nevada Test Site, and the 1954 hydrogen bomb tests at the Bikini Atoll in the South Pacific.

Dr. Martell's next position was as a radiochemist in the Chemistry Division of National Center for Atmospheric Research in Colorado. Reflecting his wide-ranging intellectual interests beyond Rocky Flats, Dr. Martell published research on such diverse topics as: "Actinides in the Environment and Their Uptake by Man" in 1975, and "Critique of Current Lung Dosimetry and Models for Radon and Its Decay Products: Occurrence Properties and Health Effects" in 1987.

Dr. Martell's special relevance to Rocky Flats is his discovery of plutonium off-site from Rocky Flats. The Martell/Poet team released their first study of off-site plutonium from Rocky Flats through the Subcommittee on Rocky Flats of the independent Colorado Committee for Environmental Information (CCEI, 1970a) in a communication to the Atomic Energy Commission. That first report to the Atomic Energy Commission in January of 1970 was entitled "Report on the Dow Rocky Flats Fire: Implications of Plutonium Releases to the Public Health and Safety"

The response of the AEC was to commission scientists P.W. Krey and E.P. Hardy from the New York Health and Safety Laboratory to conduct an independent investigation of the off-site plutonium. Their methodology and its reanalysis in 1971 by the Seed committee (Seed, 1971) were consistent with the original Martell and Poet findings. Martell et al. (1970) reported on "Fire Damage" in the journal Environment (vol. 12, pp. 14-21) which was followed by a press release from the Atomic Energy Commission.

Dr. Martell left an important legacy in his meaningful contribution to the PBS Frontline program "Secrets of a Bomb Factory." Recorded at the facility of the National Center for Atmospheric Research which overlooks Rocky Flats, Dr. Martell takes viewers through the main elements of that vitally important historical era.

The Task Force of 1975 by Colorado Governor Richard Lamm and Congressman Tim Wirth Mobilizing Citizen Inquiry:
"...this is a great case study about citizen action..."
The Lamm-Wirth Task Force is a powerful model for engagement by members of Colorado civil society. It was created by the Governor of Colorado and a U.S. Congressman from Colorado. The purpose of the Task Force was to develop an overview of risks associated with the production of nuclear weapons in close proximity to a major metropolitan area. The contrasting positions of the Task Force and the U.S. government are presented in Chapter Fourteen with a concluding commentary in historical perspective by Senator Wirth written in 2018.

After earning an undergraduate degree from Harvard University and a doctorate from Stanford University, Tim Wirth began his career in public service as a White House Fellow during the administration of President Lyndon Johnson, followed by being Deputy Assistant Secretary of Education in the Nixon Administration. As a Senator from Colorado he focused on environment issues, especially climate change. After eighteen years as an elected official, he became the first Undersecretary of State for Global Affairs from 1993 to 1997. Then he became the founding President of the UN Foundation from 1998-2013. His focus was institutional-building and public advocacy on behalf of the United Nations particularly in areas of human rights, peace, children's and women's health, population stabilization and the environment. The following quote reflects one of his deepest beliefs: "Through your support of United Nations peacekeeping, you can help to make this tantalizing word—"peace"—a reality for all the world's citizens. Together, we can spread the message that UN peacekeeping is essential. Without it, the world would be a much less stable and more violent place."

Richard D. Lamm was the three-term 38th Governor of Colorado serving from 1975 until 1987. He earned an undergraduate degree from the University of Wisconsin and a law degree from the University of California. He served the United States as a First Lieutenant in the U.S. Army. He was a tax clerk with the California Franchise Tax Board, a certified public accountant, and an attorney with the Colorado Antidiscrimination Commission. From 1969 until 1974 he was an Associate Professor of Law at the University of Denver, and a member of the Colorado House of Representatives from 1966 to 1974.

In a protest against the high cost of running for public office, he chose to walk the state in his campaign for Governor, being elected in the Fall of 1974, and re-elected in 1979 and 1983. A noted environmentalist, he followed his career in public service as Director of the Center for Public Policy and Contemporary Issues at the University of Denver.

Two quotes from Governor Lamm capture his distinctive view of the world and the place of the United States in it:

> "America does not need another political campaign based on denial and avoidance of some of our real problems. It needs a crusade to reform and renew our country and its institutions and political system."

and

> "Our globe is under new dramatic environmental pressure: our globe is warming, our ice caps melting, our glaciers receding, our coral is dying, our soils are eroding, our water tables are falling, our fisheries are being depleted, our remaining rainforests shrinking. Something is very, very wrong with our eco-system."

Visionary Creation of the 27,000 Acre Homesteader's Legacy and the Church Ranch Revolution in Historical Perspective:

"This time they've gone too far…Something needs to be done"

This statement by Marcus Church in 1973 was made to the President of Fairfield and Woods law firm in Denver, Colorado. It expressed his exasperation that plutonium known after 1970 to be coming from Rocky Flats was causing people

not to purchase beef from his beloved cattle or land he had hoped to sell. But that statement in exasperation led to the filing of a lawsuit in 1975 after he himself had been sued earlier in 1975 by someone to whom he had sold land. Marcus Church did not live to see the settlement. The lawsuit was to drag on for over 11 years from that fateful meeting in 1973. The forced sale to the U.S. government of 1,250 acres in 1951 marked a stark change for an entrepreneurial pioneer family whose stewardship of the land went back to 1857.

Catherine Lutz (2001) of High Country News recounts the history of the acquisition of a portion of the Church Ranch by the Department of Energy as told to her by Charlie McKay about what his uncle, Marcus Church, had been told in 1951 when he had been compelled to sell an initial 1,250 acres to the U.S. government.

The promise to Marcus Church was for a development boom: workers would come from all over the country and new businesses would start up and residential developments would follow. As McKay recounts: "Everybody thought it [Rocky Flats] was going to a magnet...But that didn't happen." In fact, there was a major "boom"—but it was not over new workers, business, and real estate or even Winter Wheat and Hereford cattle development but over litigation and environmental pollution and crime.

The Church Ranch Companies maintain a compelling family history which traces the homesteaders to 1857. This is a photograph of Charles McKay, the current steward of Church Ranch who made the ranch history available.

The American People Have the Right To Know the Truth With Congressman Howard Wolpe

"He stood for justice for people at home and across the world"

Howard Wolpe earned an undergraduate degree from Reed College in Portland, Oregon and a doctorate from the Massachusetts Institute of Technology. Prior to becoming a seven-term member of Congress, he taught at Western Michigan University in the field of African political studies, and been elected to the Kalamazoo City Commission and the Michigan House of Representatives. Congressman Wolpe chaired the Subcommittee on Africa of the House Foreign Affairs Committee. He was a co-manager of the legislation which placed sanctions on South Africa known as the Comprehensive Anti-apartheid Act of 1986. He was an author of the African Famine and Recovery and Development Act which included creating the African Development Fund.

During the Clinton administration Howard Wolpe was Special Envoy to the African Great Lakes Region. In that role he participated in the peace process which aimed to end civil wars in Burundi and the Democratic Republic of Congo. In 2009, President Obama appointed him Special Advisor for Africa's Great Lakes Region. He retired from the Department of State a year later. Among the many recognitions he received were the African American Institute's Star Crystal Award and the Sierra Club's Lifetime Achievement Award.

His chairmanship of pivotal Subcommittee on Investigations and Oversight of the House Science, Space and Technology committee made a significant impact on the history of Rocky Flats. This occurred in his seventh and last term in

Congress. The use of the subpoena began when institutions of the Executive branch were not cooperating with the subcommittee's investigation of Rocky Flats. The use of the subpoena transformed the documentary history of the production era. The preservation of the archives were invaluable to the preparation of this volume. The importance of the conclusions of the Wolpe subcommittee are reported in three volumes of supporting documentary evidence, testimonies under oath, and staff interviews.

U.S. Senator from Michigan Debbie Sabenow said of Howard Wolpe: "He dedicated his life to public service and brought the kind of civility to politics that is sorely missing today...He stood for justice for people at home and across the world."

Testimony Under Oath By Former FBI Agent Jon Lipsky:

"That plutonium dust is all over the site. It can potentially kill people. It is a cover-up..."

This testimony in Federal District Court in 2018 was part of an unsuccessful attempt to obtain a Preliminary Injunction against the opening of 15 trails across the Wildlife Refuge. Hallmarks of Jon Lipsky's career as a law enforcement officer have been to protect and defend the Constitution of the United States and the people of America, and to tirelessly seek out the facts as part of on-going investigations to wherever they might lead. At Rocky Flats that commitment to obtaining the "best evidence" available began when William Smith of the Criminal Investigation Division of the Environmental Protection Agency gave him in 1987 the Mary L. Walker Briefing Memo of 1986 with its assertion that elements of Rocky Flats were "patently 'illegal'."

Lipsky undertook the two-year criminal investigation under provisions of the Resource Conservation and Recovery Act and the Clean Water Act. A U.S. Magistrate approved his request for a Search Warrant. Special Agent Lipsky led the 90-agent FBI/EPA raid that lasted 18 days and seized over 1 million pages of documents.

In keeping with his commitment to the rule of law, he accepted a subpoena from the Wolpe subcommittee and testified fully despite Department of Justice lawyers assigned to challenge questions they did not want fully answered.

Fast forward to the May of 2016, Jon Lipsky was featured by 5280, Denver's Mile High Magazine as a "Rogue Agent." The translation of the meaning of "Rogue Agent" was someone who states the facts as he knows them to be true based on evidence. In the time-frame of the "Rogue Agent" article Lipsky had countered the positions of Colorado Department of Public Health and Environment that the area of the Wildlife Refuge was perfectly safe despite it having had no remediation after the 38-year production era.

He testified under oath before the jury in the Cook class action lawsuit and supported with facts that plutonium which had migrated from Rocky Flats onto plaintiffs' property constituted an "unreasonable and substantial" nuisance. This returns to the Oath he took when he became an FBI Agent: to "protect and defend." Lipsky wrote a 12-page letter to the Colorado Solid and Hazardous Waste Commission with a bottom line that: "Absent competent soil and air sampling and analysis, the remedy, with a reasonable scientific certainty, is not protective of human health and environment." His method of operation is to pursue an investigation until all the relevant evidence has been uncovered.

Frazer Lockhart, The "Accelerated Cleanup and Concept"

"You have to do the work safely before you can do the work."

Frazer Lockhart is now a Regulatory Stakeholder Interface Program Manager for the Los Alamos Legacy Cleanup contractor. In this challenging role he provides "proactive interface and coordination" with regulatory, tribal and stakeholder organizations in support of the Los Alamos Legacy Cleanup Mission. After leaving the Rocky Flats assignment he was Assistant Vice President for Environmental Programs at Stoller Newport News Nuclear which provides:

> "...complex nuclear facility management and high—consequence operations expertise with full service environmental remediation, radiological characterization, deactivation and demolition (D&D), waste management, and site closure services to federal agencies and a variety of private and public clients."

Prior to service at Rocky Flats, Mr. Lockhart had spent twenty-nine years at the Department of Energy. Among those challenging DOE assignments were to focus on high-level liquid wastes in Idaho, Hanford and Savannah River. The Service to America Medal recognition conferred by the non-profit Partnership for Public Service cited the Department of Energy as the source:

> ...the cleanup team removed more than 21 tons of weapons-usable nuclear materials; decontaminated and demolished 800 structures comprising more than 3 million square feet; drained 30,000 liters of plutonium solution; stabilized and packaged 100 tons of high-content plutonium residue; performed environmental cleanup actions at 130 sites; disposition of millions of classified items and excess property; and safely shipped more than 600,000 cubic meters of radioactive waste to a safe disposal site-enough to fill a string of railcars 90 miles long. (Samuel J. Heyman, Service to America Medal)

Other recognitions for Frazer Lockhart and his team included:

- The American Council of Engineering Companies: Grand Award Winner in the Environmental Category;
- The American Academy of Environmental Engineers: Project of the Year; and
- Secretary of Energy's 2006 Award: Project Management Improvement.

Frazer Lockhart is a retired Colonel in the U.S. Army Reserve. In his new role with the Los Alamos Legacy Cleanup Frazer Lockhart met with the Northern New Mexico Citizen's Advisory Board. He told the Board that "safety is the keystone of...performance on cleanup contracts." Further, "You have to do the work safely before you can do the work."

Mark N. Silverman and His Address to the Class of 1961 At West Point:

"...we need to develop and employ different skills than just those needed on the battlefields for which West Point so ably trains us."

Mark N. Silverman retired on June 1, 1996 after nearly 34 years of Army and federal government service. In June of 1995 he was awarded by President Clinton the Distinguished Executive Service Rank for his leadership in cleaning up Rocky Flats from 1993-1996. He was the leader of the team which created the 1994 strategic conception. He is credited by Frazer Lockhart with identifying how dysfunctional the regulatory constraints were to crafting a new approach to cleanup and environmental restoration of nuclear weapons production sites.

Three meaningful perspectives add value to understanding the life and work of Mark N. Silverman: an article in Time by Michael D. Lemonick on March 27, 1995 (Lemonick, 1995), an article in Assembly (a publication within the United States Military Academy) which Silverman co-wrote recounting the meaning of the breadth of his experiences (Silverman and Hoese, 1997), and Silverman's compelling address to the Class of 1961 at the United States Military Academy at West Point which concludes the Epilogue. The insightful Lemonick article begins with an observation which could apply to anyone who has worked in management in complex political organizations for over 30 years:

> "As a long time bureaucrat, Mark Silverman knows the usual rules of the game: cover your flanks, avoid making decisions, bury all problems under layers of paperwork."

The thrust of the Lemonick article was that Silverman knows he is sitting on a "time bomb": "Enough radioactive waste remains on the [Rocky Flats] premises to cover a football field to a depth of 20 feet."

The problem which Silverman encountered was that: "...most of what he wanted to do was against regulations." Two examples made Lemonick's point. Before Silverman could authorize the dismantling of five old guard shacks, an existing federal regulation had to be worked around that first buildings had to be offered as housing for the homeless. An existing security rule had to be streamlined because it currently took nine separate signatures before entry to security production areas could be permitted. There was a much longer list of obstacles to progress in the transition from Cold War nuclear weapons production to safe deactivation.

A more general challenge facing Silverman and his management team was that: "Of the 250 Cleanup 'milestones' set by the EPA and Colorado's Department of [Public] Health and Environment, only two dozen spelled out concrete action." A quote from Silverman underscored the transformation task: "The people who wrote these procedures had little idea of how things actually worked...They were more worried about going to jail than about plutonium risks."

In "Cleaning Up the Cold War Legacy," Silveman and Hoese (1997) highlighted that the nuclear weapons complex during World War II and the Cold War "was totally focused on the nuclear weapons production mission. It ignored what would happen when the U.S. won the Cold War."

Silverman and Hoese (1997) stated: "The story of Rocky Flats reflects the story of the Cold War itself—from the buildup on nuclear arms through the demise of the Soviet Union, including the contamination and hazardous waste left behind." They quoted Nobel Laureate Neils Bohr in 1939 that building an atomic bomb "can never be done unless you turn the entire United States into one huge factory." While the nuclear weapons complex of the United States never became "one huge factory" it did become a massive multi-state industrial complex.

Rocky Flats was originally welcomed to the Denver area in March of 1951. The then Mayor of Boulder, Colorado called the plant a "wonderful thing for

Boulder. It may bring some problems, but they are far-out-weighted by the advantages." And the Denver Post welcome characterized Rocky Flats as: "a source of satisfaction to all residents who have an abiding faith in Colorado's destiny and future greatness."

After 1989, Silverman and Hoese (1997) characterized the situation at Rocky Flats as: "...no production, no clear mission, no real deadlines, and regulatory and political turf battles only confused matters."

In 1989 the budget for Rocky Flats was less than $500 million with approximately 4,000 workers. By 1992 the budget was $800 million with almost 8,000 employees. The primary work product before Mr. Silverman became manager was "nothing got done—except for interminable studies required by regulators...and analysis, reports and debates being most of the work product." For 1994, the budget had increased to almost $900 million. It was in this era that Mark N. Silverman and Frazer Lockhart and others began the re-conceptualization of the complex cleanup.

Mark N. Silverman was laid to rest in the Cemetery at West Point. His daughter Tamra Raven described him as someone who had overcome polio as a child to graduate from West Point and serve four years in the 82nd Airborne Division. Her Father "dedicated his life to making a difference." We are indebted to his wife Trish for choosing the photograph to share.

Exemplary Scholarship In the Public Interest: Wes McKinley and Caron Balkany (2004) for The Ambushed Grand Jury and Jon Lipsky, Wes McKinley, Caron Balkany and Jacque Brever (2005) for the Citizens Investigation of 2005 (and continuing):

"This cowboy needs legal advice"

Caron Balkany provided a recorded interview on March 25, 2004 for the Maria Rogers Oral History Project. In it, she indicated that: "I've been a public interest attorney for at least two decades..." After her home in Florida was destroyed by a hurricane, she and her family relocated to Santa Fe, New Mexico. She became intrigued "with citizens there actually having an impact on something as global as nuclear weapons and nuclear waste." She spent seven years litigating against Los Alamos. Wes McKinley had come to Santa Fe to speak to Concerned Citizens for Nuclear Safety. A colleague introduced them with the comment, "This cowboy needs legal advice." As Attorney Balkany explained the dilemma which brought them together:

> I was trying to figure out how to solve his problem, which was being told by the judicial system that he couldn't talk about what happened, but knowing that he had taken an oath to uphold the law and that illegal activities had happened [had been extensively reported] behind the closed doors of the grand jury chambers which prevented him from upholding his oath. He was in quite a dilemma.

The dilemma was resolved by the exemplary scholarship commemorated in this heartfelt Appreciation. A subtitle of The Ambush Grand Jury is about catching "red-handed" the U.S. government through actions by both the Department of Justice and the Department of Energy in outright lies and extended cover-ups. Tireless use of the Freedom of Information Act with the zeal of federal prosecutors was but one inspired approach.

After the successful completion of The Ambushed Grand Jury, McKinley and Balkany went on to collaborate with Retired FBI Agent Jon Lipsky and Jacque

Brever in the Citizens Investigation of 2005 which is on-going (see Appreciation of Jon Lipsky).

Wes McKinley went on to hold elected political office in Colorado as a member of the House of Representatives from 2005 to 2013 from the 84th District in Eastern Colorado. On August 5, 2010, Patricia Calhoun (Calhoun, 2010) of Westword published an interview with Representative McKinley where they discussed the discovery of "breathable particles of plutonium" near the former Rocky Flats Plant site. In a comment on The Ambushed Grand Jury book project, McKinley said: "Since DOE is hiding its damaging data, I figured we'd just collect our own." McKinley and Balkany had the vision to understand the importance of the archives of the Wolpe Congressional investigation. They preserved the chain of custody of the archives which were such a valuable resource for this volume, and made available through the "good offices" of former FBI Agent Jon Lipsky.

This photograph is available courtesy of Eco-Ed and The Ambushed Grand Jury. McKinley and Balkany are pictured with the late Jackque Brever (June 19, 1957 - October 15, 2015) in the middle who was their colleague with Jon Lipsky on the Citizens Investigation of 2005.

The motivation for The Ambushed Grand Jury is documented on-line in terms of a series of questions:

> What do you do when you find out someone is committing a deadly crime? Usually, you go to the police. And then the government will enforce the law. But what do you do when it's the U.S. Government and its corporate contractors committing the crimes? And another part of the government is covering it all up? What happens when citizens investigate the crimes of their own government?

Jacque Brever was a Rocky Flats glovebox operator who elected to become a whistleblower to the Grand Jury. In a compelling segment of Frontline's "Secrets of a Bomb Factory," Ms. Brever recounted being called in to a 3:30 a.m. shift in an unfamiliar glovebox line. When she pulled her hand out of the glove, her face was contaminated with plutonium. Walking down the hall in shock, escorted by a co-worker, someone she did not know said "That's what you get for being a rat!"

The U.S. government discounted her credibility as a witness. But less than ten years later Rockwell International itself in a Stone False Claims Act filing against the Department of Energy confirmed the authenticity of her testimony of constant Spray Irrigation over the East Spray Fields in all weather conditions.

Patricia Calhoun, the Editor of Westword, published an interview with Jacqueline Brever in the issue of October 21-27, 1992. Describing her experience as the first Rocky Flats employee to appear before the Special Federal Grand Jury, Ms. Brever said: "I went in there, totally scared. I was scared to death…The first day was tough. Those people were looking at me. I don't think they believed me."

After testifying at length recounting incidents she had observed over her five years at the Plant, "Most jurors looked disgusted that a company could do those things." One Grand Juror followed her into the restroom and gave her a religious medal and told her that they "appreciated what she was doing." While not a religious person, Ms. Brever carried the medal for the rest of her life. She told Ms. Calhoun that there were no guidelines for her job as a "subject matter expert, plutonium operations recovery." Since there were not guidelines, she started keeping a journal and began memorializing procedural violations as well: "I got a reputation as a troublemaker." The incident which contaminated her with plutonium dust occurred on September 14, 1989, over three months after the FBI raid.

In a concluding part of the interview Ms. Brever said:

> ...I have some pretty strong feelings about no one being indicted. Some of them are still there, getting a paycheck, getting a promotion, waiting to retire at the taxpayer's expense—and they did these things. I 'm very very proud of the grand jurors. Their consciences are clear.

Major General Edward Giller and the Defense of Freedom:

"It was a long war that went on all too long"

Once again I am indebted to the scholarship of Len Ackland in Making Real Killing. Professor Ackland is an expert on the 1969 plutonium fire. He had included a passage from a House of Representatives Hearing related to the fire. The passage introduced me to a mutually knowledgeable and civil exchange between the U.S. Congress in performing their constitutional duties in the budget process, and a representative of the Atomic Energy Commission who was exceptionally well prepared to justify an allocation request for Rocky Flats. Given the lack of civility which is characteristic of recent partisan exchanges in U.S. Congress, I was curious to learn more about the wellsprings of such mutual personal respect in pursuit of the national interest and security of the United States.

As we all know, the resources of the World Wide Web and Google have become great research tools to start an inquiry. The University of Illinois had linkages to the history of Major General Edward Giller who had flown 115 combat missions over Europe during World War II including air cover for the Normandy invasion. After the war he returned to the University Illinois and earned a Ph.D. in Chemical Engineering. A little known fact in one of the articles was that the aircraft he flew during World War II (it turned out there had been 13 of them) had been named the "Millie G."

Wanting to learn about the rest of his career, since by then (2015) he would have been about 95 years of age, I looked for his Obituary. I could not find it. White pages.com had a listing for Edward and Mildred Giller in Albuquerque, New Mexico with a telephone number. He answered the telephone!

It turned out that we knew in common about a dozen U.S. government officials. It took over six months with the assistance of family members Paul and Katy Giller to arrange to spend a day with him in Albuquerque with a television crew. After earning an undergraduate degree in Civil Engineering from the University of Illinois, Edward Giller had taken an engineering position with Sinclair Oil Refinery in Houston, Texas. On his own time he took private flying lessons, earned a private flying license, and capped it off with aerial acrobatics training. Edward Giller joined the 38th Fighter Squadron as a Second Lieutenant on April 29, 1942. He transferred to the 343rd Fighter Squadron on February 2, 1943.

His son Paul Giller read on camera the log of then Major Giller's 115th combat mission which culminated in strafing runs over the Luftwaffe Brunthal Airfield which is 12 miles south of Munich, Germany. Paul took his father through the mission as it unfolded. Strafing to destroy German fighters on April 15, 1945 was being conducted at a low altitude. A shell from an anti-aircraft battery hidden in a nearby stand of trees exploded in the cockpit of Major Giller's P-51DCY-G44.

Major General Giller recounted that the explosion felt like he had been hit on the shoulder by a sledgehammer. Reaching into his flying suit, his hand came out bloody. An initial concern was that he would have to "belly the plane in" and hope that German medics would patch him up. But the engine was running fine. He told his wingman that: "We've going home." He had the presence of mind, under extreme duress, to fly one-handed on the two-hour flight across Germany, France and the English Channel and land at his home base in England.

Again on camera, Paul held up a framed case of medals from World War II. General Giller's comment was that: "It was a long war that went on all too long." The framed case contains the French Croix de Guerre and Silver Star Medals for bravery, the Purple Heart, and the Air Medal with 17 Oak Leaf Clusters (and at least a dozen others).

After World War II General Giller's service to the people of the United States covered the administrations of every President of the United States from Franklin Roosevelt to Jimmy Carter. This included being responsible for weaponization of the atom for the Atomic Energy Commission from 1967 through 1972 which covered for this study response to the plutonium fire of 1969, supervision of the preparation of the fire reports, and obtaining funding from the U.S. Congress for essential post-fire reconstruction (how he first came to the attention of this writer). Following his service for the Atomic Energy Commission, General Giller had been a member of Ambassador Herbert York's three year mission to Geneva for the Comprehensive Test Ban Treaty negotiations with the Soviet Union: "We failed!" His service concluded as a Commissioner of the Standing Consultative Commission of the US/USSR Deliberations On Compliance with SALT I, SALT II, and the ABM Treaties.

The Air Force Distinguished Service Medal was bestowed upon General Giller "for exceptionally meritorious and distinguished service in a position of great responsibility to the Government of the United States." On the occasion of his birthday on July 8, 2016, Governor Susanna Martinez of the State of New Mexico declared "Defense of Freedom Day." Both U.S. Senators from New Mexico extended their congratulations on behalf of the people of New Mexico.

Dr. LeRoy Moore On Peace Justice and Fundamental Threats To Human Life:

"...they [nuclear weapons] cannot be produced without doing violence [to]...workers and to the surrounding environment."

LeRoy Moore is a co-founder of the Rocky Mountain Peace and Justice Center which is distinctive in being non-hierarchical and inclusive. There are over 1,000 members encouraged to become part of what they choose or embrace their own vision of progressive personal and social change.

The Rocky Mountain Peace and Justice Center was founded in a spirit of unconditional nonviolence and a belief in the interconnectedness and interdependence of life on the planet. By way of illuminating more of what the Rocky Mountain Peace and Justice Center strives to accomplish, they have a section with the following elements:

- Enabling people to participate actively in democratic processes;
- Fostering a vision of a just and sustainable society;
- Providing an alternative voice to mainstream media;
- Contributing to global movements for peace, justice and ecological responsibility;
- Providing educational opportunities for the next generation who aspire to foster constructive change;
- Educating on specific issues.

An example of the seriousness of purpose behind achieving these results is the use of their legal standing to engage institutions of U.S. constitutional government on behalf of furthering the mission and vision of the organization. A recent example is utilizing the often decisive role of the Judicial branch of "separate but equal" constitutional government with its power of subpoena to engage the Executive branch over the meaning of the rule of law and the obligations imposed by the National Environmental Policy Act of 1969, in particular on the issue of the future of the Wildlife Refuge. This is s significant contribution to one of their core values: promoting democracy in America.

Finally, reflecting on the enduring contributions of LeRoy Moore, it is necessary to consider his exceptionally careful scholarship. A timeless example is his sharing the indelible contributions of Drs. Carl Johnson and Ed Martell. History has shown that they were correct in identifying that the nuclear age is unforgiving of mistakes, and that there are dangerous consequences for health, environment and safety when both the chemicals of the synthetic revolution after World War II, and ionizing radiation do not remain where they were intended to remain.

The core beliefs of LeRoy Moore include the following:

- "...ridding the planet of all nuclear weapons"
- "Nuclear weapons epitomize violence, even when they are not used."
- "...they cannot be produced without doing violence to workers and to the surrounding environment."
- "These weapons indeed pose two fundamental threats to ongoing human life: nuclear holocaust or ecological disaster."
- "...nuclear weapons present a third great danger. Their making, testing, and deployment require a secretive, centralized, non-democratic command structure."
- "...nuclear weapons have never been produced without polluting the environment and endangering workers and the public."

In one of his most influential writings, the Citizens Guide to Rocky Flats, Moore (1992) chose to quote from Martin Luther King, Jr. (King, 1997, p. 19) to represent one of his core beliefs: "A nation that continues year after year to spend more money on military defense than programs of social uplift is approaching spiritual doom." Dr. Moore published a very important book in 2019 entitled "Plutonium and People Don't Mix: Rocky Flats."

Adam Babich and the Sierra Club Lawsuit:
"Does the Sovereign Have a License to Pollute?"
Professor Adam Babich of the Tulane University Law School observed that the Department of Justice had undertaken two key initiatives in the 1980s:

- "While conducting highly publicized civil and criminal enforcement against private violators of antipollution laws, the U.S. Department of Justice has mounted a vigorous campaign to insulate federal agencies from those same laws."
- "Along the way, the department has cemented the U.S. government's reputation as one of the nation's most persistent violators of environmental statutes." (Babich, 1991, p. 28)

By way of explaining the motivation of the Department of Justice, and answering the question of "why does the government resist so adamantly?" Professor Babich's answer was "turf" and "money." He characterized both the Departments of Defense and Energy (and others) as: "It is now beyond dispute that the Departments of Defense and Energy, and some other federal agencies, have been poor stewards of public resources" (Babich, 1991, p. 28).

An added problem is that: "...the strict, joint and several, and retroactive liability provisions of the Comprehensive Environmental Response, Compensation, and Liability Act (CERCLA) took many federal agencies by surprise" (Babich, 1991, p. 28). As a practical matter, in determining whether it was the intent of Congress that the sovereign (the United States) could be sued, the courts allow suit only if Congress has "clearly and unambiguously" waived immunity. Babich noted that: "All major federal antipollution laws waive sovereign immunity (Babich, 1991, p. 29)." But as Babich notes, "the U.S. government cannot afford to clean everything at once" (Babich, 1991, p. 65).

Key issues raised by the Babich article did not mention the landmark and transformative Sierra Club case at Rocky Flats where his advocacy resulted in a significant defeat for the position of the U.S. government on behalf of both the Department of Justice and the Department of Energy. But in the vernacular of Babich (1991), the translation of "turf" was preserving Rocky Flats as a site for nuclear weapons production. The translation of "money" was "lots of money measured not in millions but in billions."

Adam Babich had been Editor-In-Chief of ELR-The Environmental Law Reporter and director of the division for publications of the nonprofit Environmental Law Institute. He had been Assistant Attorney General for the state of Colorado.

Professor Babich is best known at Rocky Flats for his service as lead counsel for the plaintiffs on three citizen suits. In 1989, before the famous RCRA case was decided in Judge Babcock's courtroom, there was a lawsuit filed on February 1, 1989 accusing Rocky Flats of violating environmental law by failing to get a permit for the Building 771 incinerator. Such a permit would make the facility subject to federal and state emissions standards. On behalf of the plaintiffs, Sierra Club spokesman A. Blakeman Early stated: "The emissions generated at Rocky Flats are more toxic to people than anything regulated by the Environmental Protection Agency." The position of Adam Babich was that the underlying legal issue was part of the RCRA litigation and would be determined there. But he also stated that:

...DOE is claiming that their plutonium-recovery process, and any other related process, is not regulated by RCRA...

To that DOE position Babich said the answer is "no."

The April 12, 1990 decision by Judge Lewis Babcock with Babish as the lead attorney (Sierra Club, 1991) noted that: "DOE does not dispute Sierra Club's factual or legal assertions." Further, "Since 1980, DOE through its operator, has been obligated...to apply for a RCRA permit for hazardous waste operations at Rocky Flats." A concluding comment from the Judge was that "DOE has advanced no cogent reason this civil determination would have 'punitive implications' relative to any ongoing criminal investigation. Such concerns are speculative."

Babich was in his continuing role as Attorney for the Sierra Club when it was Judge Babcock again who found on August 13, 1991 that the Department of Energy was in "flagrant" violation of the central permit requirement of RCRA. The Babcock Court found:

> ...the record in this case shows a constant pattern of [DOE] delay and obfuscation. [Order at 15]
>
> DOE's demonstrated attitude is that it is a governmental agency that can avoid RCRA's mandates indefinitely with impunity. [Order at 16]
>
> DOE's ongoing disregard for RCRA's linchpin permit process has been flagrant. Absent appropriate sanction, I have no credible reason to believe that DOE will comply with a two-year time requirement. Therefore, I conclude that nothing less than the threat of shutdown on noncompliance with this order will effectively enforce the requirement that DOE obtain a RCRA permit for the mixed residues now stored in violation of RCRA. [Order at 16-17]

Patricia Calhoun of Westword and the Decision To Publish the Leaked Grand Jury Report: *"Ode to Constitutionally Protected Freedoms"*

Patricia Calhoun is a co-founder of Denver's highly successful publication Westword and has served as its Editor since its inception. She writes a regular feature: "Calhoun."

Westword is distinctive for its in-depth coverage of arts, culture, restaurants, and compelling local news. It offers timely newsletters on the top news stories of the week, on music, food and drink, film and TV, things to do, marijuana, arts and culture, and happy hours. In addition to being embedded in the very fabric of Denver, Westword has been distinctive for its in-depth coverage of developments at Rocky Flats.

A defining instance of Patricia Calhoun's courage as Editor, and emblematic of her commitment to exercising freedom of the press enshrined in the U.S. Constitution, was her decision to publish the leaked Special Federal Grand Jury Report. Another example of the constructive role of Westword, beyond Patricia Calhoun's decision to publish the leaked Special Federal Grand Jury Report, is the vital role it has exercised in keeping the broader Denver communities informed about what the U.S. government did not want known based on its secrecy policies.

An illustration of the vital Westword role is provided by Eileen Welcome's contribution of July 27, 2000 "This place is a Dump." In one paragraph Eileen Welcome captured the toxic legacy of the 38-year production history:

> In the beginning, the paper bags, cardboard boxes, and drums filled with chemical and radioactive leftovers were simply pushed out the door and forgotten until someone somewhere higher up in the Atomic Energy Commission hierarchy came up with a solution. The plant didn't have time for such housekeeping chores, anyway. It had a vital mission: to produce nuclear pits, the smooth cores of plutonium or enriched uranium designed for the center of atomic bombs.

Welcome lamented that Federal law prevents Grand Jurors from telling what they know: "...still, the evidence was all there...in the soil, the water and the air." And that battle for the disclosure of the truth is going on compellingly in 2020 in Federal District Court for the District of Colorado.

Harley Alley and the Stone False Claims Act Lawsuit:
"For as long as it takes..."

On July 5, 1989 Federal District Court Judge Jim Carrigan sealed a $1 billion lawsuit filed by Hartley Alley against Rockwell International. Over a year later in November of 1990 it was unsealed. The basic allegation was that Rockwell had cheated the government. An interview with Alley by Carl Hilliard of the Associated Press was released on November 14, 1990. In what is vintage Harley Alley, he made a commitment to pursue the lawsuit "for as long as it takes." Fast forward over 17 years to 2007 when Justices of the U.S. Supreme Court considered the case of Rockwell International Corporation et al. v. United States et al. Jim Stone had been hired as an engineer by Rockwell and laid off as part of a reduction in force in 1986. He had been basically a troubleshooter with a mandate to find problems and suggest corrective action solutions to management.

In October of 1982, Stone sent a memo to Rockwell International management which shared his concerns that a proposal to mix pond sludge with cement would not work. The legal issue before the U.S. Supreme Court was whether Mr. Stone was an "original source" under the Federal False Claims Act. By then the representation by Hartley Alley had resulted in prevailing at both the Federal District Court and Court of Appeals levels. There were ancillary issues of the submission of false claims by Rockwell International to the U.S. government, and concealing environmental, health and safety problems from DOE.

Jon Lipsky interviewed Mr. Stone as the first insider who had come forward in response to the FBI criminal investigation. This was Special Agent Lipsky's recollection in 2007 of his meeting with Jim Stone in 1988:

> He was my first insider. I had a pretty good idea of how Rocky Flats was put together, but it was almost unbelievable. He was talking about stuff he knew, and he was a very intelligent man, an engineer, on a different plane than I am. He was rattling it all off, and I had to stop to ask questions. Some of the stuff I could collaborate because I'd reviewed the waste steams out there.

Lipsky recalled that Stone also had 2,300 pages of documents along with a revelation that the ductwork contained lost plutonium.

The U.S. government had first chosen not to participate in the lawsuit filed by Hartley Alley on behalf of Jim Stone. Subsequently there was a change in government policy in 1995. One of the supreme ironies is that the U.S. Supreme Court upheld the jury verdict in the amount of $4.2 million in favor of the plaintiffs but not the portion of the verdict before them that Mr. Stone was an "original source" under the law. James Stone died the next month making his case legally moot.

Hartley Alley wrote a bitter sweet letter he sent to U.S. Senators Ken Salazar and Wayne Allard of Colorado indicating Jim Stone was a:

> ...true hero in Colorado history...Had Rockwell's crimes and practices gone unchecked, there is no telling what further injuries and harm would have befallen Colorado and the health of its citizens...Mr. Stone's career was ruined because of his reporting the crimes at Rocky Flats—something he did as a patriot on behalf of the people of Colorado...

There is an important postscript about Mr. Stone's claim to FBI Special Agent Jon Lipsky in 1988 about lost plutonium in the duct work. This claim was subsequently verified, and might have resulted in him being declared to have been an "original source" under the U.S. Supreme Court ultimate decision.

In 1989 Scientech, Inc. conducted an independent review of criticality safety at Rocky Flats. Scientech, Inc. concluded that plutonium dust had been deposited in the ventilation exhaust systems of gloveboxes for the length of the production era. There was concern that a risk existed of an uncontrolled criticality incident.

The Defense Nuclear Facilities Safety Board, a Congressional oversight committee, recommended the initiation of the Duct Remediation Project. The ventilation exhaust systems were stainless steel pipes of 4 inches to 24 inches which connected the gloveboxes to high efficiency particulate air filter plenum systems which prevent the release of toxics into the environment.

The Duct Remediation Project concluded that there were 1000 kilograms of plutonium dust plus numerous other toxics in the Rocky Flats ventilation exhaust systems including fluorides, chlorides, oxides, calcium, magnesium, graphite, and oil.

Merrill G. Davidoff and the Cook Class Action Lawsuit

Merrill G. Davidoff is Chairman Emeritus and a Managing Shareholder for Berger Montague PC. He continues as Co-Chairman of the Antitrust Department with

Mr. Montague and is Chairman of the Environmental Group. In the Cook class action lawsuit, he was lead counsel and lead trial counsel for the 2005-2006 trial, and argued the appeal to the 10th Circuit Court of Appeals before Judge Neil Gorsuch, now a U.S. Supreme Court Justice. The Rocky Flats trial team, led by Mr. Davidoff, received the prestigious Public Justice Award as "Trial Lawyer of the Year."

It is a very instructive exercise to review some of the key actions by Mr. Davidoff in the lead-up to the 2005-2006 trial and in the subsequent presentation to the 10th Circuit Court of Appeals. Davidoff reported that Judge Kane accepted the stipulation of the parties "to accept a less-than-unanimous verdict (i.e. One supported by at least 8 of the 10 jurors who remained by the close of deliberation)." This made it more likely that there would be a verdict. Judge Kane also accepted the suggestion of Counsel for plaintiffs that: "...the remaining unresolved legal issues be addressed in a carefully structured pretrial process devoted to formulating the jury instructions."

This gets to the nature of the jury instructions and whether as formulated they facilitated, in a preponderance of evidence case, the position of the plaintiffs or the position of the defendants. The area of the Property Class based on the evidence did contain plutonium and it was determined to be a continuing tort. One part of the instructions on trespass by Judge Kane was that a landowner has a right "to keep his property free from contamination deposited there by others without the landowner's consent." Judge Kane further indicated that the parties had stipulated that plaintiffs had not consented to plutonium being on their properties. A key instruction from Judge Kane which favored the position of the plaintiffs was that:

> ...Plaintiffs are not required to show that plutonium is present on the class properties at any particular level or concentration, or that they suffered bodily harm because of the plutonium...

While there were two more elements of trespass what had to be proven, this first element based on the stipulations was straightforward to present to the jury. It cannot, of course, be known what was in the minds of the jurors which was persuasive. But in 2015 Judge Gorsuch of the 10th Circuit Court concluded that neither the jury instructions nor the evidence had ever been challenged.

A second major area of the actions of the attorneys for the plaintiffs which made the outcome what it became in 2016 was a critical choice after remand from the U.S. Supreme Court. Merrill Davidoff made a strategic decision to refile an amended complaint on only the state of Colorado nuisance and trespass claims. That "took off the table" the contentious issues related to the Price-Anderson Act.

Judge Gorsuch rejected the preemption defense by the attorneys for the defendants, and focused importantly on the unchallenged on appeal valid jury verdicts based on state of Colorado law. The settlement occurred the following calendar year for $375 million.

Howard Holme and the Church Cluster of Lawsuits

After earning an undergraduate degree from Stanford University and a law degree from Yale University, Howard Holme as a third-generation Denver native returned to join the law firm of Fairfield and Woods. After a visit from prominent farmer and rancher Marcus Church in 1973, the President of the firm turned the case over to young lawyer Howard Holme. Mr. Holme had been a Conscientious Objector during the Vietnam War, was a Quaker, and a strong environmentalist. His scant knowledge of Rocky Flats at the time had been based on news accounts after the 1969 fire, and having been fishing and duck hunting on nearby Church land. He gave an interview to the Maria Rogers Oral History Project on September 1, 2005. He reported in the interview that Marcus Church had gone to

the then President of Fairfield and Woods with the following lament after his "It has really gone too far…" opening comment.

> There's a big article in the newspaper that says that Rocky Flats is more contaminated than the Nevada Test Site. I've got my cattle all over it. People don't want to buy my cattle. They think they're contaminated. I've cooperated with the plant in the past, but something needs to be done.

This was the beginning. Howard Holme gave notice to the government of the United States of an intent to sue under the Federal Tort Claims Act. After the six month waiting period, a lawsuit was filed in Federal District Court for the District of Colorado. Dow Chemical and Rockwell International were named defendants under state of Colorado tort law.

Mr. Holme used the Freedom of Information Act and regular visits to the Rocky Flats Plant under federal discovery rules. He focused on the 1957 fire, the 1969 fire, accidents, incidents, and leaking of plutonium and other substances into the groundwater. The process of the lawsuit once its discovery phase got underway included hearings on soil sampling. At the time the U.S. government preferred soil samples at the depth of six to eight inches.

Federal District Court Judge Richard P. Matsch ordered pre-trial statements. That led to the submission of the 520-page statement that was not released to the public until 2018. One of the most interesting vignettes uncovered was that the 903 Pad had been named as the "Helicopter Pad."

Seven years after the lawsuit had been filed, a trial date had not been set despite repeated requests. Holme went to the 10th Circuit Court of Appeals with a motion to have Judge Matsch removed from the case. The response within a very short time by Judge Matsch was to dismiss the case because it raised political questions which under the Constitution deprived the Federal District Court of jurisdiction.

The 10th Circuit Court of Appeals disagreed, and ordered a trial in Federal District Court. That motivated the U.S. government to seek a settlement. It took three months of intense discussions in the courtroom of Judge Sherman Finesilver to reach a settlement.

From the vantage point of 2005, Howard Holme offered the following reflections:

> I was inexperienced, and if I were given that kind of case today, I would do it quite differently. What was reasonably obvious from the beginning was that the government was liable and the operators were liable. They were dealing with enormously toxic poisons, and they had handled them very carelessly.

A last word was: "I learned more about that, studied it more, in more depth than was really required for proving liability. So I would try it differently today than I tried it then…"

Len Ackland's Exemplary Scholarship In the Public Interest

Professor Len Ackland of the University of Colorado earned an undergraduate degree in History from the University of Colorado and a Master's degree in International Relations from Johns Hopkins School of Advanced International Studies. He began his career in journalism as a free-lance journalist in Vietnam in 1968.

One of the defining contributions of his long career was as Editor of the Bulletin of the Atomic Scientists. The Bulletin of the Atomic Scientists was a recipient of the National Magazine Award while he was Editor for its special issue on Chernobyl in 1987.

The original scholarship in Making a Real Killing: Rocky Flats and the Nuclear West uncovered many avenues for research which enriched this book, especially on thinking within the Atomic Energy Commission and on the aftermath of the 1969 fire in Building 776/777. Professor Ackland has been honored with a George Polk Award for Local Reporting in 1978 and a Guggenheim Fellowship in 2008.

Making a Real Killing: Rocky Flats and the Nuclear West is a model of scholarship, especially for his discovery of some of the most illuminating and well-hidden insightful vignettes from the Atomic Energy Commission era: that two staff of the Congressional Joint Committee on Atomic Energy who considered "Project Apple" (the code name for what became Rocky Flats) to be sufficiently safe that it could be located on 16th Street in Washington, D.C., or an AEC Press Release which stated that radiation was "not a factor." Patricia Nelson Limerick's review of Making a Real Killing captured the importance of this classic book: "Len Ackland's integrity and hard work reminds us how crucial energetic journalism is for a successful democracy." I have cherished being a student of Professor Ackland through his book for identifying such elements at Rocky Flats as just how far democracy and its institutions have lagged behind the development of technology.

UC Irvine Contributors

Dianna Sahhar from the School of Law Library made a major contribution to the accuracy of references to the numerous legal citations in Chapter Fifteen on the Cook class action lawsuit.

Over the period of two years, UC Irvine undergraduate students carefully went through the 19 boxes of Wolpe subcommittee subpoenaed documents to catalogue them by general topic. Most memorably, one of them came to ask me about a document he could not categorize: it turned out to be the Second (and unknown) Report of the Special Federal Grand Jury. It lacked a title page and therefore was not immediately recognizable, and is reported in Chapter Eleven.

Acknowledgements

Gratefully acknowledged are the individuals and organizations who made available images and photographs for this book including:

-The Department of Energy Office of Legacy Management,
-The Colorado Aerial Photo Service and John Young,
-Jon Lipsky for "good offices" in accessing the subpoenaed archives of the Wolpe subcommittee of the U.S. Congress and Eco-Ed and the resources of the ambushedgrandjury.com,
-Trish Silverman for the photograph of Mark Silverman,
-Barry Arthur from arkansason-line and the Arkansas Democrat-Gazette for the picture of the destroyed Titan II silo in Damascus, Arkansas,
-The New York Times for the Chinese plutonium site,
-Photojournalist Gail Fisher for the Techa River pictures in Russia,
-The Economist and the Associated Press for the photograph of the "man who saved the world."
-Major General Edward Giller and Paul Giller for welcoming a TV crew into the General's home including for the picture,
-The Department of Defense, the U.S. Air Force, and the U.S. Navy for photographs of accidents with B-52 nuclear weapons delivery systems and recovery efforts,
-Frazer Lockhart for access to photographs from Closure Legacy,and
-Merrill Davidoff for access to jury exhibits in the Cook class action lawsuit.

Sources of Inspiration

This study of Rocky Flats deriving lessons learned toward a safer 21st century nuclear world was sustained and encouraged by a number of contributors who are themselves inspirational.

John N. Gardner, a colleague since 1983, founded two organizations which are international in scope and about social justice. The first which he led for thirty years is the National Resource Center for The First-Year Experience and Students in Transition. The second is the Gardner Institute for Excellence in Undergraduate Education which he co-founded with life partner Dr. Betsy Barefoot twenty years ago to transform student success in colleges and universities throughout the Untied States. In addition to the Institute serving as publisher for this book, Professor Gardner's opening Foreword provides a blueprint for engaging college students.

Harry Ratner, a colleague since 1983, is probably the only founding staff member of both Southern California's PBS stations: KCET and KOCE. In addition to being the Editor of "Quest for Peace in the 20th and 21st Centuries" and "Toward a Sustainable 21st Century" videos, he directed all of the digital videos in "Instructional Videos Toward a Safer Nuclear World" for this book.

The opportunity to co-host Bill Moyers on his second visit to UC Irvine in 2005 (Mr. Moyers first came to UCI in the Presidential helicopter with Lyndon Johnson in 1964 for the dedication of the campus) began a fifteen-year collaboration with three warriors for successfully addressing great unsolved problems of global society: Anne Earhart, Sara Lowell and Herbert Bedolfe, III. The hallmark of their work is imagination, collaboration, and tirelessly undertaking daunting challenges toward a more sustainable 21st Century.

Both of my brothers, Scott and Rob, did something not surprising: they excelled way beyond me in Freshman English. They also made invaluable contributions to the readability of the book. Scott read an earlier version without pictures and asked: "What did the Mound area look like?" Rob read the final version of the manuscript and made an invaluable contribution to the clarity of expression of the narrative.

A cherished colleague for forty-three years, Norma Yokota has been helped in editing for a number of books over the years. A member of the Editorial staff of the MIT Press once acknowledged her by saying "every book needs a Norma." This book has been over ten years in the making for two reasons. The first reason is that all the critically important information was under the control of the U.S. government and the military-industrial contractors and therefore slow to emerge, and then only in contested circumstances through the efforts of individuals acknowledged in the Appreciations. The second reason is the contested struggle over what has been called "irreconcilable conceptions" over the future uses of the former lands of Rocky Flats in a context where 'time did not fly." This book has to have been Ms. Yokota's most challenging assignment, and all the sources of inspiration along with her are acknowledged most gratefully beyond words.

John M. Whiteley
Professor of Social Ecology, University of California, Irvine

Bibliography

Abas, B. (1989). "Rocky Flats: A Big Mistake From Day One." Bulletin of the Atomic Scientist, 45(10):19-24.

Ackland, L. (1999). Making a Real Killing: Rocky Flats and the Nuclear West. Albuquerque: University of New Mexico Press.

Adamsky, V. and Smirnov, Y. (1994). "Moscow's Biggest Bomb: the 50-Megaton Test of October 1961," Cold War International History Project Bulletin, 4(3):19-21.

AEC. (1957). Atomic Energy Commission Fire Committee Report: The Report of Investigation of Serious Incident in Building 71 [771] on September 11, 1957. Washington, D.C.: Atomic Energy Commission. Washington, D.C.: Atomic Energy Commission.

AEC. (1969). Report on Investigation of Fire Building 776-777 Rocky Flats Plant, Golden Colorado, May 11, 1969. Washington, D.C.: Atomic Energy Commission, August, 1969.

AEC. (1970). "Request by Major General Edward Giller for $78 Million for Repair for building 776/777 After the Fire of 1969 to Subcommittee of the House of Representatives Committee." October 1, 1970. Washington, D.C.: U.S. House of Representative.

AEC. (1972). "Number of Fires at Rocky Flats Plant by Years Since 1951." Letter from AEC Rocky Flats Area Office Manager Abbott to Albuquerque Operations Office Manager Donnelly, August 15, 1972. Washington, D.C.: Atomic Energy Commission.

AEC. (1974). "Investigation of a Tritium Release Occurring In Building 771 on September 3-4, 1974." Rocky Flats Repository Document IN-640. Washington, D.C.: Atomic Energy Commission.

Aguilar, J. (2015). "Critics Claim a Refuge Open to the Public is a Bad Idea. The Denver Post, October 11, 2015.

Aguilar, J. (2016). "Settlement Reached in Homeowner Lawsuit Against Rocky Flats." The Denver Post., May 19, 2016.

Aguilar, J. (2019). "Plutonium in soil sample near Rocky Flats five times higher than cleanup standard: State health officials taking a closer look after 264 picocurie-peer-gram of soil reading." The Denver Post, August 20, 2019.

Anonymous Grand Juror. (1993). "They All Knowingly Violated Environmental Laws." Los Angeles Times Magazine, August 15, 1993.

ATSDR. (2005a). The 903 Pad/Ryan's Pit Plume. Atlanta. GA: Agency for Registry, Centers for Disease Control U.S. Department of Health and Human Services.

ATSDR. (2005b). Rocky Flats Environmental Technology Site. Atlanta, GA: Centers for Disease Control U.S. Department of Health and Human Services.

ATSDR. (2010). Toxicological Profile for Plutonium. Atlanta, GA: Centers for Disease Control U.S. Department of Health and Human Services.

ATSDR. (2017). Priority List of Hazardous Substances. Agency for Toxic Substances and Disease Registry, U.S. Department of Health & Human Services. https://www.atsdr.cdc.gov/spl/resources/index.html

Babich, A. (1991). "Circumventing Environmental Laws: Does the Sovereign Have a License to Pollute?" Natural Resources and Environmental Law, 6(1):28-66, Summer.

Barker, C.J. (1974). "Contaminated Areas East of the Solar Ponds." Dow Chemical Report: Golden, CO: Dow Chemical Archives.

Bassler, D.M. (1961a). "Due to the Safety Hazard of Transporting It Is Recommended They Be Buried On the Plant Site." Letter from D.M. Bassler to E.S. Ryan, April 12, 1961. Golden, CO: Dow Chemical Archives.

Bassler, D.M. (1961b). "The Drums Should Be Buried On Site Due To the Fire Hazard." Letter from D.M. Bassler to E.S. Ryan, April 12, 1961. Golden, CO: Dow Chemical.

Beall, D. (1989). "Compliance With Environmental, Safety And Health Requirements Will Not Be Subordinated To The Plant's Production Mission." Letter from Rockwell International Chairman Donald Beall to Secretary of Energy James Watkins August 14, 1989. Pittsburgh, PA: Rockwell International.

Becker, K. (2004). Response of the Sierra Club During the 2004 Public Comment Period on the Draft Comprehensive Conservation Plan/Environment Impact Statement for the Rocky Flats National Wildlife Refuge. Golden, CO: The Sierra Club, April 21, 2004.

Biles, M. (1970). "Waste Generation Was Not Predictable." Letter to Dr. Myron C. Waddell, the Chairman of the Colorado Health Planning Council from Martin Biles, Director of the Division of Operational Safety at Rocky Flats, December 20, 1970. Golden, CO: Dow Chemical Archives.

Blush, S.M. and Heitman, T.H. (1995). Train Wreck Along the River of Money: An Evaluation of the Hanford Cleanup: A Report for the U.S. Senate Committee on Energy and Natural Resources. March, Washington, D.C.: US Senate.

Bohling, G. (2005). Kriging. University of Kansas, Lawrence, Kansas: Kansas Geological Survey.

Bowman, H.E. (1970). "One Day We Quit Burying Them." Statement of H.E. Bowman, Manufacturing Manager for Dow Chemical at Rocky Flats, and Subsequently General Manager. Golden, CO: Dow Chemical Archives.

Buckley, C. and Wu, A. (2018). "Where China Built Its Bomb, Dark Memories Haunt the Ruins." The New York Times, January 20, 2018.

Calhoun, P. (1992). "Rocky Horror" Westword, October 21-27, 1992, p. 3.

Calhoun, P. (2010) "Rocky Flats is supposed to be clean, but there's plutonium off-site, says Wes McKinley." Westword August 5, 2010.

Calkins, K.W. (1970). Internal Dow Chemical Memorandum From K.W. Calkins to Rocky Flats Plant Manager to L.M. Joshel, August 19, 1970. Golden, CO: Dow Chemical Archives.

Carson, R. (1962). Silent Spring. Boston, MA: Houghton Mifflin.

Carson, R. L. (1951). The Sea Around Us. New York: Oxford University Press.

Carson, R. L. 1955). The Edge of the Sea. Boston, MA: Houghton Mifflin.

CCEI. (1970a). "Report on the Dow Rocky Flats Fire: Implications of Plutonium Releases to the Public Health and Safety." Letter to Dr. Glenn T. Seaborg, Chairman, United States Atomic Energy Commission. Boulder, CO: Colorado Committee for Environmental Information, January 13, 1970.

CCEI. (1970b). "Plutonium Contamination in the Denver Area." Report to Governor Love and the People of Colorado. Boulder, CO: Colorado Committee for Environmental Information, February 24, 1970.

CDH. (1977). Radioactive Soil Contamination (Cesium 137 and Plutonium) in the Environment Near the Rocky Flats Nuclear Weapon Plant. CO: Colorado Department of Health.

CDH. (1990). Rocky Flats Surface Soil Survey 1970-1989. CO: Radiation Control Division, Colorado Department of Health.

Center for Public Integrity. (2017). Nuclear Negligence. Washington, D.C.: Center for Public Integrity.

Chapman. T.S. (1954). Sludge from Building 95 should Be Buried in a Shallow Trench. Golden, CO: Dow Chemical Archives.

Chapman, T.S. (1957). Releases of Smoke Up the Building 771 Smokestack Continued At Least Until Dawn. Golden, CO: Dow Chemical Archives.

ChemRisk. (1991a). Task 1 Report (R1), Identification of Chemicals and Radionuclides Used At Rocky Flats. Repository Document TW-362.

ChemRisk. (1991b). Rocky Flats Historical Investigation Interviews. Rocky Flats Repository Document RE-891. Golden, CO: Rocky Flats Archives.

ChemRisk. (1992). Reconstruction of Historical Rocky Flats Operations and Identification of Release Points. Project Tasks 3 and 4. Final Draft Report. Alameda, CA: ChemRisk, A Division of McLaren/Hart.

Chiles, J.P. and Delfiner, P. (1999). Geostatistics: Modeling Spatial Uncertainty. New York: Wiley Series in Probability and Statistics.

Church, Marcus. (1975). Freedom of Information Act Filing, Golden, CO: Marcus Church

Congressional Joint Committee on Atomic Energy. (1970). Transcript of Meeting to Discuss the Mound Area and Stream Pollution. Washington D.C.: U.S.House of Representatives, April 10,1970)

Dalton, R.J., Garb, P., Lovrich, N.P., Pierce, J.C. and Whiteley, J.M. (1999). Critical Masses: Citizens, Nuclear Weapons Production, and Environmental Destruction in the United States and Russia, Cambridge, MA: The MIT Press.

Davidoff, M. (2016). Personal communication.

deMontmollin, J.M. and Hoagland, W.R. (1961). "Analysis of the Safety Aspects of the MK 39 MOD2 Bombs Involved in B-52 G Crash Near Greensboro, North Carolina." SCDR 8-81. February 1961.

Diamond, J.M. (2005). Collapse: How Societies Choose to Fail or Succeed. New York: Penguin Books.

DOE. (1986a). Mary L. Walker Briefing Memo. Washington, D.C.: Department of Energy.

DOE. (1986b). "Comprehensive Environmental Assessment and Response Program (CEARP)-Phase 1 Installation Assessment at Rocky Flats." Golden, CO: Rocky Flats Repository Document RE-33.

DOE. (1990). Environmental Assessment For 881 Hillside (High Priority Sites) Interim Remedial Action. January 1990. Golden, CO: Rocky Flats Repository Document EN-19.

DOE. (1991a). Concerns Regarding Rocky Flats Operable Unit (881 Hillside) Interim Remedial Action French Drain Collection System.

DOE. (1991b). Phase 1 RFI/RI Work Plan-Woman Creek Drainage. Golden, CO: Rocky Flats Repository Document RF-945. August 20, 1991

DOE. (1991c). Operable Unit 6 Workplan. September 1991. Rocky Flats Repository Document FR-940. Golden, CO: Department of Energy.

DOE. (1991d). Rocky Flats Information OUs At the RFP. Rocky Flats Repository Document RF-938. Golden, CO: Department of Energy.

DOE. (1991e). Nuclear Facility Safety Committee Report On Fires During the Rockwell Era. Golden, CO: Rockwell International Corporation.

DOE. (1992). Phase III Resource Conservation and Recovery Act (RCRA) Facility Investigation/Comprehensive Environmental Response, Compensation, and Liability Act (CERCLA) Remedial Investigation (RFI/RI) of the 881 Hillside Area Operable Unit Number 1 (OU1) at the Rocky Flats Plant. Washington, D.C.: Department of Energy.

DOE. (1994). Making Contracting Work Better and Cost Less: Report of the Contract Report Team. Washington, D.C.: Department of Energy, February, 1994.

DOE. (1995). Estimating the Cold War Mortgage. Washington, D.C.: Department of Energy.

DOE. (1999). "False Statements and Claims By Fiscal Year and Date Related To Rockwell's Claim On Behalf of Earning Its Award Fee For Fiscal Year 1988." Submission to the Court in Response to a Pretrial Order in United States Ex. Rel Stone v. Rockwell, 282. F.3d 787, Washington, D.C.: Department of Energy.

DOE. (2002a). Proposed Interim Remedial Action Solar Evaporation Ponds. *www.lm.doe.gov/cercla/documents/rockyflats_docs/I101/I101-A-000287.PDF*

DOE. (2013). Wildland Fire Management Rider for the Rocky Flats, Colorado Site. July 2013, Doc. No. 504638. Washington, D.C.: Department of Energy.

DOJ. (1994). Review of the Handling By the Department of Justice (DOJ) and United States Attorney's Office (USAO) For Colorado of the Criminal Prosecution of Rockwell International Corporation (Rockwell) For Environmental Crimes At Rocky Flats Plant: Report From Mark H. Dubester and Steven Brunnel to Associate Attorney General Websten L. Hubbell. Washington, D.C.: Department of Justice, April 8, 1994.

Dow Chemical. (1954-1959). "Analyses of Plan Incinerator Ash Relevant To Assessing the Potential for Human and Ecosystem Harm." Site Survey Monthly Reports and Waste Disposal Reports. Golden, CO: Dow Chemical Archives.

Dow Chemical. (1954a). "The First Time Sealed Steel Drums Were Buried In the Mound Area." Waste Disposal Coordinating Group Monthly Report, November 2, 1954. Golden, CO: Dow Chemical Archives.

Dow Chemical. (1954b). "Efforts At Sludging Which Would Allow Shipment To the Atomic Energy Commission Site In Idaho." Waste Disposal Coordinating Group Monthly Report, November 2, 1954 and December 2, 1954. Golden, CO: Dow Chemical Archives

Dow Chemical. (1954c). "Full Report On Sludging of Oils With Bentonite Clay Unavailable." Waste Disposal Coordinating Group Monthly Report, November 2, 1954. Golden, CO: Dow Chemical Archives.

Dow Chemical. (1957a). "No Alarm was Sounded in the Superintendent's Office." Golden CO: Dow Chemical Archives.

Dow Chemical. (1957b). "Letter on Problems of Roofs In the New Buildings." December 20, 1957. Golden, CO: Dow Chemical Archives.

Dow Chemical. (1957c). Early Estimates of the Total Known and Unknown Quantities Released Was 1 Gram (.07 curies). Golden, CO: Dow Chemical Archives.

Dow Chemical. (1958). "No Word Has Been Received From ARCO In Regard To the Resumption of Shipping Contaminated Liquids To the Burial Grounds." Dow Chemical Waste Disposal Unit Progress Reports. August and September, 1958. Golden, CO: Dow Chemical Archives.

Dow Chemical. (1961). "Burning of Uranium Chips In Oil Solves the On-Site Disposal Problem." Progress Report, February 1961. Golden, CO: Dow Chemical Archives.

Dow Chemical. (1969). "Waste Management Tends To Be a Low Priority Project." Dow Chemical Report, October 24, 1969. Golden, CO: Dow Chemical Archives.

Dow Chemical. (1978). Pertinent History of Plutonium Contaminated Drum Storage Areas. Golden, CO: Dow Chemical Archives.

Drake, L.R. (1955). "Waste Was Being Generated When the Ultimate Disposition Had Not Yet Been Decided Upon." Letter from L.R. Drake to R.R. Harrison, January 5, 1955. Golden, CO: Dow Chemical Archives.

EPA. (1990). EPA Superfund Record of Decision: Interim Measures/Interim Remedial Action Plan and Decision Document 881 Hillside Area Operable Unit No. 1. www.epa.gov/superfund/sites/rods/fulltext/r0890043.pdf

EPA. (2001). Radionuclide table: Radionuclide carcinogenicity—slope factors. (Federal Guidance Report No. 13 Morbidity Risk Coefficients, in Units of Picocuries, Washington, D.C.: U.S. Environmental Protection Agency, Office of Radiation and Indoor Air. May 18, 2001.

ERDA. (1975). Response to the Lamm-Wirth Task Force. Washington, D.C.: Energy Research and Development Administration.

Erikson, K. (1991). "Radiation's Lingering Dread." Bulletin of the Atomic Scientists, 47(2):35-39.

Erikson, K. (1994). A New Species of Trouble: The Human Experience of Modern Disasters. New York, NY: W.W. Norton & Company, Inc.

Facer, G.C. (1970). Contaminated Waste Storage Burial At the Rocky Flats Plant Site. August 7, 1970. Golden, CO: Rocky Flats Repository Document HS-346, Dow Chemical Archives.

Factory Insurance Association. (1970). Final Report, Fire Inspection for U.S. Atomic Energy Commission, Rocky Flats Plant, Golden, Colorado, February, 1970: Factory Insurance Association.

Faulkner, W. (1951). Requiem for a Nun. New York: Random House.

FBI. (1989). Search Warrant Affidavit. Denver, CO: FBI Field Office.

FEDRIP. (2007). Plutonium Federal Research in Progress Database. Springfield, VA: National Technical Information Service.

Finley, B. and Graf, T. (1989). "Rocky Flats Illegally Burned, Dumped Waste, U.S. Claims." Denver Post, June 10, 1989.

Finesilver, S.G. (1989). "Instructions to the Special Federal Grand Jury." Denver, CO: Federal District Court for the District of Colorado.

Freiberg, K.J. (undated). "Chronology of the Transfer of the Contaminate Oil Drums From the 903 Pad To Building 774." Golden, CO: Dow Chemical Archives.

Frontline. (1993). Secrets of a Bomb Factory. Arlington, VA: Public Broadcasting Service.

GAO. (1988). Summary of Major Problems At DOE's Rocky Flats Plant. (October, 1988). Washington, D.C.: Government Accountability Office.

GAO. (1991). Problems Continue for Rocky Flats Solar Ponds Cleanup. (January 3, 1991). Washington, D.C.: Government Accountability Office.

GAO. (1994). Nuclear Health and Safety: Consensus on Acceptable Radiation Risk to the Public is Lacking. Washington, D.C.: Government Accountability Office, RCED-94-190.

GAO. (1999). Accelerated Closure of Rocky Flats: Status and Obstacles. RCED-99-100 04/30/99. Washington, D.C.: GAO.

GAO. (2001). Nuclear Cleanup Progress Made At Rocky Flats, But Closure By 2006 Is Unlikely, and Costs May Increase. GAO-01-284, February 28, 2001, Washington, D.C.: GAO.

GAO. (2006). Nuclear Cleanup of Rocky Flats: DOE Can Use Lessons Learned to Improve Oversight of Other Sites' Cleanup Activities. GAO-06-352, July, 2006, Washington, D.C.: GAO.

Gelb, L.H. and Betts, R.K. (1979). The Irony of Vietnam: The System Worked. Washington, D.C.: Brookings Institution Press.

Gerber, M. (1992). Legend and Legacy: Fifty Years of Defense Production At the Hanford Site. Richland, WA: Westinghouse Hanford Company.

Giller, E. (1970). Testimony Before the Boland Sub-Committee on Public Works of the House Committee on Appropriations of the U.S. House of Representatives. Washington D.C.: U.S. House of Representatives, October 1, 1970.

Graf, T. (1989). "Energy Department Freezes Millions in Bonuses to Rockwell," The Denver Post. 17 June 1989.

Groves, L.H. (1962). Now It Can Be Told: The Story of the Manhattan Project. New York: Harper.

Hahn, A. (2018). "Superior files lawsuit to block opening of Rocky Flats National Wildlife Refuge." Daily Camera, July 12, 2018.

Hammond, S.E. (1957). "Excessive Amounts of Enriched Uranium In Ash Sample From Plant Open Air Incinerator Just a Few Hundred Yards From Woman Creek Drainage." Golden, CO: Dow Chemical Archives.

Hazelton, L.D. (1962). "There Is a Fire Hazard Associated With Uranium Chips In Oil." Letter from L.D. Hazleton to Norman H. MacKay. December 6, 1962. Golden, CO: Dow Chemical Archives.

Hazle, A.J. (1985). "Interoffice Communication RE: Groundwater Contamination Sources." September 3, 1985. Golden, CO: Rocky Flats Repository Document RE-354, Dow Chemical Archives.

Hei, T.K., Wu, L-J. Lui, S-X., Vannais, D., Waldren, C., and Randers-Pherson, G. (2006). "Mutagenic Effects of a Single and an Exact Number of Particles in Mammalian Cells," pp. 3765-3770. Proceedings of the National Academy of Science Online: www.pnas.org.

Helmstedt, R.W. (1988). Aerial Photographic Analysis Comparison Report of the USDOE Rocky Flats Plant. Lockheed Engineering and Sciences. Golden, CO: Rocky Flats Repository Document EN-589.

Hobbs, F.D. (1972). "Nickel Carbonyl Disposal." September, 1972. Golden, CO: Rocky Flats Repository Document RE-933.

Holme, H. (1978). Pretrial Statement in Church Cluster of Lawsuits Civil Action No 75-M-1111, 75-M-1162, 75-M-1296 in the United States Federal District Court for the District of Colorado. Denver, CO: Fairfield and Woods.

Hsieh, A.L. (1962). Communist China's Strategy In the Nuclear Era. Englewood Cliffs, NJ: Prentice Hall. http://www.lm.doe.gov/cercla/documents/rockyflats_docs/ou01/ou01-A-000248.pdf

Iacobellis, S.I. (1989). "Rockwell International Actions Had Been Taken Under the Direction or With the Concurrence of the United States Government Acting Through the DOE." Letter From the President of Aerospace Operations for Rockwell International to Deputy Secretary of Energy W. Henson Moore August 17, 1989. Pittsburgh, PA: Rockwell International.

Illsley, C.T. and Hume, M.W. (1979). Plutonium Concentrations in Soil on Lands Adjacent to the Rocky Flat Plant. Golden, CO: Energy Systems Group, March 16, 2979.

Journel, A.G. and Huijbragts, C.J. (1978). Geostatistics. London: Academic Press.

King, Jr., Martin Luther. (1967). Conscience for Change, Canadian Broadcasting Corporation, 1967, p. 19.

Krey, P.W. and Hardy, E.P. (1970). "Plutonium in Soil Around the Rocky Flats Plant." United States Atomic Energy Commission Health and Safety Laboratory. August 1, 1970, HASL-235. New York: Atomic Energy Commission.

Krey, P.W. and Krajewski, B.T. (1972). "Plutonium Isotopic Ratios at Rocky Flats." New York: Health and Safety Laboratory-235, Atomic Energy Commission.

Lamm-Wirth Task Force. (1975). Chapter III - Hazards Associated with the Rocky Flats Plant G. Monitoring of Dangerous Material H. Other http://www.lm.doe.gov/land/sites/co/rocky_flats/closure/references/150-Lamm-Wirth%20Task%20Force%20on%20RF.pdf' .

Langell, F.H. (1957). "Eliminating Hazards Due to the Flammability of Plexiglas." Letter from F.C. Langdell, Dow Chemical Manager for Rocky Flats to Seth H. Woodruff, Jr., Manager of the Rocky Flats Area Office for the Atomic Energy Commission. December 20, 1957. Golden, CO: Dow Chemical Archives.

Langell, F.H. (1956). "Local Burial of Depleted Uranium Chips Immersed Oil Is One Possibility." Letter from F.H. Langell, Dow Chemical Manager at Rocky Flats to Gilbert C. Hoover, Manager of the Rocky Flats Area Office for the Atomic Energy Commission. June 12, 1956. Golden, CO: Dow Chemical Archives.

Lemonick, M.D. (1995). "Rocky Flats Horror Show: Why Can't Rocky Flats Plutonium Be Cleaned Up?: It's Against the Rules." Time, March 27, 1995.

Lewis, J.W. and Litai, X. (1988). China Builds the Bomb, Stanford: Stanford University Press.

Lindsay, S. and Day, J. (1989). "FBI: Flats Burned Waste Secretly," Rocky Mountain News, June 10, 1989.

Lipsky, J., McKinley, W., Brever, J. and Balkany, C. (2005). Citizens Investigation of 2005. Denver, CO: Citizens Investigation.

Lipsky, J. (2016). Personal communication.

Little, C.A. and Whicker, F.W. (1978). "Plutonium Distribution in Rocky Flats Soil." Health Physics, 34:451-457, May 1978.

Lockhart, F.R. (2006). Closure Legacy: From Weapons to Wildlife. Golden, CO: U.S. DOE Rocky Flats Project Office.

Love, C.M. (1967). "Plant Open Air Incinerator Does Not Meet Emission Visibility Standards." Letter from C.M. Love, Rocky Flats Industrial Relations Manager, to Seth R. Woodruff, Rocky Flats Area Manager for the Atomic Energy Commission. Golden, CO: Dow Chemical Archives.

Lovrich, N.P. and Whiteley, J.M. (1999). "The Governmental Response at Hanford," In R.J. Dalton, P. Garb, N.P. Lovrich, J.C. Pierce,. and J.M. Whiteley, Critical Masses: Citizens, Nuclear Weapons Production, and Environmental Destruction in the United States and Russia, Cambridge, MA: The MIT Press, 287-332.

Lutz, C. (2001). "Hot Property: A Former Nuclear Bomb Factory Gets Caught in Suburban Turf Wars." High Country News. Jan 15, 2001. https://www.hcn.org/issues/194/10202

Martell, E.A., Golfan, P. A., Kraushaar, J.J., Shea, D.W. and Williams, R.H. (1970). "Fire Damage," Environment: Science and Policy for Sustainable Development, 12:14-21.

Matthias, F.T. (1967). "Early History." Speech to the Technical Exchange Program, January 14, 1967. Hanford, WA: Hanford Engineering Works. Manhattan Engineering District.

McKay, C. (no date). Church Ranch Snapshots of History. Church Ranch, CO: Church Ranch est. 1869.

McKinley, W. and Balkany, C. (2004). The Ambushed Grand Jury, How the Justice Department Covered Up Government Nuclear Crime and How We Caught Them Red Handed. New York: The Apex Press.

Mellen, P.A. (2019). Motion for Writ of Mandamus Pursuant to 28 USCA §1361 to Locate Rocky Flats Grand Jury Documents. Civil Action, No. 19-cv-00076-RM-KLM. https://rockyflatsambushedgrandjury.com/wp-content/uploads/Motion-for-Writ-of-Mandamus.pdf.pdf

Metzger, H.P. (1972). The Atomic Establishment. New York: Simon and Schuster.

Metzger, H.P. (1975). "Burials Were Not Legal." Rocky Mountain News. March 31, 1975.

Moore, L. (1992). Citizen's Guide to Rocky Flats. Boulder, CO: Rocky Mountain Peace Center.

Moore, L. (2014). Lecture: The 'Accelerated Cleanup.' UC Irvine, 2014.

Moore, L. (2016). Personal Communication, July 2016.

Moore. L. (2019). Plutonium and People Don't Mix: Colorado's Defunct Nuclear Bomb Factory. Boulder, CO: Rocky Mountain Peace and Justice Center.

Muller, H.J. (1964). "Radiation and Heredity." American Journal of Public Health, 54(1):42-50.

National Security Archive. (2014). "Multi-Megaton Bomb Was Virtually 'Armed' When it Crashed to Earth in North Carolina, Sandia Lab Report concluded." National Security Archive Electronic Briefing Book No. 475. Washington, D.C.: National Security Archive.

Norris, R. S., Arkin, W.N., and Burr, W. (1999). "Where They Were." Bulletin of the Atomic Scientists, 55(6):30, November/December.

NRC. (1993). Pesticides in the Diets of Infants and Children. National Research Council: Washington, D.C.: National Academy Press.

NRC. (2006). Health Risks from exposure to Low Levels of Ionizing Radiation: BEIR VII Phase 2. Division on Earth and Life Studies; Board on Radiation Effects Research; Committee to Assess Health Risks from Exposure to Low Levels of Ionizing Radiation National Research Council: Washington, D.C.: National Academy Press.

Obmascik, M. (1996). "Rocky Flats Sued, Tax Payers Pay; Ex-operators' Legal Fees Hit $20 Million." The Denver Post, April, 7, 1996.

Office of Environmental Management. (1997). Linking Legacies: Connecting the Cold War Nuclear Weapons Production Process to Their Environmental Consequences. Washington, D.C.: U.S. Dept. of Energy, Department of Environmental Management.

Office of Strategic Planning and Analysis. (1995). Closing the Circle On the Splitting of the Atom. Washington, D.C.: DOE Office of Strategic Planning and Analysis.

ORAU. (2004). "ORAU Team NIOSH Dose Reconstruction Project Technical Basis Document For the Rocky Flats Plant--Site Description," www.cdc.gov/niosh/ocas/pdfs/arch/rocky2.pdf.

Owen, J.B. (1971). "Inspection of Similar Wastes Has Resulted In the Need To Repackage Nearly 100 Percent of the Containers." Letter from J.B. Owen of Dow Chemical to T.C. Jones of the Atomic Energy Commission, October 20, 1971. Golden, CO: Dow Chemical Archives.

Owen, J.B. and Steward, L.M. (1974). Environmental Inventory: A Historical Summation of Environmental Incidents Affecting Soil At Or Near the USAEC Rocky Flats Plant. January 1974. Golden, CO: Rocky Flats Repository Document RE-318.

Pankratz, H. (1989). "Rocky Flats Conspiracy Alleged: FBI Used Spy Plane." The Denver Post, June 10, 1989.

Pierce, R. (1989). "Flats Charges Include Illegal Waste Burning: Document Claims Chemicals Put in Creeks Feeding Water Supplies." Daily Camera, June 10, 1989.

Piltingsrud, C.W. (1973). "Untitled Burial Site Summary." September 28, 1973. Golden, CO: Rocky Flats Repository Document RE-547.

Poet, S.E. and Martell, E.A. (1972). "Plutonium-239 and Americium-241 Contamination in the Denver area." Health Physics, 23(4), 537-548.

Pritchett, K. (2019). "Plaintiffs in Rocky Flats National Wildlife Refuge lawsuit file brief in Support of Claim." kpritchett@prairiemountainmedia.com.

Putzier, E.A. (1970). A Summary of On-Site Radioactive Waste Disposal. April 22, 1970, Golden, CO: Rocky Flats Repository Document RE-317.

Putzier, E.A. (1977). "Deposition Describing the Plant Incinerator and Its Operation." April 27, 1977. Deposition in the 1975 Church Civil Lawsuit. Golden, CO: Dow Chemical Archives.

Putzier, E.A. (1982). The Past 30 Years at Rocky Flats Plant. Rockwell International Energy Systems Group. Work performed under United States Department of Energy Contract DEACO4-76DPO-3535. Golden CO: Rockwell International.

Quinn, J.F. (1957). "This Record Is the Best That Can Be Obtained." Letter by J.F. Quinn, Jr. to the Fire Investigation Committee. September 24, 1957. Washington, D.C.: Atomic Energy Commission.

Reno, J. (1994). "Statement of the Attorney General Concerning the Internal Report On the Rocky Flats Prosecution." Washington, D.C.: Department of Justice.

Rockwell International. (1987). Internal Report Briefing Memo. Golden, CO: Rockwell International.

Rockwell International. (1999). "Chapter and Verse of the Rockwell Submission to the Court By Date From Their Communications To DOE In Response To a Pretrial Order In the United States of America, Ex Rel. James Stone, and United States of America, Plaintiffs-Appellees and Cross Appellants, v. Rockwell International Corp., and Boeing North American, Inc. Defendants Appellants.

Rosenbaum, W. A. (1999). "The Good Lessons of Bad Experience." American Behavioral Scientist, 43(1), 74-91.

Ryan, E.S. (1953). "This Is a New Group With Little Background Knowledge of the Problems Presented By the Material to be Handled." Progress Report--Waste Disposal Unit, January, 1953 dated February 13, 1953. Golden, CO: Dow Chemical Archives.

Ryan, E.S. (1954). Letter to H.C. Anderson Dated November 23, 1954. Golden, CO: Dow Chemical Archives.

Schlosser, E. (2013). Command and Control: Nuclear Weapons, the Damascus Accident and the Illusion of Safety. New York: The Penguin Press.

Schmidt, D.W. (1994). Evaluation of Background Concentrations of Plutonium in Soils Around the Rocky Flats Plant: Technical Memorandum. Neeses, SC: Radiological Assessments Corporation, September 1994.

Schmitz, G. (1989). "Use of Officially Idled Incinerator for Toxic Waste Suspected: Feds Assume Conspiracy, Rep. Skaggs Speculates," The Denver Post, 9 June 1989.

Seastone, J. (1973). "History of Incineration and Landfill Operations." Rocky Flats Plant. Letter to E.J. Walko. October 9, 1973. Golden, CO: Rocky Flats Repository Document RF-700.

Second Citizen Special Federal Grand Jury Report. (1992). Second Citizen Special Federal Grand Jury Report. Washington, D.C.: Wolpe Committee Archives, U.S. House of Representatives.

Seed, J.R., et.al. (1971). Committee Evaluation of Plutonium Levels in Soil Within and Surrounding USAEC Installation at Rocky Flats, Colorado. Golden, CO: Rocky Flats Division, The Dow Chemical Company, July 9, 1971.

Shuster, S. (2017). "Stanislav Petrov, the Russian Officer Who Averted a Nuclear War, Feared History Repeating Itself." Time, September 19, 2017, http://time.com/4947879/stanislav-petrov-russia-nuclear-war-obituary/.

Sierra Club. (1991). Court finds DOE in "Flagrant" Violation of Hazardous Waste Requirements; Threatens Rocky Flats Shutdown." Denver, CO: Rocky Mountain Chapter, The Sierra Club, August 13, 1991.

Silverman, M.N. (1994). Rocky Flats Strategic Plan, Golden, CO: Department of Energy Rocky Flats Field Office, September 19, 1994.

Silverman, M.N. and Hoese, R.D. (1997). "Cleaning Up the Cold War Legacy." Assembly, May-June, pp. 19-23, 52-53.

Smallwood, K.S., Morrison, M.L., and Beyea, J. (1998). "Animal Burrowing Attributes Affecting Hazardous Waste Management." Environmental Management, 22(6):834.

Smith, R.E. (1975). "Buried Radioactive Material." Letter to D.J. Cockeram September 2, 1975. Golden, CO: Rocky Flats Repository Document RE-535.

Smyth, H.D. (1945). A General Account of the Development of Methods of Using Atomic Energy For Military Purposes. Princeton: Princeton University Press.

Spavin, H.A. (undated). "Isolate Component Operations in Construction of Plutonium Processing Facilities." Memorandum to John Epp of the 1957 AEC Fire Investigation Committee. Golden, CO: Dow Chemical Archives.

Special Federal Grand Jury. (1992). First "Leaked" Special Federal Grand Jury Report. Westword and the Constitution Society.

Steadman, C. (2019). Rocky Flats Subject to Another Lawsuit Over Safety." steadman@coloradocommunitymedia.com, January 15, 2019.

Steeth, P. (1989). "Investigation at Rocky Flats: A Cause for Contract Cancellation." The Denver Post, June 9, 1989.

Steward, L.M. (1973). Areas of Concern At Rocky Flats: Soil Contamination and Related Incidents-Part I: Plutonium. June 29, 1973. Golden, CO: Rocky Flats Repository Document RE-529.

Stout, J. (1974). "Storage in the Triangle Area Was Probably Minimal Until After the 1969 Fire." Memo from James Stout, Acting AEC Area Office Manager to Herman Roser AEC Albuquerque Office. October 2, 1974, Golden, CO: Dow Chemical Archives.

Subcommittee on Public Works of the House Committee on Appropriations. (1970). Public Works Appropriations, FY 1971, October 1, 1970.

Taubman, Philip. (1988). "Gromyko Says Mao Wanted Soviet A-Bomb On American G.I.s." The New York Times, February 22, 1988, p. A-1.

The Denver Post. (1989). "Rockwell Has History of Run-Ins With Regulators Every-where." The Denver Post, June 7, 1989.

The Economist. (2017). "Obituary: Stanislav Petrov Was Declared To Have Died On September 18th." The Economist, September 30, 2017, https://www.economist.com/news/obituary/21729727-man-who-saved-world-was-77-obituary-stanislav-petrov-was-declared-have-died.

Thompson, M.A. (1969). "A Serious Potential Fire Hazard." Memo from Dr. M.A. Thompson to Six Dow Health Physics and Waste Management Officials. Golden, CO: Dow Chemical Archives.

Tiger Team Report. (1989). Assessment of Environmental Conditions at the Rocky Flats Plant. Golden Colorado, August 1989. Partially released in a FOIA filing by Caron Balkany. Posted on: https://rockyflatsambushedgrandjury.com/

U.S. Senate. (1970). May 1969 Fire at Rocky Flats, Colorado. Subcommittee on Deficiencies and Supplementals of the U.S. Senate Committee on Appropriations. November 21, 1970.

Vogel, R.M. (1968a). "Environmental Contamination Is Certainly Inevitable." Health Physics Report, November, 1968. Golden, CO: Dow Chemical Archives.

Vogel, R.M. (1968b). "Potential For Environmental Contamination Is of Special Concern." Health Physics Report, December, 1968. Golden, CO: Dow Chemical Archives.

Vogel, R.M. (1970). "No Detectable Alpha Contamination Was Found In the Soil." Letter from R.M. Vogel to E.A. Putzier, June 2, 1970. Golden, CO: Dow Chemical Archives.

Wakefield, S.A. (1989). "Contemporaneous Notes of the Conversation Between the Secretary of Energy and the Chairman of Rockwell International, September 18, 1989." Washington, D.C.: Department of Energy.

Watkins, J.D. (1989a). "The Rockwell Management Plan Does Not Adequately Address the Issue of Management Problems At Rocky Flats." Letter from the Secretary of Energy to the Chairman of Rockwell International, August 25, 1989. Washington, D.C.: Department of Energy.

Watkins, J.D. (1989b). "40% of the Total Possible Award Fee Is the Proper Determination." Letter from the Secretary of Energy to the Chairman of Rockwell International Regarding the Performance Bonus "Total Award Fee." September 20, 1989. Washington, D.C.: Department of Energy.

Welsome, E. (2000). "This Place is a Dump!" Westword, July 27, 2000.

Windahl, O.M. (1952). "Annual Report 1952." Health Physics Division, Dow Chemical. Putzier Report: Summary of On Site Radioactive Waste Disposal. April 22, 1970, Golden, CO: Dow Chemical Archives.

Wirth, T. (2018). Personal Communication. December 20, 1018.

Wolpe. (1993). Environmental Crimes at the Rocky Flats Nuclear Weapons Facility. Subcommittee on Investigations and Oversight of the Committee on Science, Space, and Technology, U.S. House of Representatives. Washington, D.C.: U.S. Government Printing Office, January 3, 1993.

Yoder, R. E. (1978). "Filters Evidenced Signs of Fire and Water." Letter from Robert E. Yoder of Rockwell International to J.J. Chavez of DOE, March 1, 1978. Golden, CO: Rockwell International.

Index

Made in United States
Orlando, FL
04 November 2026

90205909R00328